# NELSON
WAmaths

UNITS ①+②

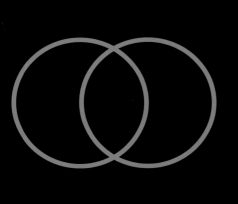
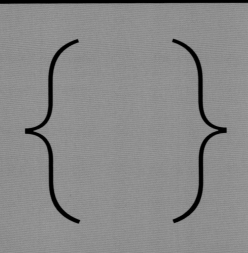
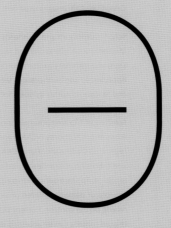

Judith Cumpsty
Dion Alfonsi
Greg Neal
Tal Ellinson

Contributing authors
Sue Garner
George Dimitriadis
Toudi Kouris
Stephen Swift
Neale Woods

mathematics methods 11

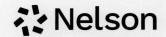

Nelson WAmaths Mathematics Methods 11
1st Edition
Judith Cumpsty
Dion Alfonsi
Greg Neal
Tal Ellinson
ISBN 9780170477543

Publisher: Dirk Strasser
Associate product manager: Cindy Huang
Project editor: Tanya Smith
Editor: Karen Chin
Series cover design: Nikita Bansal
Series text design: Alba Design (Rina Gargano)
Series designer: Nikita Bansal
Permissions researcher: Liz McShane
Content developer: Katrina Stavridis
Content manager: Alice Kane
Typeset by: Nikki M Group Pty Ltd

Any URLs contained in this publication were checked for currency during the production process. Note, however, that the publisher cannot vouch for the ongoing currency of URLs.

Acknowledgements
TI-Nspire: Images used with permission by Texas Instruments, Inc.

Casio ClassPad: Images used with permission by Shriro Australia Pty. Ltd.

School Curriculum and Standards Authority: Adapted use of 2016–2021 Mathematics Applications and Mathematics Methods examinations, marking keys, and summary examination reports, ATAR 11 and 12 Mathematics Applications and Mathematics Methods syllabuses. The School Curriculum and Standards Authority does not endorse this publication or product.

Selected VCE Examination questions are copyright Victorian Curriculum and Assessment Authority (VCAA), reproduced by permission. VCE ® is a registered trademark of the VCAA. The VCAA does not endorse this product and makes no warranties regarding the correctness or accuracy of this study resource. To the extent permitted by law, the VCAA excludes all liability for any loss or damage suffered or incurred as a result of accessing, using or relying on the content. Current VCE Study Designs, past VCE exams and related content can be accessed directly at www.vcaa.edu.au.

© 2023 Cengage Learning Australia Pty Limited

Copyright Notice
This Work is copyright. No part of this Work may be reproduced, stored in a retrieval system, or transmitted in any form or by any means without prior written permission of the Publisher. Except as permitted under the *Copyright Act 1968,* for example any fair dealing for the purposes of private study, research, criticism or review, subject to certain limitations. These limitations include: Restricting the copying to a maximum of one chapter or 10% of this book, whichever is greater; providing an appropriate notice and warning with the copies of the Work disseminated; taking all reasonable steps to limit access to these copies to people authorised to receive these copies; ensuring you hold the appropriate Licences issued by the Copyright Agency Limited ("CAL"), supply a remuneration notice to CAL and pay any required fees. For details of CAL licences and remuneration notices please contact CAL at Level 11, 66 Goulburn Street, Sydney NSW 2000,
Tel: (02) 9394 7600, Fax: (02) 9394 7601
Email: info@copyright.com.au
Website: www.copyright.com.au

For product information and technology assistance,
    in Australia call **1300 790 853**;
    in New Zealand call **0800 449 725**

For permission to use material from this text or product, please email
aust.permissions@cengage.com

National Library of Australia Cataloguing-in-Publication Data
A catalogue record for this book is available from the National Library of Australia.

Cengage Learning Australia
Level 5, 80 Dorcas Street
Southbank VIC 3006 Australia

For learning solutions, visit **cengage.com.au**

Printed in China by 1010 Printing International Limited.
1 2 3 4 5 6 7 27 26 25 24 23

Contents

To the teacher	v
About the authors	vi
Syllabus grid	vii
About this book	viii

1 Sets and combinations — 1
- Syllabus coverage — 2
- Nelson MindTap chapter resources — 2
- 1.1 Set descriptions and operations — 3
- 1.2 Combinations – counting how many ways — 13
- 1.3 Pascal's triangle and binomial expansions — 20
- Examination question analysis — 25
- Chapter summary — 28
- Cumulative examination: Calculator-free — 33
- Cumulative examination: Calculator-assumed — 34

2 Probability — 35
- Syllabus coverage — 36
- Nelson MindTap chapter resources — 36
- 2.1 The fundamentals of probability — 37
- 2.2 Probability experiments involving sets — 41
- 2.3 Probability experiments involving stages — 48
- 2.4 Probability experiments involving selections — 55
- 2.5 Conditional probability — 58
- 2.6 Independent events — 67
- 2.7 Relative frequencies and indications of independence — 73
- Examination question analysis — 77
- Chapter summary — 81
- Cumulative examination: Calculator-free — 84
- Cumulative examination: Calculator-assumed — 85

3 Linear and quadratic relationships — 87
- Syllabus coverage — 88
- Nelson MindTap chapter resources — 88
- 3.1 Sketching linear graphs — 89
- 3.2 Determining the equation of a linear graph — 101
- 3.3 Expanding — 110
- 3.4 Factorising — 116
- 3.5 Solving quadratic equations by factorising — 122
- 3.6 Solving quadratic equations by completing the square — 130
- 3.7 The quadratic formula and the discriminant — 134
- 3.8 Sketching parabolas from turning point form — 140
- 3.9 Sketching parabolas from any form — 148
- 3.10 Simultaneous equations — 157
- 3.11 Determining the equation of a parabola — 165
- 3.12 Applications of quadratic relationships — 172
- Examination question analysis — 175
- Chapter summary — 180
- Cumulative examination: Calculator-free — 184
- Cumulative examination: Calculator-assumed — 185

4 Functions and relations — 187
- Syllabus coverage — 188
- Nelson MindTap chapter resources — 188
- 4.1 Polynomials and power functions — 189
- 4.2 Solving cubic equations — 195
- 4.3 Graphing cubic functions — 199
- 4.4 Functions and relations — 206
- 4.5 Transformations of functions — 214
- 4.6 Graphs of relations — 224
- 4.7 Inverse proportion — 227
- Examination question analysis — 228
- Chapter summary — 231
- Cumulative examination: Calculator-free — 234
- Cumulative examination: Calculator-assumed — 235

5 Trigonometric functions — 236
Syllabus coverage — 237
Nelson MindTap chapter resources — 237
5.1 Right triangles and the angle of inclination — 238
5.2 Non-right triangles — 242
5.3 Radian measure, arc lengths, sectors and segments — 253
5.4 The trigonometric functions — 260
5.5 The exact values — 267
5.6 Graphs of trigonometric functions — 272
5.7 Applying trigonometric functions — 281
Examination question analysis — 284
Chapter summary — 290
Cumulative examination: Calculator-free — 295
Cumulative examination: Calculator-assumed — 297

6 Exponential functions — 299
Syllabus coverage — 300
Nelson MindTap chapter resources — 300
6.1 Index laws — 301
6.2 Fractional powers — 306
6.3 Exponential equations — 310
6.4 The exponential function $y = a^x$ — 313
6.5 Exponential growth and decay — 319
Examination question analysis — 323
Chapter summary — 327
Cumulative examination: Calculator-free — 329
Cumulative examination: Calculator-assumed — 330

7 Arithmetic and geometric sequences and series — 331
Syllabus coverage — 332
Nelson MindTap chapter resources — 332
7.1 Arithmetic sequences — 333
7.2 Modelling linear growth and decay — 343
7.3 Arithmetic series and their applications — 347
7.4 Geometric sequences — 354
7.5 Modelling exponential growth and decay — 364
7.6 Geometric series and their applications — 367
Examination question analysis — 375
Chapter summary — 379
Cumulative examination: Calculator-free — 381
Cumulative examination: Calculator-assumed — 382

8 Differential calculus — 383
Syllabus coverage — 384
Nelson MindTap chapter resources — 384
8.1 Rates of change — 385
8.2 Instantaneous rates of change and the gradient function — 392
8.3 Differentiation by first principles — 397
8.4 Differentiating polynomial functions — 400
Examination question analysis — 404
Chapter summary — 407
Cumulative examination: Calculator-free — 409
Cumulative examination: Calculator-assumed — 410

9 Applications of differential calculus — 411
Syllabus coverage — 412
Nelson MindTap chapter resources — 412
9.1 Equation of a tangent — 413
9.2 Straight line motion — 416
9.3 Stationary points — 419
9.4 Curve sketching — 425
9.5 Optimisation problems — 429
9.6 The anti-derivative — 434
9.7 Applying the anti-derivative to straight line motion — 438
Examination question analysis — 439
Chapter summary — 441
Cumulative examination: Calculator-free — 443
Cumulative examination: Calculator-assumed — 444

Answers — 446
Glossary and index — 485

To the teacher

Now there's a better way to WACE maths mastery.

Nelson WAmaths 11–12 is a new WACE mathematics series that is backed by research into the science of learning. The design and structure of the series have been informed by teacher advice and evidence-based pedagogy, with the focus on preparing WACE students for their exams and maximising their learning achievement.

- Using **backwards learning design**, this series has been built by analysing past WACE exam questions and ensuring that all theory and examples are precisely mapped to the SCSA syllabus.
- To reduce the **cognitive load** for learners, explanations are clear and concise, using the technique of **chunking** text with accompanying diagrams and infographics.
- The student book has been designed for **mastery** of the learning content.
- The exercise structure of **Recap**, **Mastery** and **Calculator-free** and **Calculator-assumed** leads students from procedural fluency to **higher-order thinking** using the learning technique of **interleaving**.
- **Calculator-free** and **Calculator-assumed** sections include exam-style questions and past SCSA exam questions.
- The cumulative structure of exercise **Recaps** and chapter-based **Cumulative examinations** is built on the learning and memory techniques of **spacing** and **retrieval**.

About the authors

Judith Cumpsty has taught Mathematics in various Western Australian schools for over thirty years, teaching at many levels, as well as having been Head of Department. She has been involved with projects such as Have Sum Fun and prepared students for OLNA. Judith has been a Team Leader for SCSA WACE marking and a proofreader for examinations published by MAWA.

Dion Alfonsi is Head of Mathematics and a Secondary Mathematics Teacher at Shenton College. In the past, he has had the roles of Years 9 & 10 Mathematics Curriculum Leader and Gifted & Talented/Academic Programs Coordinator. Dion has been a Board Member of MAWA, is a frequent presenter at the MAWA Secondary Conference and a teacher of the MAWA Problem Solving Program.

Greg Neal has taught in regional schools for over 40 years and has co-written several senior textbooks for Cengage Nelson. He has been an examination assessor, presents at conferences and has expertise with CAS technology.

Tal Ellinson is an experienced teacher across all senior secondary maths subjects. He was the Learning Specialist for Numeracy, leading maths teacher training and numeracy intervention at his previous school. He is a maths presenter for Edrolo, has consulted for education companies and presented at national conferences. Recently he has been working overseas as a data scientist in the field of artificial intelligence.

Syllabus grid

Topic		Nelson WAmaths Mathematics Methods 11 chapter
Topic 1.1: Counting and probability (18 hours)		
Combinations	1	Sets and combinations
Language of events and sets	1	Sets and combinations
Review of the fundamentals of probability	2	Probability
Conditional probability and independence	2	Probability
Topic 1.2: Functions and graphs (22 hours)		
Lines and linear relationships	3	Linear and quadratic relationships
Quadratic relationships	3	Linear and quadratic relationships
Inverse proportion	4	Functions and relations
Powers and polynomials	4	Functions and relations
Graphs of relations	4	Functions and relations
Functions	4	Functions and relations
Topic 1.3: Trigonometric functions (15 hours)		
Cosine and sine rules	5	Trigonometric functions
Circular measure and radian measure	5	Trigonometric functions
Trigonometric functions	5	Trigonometric functions
Topic 2.1: Exponential functions (10 hours)		
Indices and the index laws	6	Exponential functions
Exponential functions	6	Exponential functions
Topic 2.2: Arithmetic and geometric sequences and series (15 hours)		
Arithmetic sequences	7	Arithmetic and geometric sequences and series
Geometric sequences	7	Arithmetic and geometric sequences and series
Topic 2.3: Introduction to differential calculus (30 hours)		
Rates of change	8	Differential calculus
The concept of the derivative	8	Differential calculus
Computation of derivatives	8	Differential calculus
Properties of derivatives	8	Differential calculus
Applications of derivatives	9	Applications of differential calculus
Anti-derivatives	9	Applications of differential calculus

About this book

In each chapter

Syllabus coverage and extracts are shown at the start of the chapter along with a listing of **Nelson MindTap chapter resources**.

Important words and phrases are printed in blue and listed in the **Glossary and index** at the back of the book.

Important facts and formulas are highlighted in a shaded box.

Worked examples are explained clearly step-by-step, with the mathematical working shown on the right-hand side.

Using CAS provides clear instructions for Casio ClassPad and TI-Nspire calculators.

Exam hacks highlight valuable exam hints and common student errors.

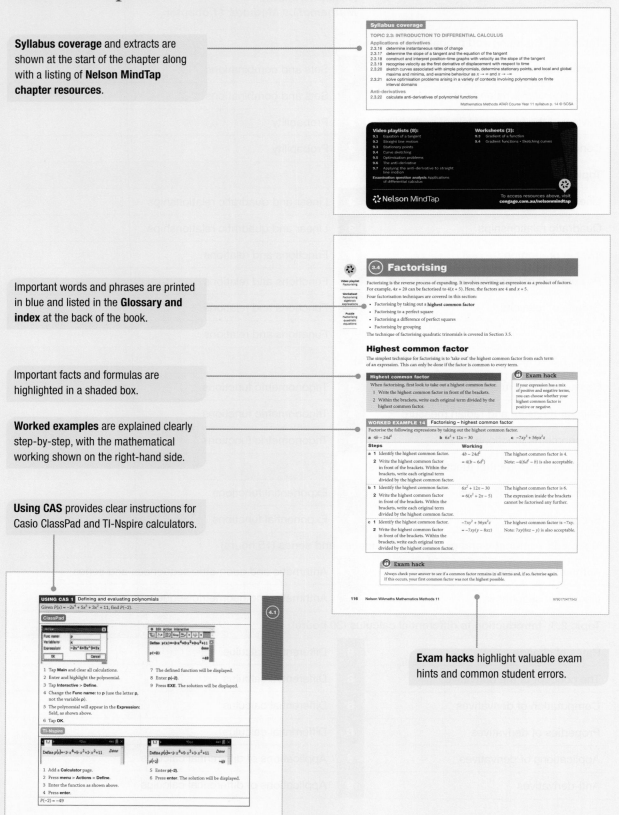

Graded exercises include **Recap**, **Mastery**, **Calculator-free** and **Calculator-assumed** questions. **Recap** questions revise skills from the previous exercise and function as lesson starters.

Mastery questions provide skill practice linked to worked examples and Using CAS, while **Calculator-free** and **Calculator-assumed** questions apply learned skills to exam-style problems with mark allocation.

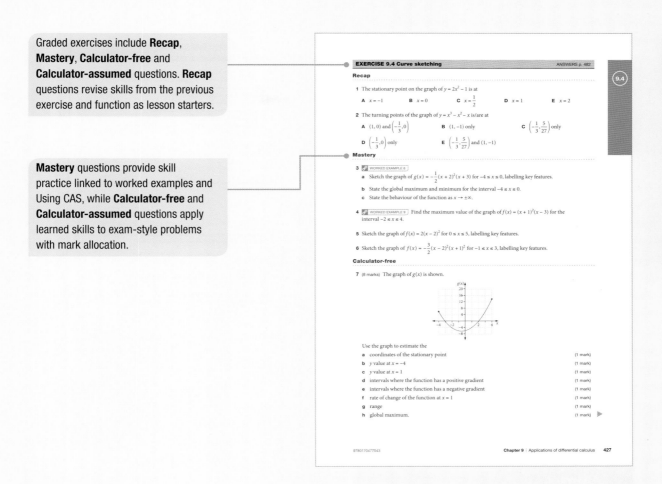

At the end of each chapter

Examination question analysis leads students through an exam-style question that exemplifies the chapter, discussing how to approach the question, providing advice on interpreting the question, common student errors, and a full worked solution with a marking key.

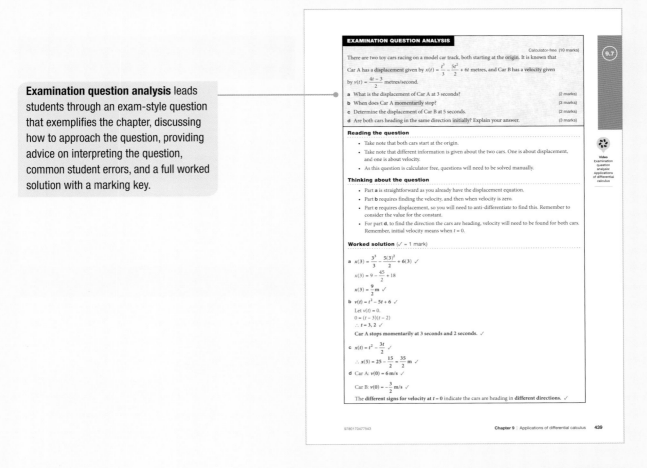

Chapter summary for easy reference.

Cumulative examination: Calculator-free and **Cumulative examination: Calculator-assumed** are mini-exams based on the format of the WACE examinations, revising work from the chapters in which they appear, as well as previous chapters.

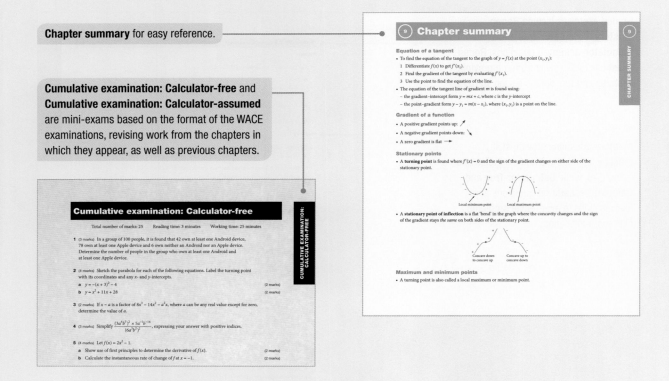

At the end of the book

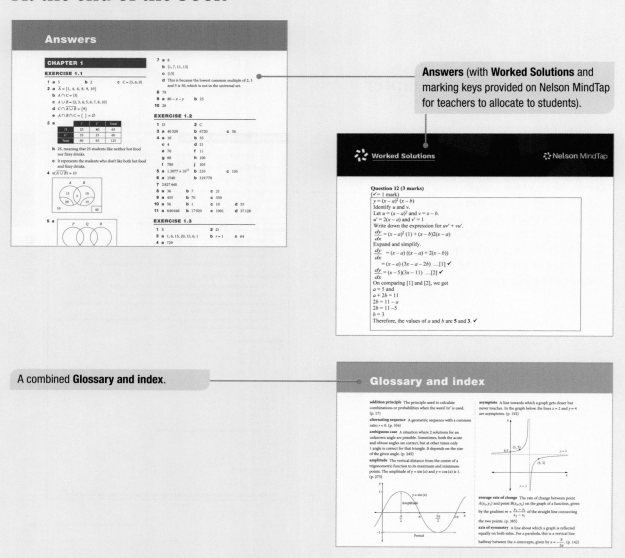

Answers (with **Worked Solutions** and marking keys provided on Nelson MindTap for teachers to allocate to students).

A combined **Glossary and index**.

Nelson MindTap

Nelson MindTap is an online learning space that provides students with tailored learning experiences. Access tools and content that make learning simpler yet smarter to help you achieve WACE maths mastery.

Nelson MindTap includes an eText with integrated interactives and online assessment.

Margin links in the student book signpost multimedia student resources found on MindTap.

Nelson MindTap for students:

- **Watch** video tutorials featuring expert teacher advice to unpack new concepts and develop your understanding.
- **Revise** using learning checks, worksheets and skillsheets to practise your skills and build your confidence.
- **Navigate** your own path, accessing the content, analytics and support as you need it.

Nelson MindTap for teachers*:

- Tailor content to different learning needs – assign directly to the student, or the whole class.
- Monitor progress using the MindTap assessment tools.
- Integrate content and assessment directly within your school's LMS for ease of access.
- Access topic tests, teaching plans and worked solutions to each exercise set.

*Complimentary access to these resources is only available to teachers who use this book as part of a class set, book hire or booklist. Contact your Cengage Education Consultant for information about access codes and conditions.

Nelson WAmaths 11–12 series

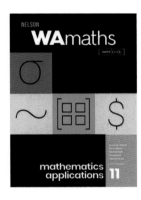
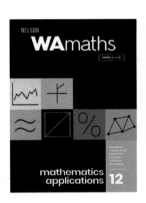
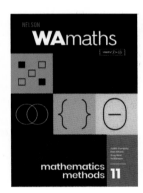
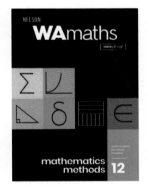

Additional credits

*modified

Chapter 2
Worked example 11 © VCAA MM2009 1Q6*
Question 1 © VCAA MM2010 2AQ21*
Question 5 © VCAA MM2015 2AQ12*
Question 6a © VCAA MM2010 2AQ14*
Question 6b © VCAA MM2006 2AQ6*
Question 6c © VCAA MM2014 2AQ11*
Question 10 © VCAA MM2006 1IQ4*
Exercise 2.4
Question 2 © VCAA MM2014 2AQ11*
Question 5 © VCAA MM2002 1IQ24*
Question 6 © VCAA MM2017N 2AQ11*
Exercise 2.5
Question 2 © VCAA MM2003 1IQ27*
Question 13 © VCAA MM2009 1Q5c*
Question 14 © VCAA MM2019 1Q3*
Question 15 © VCAA MM2015 1Q9*
Exercise 2.6
Question 1 © VCAA MM2012 2AQ13*
Question 2 © VCAA MM2012 2AQ12
Question 7 © VCAA MM2015 1Q8*
Question 8 © VCAA MM2013 2AQ10*
Question 11 © VCAA MM2008 2BQ1ai
Question 12 © VCAA MM2014 2AQ22*
Examination question analysis
© VCAA MM2013 2BQ2*
Cumulative examination: Calculator-free
Question 5 © VCAA MM2016 1Q7*
Cumulative examination: Calculator-assumed
Question 3 © VCAA MM2006 2BQ2a*

Chapter 3
Exercise 3.11
Question 10 © VCAA MM2017N 2AQ3*
Question 11 © VCAA MM2019N 2AQ8*
Cumulative examination: Calculator-assumed
Question 2 © VCAA MM2010 2BQ2*

Chapter 4
Exercise 4.2
Question 8 © VCAA MM2003 1Q1
Question 9 © VCAA MM2004 1MQ14
Question 12 © VCAA MM2006S 2AQ7
Exercise 4.3
Question 9 © VCAA MM2005 non-CAS 1IIQ6
Question 10 © VCAA MM2013 1Q9a
Question 13 © VCAA MM2006 2BQ4aic
Exercise 4.4
Question 14 © VCAA MM2016 2AQ1
Question 18 © VCAA MM2014 2AQ2
Exercise 4.5
Question 11 © VCAA MM2016 1Q3a*
Question 12 © VCAA MM2012 2BQ2a
Question 13 © VCAA MM2014 2AQ1*
Question 14 © VCAA MM2016 2AQ12*
Question 17 © VCAA MM2015 2AQ20
Exercise 4.6
Question 8 © VCAA SM2011 2AQ2
Examination question analysis
© VCAA MM2019 1Q2*
Cumulative examination: Calculator-free
Question 2 © VCAA MM2011 1Q7*
Question 3 © VCAA MM2017 1Q3
Cumulative examination: Calculator-assumed
Question 2 © VCAA MM2006 2AQ1
Question 3 © VCAA MM2015 2AQ20
Question 4 © VCAA MM2011 2AQ2
Question 6 © VCAA MM2019 2AQ13

Chapter 5
Exercise 5.6
Question 6 © VCAA MM2011 1Q3a
Question 9 © VCAA MM2012 2AQ1
Question 10 © VCAA MM2004 1IQ6
Question 12 © VCAA MM2010 2AQ1
Question 14 © VCAA MM2003 1IQ7
Examination question analysis
© VCAA MM2013 2BQ1
Cumulative examination: Calculator-free
Question 3 © VCAA MM2015 1Q4b
Question 5 © VCAA MM2015 1Q6*
Cumulative examination: Calculator-assumed
Question 2 © VCAA MM2013 2AQ1
Question 4 © VCAA MM2014 2BQ1
Question 5 © VCAA MM2017 2BQ2*

Chapter 6
Exercise 6.3
Question 8a © VCAA MM2014 1Q4
Question 8b © VCAA MM2013 1Q5b
Exercise 6.4
Question 10 © VCAA MM2017 2BQ4a*

Chapter 8
Exercise 8.1
Question 8 © VCAA MM2012 2AQ2
Question 17 © VCAA MM2009 2BQ1di
Exercise 8.4
Question 14 © VCAA MM2004 non-CAS 1IQ19
Cumulative examination: Calculator-free
Question 1 © VCAA MM2004 1IQ3*

Chapter 9
Exercise 9.3
Question 18 © VCAA MM2007 2AQ12
Question 19 © VCAA MM2018 2AQ5
Question 20 © VCAA MM2009 2AQ21
Exercise 9.4
Question 9 © VCAA MM2014 1Q5
Exercise 9.5
Question 18 © VCAA MM2012 2BQ1
Question 19 © VCAA MM2009 2BQ2
Cumulative examination: Calculator-free
Question 3 © VCAA MM2019N 2AQ3*
Cumulative examination: Calculator-assumed
Question 4 © VCAA MM2019 2AQ5
Question 5 © VCAA MM2019 2AQ6

CHAPTER 1
SETS AND COMBINATIONS

Syllabus coverage

Nelson MindTap chapter resources

1.1 Set descriptions and operations
  The language and notation of sets
  Relationships between sets
  Operations with two sets – two-way tables and Venn diagrams
  Operations with three sets – Venn diagrams

1.2 Combinations – counting how many ways
  The notion of a combination
  Using the combinations formula
  Using CAS 1: Evaluating factorials, permutations and combinations
  The multiplication and addition principles

1.3 Pascal's triangle and binomial expansions
  Deriving Pascal's triangle and its properties
  Binomial expansions and their coefficients
  Using CAS 2: Expanding binomials

Examination question analysis

Chapter summary

Cumulative examination: Calculator-free

Cumulative examination: Calculator-assumed

Syllabus coverage

TOPIC 1.1: COUNTING AND PROBABILITY

Combinations

1.1.1 understand the notion of a combination as a set of r objects taken from a set of n distinct objects

1.1.2 use the notation $\binom{n}{r}$ and the formula $\binom{n}{r} = \dfrac{n!}{r!(n-r)!}$ for the number of combinations of r objects taken from a set of n distinct objects

1.1.3 investigate Pascal's triangle and its properties to link $\binom{n}{r}$ to the binomial coefficients of the expansion of $(x+y)^n$ for small positive integers n

Language of events and sets

1.1.5 use set language and notation for events, including:
 a. $\overline{A}$ (or A') for the complement of an event A
 b. $A \cap B$ and $A \cup B$ for the intersection and union of events A and B respectively
 c. $A \cap B \cap C$ and $A \cup B \cup C$ for the intersection and union of the three events A, B and C respectively
 d. recognise mutually exclusive events.

1.1.6 use everyday occurrences to illustrate set descriptions and representations of events and set operations

Mathematics Methods ATAR Course Year 11 syllabus pp. 8–9 © SCSA

Video playlists (4):
1.1 Set descriptions and operations
1.2 Combinations – counting how many ways
1.3 Pascal's triangle and binomial expansions
Examination question analysis Sets and combinations

Worksheets (4):
1.2 Counting techniques • Factorial notation • Combinations • Combination calculations

Nelson MindTap

To access resources above, visit
cengage.com.au/nelsonmindtap

1.1 Set descriptions and operations

The language and notation of sets

In mathematics, it is quite common for us to talk about a collection of objects which have a common attribute. For example, {… −3, −2, −1, 0, 1, 2, 3, 4 …} is the collection of numbers which are integers, $\mathbb{Z}$. These collections can be referred to as **sets** where each individual number within the set is called an **element**. For example, 2 is an element of the set of integers, but 0.5 is not. We can express this mathematically as $2 \in \mathbb{Z}$ but $0.5 \notin \mathbb{Z}$, where $\in$ can be read as 'is an element of' or 'is in' and so $\notin$ means 'is not in'. However, sets are not restricted to numerical values but can be any collection of objects, qualities or quantities, and are often named using a capital letter.

Suppose we were to let A represent the set of primary colours, then we can list the elements of the set using set brackets, also called braces.

A = {red, yellow, blue}

We can see that the number of elements in set A is 3, which can be written as $n(A) = 3$, and is called the **cardinality** (or size) of the set.

A fundamental understanding of sets is important when dealing with many other topics in mathematics, such as counting techniques and probability (Chapters 1 and 2), but also functions and their domains and ranges (Chapters 3 to 6).

Video playlist
Set descriptions and operations

WORKED EXAMPLE 1 — Working with sets

Let A = {4, 8, 12, 16} and B = {1, 4, 9, 16}.

a State $n(A)$.

b State an element in A that is not an element in B.

Let C be the set of prime numbers between 1 and 10.

c List the elements of C using set notation.

Steps	Working
a Count the number of elements in set A.	$n(A) = 4$
b 1 Identify the elements in A that are also in B.	4 and 16 are in both sets.
2 State an element in A that isn't in B.	$8 \in A$, $8 \notin B$
	or
	$12 \in A$, $12 \notin B$
c Use set brackets to list the elements of set C.	C = {2, 3, 5, 7}

Relationships between sets

In Worked example 1b, we need to examine a relationship between two different sets, A and B, to identify elements that are in one but not the other. The way in which sets relate to one another are called **set operations** and allow us to express relationships such as $8 \in A$, $8 \notin B$ more concisely.

Suppose we have a **universal set** U being the set of all integers from 1 to 16, inclusive. Then

U = {1, 2, 3 … 15, 16}

With A = {4, 8, 12, 16} and B = {1, 4, 9, 16} defined as such, we can then introduce the operations of **complement**, **intersection** and **union**.

Set operation	Definition	Simple language	Notation
Complement	The **complement** of set A is a set that contains all the elements that are in the universal set but not in set A.	'not'	$\bar{A}$ or A'
	Note that you need to be familiar with both notations for the complement, so they will be used interchangeably throughout the chapters!		
Intersection	The **intersection** of sets A and B is a set that contains the elements that are in both sets.	'and'	$A \cap B$
Union	The **union** of sets A and B is obtained by combining or uniting both sets. The elements in the union are in set A or set B or both.	'(inclusive) or'	$A \cup B$

These operations can be defined for multiple sets, but we will only examine the cases in which we have two or three sets.

WORKED EXAMPLE 2 Identifying elements using set operations

Let the universal set $U = \{1, 2, 3 \ldots 15, 16\}$, $A = \{4, 8, 12, 16\}$, $B = \{1, 4, 9, 16\}$ and $C = \{2, 3, 5, 7\}$.

Use set notation to list the elements

a that are not in A

b that are in both A and B

c that are in either A or C or both

d that are in B but not A

e that are in A but neither in B nor C.

Steps		Working
a 1	Use the notation for the complement.	
2	List all the elements that are in U but not in A in set brackets.	$\bar{A} = \{1, 2, 3, 5, 6, 7, 9, 10, 11, 13, 14, 15\}$
b 1	Use the notation for the intersection of A and B.	
2	List all the elements that are common to both A and B in set brackets.	$A \cap B = \{4, 16\}$
c 1	Use the notation for the union of A and C.	
2	List all the elements that are in A or C or both in set brackets.	$A \cup C = \{2, 3, 4, 5, 7, 8, 12, 16\}$
d 1	Interpret 'but not' as 'and not'.	
2	Use the notation for the intersection of B and the complement of A.	
3	List all the elements that are in B but not in A in set brackets.	$B \cap \bar{A} = \{1, 9\}$
e 1	Interpret 'neither … nor' as 'not in either'.	
2	Use the notation for the complement of the union of B and C.	
3	List all the elements that are not in the union of B and C.	$\overline{B \cup C} = \{6, 8, 10, 11, 12, 13, 14, 15\}$
4	List all the elements that are in A.	$A = \{4, 8, 12, 16\}$
5	List all the elements that are in the intersection of the two sets in set brackets.	$(\overline{B \cup C}) \cap A = \{8, 12\}$

You may have noticed in the previous example that A and C, and B and C share no elements in common. That is, $n(A \cap C) = n(B \cap C) = 0$. The set that contains no elements is often referred to as the **empty set** and denoted as $\emptyset$. Alternatively, when two sets have no defined intersection, we can refer to them as **disjoint sets**. This particular relationship between sets will be revisited in Chapter 2: Probability.

Operations with two sets – two-way tables and Venn diagrams

Instead of relying on worded descriptions, set operations can be better visualised in displays such as **two-way tables** or **Venn diagrams**.

A two-way table can be used to show the cardinality of two sets, their complements and their four possible intersections:

- $A \cap B$: both A and B
- $A \cap \bar{B}$: A but not B
- $\bar{A} \cap B$: B but not A
- $\bar{A} \cap \bar{B}$: not A and not B.

	B	$\bar{B}$	Total
A	$n(A \cap B)$	$n(A \cap \bar{B})$	$n(A)$
$\bar{A}$	$n(\bar{A} \cap B)$	$n(\bar{A} \cap \bar{B})$	$n(\bar{A})$
Total	$n(B)$	$n(\bar{B})$	$n(U)$

From a two-way table, an obvious relationship can be seen between any set and its complement.

From this, further results such as the following can be deduced:

- $n(A \cap B) + n(A \cap \bar{B}) = n(A)$
- $n(A \cap B) + n(\bar{A} \cap B) = n(B)$

> **Complement rule**
>
> For a set A and its complement, then
> $$n(A) + n(\bar{A}) = n(U)$$
> or
> $$n(\bar{A}) = n(U) - n(A)$$
> where U is the universal set.

WORKED EXAMPLE 3 Displaying set descriptions in a two-way table

A random sample of 60 high-school students were surveyed and asked whether they took public transport to school. It was also recorded whether the students were in Middle School (Years 7–9) or Senior School (Years 10–12). Let P be the set of students who took public transport to school and M be the set of students in Middle School. It was found that 26 of the students did not take public transport. Of the 35 Middle School students, 14 took public transport.

a Copy and complete the two-way table below.

	M	M'	Total
P			
P'			
Total			60

b State $n(P' \cap M)$ and interpret the result in context of the question.

c Describe the set $P \cup M'$ in context of the question.

Steps		Working
a 1	Input the known information in the table.	<table><tr><td></td><td>M</td><td>M'</td><td>Total</td></tr><tr><td>P</td><td>14</td><td></td><td></td></tr><tr><td>P'</td><td></td><td></td><td>26</td></tr><tr><td>Total</td><td>35</td><td></td><td>60</td></tr></table>
2	Use the complement rule to complete the missing cells.	<table><tr><td></td><td>M</td><td>M'</td><td>Total</td></tr><tr><td>P</td><td>14</td><td>20</td><td>34</td></tr><tr><td>P'</td><td>21</td><td>5</td><td>26</td></tr><tr><td>Total</td><td>35</td><td>25</td><td>60</td></tr></table>
3	Check that all rows and columns add up correctly.	
b 1	Identify the correct intersection of P' and M in the table.	$n(P' \cap M) = 21$
2	State the cardinality of the set.	21 Middle School students did not take public transport.
c 1	Identify the meaning of each set.	P: took public transport M': Senior School students
2	Describe the union of the two sets.	$P \cup M'$ describes the set of students who either took public transport or are in Senior School or both.

Similarly, a Venn diagram for two sets shows the four possible relationships that can exist between the two sets and the cardinality of each set. The outer rectangle represents the universal set and the number of elements in this set is typically written in a box in the corner. Sometimes, 'bubbles' within the sets A and B are drawn to include the total in each of the two sets.

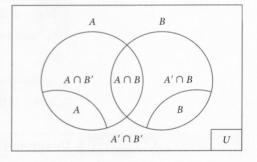

From a Venn diagram, another obvious relationship that can be seen between any two sets is the relationship between their union and their intersection. If we wanted to count all the elements in either A or B or both, we would have:

$n(A \cup B) = n(A \cap B') + n(A \cap B) + n(A' \cap B)$

but $n(A \cap B') + n(A \cap B)$ is just $n(A)$,

$n(A \cup B) = n(A) + n(A' \cap B)$

and $n(A' \cap B)$ is $n(B) - n(A \cap B)$,

$n(A \cup B) = n(A) + n(B) - n(A \cap B)$

Addition rule

For two sets A and B

$n(A \cup B) = n(A) + n(B) - n(A \cap B)$

De Morgan's laws are useful set equivalences that can also be shown using a Venn diagram.

1. $\overline{A \cap B} = \overline{A} \cup \overline{B}$

   Not both A and B is equivalent to not A or not B.

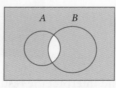

2. $\overline{A \cup B} = \overline{A} \cap \overline{B}$

   Neither A nor B is equivalent to not A and not B.

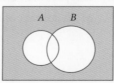

WORKED EXAMPLE 4 — Displaying set descriptions in a Venn diagram

In a random sample of 200 departing travellers at Perth International Airport, 110 were travelling alone and 35 had overweight baggage. Thirty of the travellers with overweight baggage were travelling alone. Let *A* be the set of travellers travelling alone and *B* be the set of travellers with overweight baggage.

Use a Venn diagram to determine the number of travellers who were not travelling alone and did not have overweight baggage.

Steps	Working
1 Construct a two-circle Venn diagram, labelled *A* and *B*. 2 Complete the diagram with the known information.	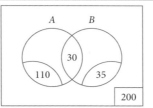
3 Identify the set that needs to be found. **Exam hack** Where appropriate, use De Morgan's laws to recognise the equivalence of the two sets.	Not travelling alone: $\bar{A}$ and: $\cap$ Did not have overweight baggage: $\bar{B}$ require: $\bar{A} \cap \bar{B}$
4 Use the addition rule to find $n(A \cup B)$.	$n(A \cup B) = n(A) + n(B) - n(A \cap B)$ $= 110 + 35 - 30$ $= 115$
5 Use the complement rule to find $n(\overline{A \cup B})$.	$n(\overline{A \cup B}) = 200 - 115 = 85$

The set operations in two-way tables and Venn diagrams

For two sets *A* and *B*	In a two-way table	In a Venn diagram
Complement of *A*		*B* | *B'* *A* | | *A'* | |	
Intersection of *A* and *B*		*B* | *B'* *A* | | *A'* | |	
Union of *A* and *B*		*B* | *B'* *A* | | *A'* | |	
Disjoint sets, *A* and *B*		*B* | *B'* *A* | 0 | *A'* | |	

Note that if *A* and *B* are disjoint sets, then $n(A \cap B) = 0$ and so $n(A \cup B) = n(A) + n(B)$.

Exam hack

In more complex set problems, you may need to use algebraic techniques to solve problems. For example, if one of the sets is unknown, let the set size be *x* and see if you can form an equation involving the known information.

Operations with three sets – Venn diagrams

When the relationships between three sets is being examined, a two-way table becomes impractical and so we rely on a three-circle Venn diagram to display the eight possible regions involving the set descriptions. The most efficient way to identify the region being described is to use the numbers 1 to 8 to arbitrarily represent each region. For example, consider the Venn diagram shown.

Let $A = \{1, 2, 6, 7\}$, $B = \{2, 3, 4, 7\}$ and $C = \{4, 5, 6, 7\}$. This numbering system can be used to identify any set description.

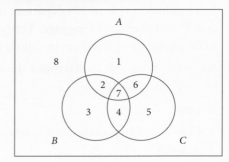

$A \cap B \cap C$ $\{7\}$

as it is the only number common to all three sets.

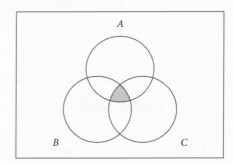

$A \cup B \cup C$ $\{1, 2, 3, 4, 5, 6, 7\}$

as it contains any number in A or B or C, or any intersection of the three.

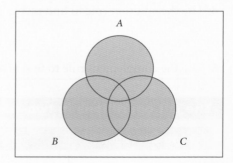

For something more complex, we can use the set operations to determine which regions to shade. For example, $A \cap B \cup \overline{C}$ would equate to

$\{1, 2, 6, 7\} \cap \{2, 3, 4, 7\} \cup \overline{\{4, 5, 6, 7\}}$

$= \{2, 7\} \cup \{1, 2, 3, 8\}$

$= \{1, 2, 3, 7, 8\}$

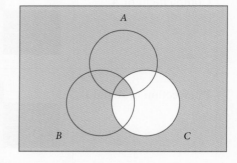

🔓 Exam hack

When identifying or representing set descriptions involving three sets, think of the ∩ ('and') and the ∪ ('or') like you would with '×' and '+' in the order of operations. If there are brackets, group those first, otherwise ∩ before ∪.

WORKED EXAMPLE 5 Identifying regions in a three-circle Venn diagram

a Draw a three-circle Venn diagram involving sets P, Q and R such that P is disjoint from Q and disjoint from R, but Q and R are not disjoint.

b Hence, shade the region representing $P \cup Q \cap R$.

Steps	Working
a Draw the diagram satisfying the conditions.	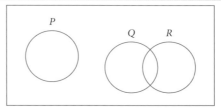
b 1 Use the numbering technique to describe the sets.	Let $P = \{1\}$, $Q = \{2, 3\}$, $R = \{3, 4\}$.
2 Use the order of set operations to simplify the set.	$P \cup Q \cap R = \{1\} \cup \{2, 3\} \cap \{3, 4\}$ $P \cup Q \cap R = \{1\} \cup \{3\}$
3 Shade the correct region.	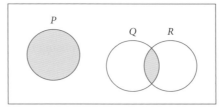

Note that if we had a situation in which each of the three sets A, B and C were disjoint from every other set, then the sets are called **pairwise disjoint**; that is, any pair of the three sets is disjoint.

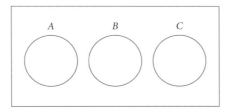

🔓 Exam hack

Be careful with three-set problems when identifying intersections between two sets from worded descriptions, as there is a difference between *A and B* (which includes $A \cap B \cap C$) and *A and B only* (which is $A \cap B \cap \bar{C}$).

WORKED EXAMPLE 6 — Solving problems using a three-circle Venn diagram

In a cohort of students, 80 study Mathematics, 60 study Physics and 70 study Modern History. Thirty of the students study both Mathematics and Physics, 25 study both Physics and Modern History, and 20 study both Mathematics and Modern History. There are only 10 students who study all three subjects.

Use a Venn diagram to determine the number of students who study exactly one of the three subjects.

Steps	Working
1 Define the three sets.	Let M be the students who study Mathematics, P be the students who study Physics and H be the students who study Modern History.
2 Construct a three-circle Venn diagram with the correct labels. 3 Complete the diagram with the known information, using 'bubbles' to indicate the totals of a region.	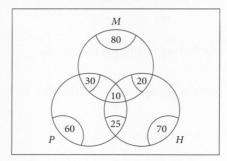
4 Complete the remaining information using the totals of each region.	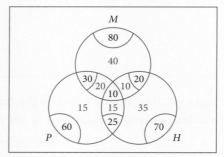
5 Identify the set that needs to be found.	Exactly one of the three subjects means • Mathematics, but not Physics or Modern History • Physics, but not Mathematics or Modern History • Modern History, but not Mathematics or Physics. Let this set be X. $n(X) = 40 + 15 + 35 = 90$
6 Write the answer.	Therefore, 90 students study exactly one of the three subjects.

EXERCISE 1.1 Set descriptions and operations

ANSWERS p. 446

Mastery

1 **WORKED EXAMPLE 1** Let $A = \{2, 3, 5, 7\}$ and $B = \{2, 4, 6, 8, 10\}$.
   a State $n(B)$.
   b State an element in A that is also an element in B.
   Let C be the set of multiples of 3 between 1 and 10.
   c List the elements of C using set notation.

2 **WORKED EXAMPLE 2** Let the universal set $U = \{1, 2, 3 \ldots 9, 10\}$, $A = \{2, 3, 5, 7\}$, $B = \{2, 4, 6, 8, 10\}$ and $C = \{3, 6, 9\}$.

Use set notation to list the elements

a that are not in A

b that are in both A and C

c that are in either A or B or both

d that are in C but in neither A nor B

e that are in A, B and C.

3 **WORKED EXAMPLE 3** A group of 125 high-school students were surveyed in the line for the school cafeteria and asked of their preferences for two types of items: hot food and fizzy drinks. Let H be the set of students who like hot food and F be the set of students who like fizzy drinks. It was found that 65 students like hot food, 60 students like fizzy drinks and 25 students like hot food and fizzy drinks.

a Copy and complete the two-way table below.

	F	F'	Total
H			
H'			
Total			125

b State $n(H' \cap F')$ and interpret the result in context of the question.

c Describe the set $H' \cup F'$ in context of the question.

4 **WORKED EXAMPLE 4** In a small company of 40 employees, there are two main projects being worked on currently: Project A and Project B. There are 20 employees working on Project A and 15 employees working on Project B. There are 5 employees who are working on both projects.

Use a Venn diagram to determine the number of employees who are not working on either of the projects.

5 **WORKED EXAMPLE 5**

a Draw a three-circle Venn diagram involving sets P, Q and R such that P is not disjoint from Q and Q is not disjoint from R, but P and R are disjoint.

b Hence, shade the region representing $(P \cup R) \cap Q$.

6 **WORKED EXAMPLE 6** In an online survey regarding hobbies conducted among 200 participants, it was found that 120 participants enjoy reading books, 90 participants enjoy watching movies and 80 participants play musical instruments. Sixty of the participants enjoy both reading books and watching movies, 20 enjoy watching movies and playing musical instruments, and 50 enjoy reading books and playing musical instruments. Ten participants were engaged with all three hobbies.

Use a Venn diagram to determine the number of participants who were engaged in at least two out of the three hobbies.

Calculator-free

7 (5 marks) Let the universal set be the integers from 1 to 15, inclusive. Let the set A contain the multiples of 2, set B contain the multiples of 3 and set C contain the multiples of 5.

  a State $n(\overline{A})$. (1 mark)

  b Determine $\overline{A \cup B \cup C}$. (2 marks)

  c List the elements of $B \cap C$. (1 mark)

  d Explain why $n(A \cap B \cap C) = 0$. (1 mark)

8 (3 marks) A small local bookstore sells fiction and non-fiction books, either in paperback or hardcover. Let F be the set of fiction books and P be the set of paperback books. The number of books currently in stock is partially shown in the two-way table below.

	P	P'	Total
F	45		
F'			25
Total		32	

If the number of fiction hardcover books in stock is a third of the number of non-fiction hardcover books, determine the total number of books in the bookstore.

9 (6 marks) At a local university, there are two main student clubs: U-Club and P-Club. In a sample of 80 students, it was found that 10 students were members of both clubs. Let x be the number of students in the sample who were members of U-Club and y be the number of students in the sample who were members of P-Club.

  a Determine a simplified algebraic expression for the number of students in the sample who were not members of either club. (3 marks)

It was later realised that the number of students who were members of P-Club is half the number of students who were members of U-Club.

  b If 45 out of the 80 were not members of either club, determine the number of students who were members of one of the two clubs, but not both. (3 marks)

Calculator-assumed

10 (2 marks) A Thai restaurant offers a menu containing 65 dishes, which includes dishes with chicken, pork or prawn. The number of chicken dishes is 15, the number of pork dishes is 15 and the number of prawn dishes is 12. If there are 25 dishes that do not contain any of the three ingredients, determine the number of dishes on the menu that contain chicken or pork, but not prawn.

1.2 Combinations – counting how many ways

The notion of a combination

Suppose there are six distinct books sitting on a shelf.

Let the six books be represented by the set {A, B, C, D, E, F}.

To understand **combination**, we first need to consider the question: in how many ways can these six books be arranged on the shelf? Without listing all the possibilities, we can instead think of it as how many options are there for each place on the shelf? Once a book is placed, it cannot occupy another place.

Place 1	Place 2	Place 3	Place 4	Place 5	Place 6
6 options	5 options	4 options	3 options	2 options	1 option

This arrangement of books on a shelf is called **permutation** and gives $6 \times 5 \times 4 \times 3 \times 2 \times 1$ ways of arranging these six books, i.e. 120 possible arrangements.

This product $6 \times 5 \times 4 \times 3 \times 2 \times 1$ can be written more efficiently as 6!, read as '6 **factorial**'.

Suppose we also want to consider how many possible arrangements of two books can be made from the six. Now with only two places, we have:

Place 1	Place 2
6 options	5 options

which gives the calculation

$$\frac{6 \times 5 \times 4 \times 3 \times 2 \times 1}{4 \times 3 \times 2 \times 1} = \frac{6!}{(6-2)!} = \frac{6!}{4!} = 30$$

The division by 4! represents the removal of 4 places from the total number of places.

> **Permutations and factorials**
>
> A **permutation** is an arrangement, or ordered selection, of n objects into r places. The formal mathematical notation for this is ^nP_r, which can be defined by the formula
>
> $$^nP_r = \frac{n!}{(n-r)!}$$
>
> When $n = r$ and the objects are distinct, then the number of possible permutations is $n!$, read as 'n **factorial**', where
>
> $$n! = n \times (n-1) \times (n-2) \times \ldots \times 1$$
>
> or
>
> $$n! = n \times (n-1)!$$
>
> This is because $0! = 1$; that is, there is only one way to arrange 0 objects – you can't arrange them!

Now consider the difference between the following two questions:

How many possible arrangements of two books can be made from six books?

and

How many possible selections of two books can be made from six books?

The first question implies there is an order to the arrangement, whereas the second question implies that order is not important, and we are only concerned with which two books we end up with.

So, suppose we consider our shelf with two places and 'remove' the concept of order.

Place 1	Place 2
6 options	5 options
there are 2 places to put the book	there is only 1 place to put the book

$$\frac{6 \times 5 \times 4 \times 3 \times 2 \times 1}{2 \times 1 \times 4 \times 3 \times 2 \times 1} = \frac{6!}{2!(6-2)!} = \frac{6!}{2!4!} = 15$$

There are 15 possible selections of two books that we can make from six books. If we think systematically about what those options are, we have:

A, B	A, C	A, D	A, E	A, F
B, C	B, D	B, E	B, F	C, D
C, E	C, F	D, E	D, F	E, F

This notion of an unordered selection is called a combination and can be derived by first thinking about selecting r objects from n objects and arranging them into r places and then removing the concept of order.

WORKED EXAMPLE 7 — Deriving the notion of combination

A family has 8 photos laid out on the kitchen table.

a How many possible arrangements of the 8 photos are there?

There are 3 spots on the wall to hang the photos.

b How many possible arrangements of 3 photos can be made from the 8 photos?

c Hence, calculate the number of ways 3 photos can be selected from the 8 photos, such that the order of selection is not important.

Steps	Working
a Use factorials to calculate the number of possible arrangements of the 8 photos.	$8! = 8 \times 7 \times 6 \times 5 \times 4 \times 3 \times 2 \times 1$ $8! = 40\,320$
b Use the understanding of a permutation to calculate the number of possible arrangements of 8 photos into 3 places, by dividing the answer in part **a** by $(8-3)!$	Place 1: 8 options, Place 2: 7 options, Place 3: 6 options $\frac{8!}{5!} = 8 \times 7 \times 6 = 336$
c Remove the concept of order from part **b** by dividing by 3!	Place 1: 8 options, there are 3 places to put the photo Place 2: 7 options, there are 2 places to put the photo Place 3: 6 options, there is only 1 place to put the photo $\frac{8!}{5!3!} = \frac{8 \times 7 \times 6}{3 \times 2 \times 1} = \frac{336}{6} = 56$

Using the combinations formula

Instead of deriving an understanding of a combination every time from the idea that it is a permutation without the order, we can generalise that idea using the combinations formula.

Combinations

A **combination** is an unordered selection of r objects from a set of n distinct objects. The formal mathematical notation for this is ^nC_r, which can be defined by the formula

$$^nC_r = {^nP_r} \div r! = \frac{n!}{r!(n-r)!}$$

This is often read as 'n choose r'.

An alternative notation for combinations is the binomial coefficient, $\binom{n}{r}$.

Some useful facts to recognise are:

- $^nC_0 = 1$, as there is only one possible way to select 0 objects from n objects. We don't select anything!
- $^nC_1 = n$, as there are n possible objects that we could select.
- $^nC_{n-1} = n$, as there are n possible objects that we could leave out.
- $^nC_n = 1$, as there is only one possible way to select n objects from n objects. We select everything!

This formula is useful to evaluate combinations without a calculator.

WORKED EXAMPLE 8 — Using the combinations formula

Use the combinations formula to calculate the following.

a $^7C_3$ **b** $^6C_4$ **c** $^{10}C_5$

Steps	Working
a 1 Substitute $n = 7$ and $r = 3$ into the combinations formula.	$\dfrac{7!}{3!4!} = \dfrac{7 \times 6 \times 5}{3 \times 2 \times 1}$
2 Express the fraction as the expanded multiplication of $^nP_r \div r!$	
3 Simplify by cancelling common factors.	$= 7 \times 5$ $= 35$
b 1 Substitute $n = 6$ and $r = 4$ into the combinations formula.	$\dfrac{6!}{4!2!} = \dfrac{6 \times 5}{2 \times 1}$
2 Express the fraction as the expanded multiplication of $^nP_r \div r!$	
3 Simplify by cancelling common factors.	$= 3 \times 5$ $= 15$
c 1 Substitute $n = 10$ and $r = 5$ into the combinations formula.	$\dfrac{10!}{5!5!} = \dfrac{10 \times 9 \times 8 \times 7 \times 6}{5 \times 4 \times 3 \times 2 \times 1}$
2 Express the fraction as the expanded multiplication of $^nP_r \div r!$	
3 Simplify by cancelling common factors.	$= 3 \times 2 \times 7 \times 6$ $= 252$

When the computations become too complicated with larger values, CAS can be used to carry out the calculations.

USING CAS 1 Evaluating factorials, permutations and combinations

Evaluate the following.

a 12! b $^{12}P_3$ c $^{12}C_3$

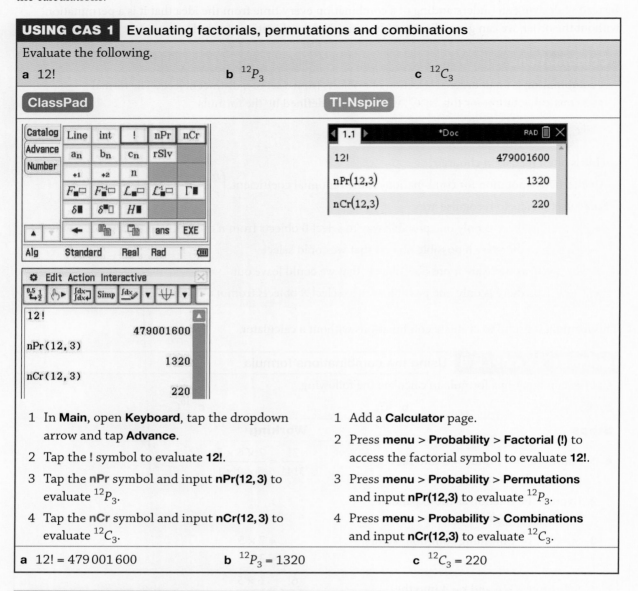

1. In **Main**, open **Keyboard**, tap the dropdown arrow and tap **Advance**.
2. Tap the **!** symbol to evaluate **12!**.
3. Tap the **nPr** symbol and input **nPr(12, 3)** to evaluate $^{12}P_3$.
4. Tap the **nCr** symbol and input **nCr(12, 3)** to evaluate $^{12}C_3$.

1. Add a **Calculator** page.
2. Press **menu > Probability > Factorial (!)** to access the factorial symbol to evaluate **12!**.
3. Press **menu > Probability > Permutations** and input **nPr(12,3)** to evaluate $^{12}P_3$.
4. Press **menu > Probability > Combinations** and input **nCr(12,3)** to evaluate $^{12}C_3$.

a 12! = 479 001 600 b $^{12}P_3$ = 1320 c $^{12}C_3$ = 220

WORKED EXAMPLE 9 Solving simple problems involving combinations

A regular deck of cards contains 52 playing cards. How many different hands are possible in a game that requires

a 5 cards in a hand
b 13 cards in a hand?

Steps	Working
a 1 Identify n and r and state the combination being calculated.	$n = 52, r = 5$
2 Use CAS, where appropriate, to evaluate the combination.	$^{52}C_5 = 2\,598\,960$
b 1 Identify n and r and state the combination being calculated.	$n = 52, r = 13$
2 Use CAS, where appropriate, to evaluate the combination.	$^{52}C_{13} = 635\,013\,559\,600$

The multiplication and addition principles

From a group of 20 students, suppose there are two subsets: 15 Year 11 students and 5 Year 12 students. If we wanted to select 5 students from the 20, then we know that there would be $^{20}C_5 = 15\,504$ different selections. However, what if a restriction was placed or further conditions were specified when making the selection? For example, the group of 5 selected must contain 3 Year 11 students AND 2 Year 12 students. Alternatively, the group can either be 5 Year 11s OR 5 Year 12s. To answer these questions involving 'AND' and 'OR', we must introduce the **multiplication** and **addition principles**.

The multiplication principle

Consider a set of R objects being selected from a set of N distinct objects. When there are two or more successive selections of $r_1, r_2 \ldots$, being made from subsets of size $n_1, n_2 \ldots$, then the total number of possible selections is calculated using the **multiplication principle**

$$^{n_1}C_{r_1} \times {}^{n_2}C_{r_2} \times \ldots$$

where $N = n_1 + n_2 + \ldots$ and $R = r_1 + r_2 + \ldots$

 Exam hack

Look for subsets and the word AND in problems involving the multiplication principle!

WORKED EXAMPLE 10 | Using the multiplication principle with combinations

For holiday homework, Freda must complete three Mathematics problems from a selection of eight, and write two English essays from a selection of six possible questions. Determine the number of possible ways that Freda can complete her holiday homework.

Steps	Working
1 Identify the two subsets and their corresponding n and r values.	Mathematics: $n = 8, r = 3$ English: $n = 6, r = 2$
2 Check for the word AND connecting the two subsets.	$^8C_3 \times {}^6C_2$
3 Establish the combinations for each subset, using the multiplication principle.	$= 56 \times 15$
4 Evaluate using CAS, where appropriate.	$= 840$

The addition principle

If the selections are being made from two or more subsets, such that r_1 objects are being selected from n_1 objects OR r_2 objects are being selected from n_2 objects and so on, then the total number of possible selections is calculated using the **addition principle**

$$^{n_1}C_{r_1} + {}^{n_2}C_{r_2} + \ldots$$

Note that if the two subsets, say A and B, contain an intersection, then the addition rule needs to be used such that

$$n(A \cup B) = n(A) + n(B) - n(A \cap B)$$

Otherwise, if they are disjoint subsets, then

$$n(A \cup B) = n(A) + n(B)$$

 Exam hack

Look for subsets and the word OR in problems involving the addition principle!

WORKED EXAMPLE 11 — Using the addition and multiplication principles with combinations

At a hotel, Elijah has the following breakfast options available to him:
- Cereals: bran, oats, granola
- Toast: white, rye or wholemeal
- Eggs: fried, poached or scrambled

Determine the number of possible breakfast selections Elijah could make if he orders

a one cereal option, one toast option and one egg option

b one cereal option or one toast option

c either one cereal option, or one toast and egg option.

Steps	Working
For each of the parts: 1 Identify whether it is an AND or OR question. 2 Establish the combinations, either using the multiplication or addition principles, or both. 3 Use CAS, where appropriate, to calculate the total number of options.	**a** AND question → multiplication principle $^3C_1 \times {}^3C_1 \times {}^3C_1$ $= 3 \times 3 \times 3$ $= 27$ **b** OR question → addition principle $^3C_1 + {}^3C_1$ $= 3 + 3$ $= 6$ **c** AND and OR question → multiplication and addition principles $^3C_1 + {}^3C_1 \times {}^3C_1$ $= 3 + 3 \times 3$ $= 3 + 9$ $= 12$

> 🔒 **Exam hack**
> Order of operations applies!

EXERCISE 1.2 Combinations – counting how many ways   ANSWERS p. 446

Recap

1 A basket of 50 toys contains 20 toys that have wheels, 18 toys that are red and 12 red toys that have wheels. The number of toys that are not red and do not have wheels is

   A 6   **B** 8   **C** 12   **D** 24   **E** 26

2 In a group of 50 students, 32 are wearing sunglasses (S), 30 are wearing hats (H) and 15 are wearing both sunglasses and hats. The value of $n(S \cap \overline{H})$ is

   A 3   **B** 15   **C** 17   **D** 30   **E** 32

Mastery

3 **WORKED EXAMPLE 7** A basketball team has 8 players.

  a How many possible ways can the 8 players be arranged in a line on the bench?

  The coach needs to choose 5 players for the match and sit them in the first five spots on the bench.

  b How many possible arrangements of 5 players can be made from the team of 8?

  c Hence, calculate the number of ways 5 players can be selected from the team of 8, such that the order of selection is not important.

4 **WORKED EXAMPLE 8** Use the combinations formula to calculate the following.

  a $^5C_2$ **b** $^7C_3$ **c** $^4C_1$ **d** $^7C_2$ **e** $^8C_4$

  f $^{11}C_{10}$ **g** $^{12}C_{10}$ **h** $^{100}C_{99}$ **i** $^{40}C_{38}$ **j** $^{15}C_{13}$

5 **Using CAS 1** Evaluate the following.

   a 15!
   b $^{15}P_2$
   c $^{15}C_2$

6 **WORKED EXAMPLE 9** Andrew has a collection of 22 books at home, all of which he hasn't read. How many different selections of books are there if he is to read

   a 3 books on the holidays
   b 8 books on the holidays?

7 **WORKED EXAMPLE 10** Carolyn, a university student, must choose 6 out of a possible 12 units of Mathematics and 4 units out of a possible 18 units of Science, over the course of her degree. How many different combinations of units are possible?

8 **WORKED EXAMPLE 11** Samantha is planning her Saturday activities from the following options:
   - Morning activity: run, cycle, swim or walk
   - Afternoon activity: watch a football match, go shopping, visit the art gallery
   - Evening activity: watch a movie, a games night at a friend's house, read a book at home

   Determine the number of possible Saturday activities Samantha could have if she

   a does one activity in the morning, one in the afternoon and one in the evening
   b does one activity in the morning or two activities in the afternoon
   c does either one morning and one afternoon activity, or two afternoon and two evening activities.

Calculator-free

9 (7 marks)

   a A squad of 15 basketball players is reduced to a final team of 12 players. Calculate the number of possible teams. (2 marks)
   b A relay team of 4 runners is selected from a group of 8 athletes. Calculate the number of different teams that are possible. (2 marks)
   c A mixed volleyball team of 6 players is selected from 7 females and 5 males. Calculate the number of possible teams that contain 3 female and 3 male players. (3 marks)

10 (7 marks) A storage box contains 8 books, of which 5 are fiction and 3 are non-fiction. Ben selects 3 books to read over the school holidays. How many selections of books are possible if Ben

   a does not mind what he selects (2 marks)
   b selects all non-fiction books (1 mark)
   c selects all fiction books (2 marks)
   d selects at least one fiction book? (2 marks)

Calculator-assumed

11 (11 marks) A student council of 12 students is to be formed from 4 Year 9 nominations, 4 Year 10 nominations, 6 Year 11 nominations and 8 Year 12 nominations. Determine the total number of possible student councils if

   a there are no restrictions (2 marks)
   b there must be an equal representation of each year group (3 marks)
   c all the Year 12s are included on the council (3 marks)
   d there are either no Year 9s or no Year 10s on the council. (3 marks)

1.3 Pascal's triangle and binomial expansions

Deriving Pascal's triangle and its properties

You may recall certain facts from your understanding of combinations, such as:

- $^nC_0 = 1$, as there is only one possible way to select 0 objects from n objects. You don't select anything!
- $^nC_1 = n$, as there are n possible objects that you could select.
- $^nC_{n-1} = n$, as there are n possible objects that you could leave out.
- $^nC_n = 1$, as there is only one possible way to select n objects from n objects. You select everything!

However, these facts are only a small part of a larger generalisation that can be made about the number of ways r objects can be selected from n distinct objects. Suppose we formed a triangle showing all these possible selections, in which the row of the triangle represents the value of n, i.e. the number of objects we are selecting from, where $n \geq 0$ and the rth entry along that row represents the number of objects we are selecting, where $0 \leq r \leq n$.

We would form a triangle such as the one below.

$$\binom{0}{0}$$
$$\binom{1}{0} \quad \binom{1}{1}$$
$$\binom{2}{0} \quad \binom{2}{1} \quad \binom{2}{2}$$
$$\binom{3}{0} \quad \binom{3}{1} \quad \binom{3}{2} \quad \binom{3}{3}$$
$$\binom{4}{0} \quad \binom{4}{1} \quad \binom{4}{2} \quad \binom{4}{3} \quad \binom{4}{4}$$

Evaluating the first five rows of this triangle, we obtain the following numerical values:

$$\begin{array}{c} 1 \\ 1 \quad 1 \\ 1 \quad 2 \quad 1 \\ 1 \quad 3 \quad 3 \quad 1 \\ 1 \quad 4 \quad 6 \quad 4 \quad 1 \end{array}$$

This is the beginning of what is called **Pascal's triangle**, representing the number of possible selections of r objects from n distinct objects. However, evidence of the triangle's existence dates back much earlier than Pascal in the 17th century and can be found in ancient Chinese mathematics of the 11th and 13th centuries, by Jia Xian and Yang Hui.

The triangle highlights some useful properties of combinations and other numerical relationships.

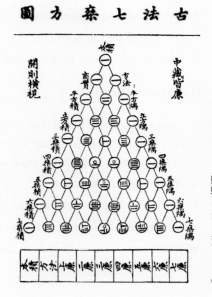

Source: Yáng Huī (楊輝), ca. 1238–1298, Public domain via Wikimedia Commons

Property	Description	Example from Pascal's triangle
The symmetry of combinations	From the triangle, we can see that in every row: • $^nC_0 = {}^nC_n$ • $^nC_1 = {}^nC_{n-1}$ More generally, $^nC_r = {}^nC_{n-r}$	$$\begin{array}{ccccccccc} & & & & 1 & & & & \\ & & & 1 & & 1 & & & \\ & & 1 & & 2 & & 1 & & \\ & 1 & & 3 & & 3 & & 1 & \\ 1 & & \boxed{4} & & 6 & & \boxed{4} & & 1 \end{array}$$ $^4C_1 = {}^4C_3$
The additive property of combinations	In any subsequent row, an entry is found by adding the two entries either side of it in the previous row. $^nC_r = {}^{n-1}C_{r-1} + {}^{n-1}C_r$	$$\begin{array}{ccccccccc} & & & & 1 & & & & \\ & & & 1 & & 1 & & & \\ & & 1 & & 2 & & 1 & & \\ & 1 & & 3 & & 3 & & 1 & \\ 1 & & 4 & & 6 & & 4 & & 1 \end{array}$$ Term 0 Term 1 Term 2 Term 3 Term 4 $1 + 3 = 4$ Add the two numbers in the row above to get the numbers in the next row. $^4C_1 = {}^3C_0 + {}^3C_1$
The total number of subsets from n objects	The number of possible **subsets** (including the empty set and the set itself) that can be made from n objects is the sum of the nth row of Pascal's triangle and is equal to 2^n.	Suppose we want to know how many possible subsets of any size can be made from three objects $\{A, B, C\}$. • $^3C_0 = 1$: the empty set, $\varnothing$ • $^3C_1 = 3$: $\{A\}, \{B\}, \{C\}$ • $^3C_2 = 3$: $\{A, B\}, \{A, C\}, \{B, C\}$ • $^3C_3 = 1$: $\{A, B, C\}$ In total, $1 + 3 + 3 + 1 = 8$, which is 2^3.

Note that a **subset**, by definition, is a set that can be formed from any number of elements of another given set, including the empty set. For example, if set B can be formed from the elements of set A, then B is a subset of A, denoted as $B \subseteq A$. Subsets can also be equal to the original set, and hence the 'line' under the symbol. If a subset B is strictly smaller than A in size, then it is called a **proper subset**, denoted as $B \subset A$.

> **Number of possible subsets**
>
> If A is a set containing n elements, then there are 2^n possible **subsets** that can be made from A, including the empty set and the set A itself.
>
> If the set A itself is excluded as a subset, then there are 2^{n-1} possible **proper subsets** that can be made from A.

WORKED EXAMPLE 12 — Investigating Pascal's triangle

Consider the first five rows of Pascal's triangle.

```
                1
              1   1
            1   2   1
          1   3   3   1
        1   4   6   4   1
```

a Determine the sixth row, i.e. when $n = 5$.

b State the value of r, $r \neq 2$ that makes this combinatorial identity true: $^5C_2 = {}^5C_r$.

c Determine the total number of possible subsets that can be formed from the set $\{A, B, C, D, E\}$.

Steps	Working
a Use the diagonal addition property of Pascal's triangle to determine the next row.	``` 1 1 1 1 2 1 1 3 3 1 1 4 6 4 1 1 5 10 10 5 1 ```
b Use the symmetrical property of Pascal's triangle to state the value of r.	$r = 5 - 2 = 3$
c 1 Count the number of elements in the set. 2 Use the row sum property (or 2^n) to calculate the number of possible subsets.	$n = 5$ $2^n = 2^5 = 32$

Binomial expansions and their coefficients

Another significant use of Pascal's triangle is in the expansion of **binomial expressions**. Consider the generalised powers of a binomial expression, $(x + y)^n$ for $n \geq 0$.

Suppose that $n = 0$, then $(x + y)^0 = 1$.

When $n = 1$, then $(x + y)^1 = x + y$, which may be better represented as $1x + 1y$.

You may also recall that when $n = 2$, then $(x + y)^2 = 1x^2 + 2xy + 1y^2$.

What do you notice?

There are a few things you should notice:

- The expansions are starting with the highest possible power of x, and subsequently losing one power at a time to the powers of y. For example, when $n = 2$,
  $$x^2y^0 \to x^1y^1 \to x^0y^2$$
- The coefficients of the expansions of $(x + y)^n$ are the terms in the nth row of Pascal's triangle.
  Why is that so?

We can think about this as a repeated selection problem involving the product of the elements of the set $\{x, y\}$. For now, we will start with $n = 2$.

Suppose we are allowed two selections, and we are allowed to select the same element twice. Let's investigate!

How many ways can you form a product of two xs?	1 way	xx
How many ways can you form a product of one x and one y?	2 ways	xy yx
How many ways can you form a product of no xs?	1 way	yy

If we were to repeat this investigation such that n = 3, that is, we can make three selections.

How many ways can you form a product of three xs?	1 way	xxx
How many ways can you form a product of two xs and one y?	3 ways	xxy
		xyx
		yxx
How many ways can you form a product of one x and two ys?	3 ways	xyy
		yxy
		yyx
How many ways can you form a product of no xs?	1 way	yyy

This would mean that the expansion of $(x + y)^3 = x^3 + 3x^2y + 3xy^2 + y^3$.

From here, we can generalise the binomial expansion of $(x + y)^n$, which is often called the **binomial theorem**.

> **The binomial theorem**
>
> For $(x + y)$, where $n \geq 0$, then
>
> $$(x + y)^n = \binom{n}{0}x^n y^0 + \binom{n}{1}x^{n-1}y^1 + \binom{n}{2}x^{n-2}y^2 + \ldots + \binom{n}{n}x^0 y^n$$
>
> and the set of values $\left\{\binom{n}{0}, \binom{n}{1}, \binom{n}{2} \ldots \binom{n}{n}\right\}$ is called the **binomial coefficients**.

 Exam hack

For every term in the expansion, the powers of x and y should sum to n.

If the value of x or the value of y (or both) is numerical, then simply substitute it in the place of the variable.

 Exam hack

Be careful when
- either x or y is a negative value, as $-k^n \neq (-k)^n$
- the x or y term has a coefficient other than 1, as $(kx)^n = k^n x^n$.

WORKED EXAMPLE 13 | Expanding powers of binomials

Use an appropriate row of Pascal's triangle to expand the following.

a $(10 + 1)^4$ **b** $(x + 1)^4$ **c** $(x - 1)^4$ **d** $(2x + 3)^4$

Steps	Working
For each of the following expansions, the fifth row ($n = 4$) of Pascal's triangle is required.	1 4 6 4 1
a 1 Apply the binomial theorem using the coefficients from Pascal's triangle.	$(10 + 1)^4$
2 Start with the highest power of 10 and decrease the power by 1 each term, multiplying by the corresponding power of 1.	$= 10^4 + 4(10)^3(1)^1 + 6(10)^2(1)^2 + 4(10)^1(1)^3 + 1^4$
3 Simplify the result.	$= 10\,000 + 4000 + 600 + 40 + 1$
	$= 14\,641$

b 1 Apply the binomial theorem using the coefficients from Pascal's triangle. $(x + 1)^4$

2 Start with the highest power of x and decrease the power by 1 each term, multiplying by the corresponding power of 1. $= x^4 + 4x^3(1)^1 + 6x^2(1)^2 + 4x^1(1)^3 + 1^4$

3 Simplify the result. $= x^4 + 4x^3 + 6x^2 + 4x + 1$

c 1 Apply the binomial theorem using the coefficients from Pascal's triangle. $(x - 1)^4$

2 Start with the highest power of x and decrease the power by 1 each term, multiplying by the corresponding power of -1. $= x^4 + 4x^3(-1)^1 + 6x^2(-1)^2 + 4x^1(-1)^3 + (-1)^4$

3 Simplify the result. $= x^4 - 4x^3 + 6x^2 - 4x + 1$

> Notice that the signs of the expansion alternate when one of the terms in the binomial is negative!

d 1 Apply the binomial theorem using the coefficients from Pascal's triangle. $(2x + 3)^4$

2 Start with the highest power of $2x$ and decrease the power by 1 each term, multiplying by the corresponding power of 3. $= (2x)^4 + 4(2x)^3(3)^1 + 6(2x)^2(3)^2 + 4(2x)^1(3)^3 + 3^4$

3 Simplify the result. $= 16x^4 + 32x^3 + 216x^2 + 216x + 81$

CAS can also be used to carry out the expansions, but in doing so, the use of Pascal's triangle becomes redundant.

USING CAS 2 Expanding binomials

Expand $(3 - 2x)^6$.

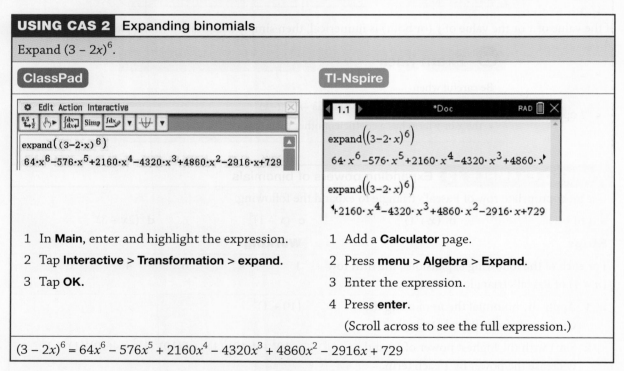

$(3 - 2x)^6 = 64x^6 - 576x^5 + 2160x^4 - 4320x^3 + 4860x^2 - 2916x + 729$

We can also use Pascal's triangle and combinations to assist in determining the coefficient of a single term in the expansion, without carrying out the whole expansion.

WORKED EXAMPLE 14 | Identify terms of a binomial expansion

Without carrying out the full expansion, determine the coefficient of the x^5 term in the expansion of $(2x - 1)^7$.

Steps

1. Set up a table with three rows.
2. In the top row, write the binomial coefficients for $n = 7$.
3. In the second row, identify the term that would have x^5.
4. In the third row, complete the corresponding power of -1.
5. Evaluate the coefficient using multiplication.

Working

1	7	21	35	35	21	7	1
$(2x)^7$	$(2x)^6$	$(2x)^5$					
		$(-1)^2$					

$21 \times 2^5 \times 1 = 21 \times 32$
$= 32(20 + 1)$
$= 640 + 32$
$= 672$

EXAMINATION QUESTION ANALYSIS

Calculator-assumed (12 marks)

A national basketball team of 12 players needs to be randomly selected from 25 qualified applicants from across the country. There are:

- 5 applicants from New South Wales
- 3 applicants from Victoria
- 4 applicants from Queensland
- 8 applicants from Western Australia
- 2 applicants from South Australia
- 2 applicants from the Australian Capital Territory
- 1 applicant from the Northern Territory.

a Determine the number of different teams that can be selected. (2 marks)

b Explain why there would be the same number of possible teams if the team was to contain 13 players instead of 12. (1 mark)

c Determine the number of different teams that can be selected that

   i include all of the applicants from the territories (2 marks)

   ii include exactly 5 applicants from Western Australia (2 marks)

   iii have at most 1 applicant from Queensland (3 marks)

   iv have at least 1 applicant from Victoria. (2 marks)

Reading the question

- Highlight the information that tells you the n and the r in the question.
- Highlight any high-order commands, such as explain why.
- Highlight any language that indicates the use of the multiplication or addition principles.

Thinking about the question

- Note that 'the territories' becomes a subset of ACT and NT, and the 'states' become another subset.
- The phrases 'at most' and 'at least' suggest the use of addition principles or the complement rules.

> **Worked solution** (✓ = 1 mark)
>
> **a** $n = 25, r = 12$ ✓
> $^{25}C_{12} = 5\,200\,300$ ✓
>
> **b** Given the symmetry of combinations (as seen in Pascal's triangle), $^nC_r = {}^nC_{n-r}$.
> If $r = 12$, then $25 - 12 = 13$ and so $^{25}C_{12} = {}^{25}C_{13}$.
>
> **uses the symmetry of Pascal's triangle to explain the equivalence** ✓
>
> or
>
> selecting 12 players from 25 applicants is the same as selecting 13 applicants to leave out of the team. So, selecting 13 players from 25 applicants is the same as selecting 12 applicants to leave out of the team. Both are equivalent in the number of ways it can be done.
>
> **uses the context of the problem to explain the equivalence** ✓
>
> **c** **i** 3 out of 3 from territories, 9 out of 22 from states
> $^3C_3 \times {}^{22}C_9 = 497\,420$
>
> **separates the set appropriately** ✓
>
> **uses the multiplication principle to obtain answer** ✓
>
> **ii** 5 out of 8 from WA, 7 out of 17 from other
> $^8C_5 \times {}^{17}C_7 = 1\,089\,088$
>
> **separates the set appropriately** ✓
>
> **uses the multiplication principle to obtain answer** ✓
>
> **iii** At most 1 applicant $\Rightarrow$ 0 or 1 applicants from Queensland
> $^4C_0 \times {}^{21}C_{12} + {}^4C_1 \times {}^{21}C_{11}$
> $= 293\,930 + 1\,410\,864$
> $= 1\,704\,794$
>
> **uses the multiplication principle correctly** ✓
>
> **shows understanding of at most one** ✓
>
> **uses the addition principle correctly** ✓
>
> **iv** At least 1 applicant – all cases subtract when there are none from Victoria.
> $^{25}C_{12} - {}^3C_0 \times {}^{22}C_{12}$
> $= 5\,200\,300 - 646\,646$
> $= 4\,553\,654$
>
> **calculates when there are none from Victoria** ✓
>
> **uses the complement to obtain answer** ✓

EXERCISE 1.3 Pascal's triangle and binomial expansions

ANSWERS p. 446

Recap

1 Dominic visits an Italian restaurant for lunch. He knows he will order either a pizza or pasta. The menu has pizzas with a choice of 3 different bases and 10 different toppings. For pasta, there is a choice of 4 different pastas with 7 different sauces.

The number of different choices Dominic has altogether is

A 4   B 7   C 28   D 30   E 58

2 A committee of 4 people is selected from 6 men and 4 women. The number of possible committees containing an equal number of men and women is

A 6   B 15   C 24   D 90   E 210

Mastery

3 WORKED EXAMPLE 12 Consider the first five rows of Pascal's triangle.

  a Determine the seventh row, i.e. when $n = 6$.
  b State the value of r, $r \neq 5$ that makes this combinatorial identity true: $^6C_5 = {}^6C_r$.
  c Determine the total number of possible subsets that can be formed from the set $\{A, B, C, D, E, F\}$.

4 WORKED EXAMPLE 13 Use an appropriate row of Pascal's triangle to expand the following.
  a $(10 - 1)^3$ b $(x + 2)^3$ c $(x - 3)^3$ d $(3x - 1)^3$

5 Using CAS 2 Expand $(3x + 5)^4$.

6 WORKED EXAMPLE 14 Without carrying out the full expansion, determine the coefficient of the
  a y^3 term in the expansion of $(3y + 2)^4$
  b x^4 term in the expansion of $(5 - x)^5$
  c x^3 term in the expansion of $(2x - 1)^4$.
  d x^4 term in the expansion of $\left(\dfrac{x}{2} - 3\right)^5$.

Calculator-free

7 (4 marks) Consider the expression $(x + y)^7$.
  a Determine the sum of the coefficients of the binomial expansion. (2 marks)
  b If $y = 2$, express $(x + y)^7$ in expanded form. (2 marks)

8 (4 marks) Consider the expression $(2x - y)^4$.
  a State the number of terms in the expansion of $(2x - y)^4$. (2 marks)
  b Determine the coefficient of the xy^3 in the expansion. (2 marks)

9 (3 marks) Prove the additive property of Pascal's triangle in the form
$$^{n+1}C_r = {}^nC_{r-1} + {}^nC_r$$
You may use the combinations formula in your proof.

Chapter summary

The language and notation of sets

Language	Meaning	Example of notation
Set	A collection of objects, denoted by a capital letter and set brackets	$A = \{1, 2, 3\}$
Element	Members of a set	$2 \in A$
Universal set	The complete set of all possible elements	$U = \{1, 2, 3, 4, 5\}$
Cardinality	The size of a set	$n(A) = 3$
Complement	The elements that are in the universal set, but not in a specific set	$\bar{A} = A' = \{4, 5\}$
Intersection	The elements that are contained in two or more sets	$A = \{1, 2, 3\}$ $B = \{3, 4\}$ $A \cap B = \{3\}$
Union	The elements that are in one or the other or both sets	$A = \{1, 2, 3\}$ $B = \{3, 4\}$ $A \cup B = \{1, 2, 3, 4\}$
Empty set	The set that contains no elements	$\emptyset = \{\ \}$
Disjoint sets	Sets that do not have an intersection	$A = \{1, 2, 3\}$ $C = \{4, 5\}$ $n(A \cap C) = 0$

Operations with two sets

Two-way tables

	B	$\bar{B}$	Total
A	$n(A \cap B)$	$n(A \cap \bar{B})$	$n(A)$
$\bar{A}$	$n(\bar{A} \cap B)$	$n(\bar{A} \cap \bar{B})$	$n(\bar{A})$
Total	$n(B)$	$n(\bar{B})$	$n(U)$

Venn diagrams

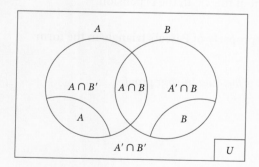

The complement rule

For a set A and its complement, then

$$n(A) + n(\overline{A}) = n(U)$$

or

$$n(\overline{A}) = n(U) - n(A)$$

Other relationships between two sets

Further results of the complement rule

- $n(A \cap B) + n(A \cap \overline{B}) = n(A)$
- $n(A \cap B) + n(\overline{A} \cap B) = n(B)$

The addition rule for two sets A and B,

$$n(A \cup B) = n(A) + n(B) - n(A \cap B)$$

De Morgan's laws are useful set equivalences that can also be shown using a Venn diagram.

1 $\overline{A \cap B} = \overline{A} \cup \overline{B}$:

   Not both A and B is equivalent to *not A or not B*.

2 $\overline{A \cup B} = \overline{A} \cap \overline{B}$:

   Neither A nor B is equivalent to *not A and not B*.

For two sets A and B	In a two-way table	In a Venn diagram
Complement of A		
Intersection of A and B		
Union of A and B		
Disjoint sets, A and B	(with 0 in $A \cap B$ cell)	
	Note that if A and B are disjoint sets, then $n(A \cap B) = 0$ and so $n(A \cup B) = n(A) + n(B)$.	

Operations with three sets

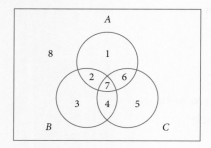

Other relationships between two sets

Let $A = \{1, 2, 6, 7\}$, $B = \{2, 3, 4, 7\}$ and $C = \{4, 5, 6, 7\}$. This numbering system can be used to identify any set description.

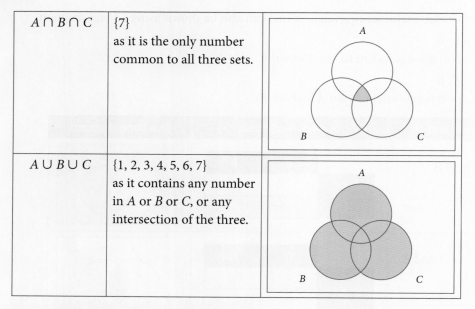

Three sets are **pairwise disjoint** if there are no intersections between any two sets.

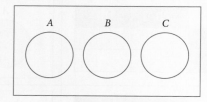

Combinations

A combination is an unordered selection, which is derived from a permutation – an ordered selection (arrangement).

A **permutation** is an arrangement of n objects into r places. The formal mathematical notation for this is ^nP_r which can be defined by the formula

$$^nP_r = \frac{n!}{(n-r)!}$$

When $n = r$ and the objects are distinct, then the number of possible permutations is $n!$, read as 'n **factorial**', where

$n! = n \times (n-1) \times (n-2) \times \ldots \times 1$

or

$n! = n \times (n-1)!$

Also, $0! = 1$.

A **combination** is an unordered selection of r objects from a set of n distinct objects. The formal mathematical notation for this is ^nC_r, which can be defined by the formula

$$^nC_r = {^nP_r} \div r! = \frac{n!}{r!(n-r)!}$$

This is often read as 'n choose r'.

An alternative notation for combinations is in the notation $\binom{n}{r}$.

Some useful facts to recognise are:
- $^nC_0 = 1$, as there is only one possible way to select 0 objects from n objects. You don't select anything!
- $^nC_1 = n$, as there are n possible objects that you could select.
- $^nC_{n-1} = n$, as there are n possible objects that you could leave out.
- $^nC_n = 1$, as there is only one possible way to select n objects from n objects. You select everything!

The multiplication principle for combinations

Consider a set of R objects being selected from a set of N distinct objects. When there are two or more successive selections of $r_1, r_2, \ldots$ being made from subsets of size $n_1, n_2, \ldots$, then the total number of possible selections is calculated using the **multiplication principle**

$$^{n_1}C_{r_1} \times {^{n_2}C_{r_2}} \times \ldots$$

where $N = n_1 + n_2 + \ldots$ and $R = r_1 + r_2 + \ldots$

The addition principle for combinations

If the selections are being made from two or more subsets, such that r_1 objects are being selected from n_1 objects OR r_2 objects are being selected from n_2 objects and so on, then the total number of possible selections is calculated using the **addition principle**

$$^{n_1}C_{r_1} + {^{n_2}C_{r_2}} + \ldots$$

Note that if the two subsets, say A and B, contain an intersection, then the addition rule needs to be used such that

$$n(A \cup B) = n(A) + n(B) - n(A \cap B)$$

Otherwise, if they are disjoint subsets, then

$$n(A \cup B) = n(A) + n(B)$$

Pascal's triangle and binomial expansions

Pascal's triangle shows the number of combinations of r objects from n distinct objects.

$$\binom{0}{0}$$

$$\binom{1}{0} \quad \binom{1}{1}$$

$$\binom{2}{0} \quad \binom{2}{1} \quad \binom{2}{2}$$

$$\binom{3}{0} \quad \binom{3}{1} \quad \binom{3}{2} \quad \binom{3}{3}$$

$$\binom{4}{0} \quad \binom{4}{1} \quad \binom{4}{2} \quad \binom{4}{3} \quad \binom{4}{4}$$

$$1$$
$$1 \quad 1$$
$$1 \quad 2 \quad 1$$
$$1 \quad 3 \quad 3 \quad 1$$
$$1 \quad 4 \quad 6 \quad 4 \quad 1$$

Properties of Pascal's triangle

Property	Description	Example from Pascal's triangle
The symmetry of combinations	From the triangle, we can see that in every row: • $^nC_0 = {^nC_n}$ • $^nC_1 = {^nC_{n-1}}$ More generally, $^nC_r = {^nC_{n-r}}$	$$\begin{array}{ccccccccc} & & & & 1 & & & & \\ & & & 1 & & 1 & & & \\ & & 1 & & 2 & & 1 & & \\ & 1 & & 3 & & 3 & & 1 & \\ 1 & & \boxed{4} & & 6 & & \boxed{4} & & 1 \end{array}$$ $$^4C_1 = {^4C_3}$$
The additive property of combinations	In any subsequent row, an entry is found by adding the two entries either side of it in the previous row. $^nC_r = {^{n-1}C_{r-1}} + {^{n-1}C_r}$	$$\begin{array}{ccccccccc} & & & & 1 & & & & \\ & & & 1 & & 1 & & & \\ & & 1 & & 2 & & 1 & & \\ & 1 & & 3 & & 3 & & 1 & \\ 1 & & 4 & & 6 & & 4 & & 1 \end{array}$$ Term 0 Term 1 Term 2 Term 3 Term 4 $1 + 3 = 4$ Add the 2 numbers in the row above to get the numbers in the next row. $$^4C_1 = {^3C_0} + {^3C_1}$$
The total number of subsets from n objects	The number of possible subsets (including the empty set and the set itself) that can be made from n objects is the sum of the nth row of Pascal's triangle and is equal to 2^n.	Suppose we wanted to know how many possible subsets of any size can be made from 3 objects $\{A, B, C\}$. • $^3C_0 = 1$: the empty set, $\varnothing$ • $^3C_1 = 3$: $\{A\}, \{B\}, \{C\}$ • $^3C_2 = 3$: $\{A, B\}, \{A, C\}, \{B, C\}$ • $^3C_3 = 1$: $\{A, B, C\}$ In total, $1 + 3 + 3 + 1 = 8$, which is 2^3.

A **subset** is a set within a set. If B is a subset of A, then $B \subseteq A$. If A is a set containing n elements, then there are 2^n possible subsets that can be made from A, including the empty set and the set A itself.

A subset is a **proper subset** if it is srictly smaller than the set itself. If B is a proper subset of A, then $B \subset A$. If the set A itself is excluded as a subset, then there are $2^n - 1$ possible proper subsets that can be made from A.

The binomial theorem

For $(x + y)^n$, where $n \geq 0$, then

$$(x + y)^n = \binom{n}{0}x^n y^0 + \binom{n}{1}x^{n-1}y^1 + \binom{n}{2}x^{n-2}y^2 + \ldots + \binom{n}{n}x^0 y^n$$

and the set of values $\left\{ \binom{n}{0}, \binom{n}{1}, \binom{n}{2} \ldots \binom{n}{n} \right\}$ is called the **binomial coefficients**.

Cumulative examination: Calculator-free

Total number of marks: 17   Reading time: 2 minutes   Working time: 17 minutes

1 (4 marks) Let the universal set be the set of 26 lowercase letters from the English alphabet.

Three sets X, Y and Z are defined as follows:

$X = \{a, e, i, o, u\}$ $Y = \{c, e, h, i, m, r, s, t, y\}$ $Z = \{a, c, e, h, i, m, s, t\}$

a List the elements of $X \cap Y \cap Z$. (1 mark)

b Determine

  i $n(\overline{X} \cap Y \cap Z)$ (1 mark)

  ii $n(\overline{X \cup Y \cup Z})$ (2 marks)

2 (6 marks)

a Evaluate $\binom{4}{3} + \binom{3}{2} + \binom{2}{1} + \binom{1}{0}$. (2 marks)

b Evaluate $\binom{7}{4}\binom{4}{2}$. (2 marks)

c State the value of n such that $^nC_4 = {}^{n+2}C_0$. (2 marks)

3 (7 marks)

a Expand and simplify $(2x - 1)^4$. (2 marks)

b Determine the sum of the coefficients of the expansion of $(a + 1)^6$. (2 marks)

c Determine the coefficient of the a^4 term in the expansion of $(a + 1)^6 + (a - 1)^6$. (3 marks)

Cumulative examination: Calculator-assumed

Total number of marks: 11   Reading time: 2 minutes   Working time: 11 minutes

1 (6 marks) A Senior School cohort contains 120 Year 11 students and 140 Year 12 students.

The Head of School wants to randomly select three students to represent the school at a local community event. Determine the number of possible groups of three students that can be formed,

   a if there are no restrictions (2 marks)

   b if two Year 12 students must be selected (2 marks)

   c if at least one Year 11 student is selected. (2 marks)

2 (5 marks) A university campus contains two cafeterias: the North Café and the South Café. A survey of 105 university students was conducted and it was found that 73 students liked the food at the North Café (N), 62 students liked the food at the South Café (S) and 31 students liked the food at the South Café, but not the food at the North Café.

   a Represent the above information in a suitable display. (2 marks)

   b Describe the set $N \cup S$ in context of the question. (1 mark)

   c Determine $n(\overline{N \cup S})$. (2 marks)

CHAPTER 2

PROBABILITY

Syllabus coverage
Nelson MindTap chapter resources

2.1 The fundamentals of probability
 The language and notation of probability
 Probability rules and links to set theory
 Experimental and theoretical probabilities

2.2 Probability experiments involving sets
 Problems involving two sets
 Mutually exclusive events
 Problems involving three sets

2.3 Probability experiments involving stages
 Problems involving two stages
 Problems involving three stages

2.4 Probability experiments involving selections
 Applying combinations to probability
 Using combinations in probability

2.5 Conditional probability
 Indicating conditionality
 Using the conditional probability formula

2.6 Independent events
 Independence using restrictions
 Successive independence
 The symmetry of independence
 Independence in context

2.7 Relative frequencies and indications of independence
 Estimates of conditional probabilities
 Possible independence of events

Examination question analysis
Chapter summary
Cumulative examination: Calculator-free
Cumulative examination: Calculator-assumed

Syllabus coverage

TOPIC 1.1: COUNTING AND PROBABILITY

Language of events and sets

1.1.4 review the concepts and language of outcomes, sample spaces, and events, as sets of outcomes
1.1.5 use set language and notation for events, including:
   a. $\bar{A}$ (or A') for the complement of an event A
   b. $A \cap B$ and $A \cup B$ for the intersection and union of events A and B respectively
   c. $A \cap B \cap C$ and $A \cup B \cup C$ for the intersection and union of the three events A, B, and C respectively
   d. recognise mutually exclusive events.
1.1.6 use everyday occurrences to illustrate set descriptions and representations of events and set operations

Review of the fundamentals of probability

1.1.7 review probability as a measure of 'the likelihood of occurrence' of an event
1.1.8 review the probability scale: $0 \leq P(A) \leq 1$ for each event A, with $P(A) = 0$ if A is an impossibility and $P(A) = 1$ if A is a certainty
1.1.9 review the rules: $P(\bar{A}) = 1 - P(A)$ and $P(A \cup B) = P(A) + P(B) - P(A \cap B)$
1.1.10 use relative frequencies obtained from data as estimates of probabilities

Conditional probability and independence

1.1.11 understand the notion of a conditional probability and recognise and use language that indicates conditionality
1.1.12 use the notation $P(A|B)$ and the formula $P(A \cap B) = P(A|B)P(B)$
1.1.13 understand the notion of independence of an event A from an event B, as defined by $P(A|B) = P(A)$
1.1.14 establish and use the formula $P(A \cap B) = P(A)P(B)$ for independent events A and B, and recognise the symmetry of independence
1.1.15 use relative frequencies obtained from data as estimates of conditional probabilities and as indications of possible independence of events

Mathematics Methods ATAR Course Year 11 syllabus pp. 8–9 © SCSA

Video playlists (8):

2.1 The fundamentals of probability
2.2 Probability experiments involving sets
2.3 Probability experiments involving stages
2.4 Probability experiments involving selections
2.5 Conditional probability
2.6 Independent events
2.7 Relative frequencies and indications of independence

Examination question analysis Probability

Worksheets (10):

2.1 Theoretical probability • Experimental probability
2.2 Venn diagrams 1 • Set operations • Venn diagrams 2 • Venn diagrams matching activity • Two-way tables
2.3 Tree diagrams 1 • Tree diagrams 2
2.5 Conditional probability

Puzzles (4):

2.1 Matching probabilities
2.2 Combined events: two-way tables
2.3 And/or problems
2.7 Conditional probability: two-way tables

Nelson MindTap

To access resources above, visit
cengage.com.au/nelsonmindtap

2.1 The fundamentals of probability

The language and notation of probability

The concept of **probability** should not be new to you – it is the mathematics of chance. Every day, we are involved with events that may or may not occur with different degrees of certainty, and so mathematicians use the concept of probability to quantify these differing degrees of certainty. Throughout this chapter, we will review the prior knowledge you should have for this topic, as well as examine some new concepts.

Fundamentals of probability

- Every probability experiment has a **sample space**, S, which is the set of all possible outcomes in the experiment.
- A single **outcome** or a group of outcomes of a probability experiment is called an **event** and is often represented using a capital letter.
- The probability of an event A occurring is denoted as:

$$P(A) = \frac{\text{number of successful outcomes for event } A}{\text{total number of possible outcomes}} = \frac{n(A)}{n(S)}$$

- Probability can be considered a function, $P(\ )$, for which the 'input' is the set of all possible events in the experiment and the 'output' is the set of numerical values, $0 \le P(A) \le 1$, where 0 represents an impossible event and 1 represents a certain event.

```
              Probability
0 ◄─────────────────────────────► 1
Impossible                    Certain
```

WORKED EXAMPLE 1 | Calculating simple probabilities

A fair spinner is divided into five sections numbered 1 to 5. Determine the probability that the spinner lands on

a an even number **b** a prime number **c** a square number.

Steps	Working
a 1 Identify the set of successful outcomes.	Let E be the set of even numbers on the spinner. $E = \{2, 4\}$
2 Use the correct notation to express the probability as a fraction of the total number of outcomes.	$P(E) = \dfrac{2}{5}$
b 1 Identify the set of successful outcomes.	Let P be the set of prime numbers on the spinner. $P = \{2, 3, 5\}$
2 Use the correct notation to express the probability as a fraction of the total number of outcomes.	$P(P) = \dfrac{3}{5}$
c 1 Identify the set of successful outcomes.	Let Q be the set of square numbers on the spinner. $Q = \{1, 4\}$
2 Use the correct notation to express the probability as a fraction of the total number of outcomes.	$P(Q) = \dfrac{2}{5}$

Video playlist
The fundamentals of probability

Worksheets
Theoretical probability
Experimental probability

Puzzle
Matching probabilities

Probability rules and links to set theory

In Chapter 1, we introduced the language and notation of set theory, which underpins much of what we talk about in probability theory. The connections can be seen in the following table.

The language of sets	The language of probability experiments	Meaning and notation
The universal set, U	The sample space, S	The set of all possible outcomes. The probability of the sample space itself occurring is certain, $P(S) = 1$.
An element	An outcome	A single member of a set, represented using the element symbol, e.g. $2 \in A$.
A set	An event	A single outcome or a group of outcomes, represented by a capital letter e.g. A.
The complement	'not'	The outcomes not in a given set A, represented by $\overline{A}$ or A'.
Complement rule: $P(\overline{A}) = 1 - P(A)$		
The intersection of sets	'and'	The intersection of sets A and B contains the outcomes that are common to both sets, represented by $A \cap B$.
The union of sets	'or' (inclusive)	The union of sets A and B contains the outcomes that are in A, in B or in both sets, represented by $A \cup B$.
The addition rule for two sets: $P(A \cup B) = P(A) + P(B) - P(A \cap B)$		

> **Exam hack**
>
> If you are not given event labels to use, choose a letter that appropriately represents the event. If it is not an obvious choice, be sure to define the events first.

WORKED EXAMPLE 2 — Using the language of 'not', 'and' and 'or'

A toy shop packages bags of marbles that contain different numbers of red, green, purple, blue and yellow marbles. A particular bag contains 10 red, 6 green, 5 purple, 9 blue and 5 yellow marbles. A marble is to be randomly selected from the bag. Determine the probability that the chosen marble is

a not purple   **b** green and purple   **c** red, yellow or blue.

Steps	Working
a 1 Count the total number of marbles.	$n(S) = 35$
2 Identify the number of purple marbles and calculate the complement.	$n(\overline{P}) = 30$
3 Use the correct notation to express the probability as a fraction of the total number of outcomes.	$P(\overline{P}) = \dfrac{30}{35}$
b 1 Recognise that a two-coloured marble is not possible.	
2 State the probability.	$P(G \cap P) = 0$
c 1 Find the total number of red, yellow and blue marbles.	$n(R \cup Y \cup B) = 24$
2 Use the correct notation to express the probability as a fraction of the total number of outcomes.	$P(R \cup Y \cup B) = \dfrac{24}{35}$

Experimental and theoretical probabilities

You may also recall there being two types of probability: **experimental probability** and **theoretical probability**. The experimental probability of an event occurring can be determined by gathering observations from an experiment with a defined number of trials, and counting the number of times a particular event has occurred out of the total number of trials. You have also heard this referred to as **relative frequency**. However, if the experiment is conducted a significantly large number of times, then the experimental probability is expected to tend towards the theoretical probability. The theoretical probability is the likelihood of the event without having to conduct the experiment, but can be calculated based on the knowledge of the situation and the number of outcomes.

For example, suppose we toss a regular coin 50 times and observe that it lands on head 20 times. Then the experimental probability of obtaining a head is P(head) = $\frac{20}{50}$ = 0.4. However, theoretically we know the probability is P(head) = $\frac{1}{2}$ = 0.5.

WORKED EXAMPLE 3 — Comparing experimental and theoretical probabilities

A regular six-sided die is rolled 40 times and the following results are obtained.

6, 6, 1, 4, 3, 5, 1, 3, 6, 2, 5, 6, 2, 3, 1, 6, 1, 1, 4, 6, 6, 2, 2, 1, 1, 5, 1, 3, 2, 6, 1, 6, 4, 2, 6, 5, 2, 2, 6, 6

Compare the experimental probability of rolling a 6 to the theoretical probability of rolling a 6.

Steps	Working
1 Count the number of times a 6 occurred.	$n(6) = 12$
2 Express the experimental probability as a fraction (and decimal for better comparison).	$P_E(6) = \frac{12}{40} = 0.3$
3 State the theoretical probability of obtaining a 6.	$P_T(6) = \frac{1}{6} = 0.1\dot{6}$
4 Compare the probabilities.	The experimental probability of rolling a 6 is 0.14 more than the theoretical probability. There were more 6s in the experiment than expected.

Experimental probability acts as an estimate of the likelihood of an event occurring, especially if the theoretical probability is unknown. This can then be used to estimate an expected number of successes in a larger number of trials. For example, if we observe 14 successes out of 20 trials, then it would be reasonable to expect 70 successes out of 100 trials.

EXERCISE 2.1 The fundamentals of probability   ANSWERS p. 447

Mastery

1 **WORKED EXAMPLE 1** A fair spinner is divided into 7 sections numbered 1 to 7. Determine the probability that the spinner lands on

   a an odd number   b a composite number   c a factor of 18.

2 **WORKED EXAMPLE 2** An educational resource shop packages bags of numbered counters that contain different numbers of counters numbered 1 to 9. A particular bag of counters contains the set
   {1, 1, 1, 2, 2, 3, 3, 3, 4, 5, 5, 5, 5, 5, 7, 7, 7, 8, 8, 9}.

   A counter is to be randomly selected from the bag. Determine the probability that the number on the chosen counter is

   a not 5   b odd and prime   c greater than 2 or less than 4.

3

a A regular six-sided die is rolled 30 times and the following results are obtained:

4, 4, 3, 3, 5, 3, 5, 6, 2, 5, 1, 1, 6, 4, 3, 3, 6, 1, 5, 1, 5, 5, 3, 1, 1, 3, 1, 3, 2, 3

Compare the experimental probability of rolling an odd number to the theoretical probability of rolling an odd number.

b A regular coin is tossed 20 times and the following results are observed:

H, H, H, T, T, H, T, H, T, H, T, T, T, H, H, T, H, T, T, T

Compare the experimental probability of obtaining a tail to the theoretical probability of obtaining a tail.

Calculator-free

4 (3 marks) Ashleigh has a biased spinner with sections numbered 1 to 5. The probability of the spinner landing on one of these numbers on any of the spins is shown in the table below.

Outcome, X	1	2	3	4	5
Probability, $P(X)$	0.25	0.22	0.3	0.18	0.05

Determine the probability that the outcome of Ashleigh's spin is

a greater than 2 (1 mark)

b even (1 mark)

c prime. (1 mark)

5 (6 marks) Over the month of June (30 days), Tony timed the number of hours each day that he spent cycling on his bicycle and recorded the times on his phone calendar. He did not cycle on three of the days. He cycled for 1 hour on six of the days, 2 hours on seven of the days, 3 hours on ten of the days and 4 hours on four of the days. If he was to randomly select a day in June on his calendar, determine the probability that he cycled for

a at least 2 hours (2 marks)

b no more than 3 hours (2 marks)

c an odd number of hours. (2 marks)

6 (5 marks) A spinner is divided into 12 equal sections. The sections are all coloured in one of the colours: red, blue, yellow, green or white. The graph below shows the frequency of the different colours in 50 spins. Use the column graph to determine how many sectors are likely to be coloured with each of the colours.

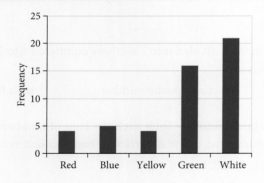

2.2 Probability experiments involving sets

Problems involving two sets

In Chapter 1, we saw that two-way tables and Venn diagrams can be used to represent the size of two sets and the relationship between their intersections. All the examples given in Chapter 1 deal with the 'number of' elements in a set; however, the two-way table and Venn diagrams can also be used to show the probability of the events, rather than just the number of elements.

	B	$\bar{B}$	Total
A	$P(A \cap B)$	$P(A \cap \bar{B})$	$P(A)$
$\bar{A}$	$P(\bar{A} \cap B)$	$P(\bar{A} \cap \bar{B})$	$P(\bar{A})$
Total	$P(B)$	$P(\bar{B})$	1

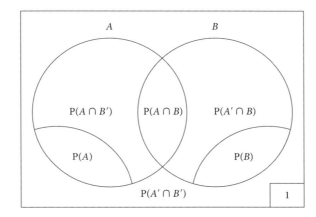

Video playlist
Probability experiments involving sets

Worksheets
Venn diagrams 1
Set operations
Venn diagrams 2
Venn diagrams matching activity
Two-way tables

Puzzle
Combined events: two-way tables

WORKED EXAMPLE 4 — Using a two-way table for probabilities

For two events A and B, $P(A \cap B) = 0.15$, $P(B) = 0.35$ and $P(A' \cap B') = 0.2$. Show the use of a two-way table to determine $P(A \cap B')$.

Steps	Working
1 Construct the two-way table and input the known probabilities, including the total of 1.	<table><tr><td></td><td>B</td><td>B'</td><td>Total</td></tr><tr><td>A</td><td>0.15</td><td></td><td></td></tr><tr><td>A'</td><td></td><td>0.2</td><td></td></tr><tr><td>Total</td><td>0.35</td><td></td><td>1</td></tr></table>
2 Complete the remaining cells using the row and column totals. Check all totals for rows and columns are consistent.	<table><tr><td></td><td>B</td><td>B'</td><td>Total</td></tr><tr><td>A</td><td>0.15</td><td>0.45</td><td>0.6</td></tr><tr><td>A'</td><td>0.2</td><td>0.2</td><td>0.4</td></tr><tr><td>Total</td><td>0.35</td><td>0.65</td><td>1</td></tr></table>
3 Identify the required value, $P(A \cap B')$.	$P(A \cap B') = 0.45$

WORKED EXAMPLE 5 — Using a Venn diagram for probabilities

For two events A and B, $P(A) = \dfrac{3}{5}$, $P(B) = \dfrac{1}{3}$ and $P(A \cap B) = \dfrac{1}{9}$. Show the use of a Venn diagram to determine $P(\overline{A \cup B})$.

Steps	Working
1 Express all probabilities with a common denominator for easy calculations.	$P(A) = \dfrac{27}{45}$ $P(B) = \dfrac{15}{45}$ $P(A \cap B) = \dfrac{5}{45}$
2 Construct the Venn diagram and input the known probabilities, including the total of 1. Note: the 'bubbles' showing the totals of $P(A)$ and $P(B)$ are not essential – you could work out $P(A \cap B')$ and $P(A' \cap B)$ instead.	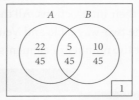
3 Use the fact that all probabilities sum to 1 to calculate the required probability.	$P(\overline{A \cup B}) = \dfrac{45}{45} - \left(\dfrac{22}{45} + \dfrac{5}{45} + \dfrac{10}{45} \right)$ $P(\overline{A \cup B}) = \dfrac{8}{45}$

These contextual problems that were originally about the number of elements in a set can now be changed to probability questions by incorporating a process of random selection into the question.

Exam hack

If you are not directed to use a suitable display, you can still choose to do so. However, sometimes a known probability rule might be enough!

WORKED EXAMPLE 6 — Solving contextual problems involving two sets

The probability that a randomly selected student from a cohort studying Art or Music is $\dfrac{3}{10}$. If $\dfrac{1}{8}$ of the students in the cohort study Music and $\dfrac{1}{4}$ study Art, then determine the probability that a randomly selected student studies both Art and Music.

Steps	Working
1 Define the events.	Let A be the event that a chosen student studies Art and M be the event that a chosen student studies Music.
2 Express the probabilities with a common denominator.	$P(A \cup M) = \dfrac{3}{10} = \dfrac{12}{40}$ $P(M) = \dfrac{1}{8} = \dfrac{5}{40}$ $P(A) = \dfrac{1}{4} = \dfrac{10}{40}$
3 Use the addition rule to calculate $P(A \cap M)$.	$P(A \cup M) = P(A) + P(M) - P(A \cap M)$ $\dfrac{12}{40} = \dfrac{5}{40} + \dfrac{10}{40} - P(A \cap M)$ $P(A \cap M) = \dfrac{3}{40}$

Mutually exclusive events

Before we apply our knowledge of sets to probability any further, it is important to redefine one more term that we saw in Chapter 1.

Suppose a regular six-sided die is to be rolled. Let the event A be the die landing on the number 6 and let the event B be the die landing on an odd number. These two events, which are both possible within the sample space, cannot occur simultaneously, meaning that the probability of the event satisfying both A and B is 0. This is an example of **mutually exclusive events**, which is the probability term for disjoint sets.

The language of sets	The language of probability experiments	Meaning and notation
Disjoint sets	**Mutually exclusive events** i.e. 'or' (exclusive)	Two sets do not share an intersection; that is, $P(A \cap B) = 0$. These events in a probability experiment cannot both occur at the same time.
The addition rule for two mutually exclusive sets: $P(A \cup B) = P(A) + P(B)$		

WORKED EXAMPLE 7 — Distinguishing between complementary and mutually exclusive events

For each of the following pairs of events, state whether they are complementary, mutually exclusive, both or neither.

a A card is drawn from a regular deck of 52 cards. Let A be the event that a red card is drawn and B be the event that a spade is drawn.

b A regular six-sided die is rolled. Let A be the event that an even number is rolled and B be the event that an odd number is rolled.

Steps	Working
a Check the following questions:	
1 Are the two events the only two possible outcomes from the experiment? If yes, then they are complementary.	There are other possible outcomes. They are not complementary.
2 Can the events occur simultaneously? If not, then they are mutually exclusive.	A card cannot be red and a spade. They are mutually exclusive. Therefore, A and B are not complementary and are mutually exclusive.
b Check the following questions:	
1 Are the two events the only two possible outcomes from the experiment? If yes, then they are complementary.	The numbers 1 to 6 are either even or odd. They are complementary.
2 Can the events occur simultaneously? If not, then they are mutually exclusive.	A number cannot be both even and odd. The events are mutually exclusive. Therefore, A and B are both complementary and mutually exclusive.

WORKED EXAMPLE 8 — Solving problems involving mutually exclusive events

Two events, A and B, are mutually exclusive such that $P(A) = \dfrac{1}{2}$ and $P(B') = \dfrac{19}{24}$. Determine

a $P(A' \cup B')$ **b** $P(A' \cap B')$

Steps	Working
1 Express all probabilities with a common denominator for easy calculations.	$P(A) = \dfrac{12}{24}$ $P(B') = \dfrac{19}{24}$
2 Construct a suitable display to represent the problem.	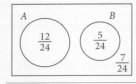
3 Use the complement rule to calculate $P(B)$.	$P(B) = 1 - \dfrac{19}{24} = \dfrac{5}{24}$
4 Complete the Venn diagram using the fact that the sum of probabilities must equal 1.	$P(\overline{A \cup B}) = 1 - \left(\dfrac{12}{24} + \dfrac{5}{24}\right) = \dfrac{7}{24}$ [Venn diagram with $A: \tfrac{12}{24}$, $B: \tfrac{5}{24}$, outside: $\tfrac{7}{24}$]
5 Recognise and use De Morgan's law $P(\overline{A} \cup \overline{B}) = P(\overline{A \cap B})$.	**a** $P(\overline{A \cap B}) = 1$, as the events are mutually exclusive.
6 Recognise and use De Morgan's law $P(\overline{A} \cap \overline{B}) = P(\overline{A \cup B})$.	**b** $P(\overline{A \cup B}) = \dfrac{7}{24}$

Problems involving three sets

We can also extend our understanding of the relationships between three sets to probability problems, using three-circle Venn diagrams as suitable displays to represent the situation, when appropriate.

WORKED EXAMPLE 9 — Calculating probabilities from three sets

Consider the sample space $S = \{1, 2, 3 \ldots 14, 15\}$. A number is to be randomly selected from this set.

Let A be the event that the number selected is even.
Let B be the event that the number selected is prime.
Let C be the event that the number selected is a multiple of 3.

Determine

a $P(A \cap B \cap C)$ **b** $P(A \cup B \cup C)$ **c** $P(A' \cap B \cup C)$

Steps	Working
List the outcomes of each event.	$A = \{2, 4, 6, 8, 10, 12, 14\}$ $B = \{2, 3, 5, 7, 11, 13\}$ $C = \{3, 6, 9, 12, 15\}$

a 1 Interpret the set description to help identify the outcomes.	The selected number must be even, prime and a multiple of 3.	
2 State the probability.	$P(A \cap B \cap C) = 0$	
b 1 Interpret the set description to help identify the outcomes.	The selected number must be even or prime or a multiple of 3. The only number not included in that set is 1.	
2 State the probability.	$P(A \cup B \cup C) = \dfrac{14}{15}$	
c 1 Interpret the set description and identify the outcomes.	The selected number must be an odd prime, or a multiple of 3. $A' \cap B = \{3, 5, 7, 11, 13\}$ $C = \{3, 6, 9, 12, 15\}$	
2 State the probability.	$P(A' \cap B \cup C) = \dfrac{9}{15}$	

WORKED EXAMPLE 10 — Solving contextual problems involving three sets

A local firm surveyed 500 of its employees about how they commute to and from work: bus (*B*), cycle (*C*) and/or train (*T*). The results of the survey found that

- 250 employees take the bus
- 150 employees take the train
- 100 employees cycle
- 80 employees take both the bus and the train
- 50 employees take both the train and cycle
- 30 employees take both the bus and cycle
- 20 employees use all three methods of transport.

If an individual was randomly selected from this group of employees, determine the probability that they do not use any of the three methods of transport.

Steps	Working
1 Construct and complete a three-circle Venn diagram to represent the problem.	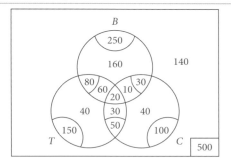
2 Use the complement rule to calculate $P(\overline{B \cup C \cup T})$.	$P(\overline{B \cup C \cup T}) = \dfrac{500 - (160 + 60 + 20 + 10 + 40 + 30 + 40)}{500}$ $= \dfrac{500 - 360}{500} = \dfrac{140}{500}$

EXERCISE 2.2 Probability experiments involving sets

ANSWERS p. 447

Recap

1 The birth month of 20 Year 11 students is shown in the table below.

Month	Jan	Feb	Mar	Apr	May	Jun	Jul	Aug	Sep	Oct	Nov	Dec
Number	3	1	2	4	0	1	0	1	2	0	1	5

The experimental probability of randomly selecting a Year 11 student with a birth month in January, February or December is

A $\dfrac{1}{4}$ B $\dfrac{3}{10}$ C $\dfrac{1}{3}$ D $\dfrac{2}{5}$ E $\dfrac{9}{20}$

2 A probability spinner is divided into 10 equal sections numbered 1 to 10. If the spinner is to be spun once, the probability that the outcome is neither prime nor even is

A 0   B $\dfrac{1}{10}$ C $\dfrac{2}{10}$ D $\dfrac{2}{5}$ E $\dfrac{4}{5}$

Mastery

3 **WORKED EXAMPLE 4** For two events A and B, $P(A \cap B) = 0.32$, $P(B') = 0.46$ and $P(A' \cap B') = 0.11$. Show the use of a two-way table to determine $P(A')$.

4 **WORKED EXAMPLE 5** For two events A and B, $P(A) = \dfrac{3}{4}$, $P(B) = \dfrac{1}{4}$ and $P(A \cap B) = \dfrac{1}{5}$. Show the use of a Venn diagram to determine $P(A' \cap B')$.

5 **WORKED EXAMPLE 6** The probability of a randomly selected student in a group of Year 11 students liking coffee is $\dfrac{9}{16}$. If $\dfrac{3}{8}$ of the students in the group like tea and $\dfrac{1}{12}$ like both coffee and tea, then determine the probability of a randomly selected student in this group liking neither coffee nor tea.

6 **WORKED EXAMPLE 7** For each of the following pairs of events, state whether they are complementary, mutually exclusive, both or neither.

   a A bag contains red, green and blue marbles. Let A be the event that a red marble is drawn from the bag and B be the event that a blue marble is drawn from the bag.

   b A regular six-sided die is rolled. Let A be the event that an even number is rolled and B be the event that a number greater than 4 is rolled.

7 **WORKED EXAMPLE 8** Two events, A and B, are mutually exclusive such that $P(A) = \dfrac{3}{4}$ and $P(A' \cap B') = \dfrac{1}{5}$. Determine $P(A \cup B')$.

8 **WORKED EXAMPLE 9** Consider the sample space $S = \{1, 2, 3 \ldots 19, 20\}$. A number is to be randomly selected from this set.

   Let A be the event that the number selected is odd.
   Let B be the event that the number selected is composite.
   Let C be the event that the number selected is a factor of 20.
   Determine

   a $P(A \cap B \cap C)$ b $P(A \cup B \cup C)$ c $P(A' \cap B' \cap C)$

9 **WORKED EXAMPLE 10** A local pizzeria surveyed 200 customers about their preferences on three controversial pizza toppings: anchovies (A), mushrooms (M) and/or pineapple (P). The results of the survey found that

- 100 individuals like anchovies on their pizza
- 80 individuals like mushrooms on their pizza
- 60 individuals like pineapple on their pizza
- 30 individuals like both anchovies and mushrooms, but not pineapple on their pizza
- 30 individuals like both mushrooms and pineapple on their pizza
- 10 individuals like both anchovies and pineapple, but not mushrooms on their pizza
- 10 individuals like all three toppings.

If an individual was randomly selected from this group of customers, determine the probability that they do not like any of the three toppings on their pizza.

Calculator-free

10 (3 marks) Two events, A and B, from a given sample space are defined such that $P(A) = 0.25$, $P(B) = 0.3$ and $P(A \cap B) = 0.12$. Determine $P(A' \cup B)$.

11 (2 marks) Two events, A and B, from a given sample space are defined such that $P(A) = \frac{1}{5}$ and $P(B) = \frac{1}{3}$. Determine

    a $P(\overline{A} \cap B)$, if $P(A \cap B) = \frac{1}{8}$ (1 mark)

    b $P(\overline{A} \cap B)$, if A and B are mutually exclusive.   (1 mark)

12 (3 marks) Two events, A and B, from a given sample space are defined such that $P(A \cup B) = \frac{2}{5}$ and $P(A \cap B) = \frac{1}{5}$. If $P(A) = 2P(B)$, calculate $P(A)$.

13 (5 marks) For two events A and B, $P(A \cap B) = p + 0.25$, $P(B') = p$, $P(A') = p - 0.15$ and $P(A \cap B') = 0.2$.

    a Determine the value of p.   (3 marks)

    b Determine $P(A' \cap B)$.   (2 marks)

14 (4 marks) For two events A and B, $P(A \cap B) = p$, $P(B') = 0.4$, $P(A) = 1.5p$ and $P(A' \cap B') = 2p$.

    a Determine the value of p.   (3 marks)

    b Determine $P(A' \cap B')$.   (1 mark)

15 (3 marks) A card is selected at random from a pack of 52 playing cards. Determine the probability of selecting

    a a queen (Q), but not a diamond (D)   (1 mark)

    b a king (K) or a heart (H).   (2 marks)

▶ **Calculator-assumed**

16 (6 marks) A drug company develops a test for a particular disease. It is found that
- the probability of the test being positive is 0.3
- the probability of a person having the disease is 0.29
- the probability of the drug testing positive and the patient not having the disease is 0.02.

a Construct and complete a suitable display to represent the probabilities in this situation. (3 marks)

b Hence, state the probability that
  i the test result is negative and the patient has the disease (1 mark)
  ii the test result is positive or the patient does not have the disease. (2 marks)

17 (7 marks) At a recent swimming carnival there were 40 competitors. The list of entries showed that
- 14 entered backstroke
- 22 entered freestyle
- 17 entered butterfly
- 5 entered all three events
- 7 entered backstroke and freestyle
- 9 entered freestyle and butterfly
- 6 entered backstroke and butterfly.

a Construct and complete a suitable display to represent the situation. (3 marks)

b Hence, state the probability that a competitor entered
  i only butterfly (1 mark)
  ii butterfly and backstroke but not freestyle (1 mark)
  iii exactly one event. (2 marks)

2.3 Probability experiments involving stages

Video playlist
Probability experiments involving stages

Worksheets
Tree diagrams 1

Tree diagrams 2

Puzzle
And/or problems

When probability experiments involve two or more successive stages, or are made up of multiple simple experiments that can be thought of as one occurring after the other, then the sample space displays can be represented using **arrays** or **tree diagrams**.

Both displays can be used for two-stage experiments, for example if a coin is tossed and then a regular die is rolled. However, for any more than two stages, we would consider using a tree diagram, or in some cases, combinations, for example if a coin is tossed three times successively.

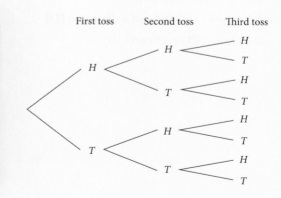

Problems involving two stages

An $m \times n$ array is a two-dimensional table in which the m events in the rows represent the outcomes of the first stage of an experiment and the n events in the columns represent the outcomes of the second stage of an experiment. The entries within the array depend on the kind of experiment being carried out. Entries can be

- ordered pairs of outcomes, for example $(H, 4)$, or
- numerical values, for example the sum of two dice, or
- combinations of letters or symbols, for example a two-letter word, AT.

	Outcomes of stage 2		
Outcomes of stage 1	...	...	...
	...	Results of both experiments	...
	...	...	...

In two-stage problems, if the same experiment is being repeated in both stages, we must take note whether there is **replacement** in the situation. For example, if a letter tile is drawn from a bag of 5 letters {A, E, I, O, U} but is not replaced, then that outcome is no longer possible in the second stage of the experiment.

Arrays, with and without replacement

An array is a table used to show the outcomes of a two-stage probability experiment, in which the outcomes of the first stage are in the rows of the table and the outcomes of the second stage are in the columns of the table.

	A	E	I	O	U
A	AA	AE	AI	AO	AU
E	EA	EE	EI	EO	EU
I	IA	IE	II	IO	IU
O	OA	OE	OI	OO	OU
U	UA	UE	UI	UO	UU

If the experiment does not involve replacement between stages, then a dash '–' is used to denote that an outcome is not possible, and this is typically seen in the diagonal of the table from the top left to the bottom right.

	A	E	I	O	U
A	–	AE	AI	AO	AU
E	EA	–	EI	EO	EU
I	IA	IE	–	IO	IU
O	OA	OE	OI	–	OU
U	UA	UE	UI	UO	–

WORKED EXAMPLE 11 — Using an array to solve a two-stage problem

Four identical balls are numbered 1, 2, 3 and 4, and are put into a box. A ball is randomly drawn from the box and not returned. A second ball is then randomly drawn from the box.

a Represent the sample space using an array.

b Determine the probability that

   i the first ball is numbered 4 and the second ball is numbered 1

   ii the sum of the numbers on the two balls is 5.

Steps	Working

a Construct a 4 × 4 array, state the possible outcomes and note that the experiment is without replacement.

	1	2	3	4
1	–	1, 2	1, 3	1, 4
2	2, 1	–	2, 3	2, 4
3	3, 1	3, 2	–	3, 4
4	4, 1	4, 2	4, 3	–

b i 1 Count the total number of outcomes.

   2 Use appropriate notation to express the probability.

$$P(4, 1) = \frac{1}{12}$$

ii 1 Modify the array to now include sums.

	1	2	3	4
1	–	3	4	5
2	3	–	5	6
3	4	5	–	7
4	5	6	7	–

   2 Use appropriate notation to express the probability.

$$P(5) = \frac{4}{12}$$

When the number of possible outcomes per stage of the experiment becomes large, or there are repeat outcomes within any given stage, an array may become tedious to draw out and so a tree diagram may be a more efficient sample space display.

A tree diagram shows the successive stages of an experiment, with each outcome per stage having a branch and the likelihood of that outcome is written as a weight on the branch. At the end of a sequence of branches, the **compound event** is noted. To calculate the probability of a compound event, the multiplication principle is used; that is, the probabilities are multiplied across the branches. If multiple compound events need to be considered as part of the answer, then the addition principle is used; that is, all the required resulting probabilities are added together.

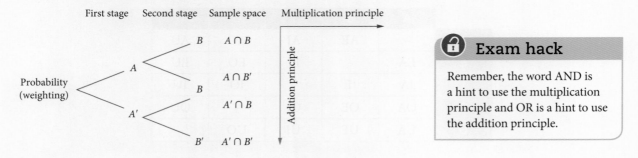

> **Exam hack**
>
> Remember, the word AND is a hint to use the multiplication principle and OR is a hint to use the addition principle.

Once again, we must take note of whether there is replacement in the situation. If there is no replacement, this will change the probability of events in future stages.

> **Exam hack**
>
> On a tree diagram, the probabilities on any set of branches MUST sum to 1.

WORKED EXAMPLE 12 | Using a tree diagram to solve a two-stage problem

Zoe has 4 pairs of black socks and 2 pairs of white socks. Her socks are randomly mixed in her drawer. Zoe takes 2 individual socks at random from the drawer in the dark.

Use a tree diagram to determine the probability that she selects a matching pair.

Steps	Working
1 Define the events of the experiment.	Let B be the event that Zoe selects a black sock and W be the event that Zoe selects a white sock.
2 Construct a two-stage tree diagram, labelling the probabilities of the first selection.	(tree diagram with first selection probabilities $\frac{8}{12}$ for B and $\frac{4}{12}$ for W, each branching to B and W on the second sock)
3 Recognise whether or not there is replacement in the situation.	The experiment does not involve replacement.
4 Given that there is no replacement, decrease the size of the sample space by 1 in the second selection and account for the first selection.	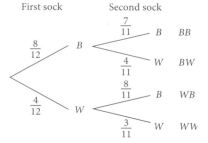
5 Use the multiplication and addition principles to calculate the probability of $BB \cup WW$.	$P(BB \cup WW) = \dfrac{8}{12} \times \dfrac{7}{11} + \dfrac{4}{12} \times \dfrac{3}{11}$ $= \dfrac{56 + 12}{132}$ $= \dfrac{68}{132}$

> **Exam hack**
>
> You don't need to waste time simplifying fractions in probability unless you have been specifically asked to do so!

Problems involving three stages

Constructing a sample space display for a three-stage experiment can be tedious, but is most efficient if it is done using a tree diagram. The properties are the same, except there is now a third stage before the sample space.

WORKED EXAMPLE 13 — Solving a three-stage problem with replacement

A barrel contains 20 balls, of which 8 are multi-coloured. Three balls are randomly selected from the barrel, with replacement. This means that a ball is selected, its colour noted, and the ball is replaced before the next ball is selected.

a Construct a tree diagram showing all possible outcomes.

b Determine the probability that 2 of the 3 balls are multi-coloured.

Steps	Working
a 1 Define C to be the event of a ball being multi-coloured.	Let C be the event that the selected ball is multi-coloured.
2 State the probabilities of C and C'.	$P(C) = \dfrac{8}{20} = 0.4$ $P(C') = 1 - 0.4 = 0.6$
3 Construct the tree diagram.	 1st ball → 2nd ball → 3rd ball 0.4 C → 0.4 C → 0.4 C : CCC 0.4 C → 0.4 C → 0.6 C' : CCC' 0.4 C → 0.6 C' → 0.4 C : CC'C 0.4 C → 0.6 C' → 0.6 C' : CC'C' 0.6 C' → 0.4 C → 0.4 C : C'CC 0.6 C' → 0.4 C → 0.6 C' : C'CC' 0.6 C' → 0.6 C' → 0.4 C : C'C'C 0.6 C' → 0.6 C' → 0.6 C' : C'C'C'
b 1 Identify the branches that satisfy the question. **2** Use the multiplication and addition principles to determine the probability.	$P(2C, 1C') = P(CCC') + P(CC'C) + P(C'CC)$ $= 3 \times 0.4^2 \times 0.6$ $= 0.288$

If the question does not specify to first construct a tree diagram, then when completing three-stage problems, we only need to consider the branches of the tree diagram that are relevant to our probability calculation.

WORKED EXAMPLE 14 — Solving a three-stage problem without replacement

A random sample of 3 cubes is selected, without replacement, from a bag containing 7 green, 5 white and 8 red cubes. Determine the probability that all 3 cubes are red, giving your answer as a simplified fraction.

Steps	Working
1 Identify whether the problem involves replacement or not.	The cubes are selected without replacement.
2 Define the event(s) in the question.	Let R be the event that a red cube is selected.
3 Decide whether the question requires a full tree diagram or not.	RRR does not require a full diagram.

4 Consider the weightings of the branches required in the tree diagram and calculate the probability.

First red: $P(R_1) = \dfrac{8}{20}$

Second red: $P(R_1 R_2) = \dfrac{8}{20} \times \dfrac{7}{19}$

Third red: $P(R_1 R_2 R_3) = \dfrac{8}{20} \times \dfrac{7}{19} \times \dfrac{6}{18}$

$P(RRR) = \dfrac{336}{6840}$

$= \dfrac{14}{285}$

EXERCISE 2.3 Probability experiments involving stages   ANSWERS p. 448

Recap

1 Two events, A and B, are mutually exclusive within a sample space. Let $P(A) = p$ and $P(B) = q$, where $0 < p < 1$ and $0 < q < 1$. The value of $P(A' \cap B')$ is

  A $(1-p)(1-q)$ **B** $1 - pq$ **C** $1 - (p+q)$
  D $2 - p - q$ **E** $1 - (p + q - pq)$

2 For two events, A and B, $P(A \cap B') = p$, $P(A) = 3p$ and $P(A \cap B) = 0.24$. The value of p is

  A 0.12 **B** 0.36 **C** 0.4 **D** 0.6 **E** 0.76

Mastery

3 WORKED EXAMPLE 11 Five identical cubes are numbered 1 to 5, and are put into a box. A cube is randomly drawn from the box, the number noted, and the cube returned to the box. A second cube is then randomly drawn from the box and its number is noted.

  a Represent the sample space using an array.
  b Determine the probability that
    i the cubes are numbered 2 and 3, in any order
    ii the product of the numbers on the two cubes is even.

4 WORKED EXAMPLE 12 A bag contains 3 red marbles and 2 green marbles. A marble is selected, its colour noted, and the marble is not replaced. A second marble is then selected.

Use a tree diagram to determine the probability of selecting 2 marbles of the same colour.

5 WORKED EXAMPLE 13 A box contains 5 red balls and 3 blue balls. John selects 3 balls from the box, replacing them after each selection.

  a Construct a tree diagram showing all possible outcomes.
  b Determine the probability that at least one of the balls that John selected is red.

6 WORKED EXAMPLE 14
  a A bag contains 4 white balls and 6 black balls. Three balls are drawn from the bag with replacement. Determine the probability that they are all black.
  b A bag contains 3 white balls and 7 yellow balls. Three balls are drawn one at a time from the bag without replacement. Determine the probability that they are all yellow.
  c A bag contains 5 red marbles and 4 blue marbles. Two marbles are drawn from the bag, without replacement, and the results are recorded. Determine the probability that the marbles are of different colours.

▶ **Calculator-free**

7 (3 marks) Two regular six-sided dice are rolled and the sum of the numbers on the uppermost faces is recorded. Calculate the probability of obtaining a sum greater than 8, or a 6 on one of the die.

8 (7 marks) A tetrahedral die has its faces numbered 1 to 4 and a regular six-sided die has its faces numbered 1 to 6. Both dice are rolled and the numbers that are face down are noted.

   a Represent the sample space using an array. (2 marks)

   b Hence, determine the probability that

      i both numbers are even (1 mark)

      ii the sum of the two numbers is odd (2 marks)

      iii the product of the two numbers is even. (2 marks)

9 (6 marks) Two six-sided dice are rolled. A fraction is formed by making the number on the first die the numerator and the number on the second die the denominator.

   a Represent the sample space using an array. (2 marks)

   b Hence, determine the probability that the fraction generated is equivalent to

      i $\frac{1}{2}$ (1 mark)

      ii $\frac{1}{3}$ (1 mark)

      iii a number greater than 1 (1 mark)

      iv a whole number. (1 mark)

10 (6 marks) At a party there are six unmarked boxes. Two boxes each have prizes, the other four boxes are empty. Two boxes are selected without replacement. Determine the probability of selecting

   a a prize box first and an empty box second (2 marks)

   b two boxes that are either both prizes or both empty (2 marks)

   c at least one box with a prize. (2 marks)

11 (3 marks) A committee of 5 Year 12 students and 6 Year 11 students elects a captain and a vice captain from its own members. Let E be the event that a Year 11 student is chosen for the role and T be the event that a Year 12 student is chosen for the role. Determine the probability of electing a captain and a vice captain of the same year level, correct to four decimal places.

12 (6 marks) Lizzie travels to work by car, bus or bicycle. The likelihood of her using each form of transport is 0.6, 0.25 and 0.15 respectively. The probability of her being late is 0.1 when she travels by car, 0.2 when she travels by bus and 0.45 when she cycles. Find the probability that Lizzie

   a is late for work (2 marks)

   b travels by car and is on time for work (2 marks)

   c travels by car or is on time for work. (2 marks)

2.4 Probability experiments involving selections

Applying combinations to probability

In cases where a tree diagram becomes too tedious to draw out, we can represent experiments involving selections using our prior knowledge of combinations from Chapter 1. Recall that when r objects are selected from a set of n distinct objects, then there are ^nC_r or $\binom{n}{r}$ ways of making the selection.

Combinations formula

$$^nC_r = \binom{n}{r} = \frac{n!}{r!(n-r)!}$$

Using combinations in probability

Combinations can be used to calculate probabilities in situations involving selections without replacement. Given that it is a probability calculation, we first need to be able to calculate the total number of possible selections that can be made. We may also then need to use the multiplication or addition principles for combinations if the problem involves AND or OR respectively.

WORKED EXAMPLE 15 — Using combinations to calculate probabilities

A mixed volleyball team of 9 players is to be randomly selected from a group of 8 girls and 5 boys. Determine the probability that the selected team contains 5 girls and 4 boys, correct to four decimal places.

Steps	Working
1 Calculate the total number of possible combinations.	Total number of possible teams: $\binom{13}{9}$
2 Identify whether the problem involves the multiplication or addition principles.	5 girls AND 4 boys → multiplication principle
3 Show the probability calculation using combinations.	$P(5G \cap 4B) = \dfrac{\binom{8}{5} \times \binom{5}{4}}{\binom{13}{9}}$
4 Evaluate using CAS, where appropriate.	$= 0.3916$

Although the multiplication and addition principles revisit the set operations of intersection and union, we can also use the complement operation in problems involving the phrase 'at least'. For example, in the problem involving 5 students being selected from 20 students (15 Year 11s and 5 Year 12s), we may want to consider how many possible selections contain *at least* one Year 11 student or *at least* two Year 12 students. This can either be done exhaustively by considering all the possible cases, or by using the complement rule and recognising that 'at least one Year 11' is equivalent to 'all possible selections' minus when there are 'no Year 11s'.

The complement principle using 'at least'

If making a selection that involves at least k objects from a set, then consider whether it is more efficient to use the complement. That is, the probability of making a selection involving at least k objects can be calculated by

P(at least k objects) = 1 − P(less than k objects)

WORKED EXAMPLE 16 — Solving combinations problems using the complement

A cricket team of 11 players is randomly selected from 7 bowlers, 8 batters and 3 wicketkeepers. Determine the probability that at least 1 wicketkeeper is chosen, correct to four decimal places.

Steps	Working
1 Calculate the total number of possible combinations.	Total number of possible teams: $\binom{18}{11}$
2 Separate the subsets into wicketkeepers and others.	3 wicketkeepers AND 15 others.
3 Recognise that at least 1 wicketkeeper is the complement of selecting 0 wicketkeepers. Use the complement and multiplication principles to establish the probability.	0 wicketkeepers means the team is formed by: $\binom{3}{0} \times \binom{15}{11}$ P(at least 1 wicketkeeper) = $1 - \dfrac{\binom{3}{0} \times \binom{15}{11}}{\binom{18}{11}}$
4 Evaluate using CAS, where appropriate.	= 0.9571

EXERCISE 2.4 Probability experiments involving selections   ANSWERS p. 448

Recap

1 Two fair coins are tossed and the number of heads that are obtained is recorded.

   The probability of obtaining no heads after both tosses is

   A 0   **B** $\dfrac{1}{4}$ **C** $\dfrac{1}{2}$ **D** $\dfrac{3}{4}$ **E** 1

2 A bag contains 6 red marbles and 4 blue marbles. Two marbles are drawn from the bag, without replacement, and the results are recorded.

   The probability that the marbles are the same colour is

   A $\dfrac{2}{45}$ **B** $\dfrac{1}{10}$ **C** $\dfrac{2}{9}$ **D** $\dfrac{7}{15}$ **E** $\dfrac{5}{9}$

Mastery

3 [WORKED EXAMPLE 15] A mixed netball team of 8 players is to be randomly selected from a group of 10 girls and 7 boys. Determine the probability that the selected team contains the same number of girls and boys, correct to four decimal places.

4 [WORKED EXAMPLE 16] A student delegation of 12 students is randomly selected from 10 Year 10s, 5 Year 11s and 5 Year 12s. Determine the probability that at least 1 Year 12 is chosen, correct to four decimal places.

Calculator-free

5 (3 marks) A bag contains 20 monetary chips, which are used to represent cash. Five chips have a value of $10 and the remainder have a value of $5.

If 4 chips are drawn at random from the bag without replacement, write an expression that can be used to calculate the probability that there will be at least 1 chip of value $10. Do not evaluate the expression.

Calculator-assumed

6 (3 marks) A bag contains 5 blue marbles and 4 red marbles. A sample of 4 marbles is taken from the bag, without replacement. Determine the probability that the proportion of blue marbles in the sample is greater than a half.

7 (9 marks) A box contains 15 batteries, of which 5 are defective. A sample of 4 batteries is selected without replacement. Determine the probability, correct to four decimal places, that the sample contains

   a 2 defective batteries (3 marks)

   b more defective batteries than functioning batteries (3 marks)

   c at least one defective battery. (3 marks)

8 (6 marks) Two mixed teams of four tennis players each select one of their team members to play off in a challenge match. Team *A* has three female players and one male player while team *B* has an equal number of male and female players.

Determine the probability that

   a both players in the challenge match are male (3 marks)

   b there is exactly one female in the challenge match. (3 marks)

9 (8 marks) Deni has 5 pencils and 3 pens in her pencil case. If she chooses 4 items at random, determine the probability, correct to four decimal places, that she picks

   a exactly 3 pencils (2 marks)

   b the same number of pencils as pens (3 marks)

   c more pencils than pens. (3 marks)

2.5 Conditional probability

Video playlist
Conditional probability

Worksheet
Conditional probability

When representing sample spaces involving the relationship between two sets, we can often encounter problems that involve restrictions or conditions placed on the sample space. These situations use **conditional probability**, which is the probability of an event occurring *given that* another event has also occurred. So, how do we know when we are dealing with conditional probability?

Indicating conditionality

An important skill in the topic of probability is to be able to identify the language that indicates conditionality, where some phrases are more obvious than others. Two key phrases that indicate conditionality are:

- if … then …
- given that

For example, if event A has occurred, then calculate the probability of event B occurring. Alternatively, calculate the probability of event B occurring, given that event A has occurred. However, it is not always as obvious as these two phrases!

> **Exam hack**
>
> To know when to use conditional probability, look for wording that implies that a restriction or condition has been placed on the original sample space.

WORKED EXAMPLE 17 — Identifying conditional statements

For each of the following statements, justify whether it is a simple probability calculation or a conditional probability calculation.

a If it is raining outside, calculate the probability that John goes for a walk.
b Calculate the probability of scoring 10 points or more.
c Find the probability that a randomly selected egg weighs more than 60 g, given that it is no more than 80 g.
d After tossing two heads on the biased coin, determine the probability that a tail is tossed.

Steps	Working
1 Identify any language that indicates conditionality. 2 If none, then it is a simple event. If conditional, justify why by identifying the restriction or condition.	a Conditional, as 'if it is raining outside' is the restriction.
	b Simple, as no restriction is being placed.
	c Conditional, as 'given that it is no more than 80 g' is the restriction.
	d Conditional, as 'tossing two heads' is the restriction.

Using the conditional probability formula

If we think about the logic of the following statement

calculate the probability of event B occurring, given that A has occurred

the probability calculation is no longer out of the total sample space.

Instead, the fraction formed to express the conditional probability needs to have the denominator representing the likelihood of event B, and the numerator representing the likelihood of both A and B having occurred.

> **Conditional probability formula**
>
> For two events A and B, the probability of A occurring given that B has occurred is calculated using
>
> $$P(A|B) = \frac{P(A \cap B)}{P(B)}$$
>
> where $A|B$ is read as 'A given B'.
>
> If A and B are mutually exclusive events, then $P(A \cap B) = 0$ and so $P(A|B) = 0$.

We can then apply this formula to any of the problem types we have examined so far: simple experiments, problems involving sets, problems involving stages and problems involving selections.

WORKED EXAMPLE 18 | Calculating conditional probabilities from a list

Consider the set of 12 letters {M, A, T, H, E, I, C, S, O, P, R, N}. A letter is to be randomly selected from this set. Calculate the probability that the letter is a vowel, given that the letter is in the word MATHEMATICS.

Steps	Working
1 Define the events.	Let V be the event that a vowel is selected and M be the event that the letter is in the word MATHEMATICS.
2 Identify the size of the restricted sample space.	$n(M) = 8$
3 Identify the size of the intersection set and state the conditional probability.	$P(V|M) = \dfrac{3}{8}$

WORKED EXAMPLE 19 | Calculating conditional probabilities from a two-way table

Two events, A and B, exist such that $P(A \cap B') = 0.04$, $P(B) = 0.6$ and $P(A') = 0.48$.

By first completing a two-way table, calculate $P(A'|B')$.

Steps	Working
1 Construct the two-way table and input the known information.	<table><tr><th></th><th>B</th><th>B'</th><th>Total</th></tr><tr><td>A</td><td></td><td>0.04</td><td></td></tr><tr><td>A'</td><td></td><td></td><td>0.48</td></tr><tr><td>Total</td><td>0.6</td><td></td><td>1</td></tr></table>
2 Complete the table using the row and column sums.	<table><tr><th></th><th>B</th><th>B'</th><th>Total</th></tr><tr><td>A</td><td>0.48</td><td>0.04</td><td>0.52</td></tr><tr><td>A'</td><td>0.12</td><td>0.36</td><td>0.48</td></tr><tr><td>Total</td><td>0.6</td><td>0.4</td><td>1</td></tr></table>
3 Use the conditional probability formula to calculate $P(A'|B')$.	$P(A'|B') = \dfrac{P(A' \cap B')}{P(B')}$ $= \dfrac{0.36}{0.4}$ $= 0.9$

Conditional probabilities from two-way tables or Venn diagrams are quite trivial when we can see the number of elements in each set, as we are simply identifying the part of the display that represents the restricted sample space and then, from that, identifying the intersection set. For example, for $P(A|B)$, we are restricting the sample space to the set shown highlighted in orange and then selecting the set shown shaded in blue.

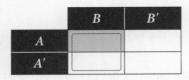
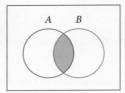

As a result, the conditional probability formula can also be written as

$$P(A|B) = \frac{n(A \cap B)}{n(B)}$$

WORKED EXAMPLE 20 — Calculating conditional probabilities from a Venn diagram

A group of 50 performers contains 18 tap dancers and 6 performers who can both sing and tap dance. All the performers either tap dance or sing. Calculate the probability of randomly selecting a tap dancer given that a singer is selected.

Steps	Working
1 Define the events appropriately.	Let T be the event that a tap dancer is chosen and S be the event that a singer is chosen.
2 Construct the Venn diagram and input the known information.	
3 Use the complement rule to complete the diagram.	$P(T \cap S') = \frac{18}{50} - \frac{6}{50} = \frac{12}{50}$ $P(T' \cap S) = 1 - \frac{6}{50} - \frac{12}{50} = \frac{32}{50}$ (Venn diagram with T containing $\frac{12}{50}$, intersection $\frac{6}{50}$, S containing $\frac{32}{50}$)
4 Use the conditional probability formula to calculate $P(T|S)$.	$P(T|S) = \frac{n(T \cap S)}{n(S)}$ $= \frac{6}{38}$

Similarly, when working with arrays, we can visualise the conditional probability $P(A|B)$ by first identifying the restricted sample space and then identifying the intersection sets.

WORKED EXAMPLE 21 — Calculating conditional probabilities from an array

A deck contains cards numbered from 1 to 5. Two cards are randomly selected without replacement.

a Construct an array to represent the sample space.
b Given that the first card selected is prime, determine the probability that the second card is also prime.
c If the product of the two cards is greater than 10, determine the probability that both cards are greater than 3.

Steps	Working
a Establish the rows and columns, as well as the entries as ordered pairs.	

	1	2	3	4	5
1	–	1, 2	1, 3	1, 4	1, 5
2	2, 1	–	2, 3	2, 4	2, 5
3	3, 1	3, 2	–	3, 4	3, 5
4	4, 1	4, 2	4, 3	–	4, 5
5	5, 1	5, 2	5, 3	5, 4	–

b 1 Highlight the outcomes that represent the restricted sample space (orange).

	1	2	3	4	5
1	–	1, 2	1, 3	1, 4	1, 5
2	2, 1	–	2, 3	2, 4	2, 5
3	3, 1	3, 2	–	3, 4	3, 5
4	4, 1	4, 2	4, 3	–	4, 5
5	5, 1	5, 2	5, 3	5, 4	–

2 Identify the intersection set (blue).

	1	2	3	4	5
1	–	1, 2	1, 3	1, 4	1, 5
2	2, 1	–	2, 3	2, 4	2, 5
3	3, 1	3, 2	–	3, 4	3, 5
4	4, 1	4, 2	4, 3	–	4, 5
5	5, 1	5, 2	5, 3	5, 4	–

3 State the conditional probability.

$$P(\text{prime} \mid \text{prime}) = \frac{6}{12}$$

c 1 Modify the array to show products.

2 Highlight the outcomes that represent the restricted sample space (orange).

	1	2	3	4	5
1	–	2	3	4	5
2	2	–	6	8	10
3	3	6	–	12	15
4	4	8	12	–	20
5	5	10	15	20	–

3 Identify the intersection set (blue).

4 State the conditional probability.

$$P(\text{card} > 3 \mid \text{product} > 10) = \frac{2}{6}$$

The benefit of tree diagrams is that they inherently display conditional probabilities on the branches of subsequent events.

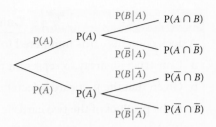

This display proves an alternative arrangement of the conditional probability formula through the multiplication principle.

The multiplication principle (conditional)

$$P(B|A) = \frac{P(A \cap B)}{P(A)} \Rightarrow P(A \cap B) = P(A) \times P(B|A)$$

Suppose in this tree diagram, we wanted to know the total probability of B. To do so, we need to consider the probability that B occurs given that either A has occurred or A has not occurred.

The law of total probability

$$P(B) = P(A \cap B) + P(\bar{A} \cap B)$$
$$P(B) = P(B|A)P(A) + P(B|\bar{A})P(\bar{A})$$

WORKED EXAMPLE 22 — Calculating conditional probabilities from a tree diagram

A team of football players find that they win 80% of their games when their star player plays and 30% of their games when he does not play. Their star player was injured at training during the week and has a 40% chance of playing on the weekend. Determine the probability that in their next game the star player plays given that they win.

Steps	**Working**
1 Define the events.	Let S be the event that the star player plays and W be the event that the team wins.
2 Construct a tree diagram.	(tree diagram showing branches S → W, W' and S' → W, W')
3 Complete the weights of the tree diagram, remembering that the sum of each set of branches must be 1.	(tree diagram with weights: 0.4 to S, then 0.8 to W, 0.2 to W'; 0.6 to S', then 0.3 to W, 0.7 to W')

4 Establish the conditional probability formula.	$P(S\mid W) = \dfrac{P(S \cap W)}{P(W)}$
5 Determine the required probabilities using the multiplication principle and law of total probability, as needed.	$P(S \cap W) = 0.4 \times 0.8 = 0.32$ $P(W) = 0.4 \times 0.8 + 0.6 \times 0.3 = 0.5$
6 State the conditional probability.	$P(S\mid W) = \dfrac{0.32}{0.5} = 0.64$

WORKED EXAMPLE 23 — Calculating probabilities using repeated conditional probabilities in a tree diagram

Beryl is the oldest hen at the farm and is a very fussy egg layer. If she lays an egg on one day, there is a probability of 0.15 that she will lay an egg the next day. If she does not lay on one day, the probability of laying an egg the next day is 0.8. On one particular Sunday, Beryl is feeling a bit indifferent and does not lay an egg. Find the probability that she lays 2 eggs in the next 3 days.

Steps	Working
1 Define the event.	Let E be the event that Beryl lays an egg.
2 Construct a tree diagram.	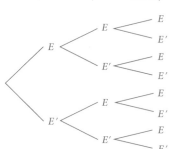
3 Complete the weights of the tree diagram, remembering that the sum of each set of branches must be 1. Since Beryl does not lay an egg on Sunday, the probabilities for Monday are $P(E) = 0.8$ and $P(E') = 0.2$.	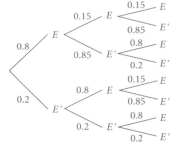
4 Determine the required probabilities using the multiplication principle and law of total probability, as needed.	$P(2 \text{ eggs}) = P(EEE') + P(EE'E) + P(E'EE)$ $= 0.8 \times 0.15 \times 0.85 + 0.8 \times 0.85 \times 8$ $+ 0.2 \times 0.8 \times 0.15$
5 State the probability.	$P(2 \text{ eggs}) = 0.67$

Combinations can also be used to solve conditional probability problems for a larger number of selections that cannot be efficiently represented in a tree diagram.

WORKED EXAMPLE 24 — Calculating conditional probabilities involving combinations

A company has 10 employees, consisting of 6 developers and 4 designers. The company plans to randomly select a project team containing 4 members. Calculate the probability that the team contains exactly 3 designers given that it has at least 1 designer.

Steps	Working
1 Define the event.	Let A be the event that the team has exactly 3 designers and B be the event that the team has at least 1 designer.
2 Calculate the number of outcomes in B using the complement rule.	$n(B) = \binom{10}{4} - \binom{6}{4} \times \binom{4}{0} = 195$
3 Interpret and calculate the number of outcomes in $A \cap B$ using the multiplication principle.	$A \cap B = A$ because exactly 3 designers is a subset of at least 1 designer.
4 Use the conditional probability formula to calculate the probability.	$n(A \cap B) = n(A) = \binom{6}{1} \times \binom{4}{3} = 24$ $$P(A \mid B) = \frac{24}{195}$$

EXERCISE 2.5 Conditional probability   ANSWERS p. 448

Recap

1 A bag contains 3 white balls and 5 black balls. Three balls are drawn from the bag without replacement. The probability that they are all black is

A $\dfrac{1}{28}$ **B** $\dfrac{27}{512}$ **C** $\dfrac{5}{28}$ **D** $\dfrac{125}{512}$ **E** $\dfrac{5}{8}$

2 A bag contains 12 bread rolls, of which 8 are white and the remainder are multigrain. Tony takes 2 bread rolls at random from the bag to eat.

The probability that at least one is a multigrain roll is

A $1 - \dfrac{2^{12}}{3^{12}}$ **B** $1 - \dfrac{^8C_2}{^{12}C_2}$ **C** $1 - \dfrac{2^{12}}{3^{12}} - 12 \times \dfrac{1}{3} \times \dfrac{2^{11}}{3^{11}}$

D $1 - \dfrac{^8C_2}{^{12}C_2} - \dfrac{^8C_1 \times {}^4C_1}{^{12}C_2}$ **E** $\dfrac{^8C_1 \times {}^4C_1}{^{12}C_2}$

Mastery

3 WORKED EXAMPLE 17 For each of the following statements, justify whether it is a simple probability calculation or a conditional probability calculation.

  a Given that the bodybuilder trained arms yesterday, calculate the probability that they will train shoulders today.

  b Calculate the probability of selecting a black marble from a bag of 4 white and 6 black marbles.

4 **WORKED EXAMPLE 18** Consider the set of 11 letters {B, I, N, O, M, A, L, E, X, P, S}. A letter is to be randomly selected from this set. Calculate the probability that the letter is a consonant, given that the letter is in the word BINOMIAL.

5 **WORKED EXAMPLE 19**

   a Two events, A and B, exist such that $P(A' \cap B) = 0.05$, $P(B') = 0.45$ and $P(A) = 0.7$. By first completing a two-way table, calculate $P(A'|B')$.

   b Two events, A and B, exist such that $P(A \cap B) = \frac{7}{24}$, $P(A \cap B') = \frac{1}{8}$ and $P(A' \cap B) = \frac{5}{12}$. By first completing a two-way table, calculate $P(A|B')$.

6 **WORKED EXAMPLE 20** A survey of 70 people shows that 55 own a smartphone, 33 own a tablet and 22 own both items. Determine the probability that a randomly selected person owns a smartphone, if it is known that the person owns a tablet.

7 **WORKED EXAMPLE 21** A deck contains cards numbered from 3 to 6. Two cards are randomly selected without replacement.

   a Construct an array to represent the sample space.
   b Given that the first card selected is even, determine the probability that the second card is also even.
   c If the sum of the two cards is less than 10, determine the probability that both the cards are odd.

8 **WORKED EXAMPLE 22** Sleepy Sam arrives at school on time on 90% of occasions when his alarm goes off in the morning but only arrives on time on 35% of occasions when his alarm does not ring. His alarm is faulty and will only ring 70% of school day mornings. Determine the probability that if Sleepy Sam arrives at school on time, his alarm has gone off that morning.

9 **WORKED EXAMPLE 23**

   a Eleanor's mum always bakes a cake on Eleanor's birthday. Her two favourite cakes are chocolate and orange. If Eleanor has a chocolate cake one year, there is a 75% chance she will get an orange cake the next year. If she has an orange cake one year, there is a 60% chance she will get a chocolate cake the next year. Eleanor has a chocolate cake on her 10th birthday. Determine the probability that Eleanor has exactly 2 chocolate birthday cakes before her 13th birthday.

   b Each afternoon, Aldo buys a piece of fruit on his way home from school. If he buys an apple one day, there is a 40% chance he will buy an apple the next day. If he buys a peach one day, there is a 70% chance he will buy a peach the next day. On Monday, Aldo buys a peach. Determine the probability that Aldo will buy a total of exactly 2 peaches on Tuesday, Wednesday and Thursday.

10 **WORKED EXAMPLE 24** Carmen has 40 different ribbons in her drawer, consisting of 15 cotton ribbons and 25 silk ribbons. She plans to wear four ribbons in her hair one day to school, but wants to randomly select them from her drawer. Calculate the probability, correct to four decimal places, that Carmen chooses exactly 2 silk ribbons given that she chooses at least 1 silk ribbon.

Calculator-free

11 (11 marks) Consider the two-way table provided showing the probabilities of events involving A and B.

	B	B'	Total
A	0.35	0.09	
A'	0.25	0.31	
Total			

    a Copy and complete the table. (1 mark)

    b Hence, calculate the following probabilities.

        i $P(A|B)$ (2 marks)

        ii $P(A|B')$ (2 marks)

        iii $P(A'|B')$ (2 marks)

        iv $P(B|A)$ (2 marks)

        v $P(B|A')$ (2 marks)

12 (8 marks) Out of a group of 150 individuals, 90 enjoy cardio exercises (C) and 60 enjoy strength training exercises (S). In the group, there are 30 individuals who do not like either type of exercise.

    a Construct a suitable display to represent this situation. (2 marks)

    b If an individual was randomly selected from the group, determine the probability that they enjoy

        i strength training, given that they also enjoy cardio (2 marks)

        ii cardio, given that they do not enjoy strength training. (2 marks)

    c Two individuals are to be randomly selected from the group, and the same person cannot be selected twice. If the first person does not like cardio, determine the probability that the second individual also does not enjoy cardio. (2 marks)

13 (4 marks) Four identical balls are numbered 1, 2, 3 and 4 and put into a box. A ball is randomly drawn from the box, and not returned to the box. A second ball is then randomly drawn from the box.

    a Construct an array to represent this situation. (2 marks)

    b Given that the sum of the numbers on the two balls is 5, determine the probability that the second ball drawn is numbered 1. (2 marks)

14 (4 marks) The only possible outcomes when a coin is tossed are a head or a tail. When an unbiased coin is tossed, the probability of tossing a head is the same as the probability of tossing a tail. Jo has four coins in her pocket; three are unbiased and one is biased. When the biased coin is tossed, the probability of tossing a head is $\frac{1}{3}$. Jo randomly selects a coin from her pocket and tosses it. Calculate the probability that she

    a tosses a tail (2 marks)

    b selects an unbiased coin, given that she tosses a tail. (2 marks)

15 (7 marks) An egg marketing company buys its eggs from farm A and farm B. Let p be the proportion of eggs that the company buys from farm A. The rest of the company's eggs come from farm B. Each day, the eggs from both farms are taken to the company's warehouse. Assume that $\frac{3}{5}$ of all eggs from farm A and $\frac{1}{5}$ of all eggs from farm B have white eggshells.

   a An egg is selected at random from the set of all eggs at the warehouse. Find, in terms of p, the probability that the egg has a white eggshell. (2 marks)

   b Another egg is selected at random from the set of all eggs at the warehouse.

      i Given that the egg has a white eggshell, find, in terms of p, the probability that it comes from farm B. (2 marks)

      ii If the probability that this egg in part **b i** comes from farm B is 0.3, find the value of p. (3 marks)

Calculator-assumed

16 (9 marks) The library conducted a survey to determine the most popular style of books. Of the 100 students surveyed:

- 60 liked fantasy
- 18 liked only romance
- 13 liked adventure and romance
- 16 liked fantasy and romance
- 26 liked exactly two of these styles
- 36 liked adventure but not romance
- 11 liked all three styles.

Determine the probability that a student chosen at random likes

   a only fantasy (2 marks)

   b fantasy or adventure or romance (2 marks)

   c none of these three styles (1 mark)

   d fantasy, given that they like romance (2 marks)

   e fantasy or adventure, given that they do not like romance. (2 marks)

2.6 Independent events

Let's recall the examples we have seen so far where the selections have been made with or without replacement. This context of replacement is a very simple representation of the idea of independence. For example, if we select a card from a regular deck of cards and then replace it before the next selection, the probabilities have been 'reset' and the first selection is not going to influence the second. However, if we select the card and do not replace it, all future probabilities change based on the type of card we selected first. This is one way to think about **independent events**: with replacement implies independence and without replacement implies dependence.

> **Independent events – definition**
>
> Two events, A and B, are independent events if the probability of one event occurring does not affect the probability of the other event occurring.

However, questions involving independence are not always going to be in the context of replacement, so we need to define the concept of independent events in a few different ways.

Independence using restrictions

Suppose we placed a restriction on a sample space such that event B occurs first and we now want to know the likelihood of event A occurring; that is, P(A|B). If the likelihood of A doesn't change given that B has occurred, then we can say that A and B are independent events.

> **Independent events (conditional rule)**
>
> For two independent events A and B, the probability of A occurring given that B has occurred remains unchanged; that is, P(A|B) = P(A).

 Exam hack

The conditional rule for independence is the most useful in problems involving sets, because the restriction can be visualised using two-way tables or Venn diagrams.

WORKED EXAMPLE 25 — Showing independence using conditional probabilities

Two events, A and B, are defined such that $P(A) = 0.35$, $P(B') = 0.2$ and $P(A' \cap B) = 0.52$. Show that A and B are independent by finding $P(A|B)$.

Steps	Working
1 Construct a suitable display for the known probabilities.	<table><tr><th></th><th>B</th><th>B'</th><th>Total</th></tr><tr><td>A</td><td></td><td></td><td>0.35</td></tr><tr><td>A'</td><td>0.52</td><td></td><td></td></tr><tr><td>Total</td><td></td><td>0.2</td><td>1</td></tr></table>
2 Complete the table using row and column sums.	<table><tr><th></th><th>B</th><th>B'</th><th>Total</th></tr><tr><td>A</td><td>0.28</td><td>0.07</td><td>0.35</td></tr><tr><td>A'</td><td>0.52</td><td>0.13</td><td>0.65</td></tr><tr><td>Total</td><td>0.8</td><td>0.2</td><td>1</td></tr></table>
3 Determine the value of P(A|B) using the conditional probability formula.	$P(A\|B) = \dfrac{0.28}{0.8} = \dfrac{28}{80}$ $= \dfrac{7}{20}$ $= 0.35$ $= P(A)$
4 Show that P(A|B) = P(A) to conclude independence.	Therefore, A and B are independent events.

Successive independence

The concept of successive independence closely relates to the idea of replacement because it is about the sequence of successive events and whether one event affected the other. A rule for successive independence can be derived from the conditional rule.

> **Independent events (successive independence rule)**
>
> Suppose that A and B were independent events; that is, A occurring does not affect the probability of B occurring and vice versa. Then the multiplication principle
>
> $P(A \cap B) = P(A) \times P(B|A)$
>
> becomes
>
> $P(A \cap B) = P(A) \times P(B)$
>
> because $P(B|A) = P(B)$.

WORKED EXAMPLE 26 — Using independence to calculate probabilities

Two independent events, A and B, exist such that $P(A') = 0.6$ and $P(A' \cap B) = 0.18$. Determine $P(A \cap B)$.

Steps	Working
1 Construct a suitable display for the known probabilities.	<table><tr><td></td><td>B</td><td>B'</td><td>Total</td></tr><tr><td>A</td><td></td><td></td><td></td></tr><tr><td>A'</td><td>0.18</td><td></td><td>0.6</td></tr><tr><td>Total</td><td></td><td></td><td>1</td></tr></table>
2 Complete as much of the table as possible using row and column sums. 3 If there is insufficient information to continue, let $P(A \cap B) = x$ and $P(B) = y$.	<table><tr><td></td><td>B</td><td>B'</td><td>Total</td></tr><tr><td>A</td><td>x</td><td></td><td>0.4</td></tr><tr><td>A'</td><td>0.18</td><td>0.42</td><td>0.6</td></tr><tr><td>Total</td><td>y</td><td></td><td>1</td></tr></table>
4 Form equations for the unknowns using the independence rule and column sum. Solve simultaneously.	$0.4y = x$ $x + 0.18 = y$ $0.4y + 0.18 = y$ $0.6y = 0.18$ $y = \dfrac{0.18}{0.6} = \dfrac{18}{60} = 0.3$ $x = 0.12$
5 State $P(A \cap B)$.	$P(A \cap B) = 0.12$

The symmetry of independence

Worked example 26 gives us insight into a very useful fact about independence. You may wonder that if it is known that A and B are independent (i.e. $P(A \cap B) = P(A)P(B) \Rightarrow x = 0.4y$), then what does $0.6y = 0.18$ suggest from the two-way table?

Looking back, $P(A') = 0.6$ and $P(B) = y$ and $P(A' \cap B) = 0.18$. So, $0.6y = 0.18$ implies that if A and B are independent, then A' and B are independent. This is one of two important facts that we can use when solving problems about independence. The first is somewhat trivial!

Property 1

The symmetry of independence

If A is independent of B, then B is independent of A.

$$P(A|B) = P(A) \Rightarrow P(B|A) = P(B)$$

Proof

By the definition of conditional probability,

$$P(A|B) = \frac{P(A \cap B)}{P(B)}$$

If $P(A|B) = P(A)$, then

$$P(A) = \frac{P(A \cap B)}{P(B)}$$

$$P(B) = \frac{P(A \cap B)}{P(A)}$$

$$P(B) = P(B|A)$$

Property 2

Independence of the complement

If A and B are independent, then A and B' are independent.

Proof

Using the law of total probability,

$$P(A) = P(A|B)P(B) + P(A|B')P(B')$$
$$P(A) = P(A \cap B) + P(A \cap B')$$

Since A and B are independent, then

$$P(A \cap B) = P(A)P(B)$$
$$P(A) = P(A)P(B) + P(A \cap B')$$

Rearranging for $P(A \cap B')$,

$$P(A \cap B') = P(A) - P(A)P(B)$$
$$= P(A) - [1 - P(B)]$$
$$= P(A)P(B')$$

Therefore, A and B' are independent.

> **Exam hack**
>
> Instead of solving problems algebraically as in Worked example 26, you can quote the properties of independence! Be sure to state the property clearly and correctly.

Independence in context

In contextual questions where we are told independent events exist, we can choose which independence rule is most appropriate to use given the context of the question. For questions involving successive independence, it is important to note the following.

> **Repetitive independent trials**
>
> If a repetitive experiment involves n independent trials with a probability of success p, then the probability of n successes is p^n.
>
> This concept will be expanded upon in Unit 3 of Mathematics Methods.

WORKED EXAMPLE 27 | Solving problems involving independence in context

Mark is competing in an archery competition. Each of his shots is independent of his previous shots and his probability of hitting a bullseye is 0.35. Determine the probability (correct to four decimal places) that Mark's fifth shot is the first bullseye he hits.

Steps	Working
1 Define the event and identify the sequence of events required.	Let B be the event that Mark hits a bullseye. $P(B'\,B'\,B'\,B'\,B)$
2 Use the complement rule to calculate $P(B')$.	$P(B') = 1 - P(B)$ $= 1 - 0.35$ $= 0.65$
3 Use the multiplication principle of repetitive independent events to calculate the required probability.	$P(B'\,B'\,B'\,B'\,B) = P(B')^4 \times P(B)$ $= 0.65^4 \times 0.35$ $= 0.0625$

EXERCISE 2.6 Independent events   ANSWERS p. 449

Recap

1 For two events, A and B, of a sample space S, it is known that $P(A \cap B) = \dfrac{2}{5}$ and $P(A \cap B') = \dfrac{3}{7}$.

  The value of $P(B'\,|\,A)$ is equal to

  A $\dfrac{6}{35}$ **B** $\dfrac{15}{35}$ **C** $\dfrac{15}{29}$ **D** $\dfrac{2}{3}$ **E** $\dfrac{29}{35}$

2 Demelza is a badminton player. If she wins a game, the probability that she will win the next game is 0.7. If she loses a game, the probability that she will lose the next game is 0.6. Demelza has just won a game. The probability that she will win exactly one of her next two games is

  A 0.33   **B** 0.35   **C** 0.42   **D** 0.49   **E** 0.82

▶ **Mastery**

3 **WORKED EXAMPLE 25** Two events, A and B, are defined such that $P(A) = 0.25$, $P(B) = 0.6$ and $P(\overline{A \cup B}) = 0.3$. Show that A and B are independent by finding $P(B|A)$.

4 **WORKED EXAMPLE 26** For two independent events A and B, if $P(A') = 0.4$ and $P(B) = 0.15$, determine $P(A \cup B')$.

5 **WORKED EXAMPLE 27** For a biased coin, the probability of obtaining a head is $\frac{1}{4}$. If the coin is tossed 4 times and each toss is independent of the previous tosses, determine the probability of obtaining a head for the first time on the last toss.

Calculator-free

6 (2 marks) For two independent events A and B, if $P(A') = 0.44$, $P(A \cap B') = 0.25$ and $P(A' \cap B) = 0.17$, determine $P(A|B')$. Justify your answer.

7 (5 marks) For two events A and B, it is known that $P(A|B) = \frac{3}{4}$ and $P(B) = \frac{1}{3}$.
   a Calculate $P(A \cap B)$. (1 mark)
   b Calculate $P(\overline{A} \cap B)$. (1 mark)
   c If A and B are independent, calculate $P(A \cup B)$. (3 marks)

Calculator-assumed

8 (4 marks) For events A and B, let $P(A \cap B) = p$, $P(A' \cap B) = p - \frac{1}{8}$ and $P(A \cap B') = \frac{3p}{5}$.
   a Explain why it is not possible for A and B to be mutually exclusive events. (1 mark)
   b Determine the value of p if A and B are independent events. (3 marks)

9 (2 marks) The weather bureau predicts a 30% chance of fog on Monday and a 40% chance of fog on Tuesday. If a fog occurring on Tuesday is independent of fog occurring on Monday, calculate the probability of fog on neither Monday nor Tuesday.

10 (3 marks) April has 3 balls that she attempts to throw into a basket. The probability of any of her throws successfully landing in the basket is 0.2 and the success of any throw is independent of the outcome on the previous throw. Calculate the probability of April landing exactly 2 balls in the basket.

11 (2 marks) Sharelle is the goal shooter for her netball team. During her matches, she has many attempts at scoring a goal. Assume that each attempt at scoring a goal is independent of any other attempt. In the long term, her scoring rate has been shown to be 80%. Determine the probability, correct to four decimal places, that she makes 8 out of 8 goals in a row during a match.

12 (3 marks) John and Rebecca are playing darts. The result of each of their throws is independent of the result of any other throw. The probability that John hits the bullseye with a single throw is $\frac{1}{4}$. The probability that Rebecca hits the bullseye with a single throw is $\frac{1}{2}$. John has four throws and Rebecca has two throws.

Determine the ratio of the probability of Rebecca hitting the bullseye at least once to the probability of John hitting the bullseye at least once.

2.7 Relative frequencies and indications of independence

In Section 2.1, we reviewed how relative frequencies obtained from data can be considered experimental probabilities and, hence, estimates for the likelihood of events occurring. We can now extend this further and use relative frequencies as estimates of conditional probabilities.

Estimates of conditional probabilities

Frequencies, that is the number of times a particular event occurs, can be represented in frequency tables, two-way tables or Venn diagrams. You should be prepared to work with any of the displays.

WORKED EXAMPLE 28 | Estimating conditional probability from data in a two-way table

A survey of 80 shoppers at the Watertown Brand Outlet determined whether people were shopping for specials or shopping for presents. Let S be the set of people shopping for specials and P be the set of people shopping for presents. Some of the results are shown in the table below.

	P	P'	Total
S			50
S'		12	
Total	45		80

a Copy and complete the two-way table.

b Estimate the probability, to four decimal places where appropriate, that a randomly selected shopper at Watertown Brand Outlet is shopping for

  i specials and presents

  ii presents, given that they are not shopping for specials

  iii specials, given that they are shopping for presents.

Steps	Working
a Use the row and column sums to complete the table.	<table><tr><td></td><td>P</td><td>P'</td><td>Total</td></tr><tr><td>S</td><td>27</td><td>23</td><td>50</td></tr><tr><td>S'</td><td>18</td><td>12</td><td>30</td></tr><tr><td>Total</td><td>45</td><td>35</td><td>80</td></tr></table>
b i 1 Use probability notation to represent the event. **2** Use the table to estimate the probability and answer to four decimal places.	$P(S \cap P) = \dfrac{27}{80}$ $= 0.3375$
ii 1 Use probability notation to represent the event. **2** Establish the conditional probability formula. **3** Use the table to estimate the probability and answer to four decimal places, where appropriate.	$P(P\|S') = \dfrac{(P \cap S')}{n(S')}$ $= \dfrac{18}{30}$ $= 0.6$
iii 1 Use probability notation to represent the event. **2** Establish the conditional probability formula. **3** Use the table to estimate the probability and answer to four decimal places, where appropriate.	$P(S\|P) = \dfrac{n(S \cap P)}{n(P)}$ $= \dfrac{27}{45}$ $= 0.6$

Video playlist
Relative frequencies and indications of independence

Puzzle
Conditional probability: two-way tables

WORKED EXAMPLE 29 — Estimating conditional probability from data in a Venn diagram

For an Outdoor Education camp, the 80 attending students could select a few activities to participate in from a list of different activities. Sixty students selected cycling (C), 35 chose bushwalking (B) and 10 chose neither of these two activities.

a Construct a Venn diagram to represent this information.

b Estimate the probability, to four decimal places where appropriate, that a randomly selected student that goes on the Outdoor Education camp

   i selects cycling, given that they also select bushwalking

   ii selects bushwalking, given that they also select cycling

   iii does not select cycling, given that they do not select bushwalking.

Steps	Working
a 1 Construct a Venn diagram for the sets B and C. 2 Input the known information. 3 Complete the Venn diagram using the known information and the addition rule.	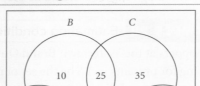 $n(B \cap C) = n(B) + n(C) - n(B \cup C)$ $= 35 + 60 - 70$ $= 25$
b 1 Use probability notation to represent the event. 2 Establish the conditional probability formula. 3 Use the table to estimate the probability and answer to four decimal places, where appropriate.	**i** $P(C \mid B) = \dfrac{P(C \cap B)}{P(B)}$ $= \dfrac{25}{35}$ $= 0.7143$
	ii $P(B \mid C) = \dfrac{P(B \cap C)}{P(C)}$ $= \dfrac{25}{60}$ $= 0.4167$
	iii $P(\overline{C} \mid \overline{B}) = \dfrac{P(\overline{C} \cap \overline{B})}{P(\overline{B})} = \dfrac{P(\overline{C \cup B})}{P(\overline{B})}$ $= \dfrac{10}{45}$ $= 0.2222$

Possible independence of events

What benefit would estimating conditional probabilities from collected data serve? As syllabus point 1.1.15 suggests, we want to be able to examine whether a conditional probability obtained from data is sufficient evidence to suggest the possible independence of two events. That is, based on the data collected, does restricting a sample space, i.e. $P(A|B)$, change the probability of an event, $P(A)$, significantly? If so, then there may be sufficient evidence to suggest the events are not independent; but if not, then there may be sufficient evidence to suggest that they are independent.

> **Exam hack**
>
> We shouldn't definitively conclude that the events are independent or dependent, because the probabilities are estimates that are coming from data, and data changes with every sample collected.

The important question is *what is a significant change in the probability in order to suggest possible independence of events?* We are not looking for $P(A|B) = P(A)$ here because relative frequencies are being used as estimates of probabilities and we cannot expect exact equivalence. Instead, we want $P(A|B) \approx P(A)$ in order to consider possible independent events.

There is no defined margin of error, as it also depends on the size of the sets and the amount of data available. For example, with larger sample sizes you can allow for a smaller margin of error because the estimates are likely to be closer to the theoretical probabilities. As a general rule, a 0.05 margin of error is a useful guide to decide whether $P(A|B)$ is close enough to $P(A)$ in order for A and B to be possibly considered as independent.

> **Using conditional probabilities for possible independence**
>
> For estimates of probabilities obtained from (experimental) data, there may be sufficient evidence to suggest events A and B are independent if
>
> $P(A|B) \approx P(A)$
>
> such that
>
> $P(A|B) = P(A) \pm 0.05$

WORKED EXAMPLE 30 | Identifying possible indications of independence

A survey of 80 shoppers at the Watertown Brand Outlet determined whether people were shopping for specials or shopping for presents. Let S be the set of people shopping for specials and P be the set of people shopping for presents. The results are shown in the table below.

	P	P'	Total
S	27	23	50
S'	18	12	30
Total	45	35	80

a Estimate the probability, to four decimal places where appropriate, that a randomly selected shopper at Watertown Brand Outlet is shopping for
  i presents
  ii presents, given that they are shopping for specials.
b Hence, justify whether there is sufficient evidence to suggest that a shopper shopping for presents is independent of a shopper shopping for specials.

Steps	Working
a **i** 1 Use probability notation to represent the event. 2 Use the table to estimate the probability and answer to four decimal places.	$P(P) = \dfrac{45}{80}$ $= 0.5625$
ii 1 Use probability notation to represent the event. 2 Establish the conditional probability formula. 3 Use the table to estimate the probability and answer to four decimal places, where appropriate.	$P(P\|S) = \dfrac{n(P \cap S)}{n(S)}$ $= \dfrac{27}{50}$ $= 0.54$
b 1 Compare the decimal representations of the probabilities by considering their difference. 2 Compare the difference to a margin of error of 0.05 and conclude whether there is a possible indication of independence.	$P(P\|S) = 0.54$ $P(P) = 0.5625$ $0.5625 - 0.54 = 0.0225$ $0.0225 < 0.05$ There is sufficient evidence to suggest that shopping for presents is independent of shopping for specials as $P(P\|S) \approx P(P)$.

These questions may also require you to comment on the validity of claims being made in context of the question.

Exam hack

Generally, answer questions to four decimal places so it makes the comparison of the probabilities easier!

WORKED EXAMPLE 31   Commenting on claims of independence

For an Outdoor Education camp, the 80 attending students could select a few activities to participate in from a list of different activities. Let the selection of cycling be C and the selection of bushwalking be B. The Venn diagram shows the number of students who participated in the different activities.

The Outdoor Education teacher claims that 'students selecting bushwalking is usually not affected by them wanting to cycle'. Use the appropriate relative frequencies to comment on the validity of the teacher's claim. Justify your answer.

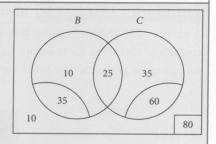

Steps	Working
1 Interpret the problem.	We are required to determine whether B and C are possible independent events.
2 Determine estimates for $P(B)$ and $P(B\|C)$ using the relative frequencies, to four decimal places.	$P(B) = \dfrac{35}{80} = 0.4375$ $P(B\|C) = \dfrac{P(B \cap C)}{P(C)} = \dfrac{25}{60}$ $= 0.4167$

3. Compare the decimal representations of the probabilities by considering their difference.

0.4375 − 0.4167 = 0.0208

0.0208 < 0.05

4. Compare the difference to a margin of error of 0.05 and comment on the validity of the claim being made, justifying your answer.

The teacher's claim appears to be valid, as there is sufficient evidence to suggest that a student selecting bushwalking is independent of them selecting cycling, as $P(B|C) \approx P(B)$.

EXAMINATION QUESTION ANALYSIS

Calculator-assumed (6 marks)

FullyFit is an international company that owns and operates many fitness centres (gyms) in several countries. At every one of FullyFit's gyms, each member agrees to have his or her fitness assessed every month by undertaking a set of exercises called S. There is a five-minute time limit on any attempt to complete S and if someone completes S in less than three minutes, they are considered 'fit'.

At FullyFit's Perth gym, it has been found that the probability that any member will complete S in less than three minutes is $\frac{5}{8}$ and the probability of any one member doing so is independent of any other member. In a particular week, three members of this gym attempt S.

a Determine the probability, correct to four decimal places, that exactly two of these three members will complete S in less than three minutes. (2 marks)

b Given that at least two of these three members complete S in less than three minutes, determine the probability, correct to four decimal places, that exactly two members complete S in less than three minutes. (2 marks)

Paula is a member of FullyFit's gym in Los Angeles. She completes S every month as required, but otherwise does not attend regularly and so her fitness level varies over many months. Paula finds that if she is 'fit' one month, the probability that she is 'fit' the next month is $\frac{3}{4}$, and if she is not 'fit' one month, the probability that she is not 'fit' the next month is $\frac{1}{2}$.

c If Paula is not 'fit' in one particular month, determine the probability that she is 'fit' for the next two months. (2 marks)

Reading the question

- Highlight any definitions that are specific to the question.
- Highlight any probabilities given.
- Highlight any rounding or accuracy commands.
- Highlight any language that indicates conditionality.

Thinking about the question

- What is the most efficient way to display the outcomes of the experiments?
- What does the word independent mean about each person?
- How many different sample spaces are there in this question?

> **Worked solution** (✓ = 1 mark)
>
> **a** Considering a tree diagram with three stages, there are three compound events that involve exactly two members completing S.
>
> $3 \times \left(\dfrac{5}{8}\right)^2 \times \left(\dfrac{3}{8}\right)$ ✓
>
> $= 0.4395$ ✓
>
> **b** Recognising that 'exactly two' is a subset of 'at least two',
>
> $P(\text{exactly 2 members complete } S \text{ in less than 3 minutes}) = \dfrac{3 \times \left(\dfrac{5}{8}\right)^2 \times \left(\dfrac{3}{8}\right)}{3 \times \left(\dfrac{5}{8}\right)^2 \times \left(\dfrac{3}{8}\right) + \left(\dfrac{5}{8}\right)^3}$ ✓
>
> $= 0.6428$ ✓
>
> **c** Given that Paula is not 'fit', the probability she is 'fit' the next month is $\dfrac{1}{2}$ and then given that she is 'fit', the probability she is 'fit' again is $\dfrac{3}{4}$.
>
> $P(\text{fit for the next two months}) = \dfrac{1}{2} \times \dfrac{3}{4}$ ✓
>
> $= \dfrac{3}{8}$ ✓

EXERCISE 2.7 Relative frequencies and indications of independence ANSWERS p. 449

Recap

1 For two independent events A and B, $P(A') = 0.6$ and $P(B') = 0.7$. The value of $P(A \cup B)$ is equal to

A 0.12   **B** 0.21   **C** 0.42   **D** 0.58   **E** 0.7

2 For two events A and B, $P(A \mid B') = \dfrac{3}{4}$, and $P(A') = P(B) = \dfrac{1}{3}$.

The two events, A and B, are independent, true or false?

Mastery

3 **WORKED EXAMPLE 28** A survey of 300 Year 11 and 12 students was conducted to examine the relationship between the enjoyment of reading books and students who study Literature. Let R be the set of students who enjoy reading books and L be the set of students who study Literature. Some of the results are shown in the table below.

	L	L'	Total
R		12	62
R'			
Total		210	300

a Copy and complete the two-way table.

b Estimate the probability, to four decimal places where appropriate, that a randomly selected Year 11 or 12 student

  i enjoys reading books and studies Literature

  ii is a Literature student who does not enjoy reading books

  iii enjoys reading books, given that they are a Literature student.

4 **WORKED EXAMPLE 29** When booking a holiday through a travel agent, 400 customers were asked whether they enjoy destinations that have beaches (B) or destinations that are mountainous (M). 250 customers stated that they enjoyed destinations that have beaches, while 210 customers stated that they enjoyed mountainous destinations. 110 customers answered that they enjoyed both.

  a Construct a Venn diagram to represent this information.
  b Estimate the probability, to four decimal places where appropriate, that a randomly selected traveller
   i enjoys destinations that have beaches, given that they enjoy mountainous destinations
   ii enjoys mountainous destinations, given that they do not enjoy destinations with beaches
   iii does not enjoy mountainous destinations, given that they enjoy destinations with beaches.

5 **WORKED EXAMPLE 30** A survey of 300 Year 11 and 12 students was conducted to examine the relationship between the enjoyment of reading books and students who study Literature. Let R be the set of students who enjoy reading books and L be the set of students that study Literature. The results are shown in the table below.

	L	L'	Total
R	50	12	62
R'	40	198	238
Total	90	210	300

  a Estimate the probability, to four decimal places where appropriate, that a randomly selected Year 11 or 12 student
   i studies Literature
   ii studies Literature, given that they enjoy reading books.
  b Hence, justify whether there is sufficient evidence to suggest that whether a student studying Literature is independent of whether a student enjoys reading books.

6 **WORKED EXAMPLE 31** When booking a holiday through a travel agent, 400 customers were asked whether they enjoy destinations that have beaches (B) or destinations that are mountainous (M). The results of the survey are shown in the Venn diagram below.

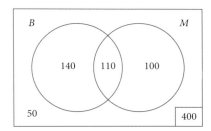

A travel agent claims that 'customers generally don't enjoy mountainous destinations if they enjoy destinations with beaches'. Use the appropriate relative frequencies to comment on the validity of the agent's claim. Justify your answer.

Calculator-assumed

7 (9 marks) In a city, there are two different modes of transport available: bus and train. A survey was conducted among 500 commuters to determine their preferred mode of transportation. Let B be the set of commuters who like taking the bus and T be the set of commuters who like taking the train. Some of the results are shown in the two-way table below.

	T	T'	Total
B	150		320
B'			
Total	250		500

    a Copy and complete the two-way table. (2 marks)

    b A commuter is randomly selected from the participants. Determine the probability, correct to four decimal places where appropriate, that the selected commuter

        i likes taking the bus (1 mark)

        ii likes taking the bus, given that they also like taking the train (2 marks)

        iii does not like taking the train, given that they do not like taking the bus. (2 marks)

    c Use the appropriate probabilities from part **b** above to comment on whether it appears that liking the bus as a mode of transportation is independent of liking the train as a mode of transport. Justify your answer. (2 marks)

8 (4 marks) Over the course of a term, Marnie's teacher records whether she arrives on time to school (T) and whether she remembered to set her alarm that morning (A). Some of the relative frequencies are shown in the table below.

	A	A'	Total
T	0.63	0.09	
T'			
Total		0.3	1

Marnie's teacher claims that she 'really needs to try to remember to set her alarm in order to improve her punctuality'. Use the appropriate relative frequencies to comment on the validity of the teacher's claim. Justify your answer.

9 (4 marks) The manager of a soccer team observes the number of times the team score the first goal in a match (F) throughout the season and whether they win the match (W). Some of the relative frequencies are shown in the table below.

	W	W'	Total
F	0.18		0.3
F'	0.07		
Total			1

The manager claims that 'the team needs to try and score first in order to win the match'. Use the appropriate relative frequencies to comment on the validity of the manager's claim. Justify your answer.

Chapter summary

The fundamentals of probability

- Every **probability** experiment has a **sample space**, which is the set of all possible outcomes in the experiment, S.
- A single **outcome** or a group of outcomes of a probability experiment is called an **event** and is often represented using a capital letter.
- The **probability** of an event A occurring is denoted as:

$$P(A) = \frac{\text{number of sucessful outcomes for event } A}{\text{total number of possible outcomes}} = \frac{n(A)}{n(S)}$$

- Probability can be considered a function, P( ), for which the 'input' is the set of all possible events in the experiment and the 'output' is the set of numerical values $0 \leq P(A) \leq 1$, where 0 represents an impossible event and 1 represents a certain event.

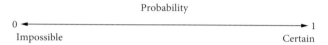

Probability experiments involving sets

- Complement rule:

$$P(\overline{A}) = 1 - P(A)$$

- The addition rule for two sets:

$$P(A \cup B) = P(A) + P(B) - P(A \cap B)$$

- For two **mutually exclusive events**:

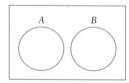

$P(A \cap B) = 0$

$P(A \cup B) = P(A) + P(B)$

Probability experiments involving stages

- An **array** is a table used to show the outcomes of a two-stage probability experiment, for which the outcomes of the first stage are in the rows of the table and the outcomes of the second stage are in the columns of the table.

	A	E	I	O	U
A	AA	AE	AI	AO	AU
E	EA	EE	EI	EO	EU
I	IA	IE	II	IO	IU
O	OA	OE	OI	OO	OU
U	UA	UE	UI	UO	UU

- If the experiment does not involve **replacement** between stages, then a dash '–' is used to denote that an outcome is not possible, and this is typically seen in the diagonal of the table from the top left to the bottom right.

	A	E	I	O	U
A	–	AE	AI	AO	AU
E	EA	–	EI	EO	EU
I	IA	IE	–	IO	IU
O	OA	OE	OI	–	OU
U	UA	UE	UI	UO	–

Tree diagrams can display multi-stage experiments and the sample space.

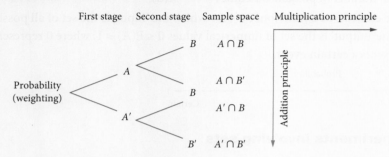

Probability experiments involving selections

- The combinations formula

$$^nC_r = \binom{n}{r} = \frac{n!}{r!(n-r)!}$$

- The complement principle using 'at least':

  The probability of making a selection involving at least k objects can be calculated by

  P(at least k objects) = 1 – P(less than k objects)

Conditional probability

- For two events A and B, the probability of A occurring given that B has occurred is calculated using:

$$P(A|B) = \frac{P(A \cap B)}{P(B)}$$

where $A|B$ is read as 'A given B'.

- If A and B are mutually exclusive events, then $P(A \cap B) = 0$ and so $P(A|B) = 0$.
- The multiplication principle for **conditional probability** states that:

$$P(B|A) = \frac{P(A \cap B)}{P(A)} \Rightarrow P(A \cap B) = P(A) \times P(B|A)$$

which is visible on a tree diagram.

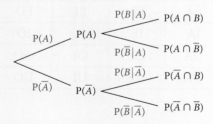

- The law of total probability states that:

  $P(B) = P(A \cap B) + P(\overline{A} \cap B)$

  $P(B) = P(B|A)P(A) + P(B|\overline{A})P(\overline{A})$

Independent events

- Two events, A and B, are **independent events** if the probability of one event occurring does not affect the probability of the other event occurring.

Conditional rule for independent events

- For two independent events A and B, the probability of A occurring given that B has occurred remains unchanged; that is, $P(A|B) = P(A)$.

The rule for successive independence

- Suppose that A and B were independent events; that is, A occurring did not affect the probability of B occurring and vice versa. Then the multiplication principle

    $P(A \cap B) = P(A)P(B|A)$

    becomes

    $P(A \cap B) = P(A)P(B)$

    because $P(B|A) = P(B)$.

Two useful properties of independence

1. The **symmetry of independence:** If A is independent of B, then B is independent of A.

    $P(A|B) = P(A) \Rightarrow P(B|A) = P(B)$

2. Independence of the complement: If A and B are independent, then A and B' are independent.

Independence in context

- If a repetitive experiment involves n independent trials with a probability of success p, then the probability of n successes is p^n.

Relative frequencies and indications of independence

- For estimates of probabilities obtained from (experimental) data, there may be sufficient evidence to suggest events A and B are independent if

    $P(A|B) \approx P(A)$

    such that

    $P(A|B) = P(A) \pm 0.05$

Cumulative examination: Calculator-free

Total number of marks: 19   Reading time: 2 minutes   Working time: 19 minutes

1 (3 marks) Consider the combinatorial equation below, where $n > 2$.

$$\binom{n}{2} \times 2! = 20$$

Determine the value of n.

2 (5 marks) Let the universal set be $U = \{-9, -7, -5, -3, -1, 1, 3, 5, 7, 9\}$.

   a State $n(U)$. (1 mark)

A number is randomly selected from the set U. Let P be the event that a prime number is selected and N be the event that a negative number is selected.

   b Explain why P and N are mutually exclusive events. (1 mark)

   c List the elements of $\overline{P} \cap \overline{N}$. (1 mark)

   d Determine the probability that a randomly selected number

      i is divisible by 3, but is neither prime nor negative (1 mark)

      ii is divisible by 3, given that it is neither prime nor negative. (1 mark)

3 (3 marks) Determine the sum of the coefficients in the expansion of $(3x + 1)^4$.

4 (3 marks) For two independent events X and Y, if $P(Y) = 0.8$ and $P(X \cup Y) = 0.86$, determine $P(X)$.

5 (5 marks) A company produces motors for refrigerators. There are two assembly lines, Line A and Line B. 5% of the motors assembled on Line A are faulty and 8% of the motors assembled on Line B are faulty. In 1 hour, 40 motors are produced from Line A and 50 motors are produced from Line B. At the end of an hour, one motor is selected at random from all the motors that have been produced during that hour.

   a Determine the probability, in the form $\dfrac{1}{b}$ where b is an integer, that the selected motor is faulty. (3 marks)

   b Determine the probability, in the form $\dfrac{1}{c}$ where c is an integer, that the selected motor was assembled on Line A, given that it is found to be faulty. (2 marks)

Cumulative examination: Calculator-assumed

Total number of marks: 28   Reading time: 3 minutes   Working time: 28 minutes

1 (10 marks) At a local high school, 150 students were surveyed about their extracurricular activities: sport (S), music (M) and art (A). The results were as follows:
- $n(S) = 80$
- $n(M) = 60$
- $n(A) = 40$
- $n(S \cap M) = 30$
- $n(M \cap A) = 20$
- $n(S \cap A) = 10$
- $n(\overline{S \cap M \cap A}) = 145$

Determine the number of students who

a   participate in sport only (2 marks)

b   participate in music or art, but not sport (2 marks)

c   participate in exactly two of the three activities (2 marks)

d   participate in at least two of the three activities (2 marks)

e   do not participate in any of the activities. (2 marks)

2 (6 marks) Consider the set of the first 10 letters of the English alphabet, {A, B, C, D, E, F, G, H, I, J}, in which there are 7 consonants and 3 vowels.

a   Determine the number of possible combinations of three letters that can be formed from this set, if

   i   there are no restrictions (1 mark)

   ii there must be at least one vowel. (2 marks)

Let A be the event that the three-letter combination contains the 'A', E be the event that the three-letter combination contains the 'E' and I be the event that the three-letter combination contains the 'I'.

b   State $n(A \cap E \cap I)$. (1 mark)

c   Determine the probability that a randomly selected combination of three letters from this set contains three vowels, given that the letter E is selected. (2 marks)

3 (4 marks) Each night, Jess goes to the gym or the pool. If she goes to the gym one night, the probability she goes to the pool the next night is 0.4, and if she goes to the pool one night, the probability she goes to the gym the next night is 0.7.

Suppose she goes to the gym one Monday night.

a   Calculate the probability that she goes to the pool for the next three nights. (2 marks)

b   Calculate the probability that she goes to the pool on Wednesday night. (2 marks)

4 (5 marks) Two events, A and B, exist such that $P(A|B') = 0.35$ and $P(A' \cap B') = 0.2$.

   a If A and B are independent events, explain why $P(A) = 0.35$. (1 mark)

   b If A and B are mutually exclusive events, determine $P(A)$ correct to four decimal places. (4 marks)

5 (3 marks) An independent organisation conducted a survey, collecting data on frequency of physical activity and ownership of gym memberships. Let F be the event that an individual exercised more than 4 times a week and G be the event that an individual had a gym membership. The results are shown in the table below.

	G	G'	Total
F	156		272
F'			
Total	330		512

After the survey, the organisation claimed that 'the number of times that someone exercises per week is largely influenced by whether they have a gym membership or not'. Show an appropriate calculation that could be used to support the organisation's claim. Justify your answer.

LINEAR AND QUADRATIC RELATIONSHIPS

CHAPTER 3

Syllabus coverage
Nelson MindTap chapter resources

3.1 Sketching linear graphs
The linear relationship $y = mx + c$
Finding the gradient
x- and y-intercepts
Sketching linear graphs
USING CAS 1: Sketching linear graphs
Horizontal and vertical lines

3.2 Determining the equation of a linear graph
Determining the equation using $y = mx + c$
Determining the equation using $y - y_1 = m(x - x_1)$
Parallel and perpendicular lines

3.3 Expanding
Expanding binomial products
Expanding a perfect square
Expanding to a difference of perfect squares
USING CAS 2: Expanding

3.4 Factorising
Highest common factor
Factorising to a perfect square
Factorising a difference of perfect squares
Factorising by grouping
USING CAS 3: Factorising

3.5 Solving quadratic equations by factorising
Factorising monic quadratic trinomials
Factorising non-monic quadratic trinomials
Solving with the null factor law

3.6 Solving quadratic equations by completing the square
Completing the square
Solving by completing the square

3.7 The quadratic formula and the discriminant
The quadratic formula
USING CAS 4: Solving quadratic equations
The discriminant
The discriminant and parameters for a, b and c
USING CAS 5: Solving with the discriminant

3.8 Sketching parabolas from turning point form
Turning point form
Sketching parabolas from turning point form
Completing the square to find turning point form

3.9 Sketching parabolas from any form
Sketching parabolas from intercept form
Sketching parabolas from general form
USING CAS 6: Sketching parabolas

3.10 Simultaneous equations
Graphical solution
Algebraic solution
Number of solutions for linear simultaneous equations
Simultaneous linear and quadratic equations
USING CAS 7: Solving simultaneous equations
USING CAS 8: Points of intersection

3.11 Determining the equation of a parabola
Identifying sufficient information
Determining the equation of a parabola

3.12 Applications of quadratic relationships
Applications of quadratic relationships

Examination question analysis
Chapter summary
Cumulative examination: Calculator-free
Cumulative examination: Calculator-assumed

Syllabus coverage

TOPIC 1.2 FUNCTIONS AND GRAPHS

Lines and linear relationships

1.2.1 recognise features of the graph of $y = mx + c$, including its linear nature, its intercepts and its slope or gradient
1.2.2 determine the equation of a straight line given sufficient information; including parallel and perpendicular lines

Quadratic relationships

1.2.3 examine examples of quadratically related variables
1.2.4 recognise features of the graphs of $y = x^2$, $y = a(x - b)^2 + c$, and $y = a(x - b)(x - c)$, including their parabolic nature, turning points, axes of symmetry and intercepts
1.2.5 solve quadratic equations, including the use of quadratic formula and completing the square
1.2.6 determine the equation of a quadratic given sufficient information
1.2.7 determine turning points and zeros of quadratics and understand the role of the discriminant
1.2.8 recognise features of the graph of the general quadratic $y = ax^2 + bx + c$

Mathematics Methods ATAR Course Year 11 syllabus p. 9 © SCSA

Video playlists (13):
3.1 Sketching linear graphs
3.2 Determining the equation of a linear graph
3.3 Expanding
3.4 Factorising
3.5 Solving quadratic equations by factorising
3.6 Solving quadratic equations by completing the square
3.7 The quadratic formula and the discriminant
3.8 Sketching parabolas from turning point form
3.9 Sketching parabolas from any form
3.10 Simultaneous equations
3.11 Determining the equation of a parabola
3.12 Applications of quadratic relationships

Examination question analysis Linear and quadratic relationships

Worksheets (8):
3.1 Graphing linear functions • x- and y-intercepts
3.3 Binomial products • Special products
3.4 Factorising algebraic expressions
3.9 Quadratic functions • Graphing quadratics • Graphing quadratic functions

Puzzles (4):
3.4 Factorising quadratic equations
3.6 Completing the square order activity
3.7 The quadratic formula
3.10 Simultaneous equations order activity

To access resources above, visit
cengage.com.au/nelsonmindtap

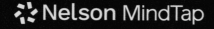

3.1 Sketching linear graphs

The linear relationship $y = mx + c$

A **linear relationship** between two variables, x and y, can be expressed in the form:

$$y = mx + c$$

where m is the gradient and c is the y-intercept.

The graph of a linear relationship is a straight line. For example, the graph of $y = 3x + 7$ can be sketched by creating a table of values, plotting some (x, y) points and drawing a line through them extending in both directions. Think of a line as representing infinitely many (x, y) points.

x	−2	−1	0	1	2
y	1	4	7	10	13

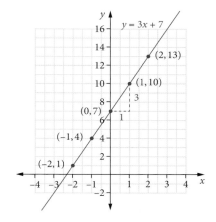

Video playlist
Sketching linear graphs

Worksheets
Graphing linear functions

x- and y-intercepts

A key feature of a linear graph is its **gradient**, which measures the 'steepness' of a line. The gradient is equal to m in the equation $y = mx + c$.

Think of the gradient as the amount by which y changes by every time x increases by 1. For example, in the graph above, y changes by +3 every time x increases by 1. This is due to the multiplication by 3 in $y = 3x + 7$.

The axis intercepts are also key features of a linear graph. The **x-intercept** is located where the line crosses the x-axis. The **y-intercept** is located where the line crosses the y-axis. The y-intercept is equal to c in the equation $y = mx + c$.

$y = mx + c$ is known as the **gradient–intercept form** of a linear equation.

> **Gradient–intercept form**
>
> The gradient–intercept form of a linear equation is $y = mx + c$.
>
> m is the gradient.
>
> c is the y-intercept.

The **degree of a polynomial** expression is the highest power on the variable. The relationship $y = mx + c$ is linear because the degree is 1 (recall that $x = x^1$). The relationship $y = ax^2 + bx + c$ is **quadratic** because the degree is 2.

Chapter 3 | Linear and quadratic relationships

Finding the gradient

The gradient (or slope) of a line is equal to the $\frac{\text{rise}}{\text{run}}$ between any two points on the line. For two points (x_1, y_1) and (x_2, y_2), gradient $= \frac{\text{rise}}{\text{run}} = \frac{y_2 - y_1}{x_2 - x_1}$.

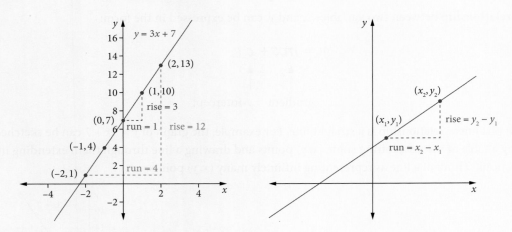

A graph with a positive gradient 'moves up' as the x values increase. A graph with a negative gradient 'moves down' as the x values increase. A graph with a zero gradient stays flat (horizontal) as the x values increase.

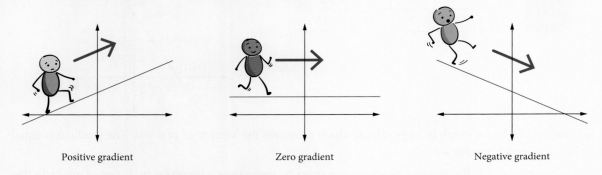

Positive gradient   Zero gradient   Negative gradient

Gradient

$$\text{gradient} = m = \frac{\text{rise}}{\text{run}} = \frac{y_2 - y_1}{x_2 - x_1}$$

where

    m is the **coefficient** of x in $y = mx + c$

    (x_1, y_1) and (x_2, y_2) are any two points on the graph.

WORKED EXAMPLE 1 — Finding the gradient

For each representation of a linear relationship, state the gradient of the graph.

a $9y + 1 = 18x$

b

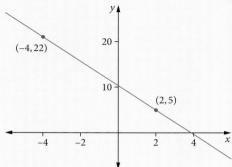

c The line which passes through the points $(5, 10)$ and $(-1, 7)$.

Steps	Working
a 1 Rearrange the equation to be in the form $y = mx + c$.	$9y + 1 = 18x$ $9y = 18x - 1$ $y = 2x - \dfrac{1}{9}$
2 In the linear equation $y = mx + c$, m represents the gradient.	$m = 2$ The gradient is 2.
b 1 The gradient equals $\dfrac{\text{rise}}{\text{run}}$. Run, left to right, is always positive. Rise is positive if the graph rises as x increases. Rise is negative if the graph falls as x increases.	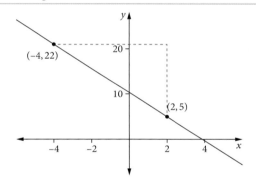
2 Calculate the rise.	As x increases, the graph falls from 22 to 5. This is a rise of -17.
3 Calculate the run.	The x values increase from -4 to 2. This is a run of 6.
4 Calculate $\dfrac{\text{rise}}{\text{run}}$.	gradient $= \dfrac{\text{rise}}{\text{run}} = -\dfrac{17}{6}$ The gradient is $-\dfrac{17}{6}$. *Note that the gradient can also be calculated using $\dfrac{y_2 - y_1}{x_2 - x_1}$.*
c 1 gradient $= \dfrac{y_2 - y_1}{x_2 - x_1}$ Choose either point to be (x_1, y_1). Choose the other point to be (x_2, y_2).	Let $(-1, 7)$ be (x_1, y_1). Let $(5, 10)$ be (x_2, y_2).
2 Substitute into $\dfrac{y_2 - y_1}{x_2 - x_1}$.	$\dfrac{y_2 - y_1}{x_2 - x_1} = \dfrac{10 - 7}{5 - (-1)}$
3 Simplify.	$= \dfrac{3}{6} = \dfrac{1}{2}$ The gradient is $\dfrac{1}{2}$.

x- and y-intercepts

The axis intercepts are key features of any graph. The y-intercept is always found by letting $x = 0$ and solving for y. The x-intercept is always found by letting $y = 0$ and solving for x.

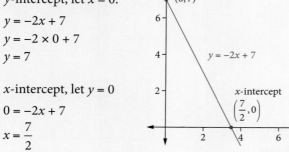

$y = -2x + 7$

y-intercept, let $x = 0$:

$y = -2x + 7$
$y = -2 \times 0 + 7$
$y = 7$

x-intercept, let $y = 0$
$0 = -2x + 7$
$x = \dfrac{7}{2}$

In the general linear equation $y = mx + c$, when $x = 0$, y equals c.

> ### x- and y-intercepts
> The x-intercept is found by letting $y = 0$ and solving for x.
> The y-intercept is found by letting $x = 0$ and solving for y.
> In $y = mx + c$, the y-intercept is always c.

WORKED EXAMPLE 2 — Finding x- and y-intercepts

Find the x- and y-intercepts of the graphs in coordinate form.

a $y = -6x - 48$ **b** $-x + 12y = 15$

Steps	Working
a 1 To find the x-intercept, let $y = 0$ and solve for x.	$y = -6x - 48$ $0 = -6x - 48$ $48 = -6x$ $x = -\dfrac{48}{6} = -8$ The x-intercept is $(-8, 0)$.
2 To find the y-intercept, let $x = 0$ and solve for y. This is the value of c in the equation $y = mx + c$.	$y = -6x - 48$ $y = -6(0) - 48$ $y = -48$ The y-intercept is $(0, -48)$.
b 1 To find the x-intercept, let $y = 0$ and solve for x.	$-x + 12y = 15$ $-x + 12(0) = 15$ $-x = 15$ $x = -15$ The x-intercept is $(-15, 0)$.
2 To find the y-intercept, let $x = 0$ and solve for y.	$-(0) + 12y = 15$ $12y = 15$ $y = \dfrac{15}{12} = \dfrac{5}{4}$ The y-intercept is $\left(0, \dfrac{5}{4}\right)$.

Sketching linear graphs

To sketch a linear graph, find the x- and y-intercepts and draw a straight line passing through them. Label the intercepts.

There are four common forms of a linear equation. Any linear equation can be rearranged into any of the forms.

The **general form** is $ax + by + c = 0$, where a, b and c are integers.

The **gradient–intercept form** is $y = mx + c$, where m is the gradient and c is the y-intercept.

The **point–gradient form** is $y - y_1 = m(x - x_1)$, where m is the gradient and (x_1, y_1) is a point on the line.

The **intercept form** is $\dfrac{x}{a} + \dfrac{y}{b} = 1$, where a is the x-intercept and b is the y-intercept.

> **Sketching linear graphs**
>
> To sketch a graph of a linear equation in any form:
> 1. Find the x-intercept by letting $y = 0$ and solving for x.
> 2. Find the y-intercept by letting $x = 0$ and solving for y.
> 3. Draw a straight line through the two axis intercepts.
> 4. Label the intercepts.

WORKED EXAMPLE 3 | Sketching linear graphs

Sketch the graphs of the following linear equations.

a $y = -9x + 45$ **b** $32x - y - 8 = 0$ **c** $y = -\dfrac{1}{4}x$

Steps	Working
a 1 Find the x-intercept by letting $y = 0$ and solving for x.	$y = -9x + 45$ $0 = -9x + 45$ $9x = 45$ $x = 5$
2 Find the y-intercept by letting $x = 0$ and solving for y.	$y = -9(0) + 45$ $y = 45$
3 Sketch the graph. Remember to label the axis intercepts.	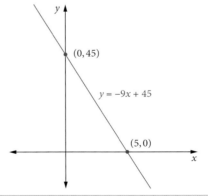

b 1 Find the x-intercept by letting $y = 0$ and solving for x.

$$32x - y - 8 = 0$$
$$32x - 0 - 8 = 0$$
$$32x = 8$$
$$x = \frac{1}{4}$$

2 Find the y-intercept by letting $x = 0$ and solving for y.

$$32(0) - y - 8 = 0$$
$$-y = 8$$
$$y = -8$$

3 Sketch the graph. Remember to label the axis intercepts.

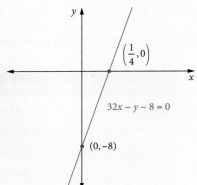

> 🔓 **Exam hack**
>
> You don't need to number every tick along the axes. Just labelling the intercepts is enough. The x-axis and y-axis can have different scales.

c 1 Find the x-intercept by letting $y = 0$ and solving for x.

$$y = -\frac{1}{4}x$$
$$0 = -\frac{1}{4}x$$
$$x = 0$$

2 Notice that the line passes through the origin.

If the x-intercept is 0, then the coordinates are $(0, 0)$ and that is the y-intercept as well. The line passes through the origin.

3 Find another point on the line so that the gradient can be seen in the graph.

Let $x = 4$.

When $x = 4$, $y = -\frac{1}{4} \times 4 = -1$.

Another point is $(4, -1)$.

4 Sketch the graph. Remember to label the axis intercept. Because the graph passes through the origin, label the extra point as well.

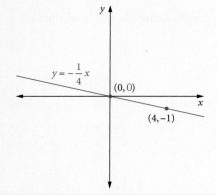

USING CAS 1 | Sketching linear graphs

Sketch the graph of $y = -\dfrac{1}{5}x + 11$. Label the axis intercepts with their coordinates.

ClassPad

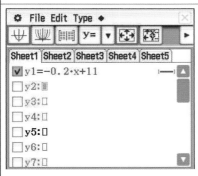

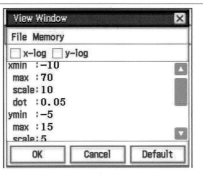

1. Tap **Menu > Graph&Table**.
2. Enter the equation as shown above and press **EXE**.
3. Tap the **Graph** tool.
4. Tap the **View Window** tool.

5. Change the minimum, maximum and scale values to those shown above and tap **OK**.

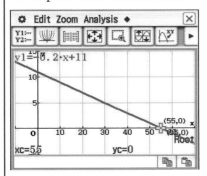

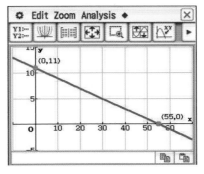

6. The graph will now be displayed (note the windows have been swapped so the graph now appears in the upper window).
7. Tap **Analysis > G-Solve > Root**.
8. The cursor will jump to the *x*-intercept.
9. Press **EXE** to save the coordinates.

10. Tap **Analysis > G-Solve > y-Intercept**.
11. The cursor will jump to the *y*-intercept.
12. Press **EXE** to save the coordinates.
13. Tap **Esc** to remove the highlight around the point.

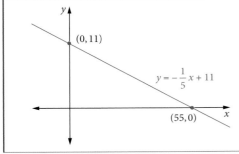

TI-Nspire

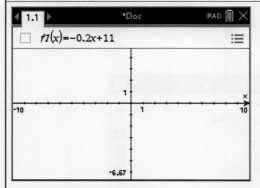

1. Add a **Graphs** application.
2. Enter the equation as shown above and press **enter**.

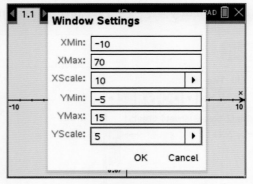

3. Press **menu > Window/Zoom > Window Settings**.
4. Change the minimum, maximum and scale values to those shown above and press **enter**.

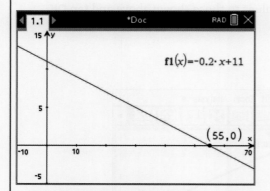

5. The graph will be displayed.
6. Press **menu > Analyze Graph > Zero**.
7. When prompted for the **lower bound?**, move the cursor to the left of the *x*-intercept and press **enter**.
8. When prompted for the **upper bound?**, move to the right of the *x*-intercept and press **enter**.
9. The coordinates will be displayed on the screen, which can then be dragged to an appropriate position.

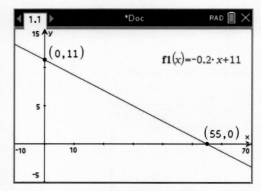

10. Press **menu > Trace**.
11. Enter **0** then press **enter** twice.
12. The coordinates will be displayed on the screen, which can then be dragged to an appropriate position.

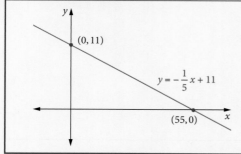

Horizontal and vertical lines

Horizontal lines

A horizontal line has the same constant y value for every value of x. It is represented by the equation $y = c$.

For example, this horizontal line passes through every point where y is 5. Its equation is $y = 5$.

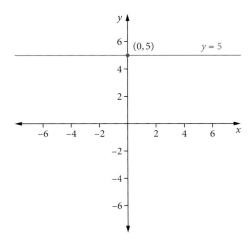

Gradient interpretation: for a horizontal line, the rise is equal to 0. Therefore, the gradient is $\frac{0}{\text{run}} = 0$ and the equation $y = mx + c$ simplifies to $y = c$.

Vertical lines

A vertical line has the same x value for every value of y. It is represented by the equation $x = d$.

For example, this vertical line passes through every point where x is -2. Its equation is $x = -2$.

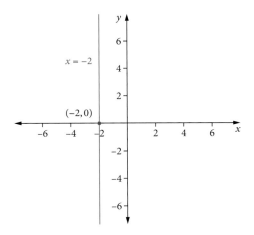

Gradient interpretation: for a vertical line, the run is equal to 0. Therefore, the gradient is $\frac{\text{rise}}{0} = $ undefined.

The form $y = mx + c$ cannot be used if m is undefined. Instead, we write the relation $x = d$.

> **Horizontal and vertical lines**
>
> A horizontal line has zero gradient and is represented by $y = c$.
> A vertical line has undefined gradient and is represented by $x = d$.
> c and d are real numbers.

WORKED EXAMPLE 4 Sketching horizontal and vertical lines

a What equation represents this graph?

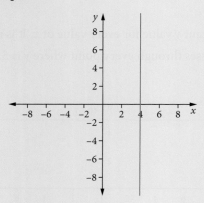

b Sketch the graph of $y = -7$.

Steps	Working
a 1 Identify which value of any coordinate is constant and which can vary.	The x value is constant. It is always equal to 4. The y value can vary.
2 Use the constant value to write the equation for the vertical line.	The equation of the line is $x = 4$.
b 1 Identify which value of any coordinate is constant and which can vary.	The y value is constant. It is always equal to -7. The x value can vary.
2 Use the constant value to sketch the graph.	A constant value of $y = -7$ refers to a horizontal line passing through a y-intercept of -7. 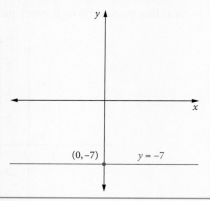

EXERCISE 3.1 Sketching linear graphs

ANSWERS p. 450

Mastery

1 **WORKED EXAMPLE 1** Find the gradient of each line represented by these linear equations.

a $y = -3x - \dfrac{2}{9}$ b $y = x$ c $4y = -x + 7$

d $2x - 11y = 0$ e $x + 17y = 34$ f $y = 12$

2 WORKED EXAMPLE 1 Find the gradient of each graph.

a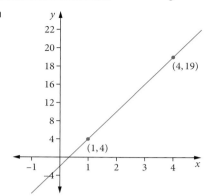

b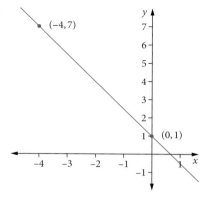

c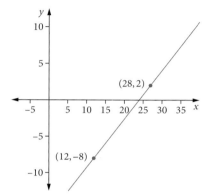

d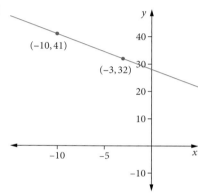

3 WORKED EXAMPLE 1 Find the gradient of the line passing through the points:
- a (4, 1) and (7, 10)
- b (−1, 15) and (3, 8)
- c (−1, −9) and (−20, −40)
- d (31, 2) and (43, 14)
- e (−27, 117) and (−11, 51)
- f (61, −5) and (5, 16)

4 WORKED EXAMPLE 2 Find the x- and y-intercepts of the graphs. Write the intercepts in coordinate form.
- a $y = 2x - 7$
- b $y = -\frac{1}{3}x + 8$
- c $2x - 10y = 16$
- d $-\frac{x}{9} + \frac{y}{4} = 1$
- e $y = x - \pi$
- f $-y + 68x + 17 = 0$

5 WORKED EXAMPLE 3 Sketch the graph of the following linear equations. Label the axis intercepts.
- a $y = 4x + 12$
- b $y = -\frac{1}{3}x - 6$
- c $y = -x + 7$
- d $y = -6x + 1$
- e $y = 7x$
- f $y = 8 - \frac{2}{9}x$

6 WORKED EXAMPLE 3 Sketch the graph of the following linear equations. Label the axis intercepts.
- a $5y - 2x = 20$
- b $x - 7y = 14$
- c $12 - 3x + 2y = 0$
- d $\frac{y}{4} - \frac{x}{11} = 2$
- e $15y + 3x = 0$
- f $\frac{1}{8}y - 7x = 4$

7 **WORKED EXAMPLE 3** Sketch the graph of the following linear equations. Label any axis intercepts.

   a $x = -3$
   b $y = -7$
   c $x = 0$
   d $y = \dfrac{5}{2}$
   e $x = \sqrt{2}$
   f $y = 54$

8 **Using CAS 1** Sketch the graph of the following linear equations. Label any axis intercepts. Note that you may need to adjust the window on the calculator to view the graph.

   a $y = -5x + 10$
   b $y = \dfrac{1}{2}x - 60$
   c $y = 20x - \dfrac{1}{3}$

9 **WORKED EXAMPLE 4** Sketch the graph of the following linear equations. Label any axis intercepts.

   a $y = -\dfrac{1}{2} + x$
   b $-20y + x = 32$
   c $y = 0$
   d $\dfrac{x}{4} + \dfrac{y}{9} = 6$
   e $x = -8$
   f $y = \dfrac{1}{7}x - 3$

Calculator-free

10 (4 marks) A line passes through the points $(3, p)$ and $(-2p, 8)$. Find the value of p if
   a the gradient of the line is 1. (2 marks)
   b the gradient of the line is 7. (2 marks)

11 (2 marks) Find the area of the rectangle bounded by the graphs of $x = -1$, $y = -2$, $x = 3$ and $y = 4$.

12 (2 marks) Consider the equation $y + 7s - x = 0$. Find the value of s if the
   a x-intercept is 35 (1 mark)
   b y-intercept is $-\dfrac{1}{2}$. (1 mark)

Calculator-assumed

13 (6 marks) Mel and her mates are planning an end-of-Year 12 road trip to Margaret River from their home town, Esperance. The distance, d km, to the destination is modelled by the linear equation $d = 711 - 85.32t$, where t is the time spent driving, in hours.
   a How far from the destination are Mel and her mates at the beginning of the drive? (1 mark)
   b How long will the crew have spent driving by the time they arrive? (2 marks)
   c Sketch the graph of d vs t to represent the drive. (2 marks)
   d What does the gradient of the graph represent? (1 mark)

3.2 Determining the equation of a linear graph

To determine the equation of a straight line, $y = mx + c$, note that there are two unknown values: m and c. Two pieces of information are needed to find two unknowns. For linear graphs, the required information is
- the gradient of the line, and
- a point on the line.

This information may arrive in different, but sufficient, forms. For example, the gradient can be calculated from two points and an axis intercept is itself a point on the line.

Sufficient information for gradient	Sufficient information for a point on the line
• a statement of the graph's gradient • (x_1, y_1) and (x_2, y_2) (use $\frac{y_2 - y_1}{x_2 - x_1}$) • the gradient of a line parallel to the line (gradients are the same) • the gradient of a line perpendicular to the line (gradients are negative reciprocals)	• the coordinates of any point on the line (if provided with two points, choose one) • the location of an axis intercept • a point of intersection with another graph

Video playlist
Determining the equation of a linear graph

Determining the equation using $y = mx + c$

The steps for finding the equation of a straight line using $y = mx + c$:
1. Find the gradient and substitute it for m.
2. Substitute the x- and y-coordinates from a point on the line into the equation. Solve for c.
3. Rewrite the full $y = mx + c$ equation using the known values for m and c.

$y = mx + c$ is known as the gradient–intercept form of the linear equation.

WORKED EXAMPLE 5 — Determining the equation using $y = mx + c$

a What is the equation of the straight line that passes through $(1, 12)$ and has a gradient of 2?

b What is the equation of the straight line that passes through the points $(-2, 5)$ and $(5, -20)$?

Steps	Working
a 1 Find the gradient and substitute it into the general equation.	The gradient is 2. $m = 2$ $y = 2x + c$
2 Substitute a point on the line into the equation.	The line passes through $(1, 12)$. When $x = 1$, $y = 12$. $12 = 2(1) + c$
3 Solve for c.	$12 = 2 + c$ $c = 10$
4 Rewrite $y = mx + c$ using the known values for m and c.	$y = 2x + 10$

b 1 Find the gradient using $\dfrac{y_2 - y_1}{x_2 - x_1}$.

$(-2, 5)$ and $(5, -20)$

$$\dfrac{y_2 - y_1}{x_2 - x_1} = \dfrac{5 - -20}{-2 - 5} = \dfrac{25}{-7} = -\dfrac{25}{7}$$

The gradient is $-\dfrac{25}{7}$.

$$m = -\dfrac{25}{7}$$

2 Substitute m into the general equation.

$$y = -\dfrac{25}{7}x + c$$

3 Substitute a point on the line into the equation. Choose either point.

The simpler point to substitute is $(-2, 5)$.

When $x = -2$, $y = 5$.

$$5 = -\dfrac{25}{7}(-2) + c$$

4 Solve for c.

$$\dfrac{35}{7} = \dfrac{50}{7} + c$$

$$c = -\dfrac{15}{7}$$

5 Rewrite $y = mx + c$ using the known values for m and c.

$$y = -\dfrac{25}{7}x - \dfrac{15}{7}$$

 Exam hack

Using the equation $y - y_1 = m(x - x_1)$ is often more efficient than using $y = mx + c$.

Determining the equation using $y - y_1 = m(x - x_1)$

The derivation of $y - y_1 = m(x - x_1)$ is as follows.

Recall that $m = \dfrac{y_2 - y_1}{x_2 - x_1}$.

Substitute (x, y) for (x_2, y_2). Think of (x, y) as a variable point that can lie anywhere on the line.

$$m = \dfrac{y - y_1}{x - x_1}$$

$$y - y_1 = m(x - x_1)$$

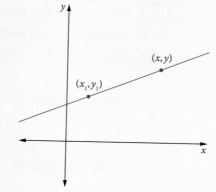

$y - y_1 = m(x - x_1)$ is known as the point–gradient form of the linear equation.

The steps for finding the equation of a straight line using $y - y_1 = m(x - x_1)$:

1 Find the gradient and substitute it for m.

2 Substitute the coordinates of a point on the line for x_1 and y_1.

3 Rearrange for y.

WORKED EXAMPLE 6 — Determining the equation using $y - y_1 = m(x - x_1)$

a What is the equation of the straight line that passes through $(1, 12)$ and has a gradient of 2?

b What is the equation of the straight line that passes through the points $(-2, 5)$ and $(5, -20)$?

Steps	Working
a 1 Find the gradient and substitute it into $y - y_1 = m(x - x_1)$.	The gradient is 2. $m = 2$ $y - y_1 = 2(x - x_1)$
2 Substitute a point on the line into the equation.	The line passes through $(1, 12)$. $x_1 = 1,\ y_1 = 12$ $y - 12 = 2(x - 1)$
3 Rearrange for y. Remember to expand and simplify.	$y - 12 = 2x - 2$ $y = 2x + 10$
b 1 Find the gradient using $\dfrac{y_2 - y_1}{x_2 - x_1}$.	$(-2, 5)$ and $(5, -20)$ $\dfrac{y_2 - y_1}{x_2 - x_1} = \dfrac{5 - (-20)}{-2 - 5} = \dfrac{25}{-7} = -\dfrac{25}{7}$ The gradient is $-\dfrac{25}{7}$. $m = -\dfrac{25}{7}$
2 Substitute m into $y - y_1 = m(x - x_1)$.	$y - y_1 = -\dfrac{25}{7}(x - x_1)$
3 Substitute a point on the line into the equation. Choose either point.	The simpler point to substitute is $(-2, 5)$. $x_1 = -2,\ y_1 = 5$ $y - 5 = -\dfrac{25}{7}(x - (-2))$ $y - 5 = -\dfrac{25}{7}(x + 2)$
4 Rearrange for y. Remember to expand and simplify.	$y - \dfrac{35}{7} = -\dfrac{25}{7}x - \dfrac{50}{7}$ $y = -\dfrac{25}{7}x - \dfrac{15}{7}$

 Exam hack

When using $y - y_1 = m(x - x_1)$ and rearranging for y, remember to expand the brackets and simplify any constant terms.

WORKED EXAMPLE 7 — Determining the equation from a graph

Find the equation of the line shown in the graph.

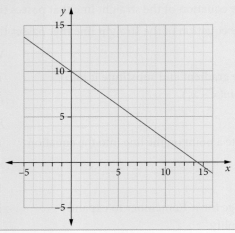

Steps	Working
1 As the y-intercept, c, and gradient, m, can be seen directly on the graph, use $y = mx + c$. Identify the y-intercept.	The y-intercept is 10. $c = 10$
2 Use the grid to find another clear point that the line passes through.	A clear point on the line is where two gridlines cross at $(8, 4)$.
3 Use this to determine the rise, run and gradient.	From the y-intercept at $(0, 10)$ to the point $(8, 4)$, there is a rise of -6 and a run of 8. $$m = \frac{\text{rise}}{\text{run}} = -\frac{6}{8} = -\frac{3}{4}$$ The gradient is $-\frac{3}{4}$.
4 Write the equation.	$y = -\frac{3}{4}x + 10$

> **Determining the equation of a straight line**
>
> To determine the equation of a straight line, find the gradient of the line and a point on the line. Then use either $y - y_1 = m(x - x_1)$ or $y = mx + c$ to find the equation.

Parallel and perpendicular lines

Parallel lines

Two lines are **parallel** if and only if they have the same gradient. They never meet and they are equidistant (equally distant) from each other.

For example, these two parallel lines both have a gradient of $\frac{1}{2}$.

If two lines $y = m_1 x + c_1$ and $y = m_2 x + c_2$ are parallel, then $m_1 = m_2$.

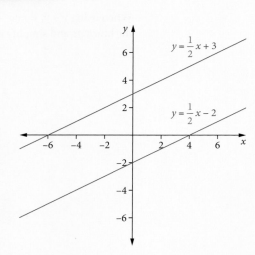

> **Parallel lines**
>
> Two parallel lines have the same gradient: $m_1 = m_2$.

Perpendicular lines

Two lines that intersect at a right angle (90°) are **perpendicular**. The gradients are negative reciprocals of each other, meaning that they multiply to −1.

Algebraically, $m_1 m_2 = -1$, which can be rearranged to $m_2 = -\dfrac{1}{m_1}$.

To gain an intuition for this formula, draw a line with a given rise and run. Then rotate the page 90° and see that the rise and run swap and a negative is introduced.

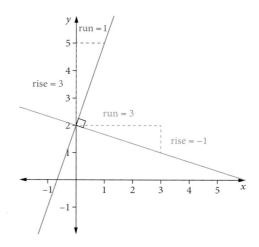

Proof

Consider two perpendicular lines AC and BC.

Let m_1 be the gradient of line AC and m_2 be the gradient of line BC.

Using $\dfrac{\text{rise}}{\text{run}}$, $m_1 = \dfrac{CD}{AD}$ and $m_2 = -\dfrac{CD}{BD}$.

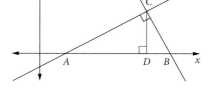

We prove similar triangles to find a relationship between the two fractions:

$\angle ADC = \angle BDC = 90°$

$\angle ACD = \angle CBD$ (using $\angle ACB = 90°$)

$\therefore \triangle ACD$ is similar to $\triangle CBD$ (by three identical angles) and $\dfrac{CD}{BD} = \dfrac{AD}{CD}$.

$\therefore m_2 = -\dfrac{CD}{BD} = -\dfrac{AD}{CD}$

$m_1 m_2 = \dfrac{CD}{AD} \times -\dfrac{AD}{CD} = -1$

The above proves the statement 'if two lines are perpendicular, then the product of their gradients is −1'. The converse statement 'if the product of the gradients of two lines is −1, then the lines are perpendicular' is also true.

A synonym for perpendicular is 'orthogonal' and the mathematical symbol is ⊥. For example, 'x-axis ⊥ y-axis' means 'the x-axis is perpendicular (orthogonal) to the y-axis'.

> **Perpendicular lines**
>
> The gradients of two perpendicular lines are the negative reciprocals of each other.
>
> $m_1 m_2 = -1$
>
> or rearranged:
>
> $m_2 = -\dfrac{1}{m_1}$

 Exam hack

Use $m_2 = -\dfrac{1}{m_1}$ to find the gradient of one line from the gradient of a perpendicular line.

It is easier to use $m_2 = -1 \div m_1$ if fractions are involved.

WORKED EXAMPLE 8 Finding the equation of parallel and perpendicular lines

a Find the equation of a line that is parallel to $y = -11x + 4$ and passes through the point $(2, 8)$.
b Find the equation of a line that is parallel to $3x - 8y = 1$ and passes through the point $(-1, 7)$.
c Find the equation of a line that is perpendicular to $y = -5x + 2$ and passes through the point $(20, -4)$.

Steps	Working
a 1 Use $m_1 = m_2$ for parallel lines, to determine the gradient.	The graph of $y = -11x + 4$ has a gradient of -11. The desired line is parallel, so it also has a gradient of -11.
2 Substitute the known gradient and a point on the line into $y - y_1 = m(x - x_1)$.	$y - y_1 = m(x - x_1)$ The gradient is -11 and the point is $(2, 8)$. $y - 8 = -11(x - 2)$
3 Rearrange.	$y - 8 = -11x + 22$ $y = -11x + 30$
b 1 Rearrange the equation into the form $y = mx + c$ to determine its gradient.	$3x - 8y = 1$ $3x - 1 = 8y$ $y = \dfrac{3}{8}x - \dfrac{1}{8}$
2 Use that $m_1 = m_2$ for parallel lines, to determine the gradient.	The graph of $3x - 8y = 1$ has a gradient of $\dfrac{3}{8}$. The desired line is parallel, so it also has a gradient of $\dfrac{3}{8}$.
3 Substitute the known gradient and a point on the line into $y - y_1 = m(x - x_1)$.	$y - y_1 = m(x - x_1)$ The gradient is $\dfrac{3}{8}$ and the point is $(-1, 7)$. $y - 7 = \dfrac{3}{8}(x - -1)$
4 Rearrange.	$y - 7 = \dfrac{3}{8}x + \dfrac{3}{8}$ $y = \dfrac{3}{8}x + \dfrac{59}{8}$
c 1 Use that $m_2 = -\dfrac{1}{m_1}$ for perpendicular lines, to determine the gradient.	The graph of $y = -5x + 2$ has a gradient of -5. The desired line is perpendicular, so it has a gradient of the negative reciprocal of -5. The gradient of the desired line is $\dfrac{1}{5}$.
2 Substitute the known gradient and a point on the line into $y - y_1 = m(x - x_1)$.	$y - y_1 = m(x - x_1)$ The gradient is $\dfrac{1}{5}$ and the point is $(20, -4)$. $y - (-4) = \dfrac{1}{5}(x - 20)$
3 Rearrange.	$y + 4 = \dfrac{1}{5}x - 4$ $y = \dfrac{1}{5}x - 8$

WORKED EXAMPLE 9 — Identifying parallel and perpendicular lines

The equations of six lines are as follows:

$L_1: 3x + y - 5 = 0$ $L_2: 4y + 2x + 6 = 0$ $L_3: -6x - 2y + 1 = 0$

$L_4: 6x + 2y - 4 = 0$ $L_5: 3x - y + 5 = 0$ $L_6: 4x + 12y + 1 = 0$

a State which, if any, of the lines are parallel.

b State which, if any, of the lines are perpendicular.

Steps	Working
a 1 Write each equation in the form $y = mx + c$ and record the gradient.	$L_1: y = -3x + 5,\ m = -3$ $L_2: y = -\dfrac{1}{2}x - \dfrac{3}{2},\ m = -\dfrac{1}{2}$ $L_3: y = -3x + \dfrac{1}{2},\ m = -3$ $L_4: y = -3x + 2,\ m = -3$ $L_5: y = 3x + 5,\ m = 3$ $L_6: y = -\dfrac{1}{3}x - \dfrac{1}{12},\ m = -\dfrac{1}{3}$
2 Identify the lines with the same gradient.	L_1, L_3 and L_4 all have a gradient of 3. L_1, L_3 and L_4 are parallel.
b Identify the pairs of lines for which the product of the gradient is –1.	$m_{L_5} \times m_{L_6} = 3 \times \left(-\dfrac{1}{3}\right) = -1$ L_5 and L_6 are perpendicular.

EXERCISE 3.2 Determining the equation of a linear graph   ANSWERS p. 451

Recap

1 Find the gradient of the line that passes through the coordinates $(5, -2)$ and $(-7, 6)$.

2 Sketch the graph of $y = -\dfrac{4}{5}x + 10$. Label the axis intercepts with their coordinates.

Mastery

3 WORKED EXAMPLES 5, 6   A line has gradient 4 and passes through the point $(3, 9)$.

   a Use $y = mx + c$ to find the equation of the line.

   b Use $y - y_1 = m(x - x_1)$ to find the equation of the line.

   c Which method is more efficient?

4 WORKED EXAMPLE 6   Find the equation of the line that

   a has gradient -1 and passes through $(3, -4)$

   b passes through $(-2, 5)$ and has gradient $-\dfrac{3}{4}$

   c passes through $(5, 3)$ and has gradient $\dfrac{5}{2}$

5. **WORKED EXAMPLE 6** Find the equation of the line with
   a gradient 2 and y-intercept of 1
   b gradient $-\dfrac{1}{3}$ and an x-intercept of $-\dfrac{2}{3}$
   c an x-intercept of 2 and gradient 2

6. **WORKED EXAMPLE 6** Find the equation of the line that passes through the following points.
   a $(3,-5)$ and $(-2,5)$
   b $(-6,4)$ and $(9,-1)$
   c $(-2,7)$ and $(3,6)$

7. **WORKED EXAMPLE 7** Find the equation of the line shown in each graph.

   a 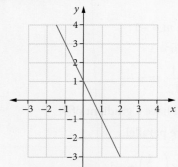   b 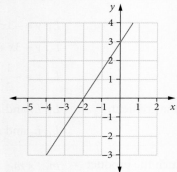   c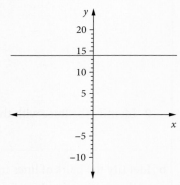

   d 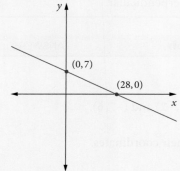   e 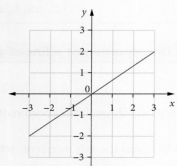   f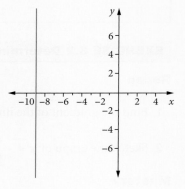

8. **WORKED EXAMPLE 8** Find the equation of the line that is parallel to
   a $y = \dfrac{1}{2}x + \dfrac{1}{4}$ and passes through $(2,5)$
   b $y = \dfrac{1}{3}x - \dfrac{2}{3}$ and passes through $(3,-1)$
   c $5x + 3y - 5 = 0$ and passes through $(5,-2)$.

9. **WORKED EXAMPLE 8** Find the equation of the line that is perpendicular to
   a $y = -\dfrac{3}{5}x + \dfrac{7}{5}$ and passes through $(1,-4)$
   b $y = 2x + 4$ and passes through $(-2,-2)$
   c $x - 3y - 1 = 0$ and passes through $(2,5)$.

10 WORKED EXAMPLES 6, 8   Find the equation of the line
   a with x-intercept -3 and y-intercept -1
   b passing through $(3, 5)$ and parallel to the x-axis
   c with gradient $-\dfrac{3}{4}$ and y-intercept -3
   d perpendicular to $y = -\dfrac{4}{3}x + \dfrac{5}{3}$ and passing through $\left(-5, \dfrac{5}{2}\right)$
   e parallel to the y-axis and passing through $(-5, 4)$
   f passing through $(1, -2)$ and $(-4, 2)$
   g passing through $(1, -7)$ and perpendicular to the y-axis
   h with x-intercept 11 and y-intercept 11
   i parallel to the x-axis and passing through $(-2, -6)$
   j with gradient $\dfrac{1}{3}$ and passing through $(-1, -2)$
   k passing through and $(-2, -6)$ perpendicular to $2x + 3y + 4 = 0$
   l with gradient -5 and intersecting $y = 8x + 1$ at $x = -1$.

11 WORKED EXAMPLE 9   The equations of six lines are:
   L_1: $2x + 3y - 5 = 0$ L_2: $3x + 2y + 6 = 0$ L_3: $5x - 3y + 2 = 0$
   L_4: $3x + 2y - 1 = 0$ L_5: $2x + 4y - 3 = 0$ L_6: $3x + 5y - 7 = 0$
   a State which, if any, of the lines are parallel.
   b State which, if any, of the lines are perpendicular.

Calculator-free

12 (4 marks) Mich the plumber earns their wage from a one-off call-out fee and an hourly rate. The graph of Mich's earnings, E, vs time, t, for a given job, is shown.

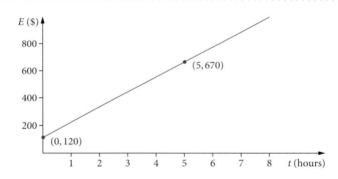

   a What is Mich's call-out fee?   (1 mark)
   b What is Mich's hourly rate?   (2 marks)
   c Write the linear equation that represents the relationship between E and t.   (1 mark)

13 (2 marks) One way to prove that a quadrilateral is a parallelogram is to show that opposite sides are parallel. If $A = (6, 5)$, $B = (1, 2)$, $C = (3, 6)$ and $D = (8, 9)$, show that $ABCD$ is a parallelogram.

Calculator-assumed

14 (2 marks) Find the gradient of the line perpendicular to the line with an x-intercept of -2 and a y-intercept of -6.

15 (4 marks) Show that the line connecting $A(0, 5)$ and $B(3, -4)$ is orthogonal to the line connecting $C(-2, 11)$ and $D(1, 12)$.

3.3 Expanding

Video playlist
Expanding

Worksheets
Binomial products
Special products

Expanding involves multiplying each **term** inside brackets by the factor outside the brackets. In the process, all brackets are removed. For example: $3(x + 5) = 3x + 15$. Expanding is the reverse process of **factorising**.

The 'terms' of an expression are the objects separated by + or − operations. For example, the expression $5x + \frac{2}{3} - 10y^2$ has three terms.

In the simplest expansion, a factor outside brackets multiplies each term within the brackets using the distributive law:

$$a \times (b + c) = a \times b + a \times c$$

This can be demonstrated geometrically. These diagrams show that $3(x + 5) = 3 \times x + 3 \times 5$.

	$x + 5$			x	5
3	area $3(x + 5)$		3	area $3x$	area 15

WORKED EXAMPLE 10 — Simple expanding

Expand these expressions and simplify if possible.

a $2(x + 3)$ **b** $-\frac{4}{5}(y - 20)$ **c** $x(1 - 8x) - 4(x + 9)$

Steps	Working
a Apply the distributive law.	$2(x + 3) = 2 \times x + 2 \times 3$ $= 2x + 6$
b 1 Apply the distributive law.	$-\frac{4}{5}(y - 20) = -\frac{4}{5} \times y - \frac{4}{5} \times (-20)$
2 Recall that the product of two negative numbers is positive and simplify the fraction multiplication.	$= -\frac{4}{5}y + 4 \times 4$ $= -\frac{4}{5}y + 16$
c 1 Apply the distributive law.	$x(1 - 8x) - 4(x + 9)$ $= x \times 1 + x \times (-8x) - 4 \times x - 4 \times 9$
2 Simplify by collecting **like terms**. Order terms from the highest to lowest degree.	$= x - 8x^2 - 4x - 36$ $= -8x^2 - 3x - 36$

Expanding binomial products

A **binomial** is an expression with two terms. For example, $x + 4$ or $3x^2 - 6$ are binomial expressions.

A **product** is the result of multiplication. For example, 27 is the product of 3 and 9, and $5x$ is the product of 5 and x.

A **binomial product** is the product of two or more binomial expressions; for example, $(3x + 2)(x - 1)$.

The binomial product $(a + b)(c + d)$ is expanded using the distributive law twice:

$$(a + b)(c + d) = a(c + d) + b(c + d)$$
$$= ac + ad + bc + bd$$

110 Nelson WAmaths Mathematics Methods 11

This can be demonstrated geometrically. These diagrams show that $(x + 3)(x + 5) = x^2 + 5x + 3x + 15$.

After expansion, collect any like terms. For example, $x^2 + 5x + 3x + 15 = x^2 + 8x + 15$.

The expansion of the product of two binomials can be remembered using the acronym FOIL.

FOIL method

First, Outer, Inner, Last

$$(a + b)(c + d) = ac + ad + bc + bd$$

F — First, O — Outer, I — Inner, L — Last

WORKED EXAMPLE 11 | Expanding binomial products

Expand and simplify these expressions.

a $(x - 5)(x + 1)$

b $(2x - 3)(4x - 7)$

Steps	Working
a 1 Use the FOIL method: $(a + b)(c + d) = ac + ad + bc + bd$	$(x - 5)(x + 1)$ First: $x \times x = x^2$ Outer: $x \times 1 = x$ Inner: $-5 \times x = -5x$ Last: $-5 \times 1 = -5$
2 Add the terms to the expression after calculating each multiplication. Check you have expanded the positives and negatives correctly.	$= x^2 + x - 5x - 5$
3 Collect like terms.	$= x^2 - 4x - 5$
b 1 Use the FOIL method: $(a + b)(c + d) = ac + ad + bc + bd$	$(2x - 3)(4x - 7)$ First: $2x \times 4x = 8x^2$ Outer: $2x \times (-7) = -14x$ Inner: $-3 \times 4x = -12x$ Last: $-3 \times (-7) = 21$
2 Add the terms to the expression after calculating each multiplication. Check you have expanded the positives and negatives correctly.	$= 8x^2 - 14x - 12x + 21$
3 Collect like terms.	$= 8x^2 - 26x + 21$

Expanding a perfect square

Some expansion patterns are so common that it is worth memorising each as its own rule. The first is the expansion of a **perfect square**, $(a + b)^2$ or $(a - b)^2$:

$$(a + b)^2 = (a + b)(a + b)$$
$$= a^2 + ab + ba + b^2$$
$$= a^2 + 2ab + b^2$$

$$(a - b)^2 = (a - b)(a - b)$$
$$= a^2 - ab - ba + b^2$$
$$= a^2 - 2ab + b^2$$

> **Expanding perfect squares**
> $(a + b)^2 = a^2 + 2ab + b^2$
> $(a - b)^2 = a^2 - 2ab + b^2$

WORKED EXAMPLE 12 — Perfect square expansion

Expand the following expressions.

a $(x + 5)^2$

b $(2x - 3)^2$

Steps	Working	
a 1 While the expression can be expanded using FOIL, recognise the perfect square expansion pattern to do it more efficiently. Note the first term and the second term. Recall $(a + b)^2 = a^2 + 2ab + b^2$.	$(x + 5)^2$	The first term is x. The second term is 5.
2 Read the pattern as: • first term squared • twice the product of the two terms • second term squared.	$= x^2 + 2 \times x \times 5 + 5^2$ $= x^2 + 10x + 25$	
b 1 While the expression can be expanded using FOIL, recognise the perfect square expansion pattern to do it more efficiently. Note the first term and the second term. Recall $(a - b)^2 = a^2 - 2ab + b^2$.	$(2x - 3)^2$	The first term is $2x$. The second term is -3.
2 Read the pattern as: • first term squared • twice the product of the two terms • second term squared.	$= (2x)^2 + 2(2x)(-3) + (-3)^2$ $= 4x^2 - 12x + 9$	

Expanding to a difference of perfect squares

Another pattern worth memorising is the expansion of $(a - b)(a + b)$ to arrive at a **difference of perfect squares**, $a^2 - b^2$.

$(a - b)(a + b)$
$= a^2 + ab - ba - b^2$
$= a^2 - b^2$

> **Expanding to a difference of perfect squares**
> $(a - b)(a + b) = a^2 - b^2$

WORKED EXAMPLE 13 | Expanding to a difference of perfect squares

Expand the following expressions.
a $(x - 6)(x + 6)$
b $(3x + 5)(3x - 5)$

Steps	Working	
a 1 While the expression can be expanded using FOIL, recognise the expansion to a difference of perfect squares to do it more efficiently.	$(x - 6)(x + 6)$	a is x. b is 6.
2 Note a and b. Recall $(a - b)(a + b) = a^2 - b^2$.		
3 Read the pattern as a squared minus b squared.	$= x^2 - 6^2$ $= x^2 - 36$	
b 1 While the expression can be expanded using FOIL, recognise the expansion to a difference of perfect squares to do it more efficiently.	$(3x + 5)(3x - 5)$	a is $3x$. b is 5.
2 Note a and b. Recall $(a - b)(a + b) = a^2 - b^2$.		
3 Read the pattern as a squared minus b squared.	$= (3x)^2 - 5^2$ $= 9x^2 - 25$	

USING CAS 2 | Expanding

Expand $2(x - 3) - (x - 7)(x + 7) - (x - 9)^2$ and simplify.

ClassPad

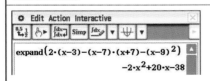

1. Enter and highlight the expression.
2. Tap **Interactive > Transformation > Expand**.
3. Keep the defaults settings in the dialogue box and tap **OK**.

TI-Nspire

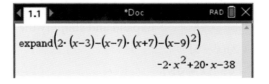

1. Press **menu > Algebra > Expand**.
2. Enter the expression and press **enter**.

The answer is $-2x^2 + 20x - 38$.

EXERCISE 3.3 Expanding

ANSWERS p. 452

Recap

1 The equation of the line with gradient $-\dfrac{4}{5}$ and passing through $(2, -4)$ is

    A $4x + 5y + 12 = 0$ **B** $4x + 5y - 28 = 0$ **C** $4x + 5y - 12 = 0$
    D $4x - 5y - 12 = 0$ **E** $4x - 5y - 4 = 0$

2 Find the equation of the line that passes through $(-18, 2)$ and is perpendicular to $y = 6x + 11$.

Mastery

3 WORKED EXAMPLE 10 Expand and simplify the following expressions.
    a $2(2x + 3)$ **b** $-4(2x - 3)$
    c $7(5 - x)$ **d** $x(x - 8)$
    e $11(x - 8) - 2(x + 9)$ **f** $x(x + 4) - x(10 - x)$

4 WORKED EXAMPLE 11 Expand and simplify the following expressions.
    a $(x + 2)(x + 3)$ **b** $(x - 4)(x + 5)$
    c $(x - 8)(x - 6)$ **d** $(x + 1)(x - 11)$
    e $(x + 7)(x + 2) + (x - 9)(x - 3)$ **f** $(x + 4)(x - 6) - (x - 3)(x + 6)$

5 WORKED EXAMPLE 11 Expand and simplify the following expressions.
    a $(3x + 2)(2x + 3)$ **b** $(2x + 5)(2x - 1)$
    c $(6x - 1)(x + 7)$ **d** $(3x - 1)(5x - 4)$
    e $(2x + 9)(x - 8) - x(x - 14)$ **f** $(3x - 1)(x - 5) - (2x - 11)(2x + 4)$

6 WORKED EXAMPLE 12 Expand and simplify the following expressions.
    a $(x + 7)^2$ **b** $(x - 8)^2$
    c $(2x + 3)^2$ **d** $(3x - 5)^2$
    e $(x - 1)^2 + (2x - 1)^2$ **f** $(p - q)^2 - (p + q)^2$
    g $\left(x + \dfrac{1}{2}\right)^2$ **h** $\left(x - \dfrac{5}{2}\right)^2$

7 WORKED EXAMPLE 13 Expand and simplify the following expressions.
    a $(x + 3)(x - 3)$ **b** $(x - 7)(x + 7)$
    c $\left(x - \sqrt{5}\right)\left(x + \sqrt{5}\right)$ **d** $(2x + 1)(2x - 1)$
    e $\left(4x + \sqrt{3}\right)\left(4x - \sqrt{3}\right)$ **f** $\left(8x + \dfrac{1}{9}\right)\left(8x - \dfrac{1}{9}\right)$

8 Using CAS 2 Expand and simplify the following expressions.
    a $5(x - 1) - (x + 3)(x - 3) + (4x - 3)^2$ **b** $-x(x + 8) + (2x - 9)(2x + 9) + (7x - 5)^2$

9 **WORKED EXAMPLES 10–13** Expand and simplify the following expressions.

a $(m - 5n)^2$

b $3(2x + 1)$

c $\left(7p + \dfrac{1}{4}\right)\left(7p - \dfrac{1}{4}\right)$

d $4(x + 1)(x - 6)$

e $(2t - 7)(3t - 9)$

f $-4(2x - 3)^2$

g $(x + 2)(2x - 3y + 7)$

h $(xy + 11v)(xy - 11v)$

10 **WORKED EXAMPLE 12** By guess and check, determine whether the following are expansions of perfect squares.

a $x^2 + 8x + 16$

b $x^2 - 10x + 25$

c $9x^2 + 6x + 1$

d $x^2 - 12x - 36$

Calculator-free

11 (3 marks)

a If $x^2 - 12x + t$ is a perfect square for all x, find t. (1 mark)

b If $px^2 + 28x + 4$ is a perfect square for all x, find p. (1 mark)

c If $25x^2 + qx + 49$ is a perfect square for all x, find the possible values of q. (1 mark)

12 (3 marks)

a Show that $(x - y)(x^2 + xy + y^2) = x^3 - y^3$. (2 marks)

b Hence, find the expansion of $(x - 3)(x^2 + 3x + 9)$. (1 mark)

Calculator-assumed

13 (4 marks) The dimensions of a rectangular garden bed are modelled as follows, where units are measured in metres.

a Find an expression for the area of the garden bed in expanded form. (1 mark)

b What is the area of the garden bed if $x = 17$ m? (1 mark)

c What is the area of the garden bed if $x = 5.001$ cm? (1 mark)

d Write an equation whose solution is the x value corresponding to an area of 100 m². Do not solve the equation. (1 mark)

3.4 Factorising

Video playlist
Factorising

Worksheet
Factorising algebraic expressions

Puzzle
Factorising quadratic equations

Factorising is the reverse process of expanding. It involves rewriting an expression as a product of factors. For example, $4x + 20$ can be factorised to $4(x + 5)$. Here, the factors are 4 and $x + 5$.

Four factorisation techniques are covered in this section:
- Factorising by taking out a **highest common factor**
- Factorising to a perfect square
- Factorising a difference of perfect squares
- Factorising by grouping

The technique of factorising quadratic trinomials is covered in Section 3.5.

Highest common factor

The simplest technique for factorising is to 'take out' the highest common factor from each term of an expression. This can only be done if the factor is common to every term.

Highest common factor

When factorising, first look to take out a highest common factor.
1. Write the highest common factor in front of the brackets.
2. Within the brackets, write each original term divided by the highest common factor.

Exam hack

If your expression has a mix of positive and negative terms, you can choose whether your highest common factor is positive or negative.

WORKED EXAMPLE 14 — Factorising – highest common factor

Factorise the following expressions by taking out the highest common factor.

a $4b - 24d^2$ **b** $6x^2 + 12x - 30$ **c** $-7xy^2 + 56yx^2z$

Steps	Working	
a 1 Identify the highest common factor.	$4b - 24d^2$	The highest common factor is 4.
2 Write the highest common factor in front of the brackets. Within the brackets, write each original term divided by the highest common factor.	$= 4(b - 6d^2)$	Note: $-4(6d^2 - b)$ is also acceptable.
b 1 Identify the highest common factor.	$6x^2 + 12x - 30$	The highest common factor is 6.
2 Write the highest common factor in front of the brackets. Within the brackets, write each original term divided by the highest common factor.	$= 6(x^2 + 2x - 5)$	The expression inside the brackets cannot be factorised any further.
c 1 Identify the highest common factor.	$-7xy^2 + 56yx^2z$	The highest common factor is $-7xy$.
2 Write the highest common factor in front of the brackets. Within the brackets, write each original term divided by the highest common factor.	$= -7xy(y - 8xz)$	Note: $7xy(8xz - y)$ is also acceptable.

Exam hack

Always check your answer to see if a common factor remains in all terms and, if so, factorise again. If this occurs, your first common factor was not the highest possible.

Factorising to a perfect square

By expanding:

$(a + b)^2 = (a + b)(a + b) = a^2 + ab + ba + b^2$
$\qquad = a^2 + 2ab + b^2$

$(a - b)^2 = (a - b)(a - b) = a^2 - ab - ba + b^2$
$\qquad = a^2 - 2ab + b^2$

Because factorising is the reverse process of expanding, we can reverse the above and conclude that $a^2 + 2ab + b^2$ factorises to $(a + b)^2$ and $a^2 - 2ab + b^2$ factorises to $(a - b)^2$.

> **Factorising to a perfect square**
> $a^2 + 2ab + b^2 = (a + b)^2$
> $a^2 - 2ab + b^2 = (a - b)^2$

To find such factorisations, look for situations where the middle term is double the product of the square roots of the first and third terms.

WORKED EXAMPLE 15 | Factorising to a perfect square

Factorise the following expressions.

a $x^2 + 18x + 81$

b $8x^2 - 40x + 50$

Steps	Working	
a 1 Check if perfect square factorisation is possible. Note the square root of the first term and the square root of the third term.	$x^2 + 18x + 81$ $2(x)(9) = 18x$	The square root of the first term is x. The square root of the third term is 9. The middle term is double the product of the x and 9. Perfect square factorisation is possible.
2 Recall $a^2 + 2ab + b^2 = (a + b)^2$.	$= (x + 9)^2$	
b 1 When factorising, always check first if there is a common factor of all the terms.	$8x^2 - 40x + 50$	2 is the highest common factor.
2 Take out the common factor.	$= 2(4x^2 - 20x + 25)$	
3 Consider the expression within the brackets. Check if perfect square factorisation is possible. Note the square root of the first term and the square root of the third term.	$4x^2 - 20x + 25$ $2(2x)(-5) = -20x$	The square root of the first term is $2x$. The square root of the third term is 5 (or -5). The middle term is double the product of the $2x$ and -5. Perfect square factorisation is possible.
4 Recall $a^2 - 2ab + b^2 = (a - b)^2$.	$= (2x - 5)^2$	
5 Write the full factorisation of the original expression.	$8x^2 - 40x + 50$ $= 2(4x^2 - 20x + 25)$ $= 2(2x - 5)^2$	

Factorising a difference of perfect squares

By expanding:

$(a - b)(a + b) = a^2 + ab - ba - b^2$
$= a^2 - b^2$

Because factorising is the reverse process of expanding, we can conclude that $a^2 - b^2$ factorises to $(a - b)(a + b)$.

Factorising a difference of perfect squares
$a^2 - b^2 = (a - b)(a + b)$

 Exam hack

There is no factorisation of $a^2 + b^2$ in the real number system.

WORKED EXAMPLE 16 Factorising a difference of perfect squares

Factorise the following expressions.

a $x^2 - 64$ **b** $x^2 - 5$ **c** $162u^2 - 98v^6$ **d** $(x - 3)^2 - 24$

Steps	Working	
a Recall that $a^2 - b^2 = (a - b)(a + b)$.	$x^2 - 64$	The square root of the first term is x.
a is the square root of the first term and b is the square root of the second term.	$= (x - 8)(x + 8)$	The square root of the second term is 8.
b Recall that $a^2 - b^2 = (a - b)(a + b)$.	$x^2 - 5$	The square root of the first term is x.
a is the square root of the first term and b is the square root of the second term.	$= (x - \sqrt{5})(x + \sqrt{5})$	The square root of the second term is $\sqrt{5}$.
c 1 When factorising, always check first if there is a common factor of all the terms.	$162u^2 - 98v^6$	2 is the highest common factor.
2 Take out the common factor.	$= 2(81u^2 - 49v^6)$	
3 Consider the expression within the brackets.	$81u^2 - 49v^6$	The square root of the first term is $9u$.
Recall that $a^2 - b^2 = (a - b)(a + b)$. a is the square root of the first term and b is the square root of the second term.	$= (9u - 7v^3)(9u + 7v^3)$	The square root of the second term is $7v^3$ because $(v^3)^2 = v^6$.
4 Write the full factorisation of the original expression.	$162u^2 - 98v^6$ $= 2(81u^2 - 49v^6)$ $= 2(9u - 7v^3)(9u + 7v^3)$	
d Recall that $a^2 - b^2 = (a - b)(a + b)$.	$(x - 3)^2 - 24$	
a is the square root of the first term and b is the square root of the second term.	The square root of the first term is $x - 3$.	
	The square root of the second term is $\sqrt{24}$.	
	$\sqrt{24} = \sqrt{4 \times 6} = 2\sqrt{6}$	
	$\therefore$ the factorisation is $(x - 3 - 2\sqrt{6})(x - 3 + 2\sqrt{6})$.	

Factorising by grouping

Factorising by grouping is a technique for factorising four or more terms, where there is not a common factor between all terms. For example:

$xy + 8y + 5x + 40$ ← 4 terms with no common factor, but two terms with a common factor of y and two terms with a common factor of 5
$= y(x + 8) + 5(x + 8)$ ← 2 terms with a common factor of $x + 8$
$= (x + 8)(y + 5)$ ← 1 fully factorised expression

> **Factorising by grouping**
>
> To factorise by grouping:
>
> 1 Separate the terms into groups with their own common factor and factorise each group.
>
>   *Choose a grouping so that, after factorising, the expression in brackets is common to all groups.*
>
> 2 Use the expression in brackets as a new common factor and factorise again.

Factorising by grouping is not always possible and is a less common method of factorising.

WORKED EXAMPLE 17 — Factorising by grouping

Factorise the following expressions.

a $x^2 - 4x + 28 - 7x$

b $6p + 3q - 2pq - q^2$

Steps	Working
a 1 Separate the terms into groups that have a common factor. Choose a grouping so that, after factorising, the expression in brackets will be common to all groups.	$x^2 - 4x + 28 - 7x$ Group x^2 and $-4x$ (taking out an x, the expression in brackets will be $x - 4$). Group $-7x$ and 28 (taking out -7, the expression in brackets will be $x - 4$).
2 Factorise each group individually.	$x^2 - 4x + 28 - 7x$ $= x^2 - 4x - 7x + 28$ $= x(x - 4) - 7(x - 4)$
3 Use the expression in brackets as a new common factor and factorise again.	$= (x - 4)(x - 7)$
b 1 Separate the terms into groups that have a common factor. Choose a grouping so that, after factorising, the expression in brackets will be common to all groups.	$6p + 3q - 2pq - q^2$ Group $6p$ and $3q$ (taking out 3, the expression in brackets will be $2p + q$). Group $-2pq$ and $-q^2$ (taking out $-q$, the expression in brackets will be $2p + q$).
2 Factorise each group individually.	$6p + 3q - 2pq - q^2$ $= 3(2p + q) - q(2p + q)$
3 Use the expression in brackets as a new common factor and factorise again.	$= (2p + q)(3 - q)$

USING CAS 3 — Factorising

Factorise $36x^2 - 108x + 81$.

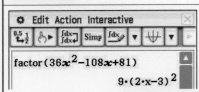

ClassPad

1. Enter and highlight the expression.
2. Tap **Interactive > Transformation > factor > factor**.
3. Keep the defaults settings in the dialogue box and tap **OK**.

TI-Nspire

1. Press **menu > Algebra > Factor**.
2. Enter the expression and press **enter**.

The answer is $9(2x - 3)^2$.

EXERCISE 3.4 Factorising

ANSWERS p. 452

Recap

1. What are the x- and y-intercepts of the graph of $y = \dfrac{5}{6}x - 25$?

2. Expand the following expressions.
   - **a** $4x(5x - 1)$
   - **b** $(x - 8)^2$
   - **c** $(x - 3)(x + 3)$
   - **d** $(2x + 7)^2$

Mastery

3. **WORKED EXAMPLE 14** Factorise the following expressions by taking out the highest common factor.
   - **a** $2x + 16$
   - **b** $4x - 12$
   - **c** $9b - 36b^2$
   - **d** $-4xy + 10x + 8$
   - **e** $12x^2 - 24x^3$
   - **f** $7x^2 - 14x + 56$
   - **g** $15dy - 40d^2 + 5d$
   - **h** $-6x^2yz + 30xy^2z$

4. **WORKED EXAMPLE 15** Factorise the following expressions.
   - **a** $x^2 + 12x + 36$
   - **b** $x^2 + 18x + 81$
   - **c** $x^2 - 8x + 16$
   - **d** $x^2 - 22x + 121$
   - **e** $49x^2 - 14x + 1$
   - **f** $9x^2 - 30x + 25$
   - **g** $8x^2 - 48x + 72$
   - **h** $32x^2 - 16x + 2$

5. **WORKED EXAMPLE 16** Factorise the following expressions.
   - **a** $x^2 - 25$
   - **b** $x^2 - 49$
   - **c** $t^2 - 36$
   - **d** $x^2 - 6$
   - **e** $p^2 - 7$
   - **f** $(x + 2)^2 - 81$
   - **g** $(x - 7)^2 - 5$
   - **h** $12u^2 - 27p^6$

6 **WORKED EXAMPLE 17** Factorise the following expressions by grouping.

a $x^2 + x - 4x - 4$
b $x^2 - 2x - 5x + 10$
c $x^2 + 6x - 8x - 48$
d $3x^2 + 2x - 15x - 10$
e $3x^2 - 6x - 6x + 12$
f $4p^2 - 8pq + pq - 2q^2$
g $4x^3 - 16x^2 + 4x - 16$
h $a^2x^3 - 2ax^2 - 9a^2x + 18a$

7 **WORKED EXAMPLES 14–17** Factorise the following expressions. If they cannot be factorised, rewrite the unchanged expression.

a $x^2 - 16$
b $3x^2 + 6x + 3$
c $(x - 8)(x + 7)$
d $(x - 3)(2x + 4)$
e $4x^2 - 28x + 49$
f $-24x + 72xy$
g $36m^2 - 1$
h $x^2 + 9$
i $p^2 + 5p - 7p - 35$
j $(u - 1)^2 - 13$
k $5x^2 - 2x + 40x - 16$
l $2x^2 - 18a^2$

8 **Using CAS 3** Factorise the following expressions.

a $80x^2 - 360x + 405$
b $3x^2 - 3p^2q^2$
c $-7x^2 + 28x + 420$

Calculator-free

9 (4 marks) Factorise and simplify the following expressions.

a $(x + 5)^2 - 14(x + 5) + 49$ (2 marks)
b $(x - 11)^2 - (x + 11)^2$ (2 marks)

Calculator-assumed

10 (2 marks) The product of Jordan's age and her sister's age is represented by $x^2 + 9x$, where x represents her sister's age. How much older is Jordan than her sister?

11 (4 marks) The area of a right triangle, in mm^2, is represented by $2t^2 - 10t + \dfrac{25}{2}$, where $t > 0$.

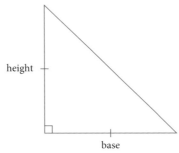

a Show that the product of the base and height equals $(2t - 5)^2$. (2 marks)
b If the triangle is isosceles and the length of the base is 17 mm, find the value of t. (2 marks)

Video playlist
Solving quadratic equations by factorising

3.5 Solving quadratic equations by factorising

A **quadratic equation** is any equation that can be rearranged into the form $ax^2 + bx + c = 0$.

A **quadratic trinomial** is the expression $ax^2 + bx + c$ on the left-hand side of the equation. The prefix 'tri' refers to the three terms.

The goal of factorising quadratic trinomials is to write the expression as a product of two linear factors, for example:

$x^2 - 7x + 10$
$= (x - 5)(x - 2)$

$x - 5$ and $x - 2$ are linear factors because their degree is 1.

> 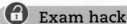 **Exam hack**
>
> An equation has an equal sign.
> An expression has no equal sign.
> Quadratic equations can be solved.
> **Quadratic expressions** can be factorised.

Factorising monic quadratic trinomials

In a **monic quadratic trinomial**, the coefficient of the x^2 term is equal to 1. The general form is $x^2 + bx + c$.

The factorisation of a monic quadratic trinomial uses a pattern found by expanding the factorised form:

$(x + u)(x + v)$
$= x^2 + vx + ux + uv$
$= x^2 + (u + v)x + uv$
$= x^2 + (\text{SUM})x + (\text{PRODUCT})$
$= x^2 + bx + c$

In other words, to factorise a monic quadratic trinomial, seek two numbers whose sum is b and whose product is c. The following methods provide structure for this search.

Method 1 – Factorising by the cross method

To factorise any quadratic trinomial by the cross method:

1. Draw a cross and write a factor pair of the ax^2 term on the left side. If $a = 1$, use x and x.
2. Write a factor pair of c on the right side.
3. Check if the choice of factor pairs is correct by multiplying along each diagonal and adding the resultant terms together. The sum should equal bx.
4. Once the correct cross is found, write each linear factor by reading horizontally.

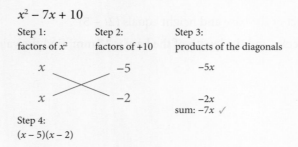

Some useful tips:
- Start by writing all possible factor pairs of c, including all positive and negative options.
- Draw a new cross for a new attempt.
- Check the factorised answer by drawing FOIL lines and imagining the expansion.

WORKED EXAMPLE 18 | Factorising monic quadratic trinomials – cross method

Factorise the following expressions.

a $x^2 + 7x + 12$

b $3x^2 - 39x + 126$

Steps	Working
a 1 Write all the factor pairs of c, including positive and negative variations.	$x^2 + 7x + 12$ 1, 12 2, 6 3, 4 –1, –12 –2, –6 –3, –4
2 Set up the cross. Because the expression is monic, use x and x on the left side. Try different factor pairs of c.	Step 1: factors of x^2 Step 2: factors of +12 x +3 ╳ x +4
3 Calculate the products of the diagonals and check they sum to bx. If not, try a different factor pair.	Step 3: products of the diagonals x +3 3x ╳ x +4 4x sum: 7x ✓
4 Write the linear factors by reading horizontally.	$x^2 + 7x + 12$ $= (x + 3)(x + 4)$ Step 4: read horizontally
b 1 Whenever factorising, first check if there is a common factor in all the terms.	$3x^2 - 39x + 126$ $= 3(x^2 - 13x + 42)$
2 Write all the factor pairs of c, including positive and negative variations.	Factorise $x^2 - 13x + 42$. 1, 42 2, 21 3, 14 6, 7 –1, –42 –2, –21 –3, –14 –6, –7
3 Set up the cross. Because the expression is monic, use x and x on the left side. Try different factor pairs of c. Calculate the products of the diagonals and check they sum to bx. If not, try a different factor pair.	Step 1: factors of x^2 Step 2: factors of +42 Step 3: products of the diagonals x –6 –6x ╳ x –7 –7x sum: –13x ✓
4 Write the linear factors by reading horizontally.	$x^2 - 13x + 42$ $= (x - 6)(x - 7)$ Step 4: read horizontally
5 Write the full factorisation of the original expression.	$3x^2 - 39x + 126$ $= 3(x - 6)(x - 7)$

 Exam hack

If a is negative in $ax^2 + bx + c$, factorise by taking out a negative from the entire expression.

Method 2 – Factorising by grouping

To factorise a monic quadratic trinomial, $x^2 + bx + c$, by grouping, split bx into two terms whose coefficients multiply to give c.

$x^2 - 7x + 10$
$= x^2 - 5x - 2x + 10$ -5 and -2 are chosen because their product is $+10$
$= x(x - 5) - 2(x - 5)$ using the grouping technique from Section 3.4
$= (x - 5)(x - 2)$

WORKED EXAMPLE 19	Factorising monic quadratic trinomials – grouping method

Factorise the following expressions.

a $x^2 - x - 30$
b $-4x^2 + 48x - 44$

Steps	**Working**
a 1 Write all the factor pairs of c, including positive and negative variations.	$x^2 - x - 30$ $-1, 30$ $-2, 15$ $-3, 10$ $-5, 6$ $1, -30$ $2, -15$ $3, -10$ $5, -6$
2 Choose the pair whose values sum to b.	5 and -6 sum to give -1.
3 Split bx using these two values.	$x^2 - x - 30$ $= x^2 + 5x - 6x - 30$
4 Factorise by grouping.	$= x(x + 5) - 6(x + 5)$ $= (x + 5)(x - 6)$
b 1 Whenever factorising, first check if there is a common factor in all the terms.	$-4x^2 + 48x - 44$ $= -4(x^2 - 12x + 11)$
2 Write all the factor pairs of c, including positive and negative variations.	Factorise $x^2 - 12x + 11$. $1, 11$ $-1, -11$
3 Choose the pair whose values sum to b.	-1 and -11 sum to give -12.
4 Split bx using these two values.	$x^2 - 12x + 11$ $= x^2 - 11x - x + 11$
5 Factorise by grouping.	$= x(x - 11) - (x - 11)$ $= (x - 11)(x - 1)$
6 Write the full factorisation of the original expression.	$-4x^2 + 48x - 44$ $= -4(x - 11)(x - 1)$

 Exam hack

After a lot of practice with expanding and factorising monic trinomials, you may be able to go to the factorised form in one step. Write $(x \_\_\_\_)(x \_\_\_\_)$, then fill in the missing terms by imagining the expansion using FOIL.

Factorising non-monic quadratic trinomials

Method 1 – Factorising by the cross method

The steps for factorising any quadratic trinomial, $ax^2 + bx + c$, using the cross method, are described on page 122. If the expression is non-monic, then the left side of the cross must show a factor pair of ax^2.

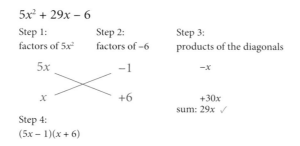

$5x^2 + 29x - 6$

Step 1: factors of $5x^2$
Step 2: factors of -6
Step 3: products of the diagonals

$5x$ -1 $-x$
x $+6$ $+30x$
                     sum: $29x$ ✓

Step 4: $(5x - 1)(x + 6)$

WORKED EXAMPLE 20 — Factorising non-monic quadratic trinomials – cross method

Factorise $2x^2 + 15x + 22$.

Steps	Working
1 Write all the factor pairs of c, including positive and negative variations.	$2x^2 + 15x + 22$ 1, 22 2, 11 $-1, -22$ $-2, -11$
2 Write the positive factor pairs of a (if a is negative, factorise by taking out -1 from the entire expression before this step).	1, 2
3 Set up the cross. Because the expression is non-monic, use $2x$ and x on the left side. Try different factor pairs of c. Calculate the products of the diagonals and check they sum to bx. If not, try a different factor pair.	Step 1: factors of $2x^2$ Step 2: factors of $+22$ Step 3: products of the diagonals $2x$ $+11$ $11x$ x $+2$ $4x$ sum: $15x$ ✓
4 Write the linear factors by reading horizontally.	$2x^2 + 15x + 22$ $= (2x + 11)(x + 2)$ Step 4: read horizontally

Method 2 – Factorising by grouping

To factorise any quadratic trinomial, $ax^2 + bx + c$, by grouping, split bx into two terms whose coefficients multiply to give the product ac.

$5x^2 + 29x - 6$
$= 5x^2 + 30x - x - 6$ $+30$ and -1 are chosen because their product equals 5×-6
$= 5x(x + 6) - (x + 6)$ using the grouping technique from Section 3.4
$= (5x - 1)(x + 6)$

The rationale behind this method is as follows:

$(px + u)(qx + v)$
$= pqx^2 + (uq + pv)x + uv$
$= ax^2 + bx + c$

An inspection shows that b is broken up into two terms (uq and pv) whose product is equal to the product ac.

WORKED EXAMPLE 21 — Factorising non-monic quadratic trinomials – grouping method

Factorise $3x^2 - 17x + 24$.

Steps	Working
1 Calculate the product ac.	$3x^2 - 17x + 24$ $ac = 72$
2 Choose a factor pair of this product whose sum equals b.	A factor pair of 72 is $-9, -8$. These numbers sum to -17.
3 Split bx using these two values.	$3x^2 - 17x + 24$ $= 3x^2 - 9x - 8x + 24$
4 Factorise by grouping.	$= 3x(x - 3) - 8(x - 3)$ $= (3x - 8)(x - 3)$

Factorising quadratic trinomials

A quadratic trinomial is an expression in the form $ax^2 + bx + c$.

Quadratic trinomials cannot always be factorised, but when they can, the cross method or the grouping method can be used. When $a = 1$, the trinomial ($x^2 + bx + c$) is monic.

Solving with the null factor law

A quadratic equation $ax^2 + bx + c = 0$ can be solved using the **null factor law** after the expression $ax^2 + bx + c$ has been factorised.

The null factor law

Consider two unknown values, m and n. If $m \times n$ is equal to zero, what can be deduced about m and n? Think closely and notice that either m or n must equal 0.

Null factor law

If $mn = 0$, then $m = 0$ or $n = 0$.

For example:

$x^2 + 6x - 16 = 0$
$(x - 2)(x + 8) = 0$

The equation is in the form $m \times n = 0$, where m and n are the linear factors $x - 2$ and $x + 8$, respectively. By the null factor law:

$x - 2 = 0$ or $x + 8 = 0$
$x = 2$ or $x = -8$

The solutions to $(x - 2)(x + 8) = 0$ are *any* values of x that make the expression on the left-hand side equal the right-hand side. Therefore, *both* $x = 2$ and $x = -8$ are solutions.

In Chapter 4, the null factor law is also used to solve equations with higher degree polynomials..

Other methods for solving quadratic equations include completing the square and using the quadratic formula, which are covered in Sections 3.6 and 3.7.

WORKED EXAMPLE 22 | Solving using the null factor law

Solve the following quadratic equations.

a $(x - 5)(x + 2) = 0$

b $x(2x + 3) = 0$

c $x^2 - 9x - 36 = 0$

d $6x^2 - 14x = -8$

Steps	Working
a The equation is factorised and is in the form = 0. Apply the null factor law.	$(x - 5)(x + 2) = 0$ $x - 5 = 0$ or $x + 2 = 0$ $x = 5$ $x = -2$ The solutions are $x = 5, -2$.
b The equation is factorised and is in the form = 0. Apply the null factor law.	$x(2x + 3) = 0$ $x = 0$ or $2x + 3 = 0$ $x = -\dfrac{3}{2}$ The solutions are $x = -\dfrac{3}{2}, 0$.
c 1 The equation is in the form = 0. Factorise the quadratic expression. 2 Apply the null factor law.	$x^2 - 9x - 36 = 0$ $(x - 12)(x + 3) = 0$ $x - 12 = 0$ or $x + 3 = 0$ $x = 12$ $x = -3$ The solutions are $x = -3, 12$.

d 1 Rearrange the equation to be in the form = 0.
  2 Divide both sides by 2.
  3 Factorise the quadratic expression.
  4 Apply the null factor law.
	$6x^2 - 14x = -8$
	$6x^2 - 14x + 8 = 0$
	$3x^2 - 7x + 4 = 0$
	$(3x - 4)(x - 1) = 0$
	$3x - 4 = 0$ or $x - 1 = 0$
	$x = \dfrac{4}{3}$ $x = 1$
	The solutions are $x = 1, \dfrac{4}{3}$.

Exam hack

The null factor law can only be used when the right-hand side of the equation equals zero.

EXERCISE 3.5 Solving quadratic equations by factorising

ANSWERS p. 452

Recap

1 Write the value of b in $ax^2 + bx + c$ for the expanded form of each of these expressions.

a $(x + 9)(x + 1)$ **b** $(x + 2)(x - 8)$ **c** $(x - 7)(x - 4)$

2 Factorise $x^2 - 16x + 64$.

Mastery

3 WORKED EXAMPLES 18, 19 Write down all the factor pairs of each number.

a 12 **b** 20 **c** −15 **d** 24
e −30 **f** −18 **g** 72 **h** −21
i 16 **j** −5 **k** −28 **l** 45

4 WORKED EXAMPLES 18, 19 Factorise the following expressions.

a $x^2 + 5x + 6$ **b** $x^2 + 11x + 28$ **c** $x^2 + 14x + 24$
d $x^2 - 10x + 24$ **e** $x^2 + 8x - 9$ **f** $x^2 - 3x - 40$
g $x^2 + 17x - 60$ **h** $x^2 - 12x + 32$ **i** $x^2 - 2x - 35$

5 WORKED EXAMPLES 18, 19 Factorise the following expressions.

a $2x^2 + 14x - 36$ **b** $4x^2 - 68x + 240$
c $-5x^2 - 80x + 75$ **d** $-3x^2 + 15x + 72$

6 WORKED EXAMPLES 20, 21 Factorise the following expressions.

a $2x^2 + 3x + 1$ **b** $3x^2 - 13x + 12$ **c** $2x^2 + 19x - 10$
d $5x^2 - 11x + 6$ **e** $4x^2 + 3x - 10$ **f** $6x^2 - 19x + 10$

7 WORKED EXAMPLE 22 Solve for x in these quadratic equations.

a $(x + 3)(x - 1) = 0$ **b** $(x - 1)(x + 8) = 0$ **c** $x(x - 6) = 0$
d $(x - 2)(4x + 1) = 0$ **e** $(3x + 2)(x + 5) = 0$ **f** $(7x - 4)(9x - 1) = 0$

8 WORKED EXAMPLE 22 Solve for x in these quadratic equations.

a $x^2 + 7x + 6 = 0$ **b** $x^2 + 2x - 63 = 0$ **c** $x^2 - 12x + 20 = 0$
d $x^2 - x = 30$ **e** $3x^2 - 21x = -36$ **f** $-2x^2 + 14x + 16 = 0$

9 WORKED EXAMPLE 22 Solve for x in these quadratic equations.

a $3x^2 - 11x - 4 = 0$ **b** $2x^2 + 13x + 15 = 0$ **c** $6x^2 - 11x - 2 = 0$

10 WORKED EXAMPLE 22 Solve for x in these quadratic equations.

a $x^2 + 28x + 75 = 0$ **b** $x^2 + 6x + 9 = 0$ **c** $-4x - 96 = -x^2$
d $x^2 - 81 = 0$ **e** $x^2 - 18x + 45 = 0$ **f** $4x^2 - 8x = 0$
g $x^2 + x = 42$ **h** $4x^2 + 19x - 5 = 0$ **i** $x^2 - 36 = -9x$
j $x^2 - 5 = 0$ **k** $-3x^2 - 24x + 99 = 0$ **l** $4x^2 - 4x - 3 = 0$

Calculator-free

11 (4 marks) A rectangle has width $(x - 3)$ m and length $(x + 6)$ m.

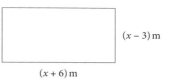

a Show that the area of the rectangle equals $x^2 + 3x - 18$ m². (1 mark)

b If the area of the rectangle is 36 m², find the value of x and, hence, the length and width of the rectangle. (3 marks)

12 (5 marks) The price of a melon is $2 more than the price of a mango. Let x be the price of a mango.

a Write an expression for the price of a melon in terms of x. (1 mark)

Amir wants to spend $12 on one type of fruit.

b Write an expression for the number of mangos Amir can buy for $12, in terms of x. (1 mark)

c Write an expression for the number of melons Amir can buy for $12, in terms of x. (1 mark)

d If, for $12, Amir can buy 3 more mangos than he can buy melons, find the price of each fruit. (2 marks)

Calculator-assumed

13 (5 marks) A ball is projected vertically upwards. At time t seconds, its height, s metres, is modelled by the equation $s = 16t - 5t^2 + 2$.

a What is the initial height? (1 mark)

b Find the two times that the ball is at a height of 14 m. (2 marks)

c How many seconds does it take for the ball to fall back to the initial height? (2 marks)

14 (5 marks) The triangular number sequence 1, 3, 6, 10, 15… counts the number of dots in consecutive equilateral triangles, where the number of rows increases by one each time.

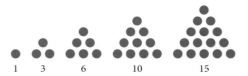

The formula for the number of dots, d, in the nth triangle is $d = \dfrac{n(n + 1)}{2}$. For example, when $n = 4$, $d = 10$.

a How many dots are in the 20th triangle? (1 mark)

A triangle at an unknown position has 36 dots.

b i Find a quadratic equation, in terms of n, to represent this idea. Write the equation in the form $n(n + 1) = k$, where k is a positive integer. (1 mark)

ii Rearrange the above equation to be in the form $an^2 + bn + c = 0$, where a, b and c are integers. (1 mark)

iii Determine the position of the triangle with 36 dots by solving the equation for n. (2 marks)

Video playlist
Solving quadratic equations by completing the square

Puzzle
Completing the square order activity

3.6 Solving quadratic equations by completing the square

We have, thus far, learnt to solve quadratic equations by factorising and then applying the null factor law. However, the trinomial in $ax^2 + bx + c = 0$ form is not always easily factorised. If solutions exist, they can always be found by **completing the square** and solving for x.

Completing the square

Completing the square manufactures a perfect square within a quadratic expression, as follows:

$$x^2 + bx + c$$
$$= x^2 + bx + \left(\frac{b}{2}\right)^2 - \left(\frac{b}{2}\right)^2 + c$$
$$= \underbrace{\left(x + \frac{b}{2}\right)^2}_{\text{Perfect square}} - \left(\frac{b^2}{4} - c\right)$$

Completing the square on $x^2 + bx + c$

To complete the square:

1. Halve b and square the result. $\quad \left(\dfrac{b}{2}\right)^2$

2. Add this value after the x term. Subtract as well to keep the expression unchanged. $\quad x^2 + bx + \left(\dfrac{b}{2}\right)^2 - \left(\dfrac{b}{2}\right)^2 + c$

3. Place brackets around the first three terms and simplify the terms outside the brackets. $\quad \left(x^2 + bx + \left(\dfrac{b}{2}\right)^2\right) - d$

4. Factorise to a perfect square. The second term in the linear factor is half of b. $\quad \left(x + \dfrac{b}{2}\right)^2 - d$

To complete the square on $ax^2 + bx + c$, where $a \neq 1$, first take out a factor of a from all three terms. Then complete the square within the brackets:

$$ax^2 + bx + c$$
$$= a\left(x^2 + \frac{b}{a}x + \frac{c}{a}\right)$$
$$= a\left(x^2 + \frac{b}{a}x + \left(\frac{b}{2a}\right)^2 - \left(\frac{b}{2a}\right)^2 + \frac{c}{a}\right)$$
$$= a\left(\left(x + \frac{b}{2a}\right)^2 + \left(\frac{c}{a} - \frac{b^2}{4a^2}\right)\right)$$
$$= a\left(x + \frac{b}{2a}\right)^2 + \left(c - \frac{b^2}{4a}\right)$$

WORKED EXAMPLE 23 — Completing the square where $a = 1$

Express $x^2 - 4x + 7$ in the form $a(x - h)^2 + k$ by first completing the square.

Steps	Working
1 Halve the coefficient of the x term, then square it.	The x term is $-4x$. Half of -4 is -2. -2 squared is 4.
2 Add this value after the x term. Subtract this value as well to keep the expression unchanged.	$x^2 - 4x + 7$ $= x^2 - 4x + 4 - 4 + 7$
3 Place brackets around the first three terms and simplify the remaining constant terms.	$= (x^2 - 4x + 4) + 3$
4 Factorise the first three terms using $(a + b)^2 = a^2 + 2ab + b^2$. The second term in the brackets will be half the x term.	The x term is $-4x$. Half of -4 is -2. $(x^2 - 4x + 4) + 3$ $= (x - 2)^2 + 3$

WORKED EXAMPLE 24 — Completing the square where $a \neq 1$

Express $2x^2 + 3x + 8$ in the form $a(x - h)^2 + k$ by completing the square.

Steps	Working	
1 As $a \neq 1$, first take out a factor of a from all three terms.	$2x^2 + 3x + 8$	
2 Halve the coefficient of the x term, then square it.	$= 2\left(x^2 + \dfrac{3}{2}x + 4\right)$	The x term is $\dfrac{3}{2}x$. Half of $\dfrac{3}{2}$ is $\dfrac{3}{4}$. $\dfrac{3}{4}$ squared is $\dfrac{9}{16}$.
3 Add this value after the x term. Subtract this value as well to keep the expression unchanged.	$2\left(x^2 + \dfrac{3}{2}x + 4\right)$ $= 2\left(x^2 + \dfrac{3}{2}x + \dfrac{9}{16} - \dfrac{9}{16} + 4\right)$	
4 Place brackets around the first three terms and simplify the remaining constant terms.	$= 2\left[\left(x^2 + \dfrac{3}{2}x + \dfrac{9}{16}\right) - \dfrac{9}{16} + \dfrac{64}{16}\right]$ $= 2\left[\left(x^2 + \dfrac{3}{2}x + \dfrac{9}{16}\right) + \dfrac{55}{16}\right]$	
5 Factorise the first three terms using $(a + b)^2 = a^2 + 2ab + b^2$. The second term in the brackets will be half the x term.	$= 2\left[\left(x + \dfrac{3}{4}\right)^2 + \dfrac{55}{16}\right]$	The x term is $\dfrac{3}{2}x$. Half of $\dfrac{3}{2}$ is $\dfrac{3}{4}$.
6 Expand the a term, so the answer is in the form $a(x - h)^2 + k$.	$= 2\left(x + \dfrac{3}{4}\right)^2 + \dfrac{55}{8}$	

Solving by completing the square

An expression in the form $\left(x + \dfrac{b}{2}\right)^2 = p$ can be solved by taking the square root of both sides, if $p > 0$, and rearranging for x. The next step is $x + \dfrac{b}{2} = \pm\sqrt{p}$ because p has both a positive and negative square root. On its own, the symbol $\sqrt{}$ refers only to the positive square root.

If $p < 0$, there is no solution as negative numbers have no square root in the real number system.

WORKED EXAMPLE 25 — Solving for x by first completing the square

Solve the following equations for x. If there are no solutions, write *no real solutions*.

a $2x^2 - 12x = -1$

b $x^2 - 6x + 109 = 0$

Steps	Working	
a 1 Write the equation in general form $ax^2 + bx + c = 0$.	$2x^2 - 12x + 1 = 0$	
2 Divide both sides by the coefficient of x^2. Recall that $\dfrac{0}{a} = 0$.	$x^2 - 6x + \dfrac{1}{2} = 0$	
3 Halve the coefficient of the x term, then square it.	$x^2 - 6x + 9 - 9 + \dfrac{1}{2} = 0$	The x term is $-6x$. Half of -6 is -3. -3 squared is 9.
4 Add this value after the x term. Subtract this value as well to keep the expression unchanged.		
5 Place brackets around the first three terms and simplify the remaining constant terms.	$(x^2 - 6x + 9) - 9 + \dfrac{1}{2} = 0$	
6 Factorise the first three terms using $(a + b)^2 = a^2 + 2ab + b^2$. The second term in the brackets will be half the x term.	$(x - 3)^2 - \dfrac{17}{2} = 0$	The x term is $-6x$. Half of -6 is -3.
7 Rearrange to get x by itself.	$(x - 3)^2 - \dfrac{17}{2} = 0$	
	$(x - 3)^2 = \dfrac{17}{2}$	
Exam hack — Remember the $\pm$ in front of the $\sqrt{}$.	$x - 3 = \pm\sqrt{\dfrac{17}{2}}$	
	$x - 3 = \pm\sqrt{\dfrac{17}{2}} \times \dfrac{\sqrt{2}}{\sqrt{2}} = \pm\dfrac{\sqrt{34}}{2}$	
	$x = 3 \pm \dfrac{\sqrt{34}}{2}$	
b 1 Complete the square.	$x^2 - 6x + 109 = 0$	
	$(x^2 - 6x + 9) - 9 + 109 = 0$	
	$(x - 3)^2 + 100 = 0$	
2 Rearrange to get x by itself.	$(x - 3)^2 = -100$	
	$(x - 3)^2 = \pm\sqrt{-100}$	
	There is no square root of -100 in the real number system.	
	'*no real solutions*'	

EXERCISE 3.6 Solving quadratic equations by completing the square

ANSWERS p. 453

Recap

1 Factorise the following expressions.

   a $x^2 + 6x + 9$ **b** $x^2 - 18x + 81$ **c** $4x^2 - 40x + 100$

2 Solve the following equations for x.

   a $(x-8)(x+2) = 0$ **b** $(2x+3)(x-9) = 0$ **c** $x\left(x + \dfrac{3}{2}\right) = 0$

Mastery

3 **WORKED EXAMPLE 23** Write each of these expressions in the form $(x-h)^2 + k$ by completing the square.

   a $x^2 + 6x + 17$ **b** $x^2 + 8x + 14$ **c** $x^2 - 2x - \dfrac{5}{2}$

   d $x^2 + x - 1$ **e** $x^2 - 5x + 7$ **f** $x^2 - 3x - 11$

4 **WORKED EXAMPLE 24** Write each of these expressions in the form $a(x-h)^2 + k$ by completing the square.

   a $2x^2 - 20x + 54$ **b** $3x^2 - 6x - 7$ **c** $-2x^2 - 36x - 164$

   d $-4x^2 + 32x - 70$ **e** $2x^2 - 14x + 22$ **f** $-5x^2 - 15x - 14$

5 **WORKED EXAMPLE 25** Solve the following equations for x. If there are no solutions, write *no real solutions*.

   a $(x-12)^2 - 36 = 0$ **b** $(x+5)^2 - 1 = 0$ **c** $(x+9)^2 + 16 = 0$

   d $\left(x + \dfrac{11}{2}\right)^2 - 81 = 0$ **e** $3(x-4)^2 - 21 = 0$ **f** $\left(x + \dfrac{5}{2}\right)^2 - 24 = 0$

6 **WORKED EXAMPLE 25** Solve the following equations for x. If there are no solutions, write *no real solutions*.

   a $x^2 + 6x - 10 = 0$ **b** $x^2 - 4x + 5 = 0$ **c** $x^2 + 7x - 3 = 0$

   d $3x^2 - 30x - 33 = 0$ **e** $x^2 - x + 11 = 0$ **f** $-x^2 - 5x + 5 = 0$

Calculator-free

7 (3 marks) Consider the equation $x^2 + bx + c = 0$.

   a By completing the square, rewrite the equation in the form $\left(x + \dfrac{b}{2}\right)^2 - p = 0$.   (1 mark)

   b Hence, state the condition, in terms of b and c, for there to be no solution to the equation.   (2 marks)

Calculator-assumed

8 (4 marks)

   a Write the expression $-2x^2 + 10x - 5$ in the form $a(x-h)^2 + k$ where a, h and k are real numbers.   (2 marks)

   b Hence, solve the equation $-2x^2 + 10x - 5 = 0$ for x.   (2 marks)

Video playlist
The quadratic formula and the discriminant

Puzzle
The quadratic formula

3.7 The quadratic formula and the discriminant

The quadratic formula

The **quadratic formula** can be used to solve any quadratic equation in the form $ax^2 + bx + c = 0$. It is a shortcut for solving an equation by completing the square. The derivation comes from completing the square and rearranging for x.

$$ax^2 + bx + c = 0$$

$$x^2 + \frac{b}{a}x + \frac{c}{a} = 0$$

$$x^2 + \frac{b}{a}x + \frac{b^2}{4a^2} - \frac{b^2}{4a^2} + \frac{c}{a} = 0$$

$$\left(x + \frac{b}{2a}\right)^2 - \frac{b^2}{4a^2} + \frac{c}{a} = 0$$

$$\left(x + \frac{b}{2a}\right)^2 = \frac{b^2 - 4ac}{4a^2}$$

$$x = -\frac{b}{2a} \pm \sqrt{\frac{b^2 - 4ac}{4a^2}}$$

$$x = \frac{-b \pm \sqrt{b^2 - 4ac}}{2a}$$

> **The quadratic formula**
>
> For a quadratic equation $ax^2 + bx + c = 0$, the solutions are
> $$x = \frac{-b \pm \sqrt{b^2 - 4ac}}{2a}$$

The $\pm$ symbol denotes two solutions: $x = \frac{-b + \sqrt{b^2 - 4ac}}{2a}$ and $x = \frac{-b - \sqrt{b^2 - 4ac}}{2a}$.

To solve quadratic equations using the quadratic formula:

1. Write the equation in general form: $ax^2 + bx + c = 0$.
2. Identify the values of a, b and c and substitute into the quadratic formula.
3. Simplify.

WORKED EXAMPLE 26 — Using the quadratic formula

Solve these equations using the quadratic formula.

a $2x^2 = 2x + 3$

b $-3x^2 + x - 7 = 0$

Steps	Working
a 1 Write in general form: $ax^2 + bx + c = 0$.	$2x^2 = 2x + 3$ $2x^2 - 2x - 3 = 0$
2 Identify a, b and c and substitute into the quadratic formula: $$x = \frac{-b \pm \sqrt{b^2 - 4ac}}{2a}$$	$a = 2,\ b = -2,\ c = -3$ $$x = \frac{-(-2) \pm \sqrt{(-2)^2 - 4(2)(-3)}}{2(2)}$$
3 Simplify.	$= \dfrac{2 \pm \sqrt{28}}{4}$ $= \dfrac{2 \pm \sqrt{4 \times 7}}{4}$ $= \dfrac{2 \pm 2\sqrt{7}}{4}$ $= \dfrac{1 \pm \sqrt{7}}{2}$ $x = \dfrac{1 + \sqrt{7}}{2},\ \dfrac{1 - \sqrt{7}}{2}$
b 1 Identify a, b and c and substitute into the quadratic formula: $$x = \frac{-b \pm \sqrt{b^2 - 4ac}}{2a}$$	$-3x^2 + x - 7 = 0$ $a = -3,\ b = 1,\ c = -7$ $$x = \frac{-1 \pm \sqrt{(1)^2 - 4(-3)(-7)}}{2(-3)}$$
2 Simplify. If you encounter the square root of a negative number, stop, as there is no square root of a negative number in the real number system. **Exam hack** A solution to a quadratic equation is also known as a '**root**' of the equation.	$= \dfrac{-1 \pm \sqrt{1 - 84}}{-6}$ $= \dfrac{-1 \pm \sqrt{-83}}{-6}$ Stop simplifying. There is no real number square root of −83. Therefore, there are no real solutions to this quadratic equation.

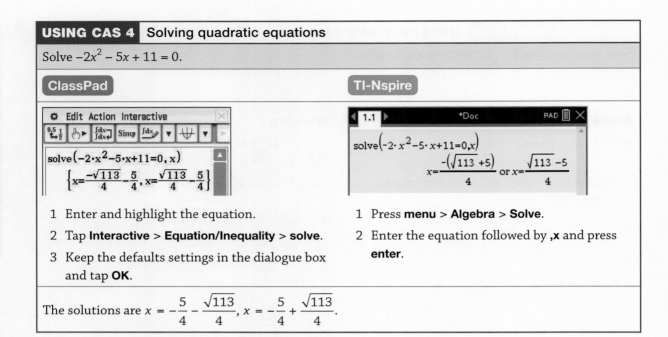

The solutions are $x = -\dfrac{5}{4} - \dfrac{\sqrt{113}}{4}$, $x = -\dfrac{5}{4} + \dfrac{\sqrt{113}}{4}$.

The discriminant

There are not always two solutions to a quadratic equation. The number of solutions depends on the value of $b^2 - 4ac$, which appears under the square root in the quadratic formula $x = \dfrac{-b \pm \sqrt{b^2 - 4ac}}{2a}$.

- When $b^2 - 4ac$ is positive, there are two distinct solutions due to the $\pm$.
- When $b^2 - 4ac$ is zero, there is one solution because $\pm\sqrt{0} = 0$ and the quadratic formula simplifies to $x = \dfrac{-b}{2a}$.
- When $b^2 - 4ac$ is negative, there are no solutions because $\sqrt{b^2 - 4ac}$ is the square root of a negative number, which has no value in the real number system.

The expression $b^2 - 4ac$ is so useful that it has its own name: the **discriminant**. It is represented by the Greek letter delta, Δ.

> **The discriminant**
>
> For a quadratic equation $ax^2 + bx + c = 0$, the discriminant is $\Delta = b^2 - 4ac$.
> - If $\Delta > 0$, the quadratic equation has two distinct real solutions.
> - If $\Delta = 0$, the quadratic equation has one real solution.
> - If $\Delta < 0$, the quadratic equation has no real solutions.

If Δ is a perfect square, then $\sqrt{\Delta}$ simplifies to a non-negative integer and any solutions will be rational. If Δ is not a perfect square, any solutions will be irrational.

WORKED EXAMPLE 27 — The discriminant

For each of these quadratic equations, use the discriminant to
  i determine the number of solutions in the real number system
  ii if there are real solutions, state whether they are rational or irrational.

a $-x^2 + 5x + 6$

b $x^2 - 7x + \dfrac{49}{4}$

Steps			Working
a	**i**	1 Identify a, b and c.	$-x^2 + 5x + 6$ $a = -1,\ b = 5,\ c = 6$
		2 Substitute a, b and c into the expression for the discriminant and simplify.	$\Delta = b^2 - 4ac$ $= (5)^2 - 4(-1)(6)$ $= 25 + 24$ $= 49$
		3 Interpret the discriminant for the number of solutions.	$\Delta > 0$ Therefore, there are two real solutions to the quadratic equation.
	ii	Interpret the discriminant for the type of solutions.	$\Delta = 49$ The discriminant is a perfect square. Therefore, the solutions will be rational.
b	**i**	1 Identify a, b and c.	$x^2 - 7x + \dfrac{49}{4}$ $a = 1,\ b = -7,\ c = \dfrac{49}{4}$
		2 Substitute a, b and c into the expression for the discriminant and simplify.	$\Delta = b^2 - 4ac$ $= (-7)^2 - 4(1)\left(\dfrac{49}{4}\right)$ $= 49 - 49$ $= 0$
		3 Interpret the discriminant for the number of solutions.	$\Delta = 0$ Therefore, there is one solution to the quadratic equation.
	ii	Interpret the discriminant for the type of solutions.	$\Delta = 0$ The discriminant is a perfect square. Therefore, the solution will be rational.

The discriminant and parameters for a, b and c

The number of solutions to a quadratic equation $ax^2 + bx + c = 0$ depends on the values of a, b and c. If unknown, these values are called **parameters**. To determine which value(s) of the parameters lead to zero, one or two solutions, use the discriminant.

WORKED EXAMPLE 28 — The discriminant and parameters for a, b and c

a Find the value(s) of k for which $5x^2 - 2x + k = 0$ has one solution.

b Find the value(s) of p for which $px^2 + 9x - 1 = 0$ has no solutions.

Steps	Working
a 1 Find the discriminant.	$5x^2 - 2x + k = 0$ $\Delta = b^2 - 4ac$ $= (-2)^2 - 4(5)(k)$ $= 4 - 20k$
2 Write the condition on the discriminant for the desired number of solutions.	For one solution, $\Delta = 0$, therefore, $4 - 20k = 0$.
3 Solve for the unknown parameter.	$4 - 20k = 0$ $4 = 20k$ $k = \dfrac{1}{5}$
4 Interpret the result.	There will be one solution to the equation $5x^2 - 2x + k = 0$ when $k = \dfrac{1}{5}$.
b 1 Find the discriminant.	$px^2 + 9x - 1 = 0$ $\Delta = b^2 - 4ac$ $= (9)^2 - 4(p)(-1)$ $= 81 + 4p$
2 Write the condition on the discriminant for the desired number of solutions.	For no solutions, $\Delta < 0$, therefore, $81 + 4p < 0$.
3 Solve for the unknown parameter.	$81 + 4p < 0$ $4p < -81$ $p < -\dfrac{81}{4}$
4 Interpret the result.	There will be no solutions to the equation $px^2 + 9x - 1 = 0$ when $p < -\dfrac{81}{4}$.

USING CAS 5 — Solving with the discriminant

Find the value(s) of m for which $5x^2 - mx - x + 7 = 0$ has two solutions.

1 Rearrange to be in the form $ax^2 + bx + c$.	$5x^2 - (m + 1)x + 7 = 0$
2 Find the discriminant.	$\Delta = b^2 - 4ac$ $= (-(m+1))^2 - 4(5)(7)$ $= (m+1)^2 - 140$
3 Write the condition on the discriminant for the desired number of solutions.	For two solutions, $\Delta > 0$. $(m+1)^2 - 140 > 0$

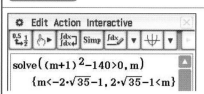

ClassPad	TI-Nspire
1 Enter and highlight the inequation. 2 Tap **Interactive** > **Equation/Inequality** > **solve**. 3 In the dialogue box, change the **Variable:** field to **m** and tap **OK**.	1 Press **menu** > **Algebra** > **Solve**. 2 Enter the inequation followed by **,m** and press **enter**.

The solutions are $m < -1 - 2\sqrt{35}$, $m > -1 + 2\sqrt{35}$.

EXERCISE 3.7 The quadratic formula and the discriminant ANSWERS p. 453

Recap

1 a Write $x^2 - 10x + 3$ in the form $(x - h)^2 + k$ by completing the square.

b Hence, solve the equation $x^2 - 10x + 3 = 0$.

2 Find the gradient of this line.

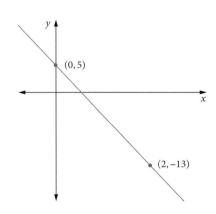

Mastery

3 WORKED EXAMPLE 26 For each of these equations

　　i write the values of a, b and c, where $ax^2 + bx + c = 0$

　　ii hence, solve the equation using the quadratic formula.

a $x^2 + 4x - 11 = 0$　　**b** $4x^2 - 5x - 3 = 0$　　**c** $-3x^2 + 6x + 2 = 0$

d $-5x^2 + 10x - 3 = 0$　　**e** $-x^2 - x + 11 = 0$　　**f** $2x^2 + 5x - 3 = 0$

g $-4x^2 - 6x + 1 = 0$　　**h** $2x^2 + 7x - 4 = 0$　　**i** $11x^2 + 9x - 1 = 0$

4 Using CAS 4 Solve the following quadratic equations.

a $-x^2 - 10x - 7 = 0$　　**b** $6x^2 - 11x = 2$　　**c** $3x^2 - 9x - 4 = 0$

5 WORKED EXAMPLE 27 For each of the following equations

　　i find the discriminant

　　ii hence, state the number of solutions in the real number system.

a $-2x^2 + 9x - 2 = 0$　　**b** $7x^2 - 5x = 0$　　**c** $3x^2 + x + 10 = 0$

d $4x^2 - 12x + 9 = 0$　　**e** $6x^2 - x + 2 = 0$　　**f** $3x^2 + 2x + 7 = 0$

g $-x^2 + 6x - 9 = 0$　　**h** $5x^2 - 2x - 5 = 0$　　**i** $-6x^2 - 4x - 5 = 0$

6 WORKED EXAMPLE 28

  a Find the value(s) of k for which $x^2 - 7x + k = 0$ has one solution.
  b Find the value(s) of p for which $px^2 + 3x + 1 = 0$ has two solutions.
  c Find the value(s) of m for which $-6x^2 - 2x + 7m = 0$ has no solutions.

7 Using CAS 5

  a Find the value(s) of q for which $x^2 + qx + 30 = 0$ has one solution.
  b Find the value(s) of r for which $5x^2 + 4rx + 1 = 0$ has two solutions.
  c Find the value(s) of d for which $10x^2 - dx + 2 - d = 0$ has no solutions.

Calculator-free

8 (5 marks) Consider the equation $x^2 - 3x = t$.

  a Use the quadratic formula to solve for x in terms of t. (2 marks)
  b Find the solution(s) for x when $t = 1$. (1 mark)
  c Find the value of t that produces the solutions $x = -1$ and 4. (2 marks)

Calculator-assumed

9 (4 marks) Consider the equation $-3x^2 + px - p + 3 = 0$.

  a Find the discriminant of the equation in terms of p. (2 marks)
  b Hence, show that the equation will always have two solutions. (2 marks)

3.8 Sketching parabolas from turning point form

A **quadratic relationship** between two variables, x and y, can be expressed by the function $y = ax^2 + bx + c$. The graph of a quadratic relationship is called a **parabola**.

The **quadratic function** $y = ax^2 + bx + c$ may be transposed into two other forms.

- General form: $y = ax^2 + bx + c$
- Intercept form: $y = a(x - u)(x - v)$
- Turning point form: $y = a(x - h)^2 + k$

In this section, we will learn how to sketch quadratic relationships in turning point form.

Turning point form

Any parabola can be expressed as a transformation of the basic parabola $y = x^2$.

The graph of $y = x^2$ can be determined from a table of values. Think of the curve as representing infinitely many (x, y) points.

x	−6	−4	−2	0	2	4	6
y	36	16	4	0	4	16	36

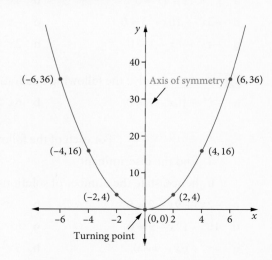

Equation	Transformation from $y = x^2$	Example parabolas
$y = ax^2$, where $a > 0$	vertical dilation by a scale factor of a If $a > 1$, every y value is dilated away from the x-axis. If $0 < a < 1$, every y value is dilated towards the x-axis.	
$y = -x^2$	reflection in the x-axis	
$y = (x - h)^2$	horizontal translation of h units If $h > 0$, the translation is to the right. If $h < 0$, the translation is to the left.	
$y = x^2 + k$	vertical translation of k units If $k > 0$, the translation is up. If $k < 0$, the translation is down.	

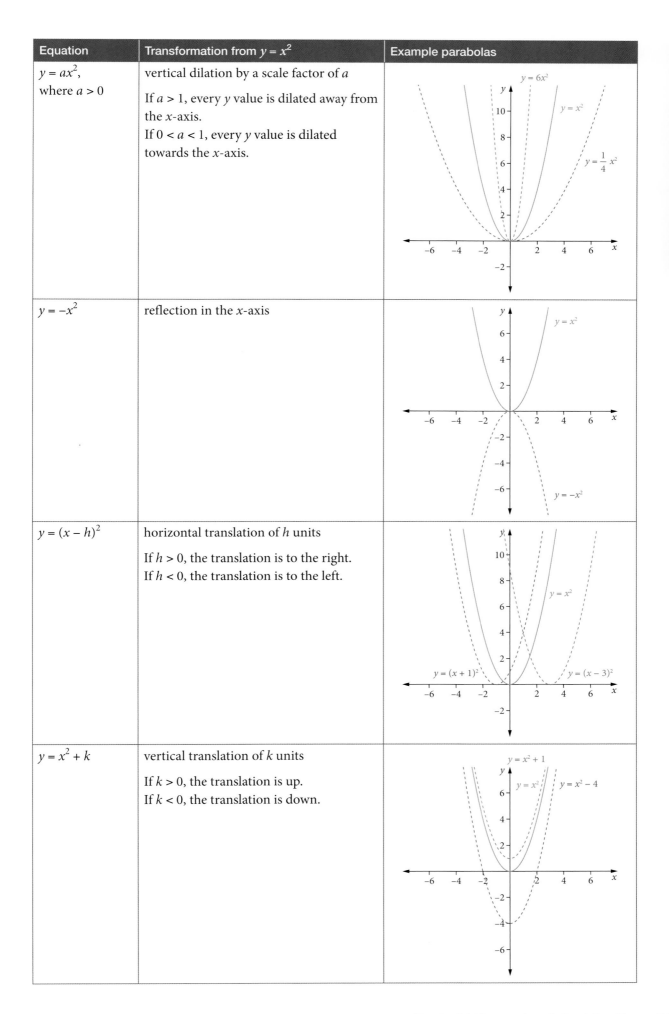

The transformations are all visible in turning point form: $y = a(x - h)^2 + k$. The **turning point**, or vertex, of a parabola is (h, k). If $a > 0$, the turning point is a minimum. If $a < 0$, the turning point is a maximum.

The **axis of symmetry** is an imaginary vertical line, $x = h$, running through the turning point. An axis of symmetry is a line about which a shape is symmetrical.

> **Turning point form**
>
> The equation of a parabola can be expressed in turning point form: $y = a(x - h)^2 + k$.
> It contains the following information:
> - the coordinates of the turning point: (h, k)
> - the nature of the turning point and the shape of the graph:
>   - turning point is a minimum if $a > 0$
>   - turning point is a maximum if $a < 0$
> - the axis of symmetry: $x = h$.

> **Turning point form – transformations**
>
> The graph of $y = a(x - h)^2 + k$ can be obtained from the graph of $y = x^2$ with the following sequence of transformations:
> 1. vertical dilation by a scale factor of $|a|$
> 2. if a is negative, a reflection in the x-axis
> 3. horizontal translation h units right (if $h > 0$) or $|h|$ units left (if $h < 0$)
> 4. vertical translation k units up (if $k > 0$) or $|k|$ units down (if $k < 0$).

 Exam hack

If h is negative, the translation is to the left, but this will show as a *positive* number in the turning point form. For example, if $h = -3$, then $y = a(x - (-3))^2 + k = a(x + 3)^2 + k$.

WORKED EXAMPLE 29 | Interpreting turning point form

For the parabola represented by each quadratic function, find
- i the coordinates of the turning point
- ii whether the turning point is a maximum or a minimum
- iii the axis of symmetry.

a $y = 5(x - 3)^2 + 7$

b $y = -2(x + 1)^2 - \dfrac{11}{2}$

Steps	Working
a i Find h and k and write them in coordinate form.	The turning point form is $y = a(x - h)^2 + k$. $y = 5(x - 3)^2 + 7$ $h = 3, k = 7$ The turning point is $(3, 7)$.
ii Determine whether there has been a reflection in the x-axis using the sign of a.	The value of a is 5, which is positive. There has not been a reflection in the x-axis. Therefore, the turning point $(3, 7)$ is a minimum.
iii The axis of symmetry is the line going through the x-coordinate of the turning point.	The x-coordinate of the turning point is 3. The axis of symmetry is $x = 3$.

b i Find h and k and write them in coordinate form.

The turning point form is $y = a(x - h)^2 + k$.
$$y = -2(x + 1)^2 - \frac{11}{2}$$
$$= -2(x - -1)^2 + -\frac{11}{2}$$
$$h = -1, k = -\frac{11}{2}$$
The turning point is $\left(-1, -\frac{11}{2}\right)$.

ii Determine whether there has been a reflection in the x-axis using the sign of a.

The value of a is -2, which is negative. There has been a reflection in the x-axis.

Therefore, the turning point $\left(-1, -\frac{11}{2}\right)$ is a maximum.

iii The axis of symmetry is the line going through the x-coordinate of the turning point.

The x-coordinate of the turning point is -1.
The axis of symmetry is $x = -1$.

> **Exam hack**
> Always check if a is positive or negative. A happy curve ☺ comes from a positive a. A sad curve ☹ comes from a negative a.

Sketching parabolas from turning point form

To sketch the parabola, $y = a(x - h)^2 + k$, find these three key features and label them on the graph:
- the turning point
- the y-intercept
- any x-intercepts, if they exist.

The turning point

1 Read the coordinates of the turning point (h, k) from the equation.
2 Note the shape of the graph. If a is positive, the turning point is a minimum. If a is negative, the turning point is a maximum.

The y-intercept

1 Let $x = 0$. Write $y = a(0 - h)^2 + k$, using the specific values for a, h and k.
2 Simplify the right-hand side to find the value of y.

The x-intercepts

1 Let $y = 0$. Write $a(x - h)^2 + k = 0$, using the specific values for a, h and k.
2 Solve for x.

WORKED EXAMPLE 30 — Finding x-intercepts in turning point form

Find the x-intercept(s) of a parabola with the equation

a $y = 4\left(x - \dfrac{1}{2}\right)^2 - 7$

b $y = 4(x - 2)^2 + 12$

Steps	Working
a 1 To find x-intercepts, let $y = 0$.	$4\left(x - \dfrac{1}{2}\right)^2 - 7 = 0$
2 Rearrange to get the $(x - h)^2$ expression by itself.	$4\left(x - \dfrac{1}{2}\right)^2 = 7$ $\left(x - \dfrac{1}{2}\right)^2 = \dfrac{7}{4}$
3 Take the square root of both sides. Remember to include the + and the −.	$x - \dfrac{1}{2} = \pm\sqrt{\dfrac{7}{4}}$

4 Simplify and rearrange to get x by itself.	$x - \dfrac{1}{2} = \pm\dfrac{\sqrt{7}}{2}$
	  $x = \dfrac{1 \pm \sqrt{7}}{2}$
	  There are two x-intercepts:
	  $x = \dfrac{1 + \sqrt{7}}{2}$ and $x = \dfrac{1 - \sqrt{7}}{2}$.

b 1 To find x-intercepts, let $y = 0$.	$4(x - 2)^2 + 12 = 0$
2 Rearrange to get the $(x - h)^2$ expression by itself.	$4(x - 2)^2 = -12$ $(x - 2)^2 = -3$
3 Take the square root of both sides. Remember to include the + and the −.	$x - 2 = \pm\sqrt{-3}$
4 Stop solving as there is no square root of a negative number in $\mathbb{R}$. There are no real solutions to this equation.	There is no real value for $\sqrt{-3}$. Therefore, this parabola has no x-intercepts.

🔓 Exam hack

The x-intercepts of any graph can always be found by letting $y = 0$ and solving for x.

Nelson WAmaths Mathematics Methods 11

WORKED EXAMPLE 31 Sketching parabolas from turning point form

For the parabola of each equation
  i state the coordinates of the turning point and whether it is a maximum or minimum
  ii find the y-intercept
  iii find any x-intercepts, if they exist
  iv sketch the graph.

a $y = \dfrac{3}{4}(x+1)^2 - 3$ **b** $y = -(x-4)^2 - 7$

Steps	Working
a i 1 Find h and k and write them in coordinate form.	$y = \dfrac{3}{4}(x+1)^2 - 3$ $h = -1, k = -3$ The turning point is $(-1, -3)$.
2 Interpret the sign of a.	a is positive. There has not been a reflection in the x-axis. Therefore, the turning point is a minimum.
ii 1 To find the y-intercept, let $x = 0$.	$y = \dfrac{3}{4}(0+1)^2 - 3$
2 Simplify the right-hand side.	$y = \dfrac{3}{4} - 3$ $= \dfrac{3}{4} - \dfrac{12}{4}$ $= -\dfrac{9}{4}$ The y-intercept is $\left(0, -\dfrac{9}{4}\right)$.
iii 1 To find any x-intercepts, let $y = 0$.	$\dfrac{3}{4}(x+1)^2 - 3 = 0$
2 Solve for x by rearranging. Include the $\pm$ in front of the $\sqrt{\ }$.	$\dfrac{3}{4}(x+1)^2 = 3$ $(x+1)^2 = 4$ $(x+1) = \pm\sqrt{4}$ $x + 1 = \pm 2$ $x = -1 \pm 2$ $x = -3, 1$ The x-intercepts are $(-3, 0)$ and $(1, 0)$.
iv Label the turning point, the y-intercept and the x-intercepts. Note the shape of the graph.	The graph has not been reflected in the x-axis. 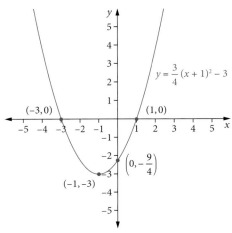

b i 1 Find h and k and write them in coordinate form.		$y = -(x-4)^2 - 7$ $h = 4, k = -7$ The turning point is $(4, -7)$.
2 Interpret the sign of a.		a is negative. There has been a reflection in the x-axis. Therefore, the turning point is a maximum.
ii 1 To find the y-intercept, let $x = 0$.		$y = -(0-4)^2 - 7$
2 Simplify the right-hand side.		$y = -16 - 7$ The y-intercept is $(0, -23)$.
iii 1 To find any x-intercepts, let $y = 0$.		$-(x-4)^2 - 7 = 0$
2 Solve for x by rearranging. Include the $\pm$ in front of the $\sqrt{\ }$.		$(x-4)^2 = -7$ $x - 4 = \pm\sqrt{-7}$
3 Stop solving as there is no square root of a negative number in $\mathbb{R}$. There are no solutions to this equation.		There is no real value for $\sqrt{-7}$. Therefore, this parabola has no x-intercepts.
iv Label the turning point and the y-intercept. Note the shape of the graph.		The graph has been reflected in the x-axis.

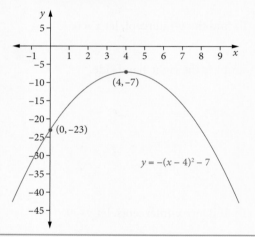

Completing the square to find turning point form

A quadratic function $y = ax^2 + bx + c$ can be transposed into turning point form by completing the square.

WORKED EXAMPLE 32 — Completing the square to find turning point form

Consider the equation $y = x^2 - 10x + 2$.

a Complete the square to find the turning point form.
b Hence, state the coordinates of the turning point.

Steps	Working
a 1 Halve the coefficient of the x term, then square it. Add and subtract this number.	Half of -10 is -5. -5 squared is 25. $y = x^2 - 10x + 25 - 25 + 2$
2 Place brackets around the first three terms and simplify the remaining constant terms.	$= (x^2 - 10x + 25) - 23$
3 Factorise the first three terms. The second term in the brackets will be half the x term.	$y = (x-5)^2 - 23$
b Find h and k and write them in coordinate form.	The turning point form is $y = a(x-h)^2 + k$. $h = 5, k = -23$ The turning point is $(5, -23)$.

EXERCISE 3.8 Sketching parabolas from turning point form

ANSWERS p. 454

Recap

1 Solve the equation $-x^2 - 7x + 11 = 0$ using the quadratic formula.

2 Consider the equation $0 = -2x^2 + 8x - 11$.
   a Find the discriminant.
   b Hence, state the number of solutions to the equation, giving a reason.

Mastery

3 **WORKED EXAMPLE 29** For each parabola, find
   i the coordinates of the turning point
   ii whether the turning point is a maximum or a minimum
   iii the axis of symmetry.
   a $y = 2(x - 7)^2 + 5$
   b $y = (x + 7)^2 + 32$
   c $y = 4(x - 11)^2 - 12$
   d $y = -3(x + 5)^2 - 8$
   e $y = 5(x - 3)^2 + 4$
   f $y = -2(x + 50)^2 - 26$

4 **WORKED EXAMPLE 30** Find the x-intercepts of each parabola.
   a $y = (x + 6)^2 - 49$
   b $y = 3(x - 5)^2 - 27$
   c $y = 3(x - 11)^2 - 4$
   d $y = 2(x - 1)^2 + 12$
   e $y = -3(x - 2)^2 + 75$
   f $y = -5(x - 10)^2 + 9$

5 **WORKED EXAMPLE 31** For each parabola
   i state the coordinates of the turning point and whether it is a maximum or minimum
   ii find the y-intercept
   iii find any x-intercepts, if they exist
   iv sketch the graph.
   a $y = (x + 3)^2 - 25$
   b $y = 2(x + 1)^2 - 32$
   c $y = -(x - 6)^2 + 100$
   d $y = -4(x + 1)^2 + 17$
   e $y = -2(x - 5)^2 - 8$
   f $y = 4(x - 1)^2 - 6$
   g $y = 5(x - 5)^2 + 45$
   h $y = 2(x - 4)^2 - 72$
   i $y = -6(x + 3)^2 + 7$

6 **WORKED EXAMPLES 31–32** For each quadratic relationship
   i write the equation in turning point form
   ii sketch the parabola, labelling all key features.
   a $y = x^2 + 2x - 3$
   b $y = -x^2 - 10x - 16$
   c $y = 3x^2 + 18x - 405$
   d $y = x^2 - 10x + 74$
   e $y = -x^2 + 6x + 1$
   f $y = 2x^2 - 16x + 20$

Calculator-free

7 (6 marks) The sum of the base, b, and height, h, of a triangle is 12 cm.

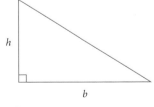

   a Write an expression for the height of the triangle in terms of its base. (1 mark)
   b Hence, find an equation for the area of the triangle in terms of b. (1 mark)
   c Write the equation for the area of the triangle in turning point form. (2 marks)
   d Hence, state the maximum area of the triangle and the base length for which this occurs. (2 marks)

8 (3 marks) Consider the graph of the equation $y = a(x - h)^2 + k$.

   a Find the x-intercepts of the parabola in the form $x = m \pm \sqrt{n}$. (1 mark)

   b Under what condition, in terms of k and a, will there be no x-intercepts? (1 mark)

   c Under what condition, in terms of k and a, will there be one x-intercept? (1 mark)

Calculator-assumed

9 (7 marks) An archway over a bridge is modelled by the equation $y = -x^2 + 26x - 25$. y is the height of the archway above the bridge and x is the horizontal distance from a fixed reference point. Distances are measured in metres.

   a Rewrite the equation in turning point form. (2 marks)

   b The archway is attached to both ends of the bridge, at the x-intercepts of the graph. How wide is the bridge? (2 marks)

   c How tall is the archway? (1 mark)

   d Sketch the graph of the archway for non-negative values of y. (2 marks)

3.9 Sketching parabolas from any form

Video playlist
Sketching parabolas from any form

Worksheets
Quadratic functions

Graphing quadratics

Graphing quadratic functions

A quadratic relationship can be expressed in three forms:
- Intercept form: $y = a(x - u)(x - v)$
- Turning point form: $y = a(x - h)^2 + k$
- General form: $y = ax^2 + bx + c$

Sketching parabolas

To sketch a parabola from any of the forms, find these three key features and label them on the graph:
- the y-intercept
- any x-intercepts, if they exist
- the turning point.

The techniques for sketching parabolas from turning point form were covered in Section 3.8.

Sketching parabolas from intercept form

$y = a(x - u)(x - v)$

The y-intercept

1. Let $x = 0$.
2. Solve for y by simplifying the right-hand side.

The x-intercepts

1. Let $y = 0$.
2. Solve for x using the null factor law: $x = u, v$.

The turning point

1. Find the x-coordinate of the turning point. It is halfway between the two x-intercepts and is equal to their average: $\dfrac{u + v}{2}$.

2. Substitute the x-coordinate into the equation of the parabola to find the y-coordinate of the turning point.

148   Nelson WAmaths Mathematics Methods 11

WORKED EXAMPLE 33 — Sketching parabolas from intercept form

Consider the parabola with the equation $y = -2(x - 3)(x + 6)$.

a Find the y-intercept.
b Find any x-intercepts, if they exist.
c State the coordinates of the turning point and whether it is a maximum or minimum.
d Sketch the graph.

Steps	Working
a To find the y-intercept, let $x = 0$. Simplify the right-hand side.	$y = -2(0 - 3)(0 + 6)$ $y = -2(-3)(6)$ $= 36$ The y-intercept is $(0, 36)$.
b To find any x-intercepts, let $y = 0$. Use the null factor law to solve for x.	$-2(x - 3)(x + 6) = 0$ $(x - 3)(x + 6) = 0$ $x = 3, -6$ The x-intercepts are $(3, 0)$ and $(-6, 0)$.
c 1 The x-coordinate of the turning point is halfway between the x-intercepts.	$\dfrac{3 + (-6)}{2} = -\dfrac{3}{2}$ The x-coordinate of the turning point is $-\dfrac{3}{2}$.
2 Substitute the x-coordinate of the turning point into the initial quadratic function to find the corresponding y-coordinate.	$y = -2\left(-\dfrac{3}{2} - 3\right)\left(-\dfrac{3}{2} + 6\right)$ $= -2\left(-\dfrac{9}{2}\right)\left(\dfrac{9}{2}\right)$ $= \dfrac{81}{2}$ The y-coordinate of the turning point is $\dfrac{81}{2}$. The coordinates of the turning point are $\left(-\dfrac{3}{2}, \dfrac{81}{2}\right)$.
3 Interpret the sign of a.	a is negative. There has been a reflection in the x-axis. Therefore, the turning point is a maximum.
d Label the turning point, the y-intercept and the x-intercepts. Note the shape of the graph.	The graph has been reflected in the x-axis.

Exam hack

The solutions to 'quadratic expression = 0' are the x-intercepts of 'y = quadratic expression'. They are also called 'roots' or 'zeros'.

Sketching parabolas from general form

$y = ax^2 + bx + c$ is the general form of a quadratic relationship.

The y-intercept

1. Let $x = 0$.
2. Solve for y by simplifying the right-hand side. This simplifies to $y = c$.

The x-intercepts

1. Let $y = 0$.
2. Solve for x with one of the following methods:
   - factorise into intercept form and solve using the null factor law
   - solve using the quadratic formula
   - complete the square and rearrange for x.

Exam hack

When solving for x in general form, first try to factorise and solve with the null factor law. If it is possible, it is the easiest method.

The turning point

1. Find the x-coordinate of the turning point with one of these methods:
   - $x = -\dfrac{b}{2a}$, the axis of symmetry
   - halfway between the x-intercepts (if they have already been found).

2. Substitute the x-coordinate into the equation of the parabola to find the y-coordinate of the turning point. The x-coordinate of the turning point has the same value as the axis of symmetry, $x = -\dfrac{b}{2a}$, which has been derived from the quadratic formula. It is halfway between the x-intercepts: $x = \dfrac{-b}{2a} \pm \dfrac{\sqrt{b^2 - 4ac}}{2a}$.

WORKED EXAMPLE 34 | Sketching parabolas from general form

For each parabola
  i   find the y-intercept
  ii find any x-intercepts, if they exist
  iii state the coordinates of the turning point and whether it is a maximum or minimum
  iv sketch the graph.

a $y = 3x^2 - 6x - 72$

b $y = -x^2 - 10x + 6$

Steps	Working
a i 1 To find the y-intercept, let $x = 0$.	$y = 3x^2 - 6x - 72$ $y = 3(0) - 6(0) - 72$
2 Simplify the right-hand side.	$= -72$ The y-intercept is $(0, -72)$.
ii 1 To find any x-intercepts, let $y = 0$.	$3x^2 - 6x - 72 = 0$
2 If there is a common factor in all terms, divide it out.	$x^2 - 2x - 24 = 0$
3 Solve for x. First, try to factorise. If it is not possible, use the quadratic formula instead.	$(x + 4)(x - 6) = 0$ $x = -4, 6$ The x-intercepts are $(-4, 0)$ and $(6, 0)$.

iii
1. The x-coordinate of the turning point is $-\dfrac{b}{2a}$.

   $-\dfrac{b}{2a} = \dfrac{-(-6)}{2 \times 3} = 1$

   The x-coordinate of the turning point is 1.

2. Substitute the x-coordinate of the turning point into the initial equation to find the corresponding y-coordinate.

   $y = 3(1)^2 - 6(1) - 72$
   $= 3 - 6 - 72$
   $= -75$

   The coordinates of the turning point are $(1, -75)$.

3. Interpret the sign of a.

   a is positive. There has not been a reflection in the x-axis. Therefore, the turning point is a minimum.

iv Label the turning point, the y-intercept, and the x-intercepts. Note the shape of the graph.

The graph has not been reflected in the x-axis.

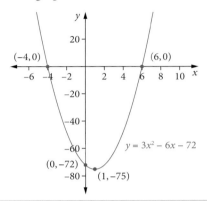

b i
1. To find the y-intercept, let $x = 0$.

   $y = -x^2 - 10x + 6$
   $y = -(0)^2 - 10(0) + 6$

2. Simplify the right-hand side

   $= 6$

   The y-intercept is $(0, 6)$.

ii
1. To find any x-intercepts, let $y = 0$.
2. If $a < 0$, multiply both sides by -1.
3. Solve for x. First, try to factorise. If it is not possible, use the quadratic formula instead.

   $-x^2 - 10x + 6 = 0$
   $x^2 + 10x - 6 = 0$

   There is no simple factorisation. Use the quadratic formula.

   $a = 1, b = 10, c = -6$

   $x = \dfrac{-10 \pm \sqrt{10^2 - 4(1)(-6)}}{2(1)}$

   $= \dfrac{-10 \pm \sqrt{124}}{2}$

   $= \dfrac{-10 \pm 2\sqrt{31}}{2}$

   $= -5 \pm \sqrt{31}$

   The x-intercepts are $(-5 - \sqrt{31}, 0)$ and $(-5 + \sqrt{31}, 0)$.

iii
1. The x-coordinate of the turning point is $-\dfrac{b}{2a}$.

   $-\dfrac{b}{2a} = -\dfrac{-10}{2 \times -1} = -5$

   The x-coordinate of the turning point is -5.

2. Substitute the x-coordinate of the turning point into the initial equation to find the corresponding y-coordinate.

   $y = -(-5)^2 - 10(-5) + 6$
   $= -25 + 50 + 6 = 31$

   The coordinates of the turning point are $(-5, 31)$.

3. Interpret the sign of a.

   a is negative. There has been a reflection in the x-axis. Therefore, the turning point is a maximum.

iv Label the turning point, the *y*-intercept and the *x*-intercepts. Note the shape of the graph.

The graph has been reflected in the *x*-axis.

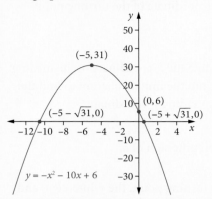

🔓 Exam hack

The *x*-intercepts of $y = ax^2 + bx + c$ can be found through the quadratic formula or solving by completing the square. The quadratic formula is often easier.

USING CAS 6 Sketching parabolas

Sketch the graph of $f(x) = -x^2 - x + 7$

ClassPad

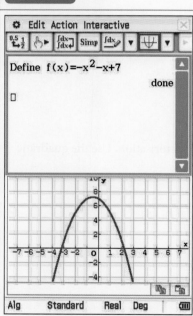

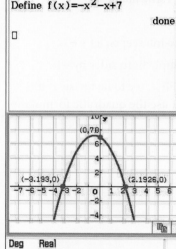

1. Enter and highlight the expression **–*x*² – *x* + 7**.
2. Tap **Interactive > Define**.
3. Keep the defaults settings in the dialogue box and tap **OK**.
4. Tap on the **Graph** tool to open the graph window.
5. Highlight either $f(x)$ or $-x^2 - x + 7$ and drag it into the graph window.
6. Adjust the window settings to view the top of the graph.

7. To find the first *x*-intercept, tap **Analysis > G-Solve > Root** and press **EXE** to paste the coordinates on the screen.
8. To find the second *x*-intercept, press the **right arrow** and press **EXE** to paste the coordinates on the screen.
9. To find the *y*-intercept, tap **Analysis > G-Solve > y-intercept** and press **EXE** to paste the coordinates on the screen.

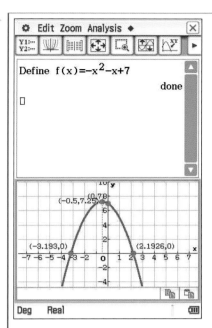

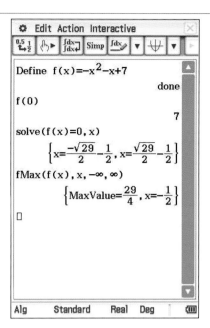

10 To find the turning point, tap **Interactive** > **G-Solve** > **fMax** and press **EXE** to paste the coordinates on the screen.

11 Tap to highlight the **Main** application in the upper window and tap **Resize**.

12 To find the y-intercept, enter **f(0)** and tap **EXE**.

13 To find the x-intercepts, enter and highlight **f(x) = 0** and tap **Interactive** > **Equation/Inequality** > **solve**.

14 Keep the defaults settings in the dialogue box and tap **OK**.

15 To find the coordinates of the turning point, enter and highlight **f(x)** and tap **Interactive** > **Calculation** > **fMin/fMax** > **fMax**.

16 Keep the defaults settings in the dialogue box and tap **OK**.

The coordinates are:

x-intercepts: $\left(\dfrac{-1-\sqrt{29}}{2}, 0\right), \left(\dfrac{-1+\sqrt{29}}{2}, 0\right)$

y-intercept: $(0, 7)$

turning point: $\left(-\dfrac{1}{2}, \dfrac{29}{4}\right)$

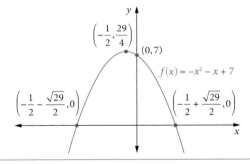

TI-Nspire

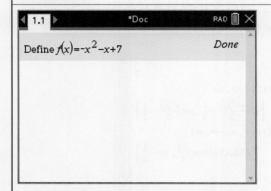

1. In a **Calculator** page, press **menu > Actions > Define**.
2. Enter the function as shown above and press **enter**.

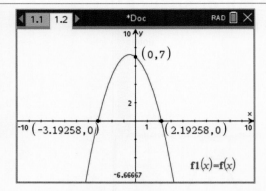

3. Add a **Graphs** page.
4. In the **Graph Entry Line**, enter **f(x)** and press **enter**.
5. Adjust the window settings to view the top of the graph.
6. To find the first x-intercept, press **menu > Analyze Graph > Zero**.
7. When prompted for the **lower bound?** and the **upper bound?**, click to the left then the right of the intercept. Repeat for the second x-intercept.
8. To find the y-intercept, press **menu > Trace**.
9. Enter **0** and press **enter** twice.

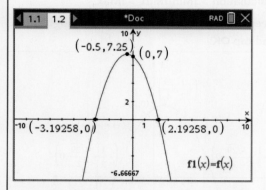

10. To find the turning point, press **menu > Analyze Graph > Maximum**.
11. When prompted for the **lower bound?** and the **upper bound?**, click to the left then the right of the turning point.

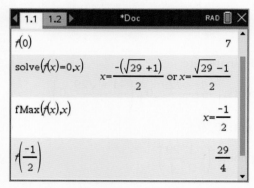

12. Return to the **Calculator** page.
13. To find the y-intercept, enter **f (0)** and press **enter**.
14. To find the x-intercepts, press **menu > Algebra > solve**.
15. Enter **f(x)=0,x** and press **enter**.
16. To find the x-coordinate of the turning point, press **menu > Calculus > Function Maximum**.
17. Enter **f(x),x** and press **enter**.
18. To find the y-coordinate of the turning point, substitute the answer into **f(x)** and press **enter**.

The coordinates are:

x-intercepts: $\left(\dfrac{-1-\sqrt{29}}{2}, 0\right), \left(\dfrac{-1+\sqrt{29}}{2}, 0\right)$

y-intercept: $(0, 7)$

turning point: $\left(-\dfrac{1}{2}, \dfrac{29}{4}\right)$

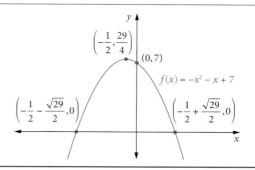

EXERCISE 3.9 Sketching parabolas from any form

ANSWERS p. 456

Recap

1 Find the turning point of $y = -3(x + 1)^2 + 14$ and state whether it is a maximum or a minimum.

2 Find the x-intercepts of $y = 4(x + 3)^2 - 80$.

Mastery

3 **WORKED EXAMPLE 33** For each parabola
   i   find the y-intercept
   ii find any x-intercepts
   iii find the x value of the turning point
   iv find the y value of the turning point
   v   sketch the graph, labelling all key features.

   a $y = (x + 5)(x - 15)$
   b $y = -(x + 4)(x - 8)$
   c $y = (x + 3)(x - 4)$
   d $y = (2x - 5)(x + 3)$
   e $y = -2(x + 3)(x - 12)$
   f $y = 3(x - 14)(x - 2)$

4 **WORKED EXAMPLE 34** For each parabola
   i   find the y-intercept
   ii find any x-intercepts
   iii find the x value of the turning point
   iv find the y value of the turning point
   v   sketch the graph, labelling all key features.

   a $y = x^2 + 2x - 8$
   b $y = -x^2 + 36$
   c $y = -4x^2 - 8x - 11$
   d $y = -x^2 - 12x - 32$
   e $y = 2x^2 + 6x$
   f $y = 3x^2 - 12x - 63$

5 **WORKED EXAMPLE 34** For each parabola
   i   find the y-intercept
   ii find any x-intercepts
   iii find the x value of the turning point
   iv find the y value of the turning point
   v   sketch the graph, labelling all key features.

   a $y = x^2 - 6x - 2$
   b $y = -x^2 - 4x + 8$
   c $y = x^2 + 6x - 10$
   d $y = 4x^2 + 3x + 1$
   e $y = -x^2 + 18x + 4$
   f $y = -3x^2 + x + 1$

6 **WORKED EXAMPLES 31, 33–34** Sketch each graph, labelling all key features.

a $y = (x - 8)(x + 3)$
b $y = -x^2 - 5x - 9$
c $y = 2(x + 6)^2 - 12$
d $y = (3x + 2)(2x - 1)$
e $y = 2(x + 7)^2 - 18$
f $y = -4x(x - 14)$
g $y = x^2 - 10x - 11$
h $y = -4(x - 5)^2 - 64$
i $y = -x^2 + 7x + 3$
j $y = 2(x - 3)^2 - 4$
k $y = -2(x + 5)(x - 5)$
l $y = 3x^2 - 14x - 5$

7 **Using CAS 6** Sketch each graph, labelling all key features.

a $y = (x - 8)^2 - 15$
b $y = -3(x - 9)(x + 2)$
c $y = 5x^2 - x + 12$

Calculator-free

8 (4 marks) Consider the equation $y = 2(x - 7)(x + 5)$.

a Express the equation in general form: $y = ax^2 + bx + c$. (1 mark)
b Express the equation in turning point form: $y = a(x - h)^2 + k$. (1 mark)
c Sketch the graph, labelling all key features with their coordinates. (2 marks)

9 (3 marks) Sketch the graph of $y = -x^2 + 6x + 6$, labelling all key features with their coordinates.

Calculator-assumed

10 (5 marks) A diver dives from a platform into the water. The arc of the dive is modelled by the equation $h = -7d^2 + 14d + 23$, where $d, h \geq 0$. d represents the horizontal distance from the platform and h represents the height of the diver above the water. Distances are measure in metres.

a What is the starting height of the diver above the water? (1 mark)
b What horizontal distance from the platform does the diver hit the water, correct to two decimal places? (2 marks)
c Sketch the graph where $d, h \geq 0$, labelling all key features with their coordinates (if necessary, to two decimal places). (2 marks)

11 (4 marks) The Lhotse is the fourth highest mountain in the world. The trek up the south face (one of the steepest climbs in the world) is modelled by the equation $y = -\dfrac{32}{50625} x(x - 4500)$, for $0 \leq x \leq h$, where h is the x value of the turning point, y represents the vertical height above base camp in metres, and x represents the horizontal distance from base camp in metres.

a Find the turning point of the graph. (1 mark)
b If the base camp is 5316 m above sea level, how tall is Lhotse? (1 mark)
c Sketch the graph for $0 \leq x \leq h$, labelling all key features with their coordinates. (2 marks)

3.10 Simultaneous equations

Simultaneous equations are a set of equations with shared variables for which a common solution is sought. The solution is a set of values that satisfies all equations simultaneously. Simultaneous equations are also known as a system of equations.

Simultaneous equations can be solved algebraically or graphically. Graphically, the solution is the point(s) of intersection of the graphs.

Graphical solution

Consider the lines of $y = 1 - x$ and $y = 2x + 4$.

The graphs have a point of intersection at $(-1, 2)$.
The solution is $x = -1$, $y = 2$.

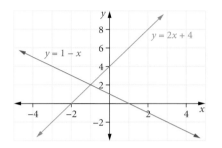

The graphs don't have to be linear. Consider the line $y = 1 - 2x$ and the curve $y = x^2 + 1$.

They have two points of intersection, at $(0, 1)$ and $(-2, 5)$.
There are two solutions: $x = 0$, $y = 1$ and $x = -2$, $y = 5$.

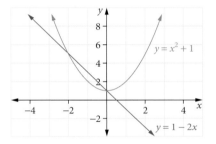

Video playlist
Simultaneous equations

Puzzle
Simultaneous equations order activity

WORKED EXAMPLE 35	Simultaneous equations using graphs

Solve the simultaneous equations $x + 2y - 10 = 0$ and $y = 3x - 2$ by sketching the two lines.

Steps	Working
1 Graph $x + 2y - 10 = 0$ and $y = 3x - 2$.	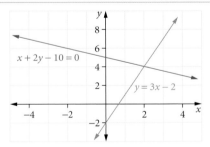
2 Read the solution point from the graph.	The point of intersection is $(2, 4)$. So, the solution is $x = 2$, $y = 4$.

> 🔒 **Exam hack**
>
> Check your solution by substituting it back into the two equations for x and y.

Algebraic solution

There are two algebraic methods for solving simultaneous linear equations.

The **elimination method** adds or subtracts equations to eliminate x or y. One or more equations may first need to be multiplied by a constant to make elimination possible.

The **substitution method** substitutes an expression from one equation into the other to create an equation with just one variable.

WORKED EXAMPLE 36 Solving by elimination

a Solve the simultaneous equations $x + 2y = 10$ and $-3x + y = -2$.

b Solve the simultaneous equations $2x + 7y = 4$ and $-5x + 3y = -51$.

Steps	Working
a 1 Multiply an equation by a constant to get matching terms for one of the variables. Identify each equation by labelling it with a number.	$x + 2y = 10$ [1] $-3x + y = -2$ [2] To match y terms, multiply [2] by 2. $-6x + 2y = -4$ [3]
2 Add or subtract the equations to eliminate one of the variables.	[1] minus [3]: $7x = 14$ $x = 2$
3 Substitute the known value into the 'easier' equation to find the remaining unknown.	$x \rightarrow$ [1] $2 + 2y = 10$ $2y = 8$ $y = 4$
4 Write the solution for both variables.	$x = 2, y = 4$
b 1 Multiply both equation by a constant to get matching terms for one of the variables. Identify each equation by labelling it with a number.	$2x + 7y = 4$ [1] $-5x + 3y = -51$ [2] To match x terms, multiply [1] by 5 and [2] by 2. $10x + 35y = 20$ [3] $-10x + 6y = -102$ [4]
2 Add or subtract the equations to eliminate one of the variables.	[3] + [4]: $41y = -82$ $y = -2$
3 Substitute the known value into the 'easier' equation to find the remaining unknown.	$y \rightarrow$ [1] $2x + 7(-2) = 4$ $2x - 14 = 4$ $x = 9$
4 Write the solution for both variables.	$x = 9, y = -2$

 Exam hack

If you multiply equation [1] by a constant to produce equation [3], for example 4 × [1] = [3], then never add/subtract [1] and [3]. These equations represent the same information, so combining them won't produce anything useful.

WORKED EXAMPLE 37 | Solving by substitution

Solve the simultaneous equations $y = 5x - 11$ and $-4x - y = 8$.

Steps	Working
1 Substitute the expression for one variable into the second equation. This will produce an equation in only one variable.	$y = 5x - 11$ [1] $-4x - y = 8$ [2] Sub [1] into [2]: $-4x - (5x - 11) = 8$
2 Solve the equation.	$-9x + 11 = 8$ $-9x = -3$ $x = \dfrac{1}{3}$
3 Substitute the known value into the 'easier' equation to find the remaining unknown.	$x \to$ [1] $y = 5\left(\dfrac{1}{3}\right) - 11$ $y = -\dfrac{28}{3}$
4 Write the solution for both variables.	$x = \dfrac{1}{3}, y = -\dfrac{28}{3}$

 Exam hack

After finding one unknown, remember to substitute it back to find the other.

Number of solutions for linear simultaneous equations

The number of solutions for simultaneous linear equations indicates how many times the lines intersect.

There are three possibilities:

- one unique solution (different gradients)

- no solutions (parallel lines: same gradient, different y-intercepts)

- infinite solutions (the same line: same gradient, same y-intercepts).

WORKED EXAMPLE 38 Number of solutions for linear simultaneous equations

a How many points of intersection are there between $y = 2x + 4$ and $y = -5x - 9$?
b How many points of intersection are there between $y = -4x - 9$ and $3y + 12x = 5$?
c Find the value(s) of m such that the simultaneous equations $mx + 2y = 20$ and $(m + 1)x - 2y = 10$ have no solution, where m is a real constant.

Steps	Working
a Compare the gradients of the linear graphs and interpret the result.	$m_1 = 2$, $m_2 = -5$ The gradients of the two lines are different. Therefore, there is one point of intersection.
b 1 Rearrange both equations into the form $y = mx + c$.	$y = -4x - 9$ is in the correct form. $3y + 12x = 5$ becomes $y = -4x + \dfrac{5}{3}$.
2 Compare the gradients of the linear graphs and interpret the result.	$m_1 = -4$, $m_2 = -4$ The gradients of the two lines are the same. There will either be no points of intersection or infinite points of intersection.
3 Compare the y-intercepts of the linear graphs and interpret the result.	$c_1 = -9$, $c_2 = \dfrac{5}{3}$ The y-intercepts are different. There are no points of intersection.
c 1 Find the gradient of both linear equations by rearranging the equations in the form $y = mx + c$.	$mx + 2y = 20$ $2y = -mx + 20$ $y = -\dfrac{m}{2}x + 10$ gradient $= -\dfrac{m}{2}$ $(m + 1)x - 2y = 10$ $(m + 1)x - 10 = 2y$ $y = \dfrac{m+1}{2}x - 5$ gradient $= \dfrac{m+1}{2}$
2 If the equations have no solution, then the lines must be parallel, so their gradients must be equal. Equate the gradients.	$-\dfrac{m}{2} = \dfrac{m+1}{2}$ $-m = m + 1$ $-2m = 1$ $m = -\dfrac{1}{2}$
3 However, to prevent them being the same line, their y-intercepts must be different. Substitute $m = -\dfrac{1}{2}$ to check that they are NOT the same line.	$y = -\dfrac{m}{2}x + 10$, when $m = -\dfrac{1}{2}$: $y = \dfrac{1}{4}x + 10$ \[1\] $y = \dfrac{m+1}{2}x - 5$, when $m = -\dfrac{1}{2}$: $y = \dfrac{1}{4}x - 5$ \[2\] Different y-intercepts (10 and -5) indicate the lines are not identical.
4 Write the answer.	$m = -\dfrac{1}{2}$

Simultaneous linear and quadratic equations

To solve simultaneous equations involving mixed linear and quadratic equations, use substitution. The number of intersection points between a parabola and a linear line (0, 1 or 2) can be determined using the discriminant.

WORKED EXAMPLE 39 Simultaneous linear and quadratic equations

a Find any points of intersection between the linear line $y = 2x + 2$ and the parabola $y = -3x^2 + 3$.
b How many points of intersection are there between $y = 9x^2 - 10x$ and $y = 20x - 25$?

Steps	Working
a 1 Points of intersection occur at the solution of the simultaneous equations. Substitute the y value of one equation into the other.	$y = 2x + 2$ [1] $y = -3x^2 + 3$ [2] $2x + 2 = -3x^2 + 3$ Solutions to this equation are the x values of any points of intersection.
2 Rearrange into the form $ax^2 + bx + c = 0$ and find the discriminant. Interpret the discriminant.	$3x^2 + 2x - 1 = 0$ $\Delta = 4 - 4(3)(-1) = 16$ The discriminant is positive, therefore there are two solutions.
3 Solve for x.	$3x^2 + 2x - 1 = 0$ $(3x - 1)(x + 1) = 0$ $x = \dfrac{1}{3}$ and $x = -1$
4 Substitute the known values into the 'easier' equation to determine the corresponding y values.	$x = \dfrac{1}{3}$ into [1]: $y = 2\left(\dfrac{1}{3}\right) + 2$ $= \dfrac{8}{3}$ $x = -1$ into [1]: $y = 2(-1) + 2$ $= 0$
5 Write each solution pair in coordinate form.	The points of intersection are $\left(\dfrac{1}{3}, \dfrac{8}{3}\right)$ and $(-1, 0)$.
b 1 Points of intersection occur at the solution of the simultaneous equations. Substitute the y value of one equation into the other.	$y = 9x^2 - 10x$ [1] $y = 20x - 25$ [2] $9x^2 - 10x = 20x - 25$ Solutions to this equation are the x values of any points of intersection.
2 Rearrange into the form $ax^2 + bx + c = 0$ and find the discriminant.	$9x^2 - 30x + 25 = 0$ $\Delta = 900 - 4(25)(9) = 0$
3 Interpret the discriminant.	The discriminant is 0. There is one solution to the equation $9x^2 - 10x = 20x - 25$. Therefore, there is one point of intersection between the graphs.

> **🔒 Exam hack**
>
> If there are two intersection points, find the x and y values for both points.

USING CAS 7 Solving simultaneous equations

Solve this pair of simultaneous equations.

$4x - 5y = 53$

$y = -\dfrac{5}{2}x - 4$

ClassPad

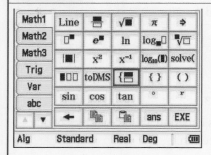 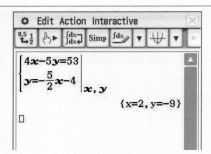

1. In **Main**, press **Keyboard** > **Math1**.
2. Tap on the **Simultaneous Equations** template. If solving more than 2 equations, tap multiple times to add more equations.
3. Enter the two equations into the template as shown above.
4. At the end of the template, enter the variables **x,y** and press **EXE**.

TI-Nspire

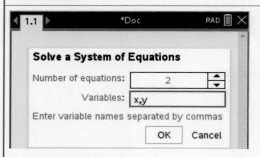 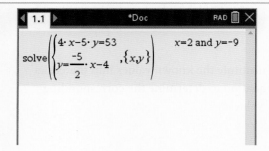

1. Add a **Calculator** page.
2. Press **menu** > **Algebra** > **Solve Simultaneous Equations** > **Solve System of Equations**.
3. Keep the defaults settings in the dialogue box as shown above and press **enter**.
4. Enter the equations into the template as shown above and press **enter**.

The solutions are $x = 2$ and $y = -9$.

USING CAS 8 | Points of intersection

Sketch the graphs of $y = 2x + 2$ and $3x^2 + y = 3$ and find their point(s) of intersection.

ClassPad

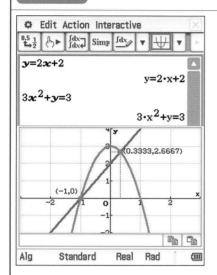

1. In the **Main** menu, enter the equations $y = 2x + 2$ and $3x^2 + y = 3$.
   Using **Main** menu is preferable to **Graph&Table**, as we don't have to make y the subject before we enter the equation.
2. Tap **Graph** to open the Graph window.
3. Highlight and drag each equation into the **Graph** window below.
4. Adjust the viewing window for a suitable view.
5. In the lower screen, tap **Analysis** > **G-Solve** > **Intersection**, pressing the right arrow to move to the next point.

TI-Nspire

1. Add a **Graphs** page.
2. Enter **2x + 2** then press **enter** to display the first graph.
3. Press **menu** > **Graph Entry/Edit** > **Relation**.
4. Enter **3x² + y = 3** then press **enter** to display the second graph.
5. Press **menu** > **Analyze Graph** > **Intersection**.
6. When prompted for the **lower bound?**, move the cursor to the left of the first point of intersection and press **enter**.
7. When prompted for the **upper bound?**, move the cursor to the right of the first point of intersection and press **enter**.
8. Repeat steps 6 and 7 for the second point of intersection.
9. Adjust the window settings to suit.

Algebraic solution

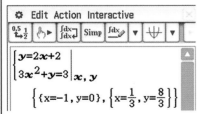

1. In **Main**, open the **Keyboard** and select the **Math1** soft keyboard.
2. Tap on the **simultaneous equations** template.
3. Enter the equations and variables as shown above, then press **EXE**.

Algebraic solution

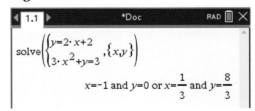

1. Add a **Calculator** page.
2. Press **menu** > **Algebra** > **Solve System of Equations** > **Solve System of Equations**.
3. Keep the default settings and select **OK**.
4. In the template, enter the equations as shown above and press **enter**.

The points of intersection are $(-1, 0)$ and $\left(\dfrac{1}{3}, \dfrac{8}{3}\right)$.

> **Simultaneous equations**
> - A solution to simultaneous equations is a point of intersection on the graphs.
> - A system of linear equations can be solved algebraically using elimination or substitution.
> - Two lines have one point of intersection if their gradients are different.
> - Two lines have no points intersection if their gradients are the same and they have different y-intercepts.
> - Two lines have infinite points intersection if their gradients are the same and they have the same y-intercept.
> - Simultaneous linear and quadratic equations can be solved using substitution.

EXERCISE 3.10 Simultaneous equations ANSWERS p. 461

Recap

1. Find the coordinates of the turning point of the parabola $y = -x^2 + 10x - 8$.

2. Sketch the graph of $4y + 5x = 10$.

Mastery

3. **WORKED EXAMPLE 35** Solve the simultaneous equations by sketching the linear lines.
   a. $y = 2x + 1$ and $y = -x - 2$
   b. $y = x + 4$ and $y = -x - 6$
   c. $x + 2y = 8$ and $y = x - 2$

4. **WORKED EXAMPLE 36** Solve each pair of simultaneous equations using elimination.
   a. $3x + 2y = 12$
      $4x - 2y = 2$
   b. $3x - 4y = 8$
      $2x + y = 9$
   c. $3x - 5y = 15$
      $-4x + 2y = -20$
   d. $x + y = 1$
      $2x + y = 5$
   e. $x + y = 10$
      $2x - 2y = 5$
   f. $-7y + 2x = -4$
      $-2x + 3y = 12$

5. **WORKED EXAMPLE 37** Solve each pair of simultaneous equations using substitution.
   a. $4x + 5y = -12$
      $y = -x - 4$
   b. $-5x - 3y = 36$
      $y = -x - 8$
   c. $-7x - 1y = 66$
      $y = 2x + 24$
   d. $4x - 5y = 55$
      $y = 2x - 23$
   e. $-5x - 3y = 36$
      $y = \dfrac{2}{3}x - 5$
   f. $2x + 4y = -8$
      $y = -\dfrac{3}{4}x - \dfrac{7}{2}$

6. **WORKED EXAMPLE 38** How many points of intersection are there between these lines?
   a. $y = 5x + 3$
      $y = -5x - 11$
   b. $y = -2x - 6$
      $y = -2x - 12$
   c. $y = x + 1$
      $y = x + 1$
   d. $4y + x = 7$
      $-8y - 2x + 14 = 0$
   e. $3y + 9x = 1$
      $-y + 3x = 2$
   f. $5y + 2x = 8$
      $2y + 5x = 8$

7. **WORKED EXAMPLE 39** Find any point(s) of intersection between each parabola and line.
   a. $y = x^2 - x - 3$
      $y = 3x + 9$
   b. $y = -x^2 - x - 8$
      $y = -x - 12$
   c. $y = -x^2 - 6$
      $y = -2x - 5$
   d. $y = -x^2 + x - 7$
      $y = -3x - 3$
   e. $y = x^2 + x + 2$
      $y = 2 - x$
   f. $y = x^2 - x - 4$
      $y = 8 - 5x$

8 **Using CAS 7** Solve each system of equations.

a for x and y
$-3x + 11y = 9$
$y = 6x - 1$

b for a, b, c and d
$b = ca + d$
$a = c + 2d$
$c = d$
$a = 2b - 1$

9 **Using CAS 8** Find the point(s) of intersection between $y = -x^2 - 5x + 12$ and $y = -3x + 6$ to two decimal places.

Calculator-free

10 (3 marks) For which value(s) of m do the simultaneous linear equations $mx + 7y = 12$ and $7x + my = m$ have one solution?

11 (3 marks) For which values of k and m do the simultaneous linear equations $2y + (m - 1)x = 2$ and $my + 3x = k$ have infinitely many solutions?

Calculator-assumed

12 (4 marks) Consider the parabola $y = x^2 + k$ and the line $y = kx - 3$. For what values of k are there

a no points of intersection? (2 marks)

b two points of intersection? (1 mark)

c one point of intersection? (1 mark)

3.11 Determining the equation of a parabola

To determine the equation of a parabola, use features of the graph to find any unknown parameters in the equation. Features may include axes intercepts, the turning point or any point on the graph.

Each quadratic form has 3 unknowns which, when found, determine the complete equation of the parabola.

- Intercept form: $y = a(x - u)(x - v)$ ← find a, u and v
- Turning point form: $y = a(x - h)^2 + k$ ← find a, h and k
- General form: $y = ax^2 + bx + c$ ← find a, b and c

These equations are different in *form* only. Expanding or factorising can transition between the forms.

Identifying sufficient information

To find n unknowns, n pieces of information are required. When determining the equation of a parabola, ask 'how many unknowns are there?' and 'how many pieces of information do I need to find?'

> **WORKED EXAMPLE 40** | **Identifying sufficient information**
>
> For each quadratic function below
> - i state which unknowns need to be found to determine the complete equation of the parabola
> - ii determine how many pieces of information are needed to find the unknown parameters
> - iii give an example of the information that could be used for each unknown.
>
> **a** $y = (x - 2)(x - v)$ **b** $y = a(x - 9)^2 + k$
> **c** $y = x^2 + bx + c$ **d** $y = -(x - h)^2 + k$

Steps	Working
a i Identify the letters that are not x or y.	$y = (x - 2)(x - v)$ There is one unknown: v.
ii n unknowns require n pieces of information.	One piece of information is required.
iii State information that could be used to find the unknown(s).	Another x-intercept (or another point on the graph).
b i Identify the letters that are not x or y.	$y = a(x - 9)^2 + k$ There are two unknowns: a and k.
ii n unknowns require n pieces of information.	Two pieces of information are required.
iii State information that could be used to find the unknown(s).	Two points on the graph (or one point and the y value of the turning point).
c i Identify the letters that are not x or y.	$y = x^2 + bx + c$ There are two unknowns: b and c.
ii n unknowns require n pieces of information.	Two pieces of information are required.
iii State information that could be used to find the unknown(s).	Two points on the graph.
d i Identify the letters that are not x or y.	$y = -(x - h)^2 + k$ There are two unknowns: h and k.
ii n unknowns require n pieces of information.	Two pieces of information are required.
iii State information that could be used to find the unknown(s).	The turning point (or two points on the graph). Note that the turning point is two pieces of information.

The coordinates of a turning point provide two pieces of information: a point on the graph and the fact that it is a turning point. Thus, the coordinates of a turning point allow discovery of two unknowns, for example, h and k in $y = a(x - h)^2 + k$.

Determining the equation of a parabola

To determine the equation of a parabola, choose the form that best suits the information provided.

Form		Use it when provided with:
Intercept form	$y = a(x - u)(x - v)$	two x-intercepts and another point on the parabola (or one x-intercept and two points on the parabola)
Turning point form	$y = a(x - h)^2 + k$	the turning point and another point on the parabola
General form	$y = ax^2 + bx + c$	all other scenarios; for example, three points on the parabola

Repeated solutions

A parabola has one x-intercept if and only if the x-axis is 'tangent to' the graph at that point, meaning that they just touch. Such an x-intercept is called a **repeated solution**. For a repeated solution $x = u$, the equation is $y = a(x - u)^2$.

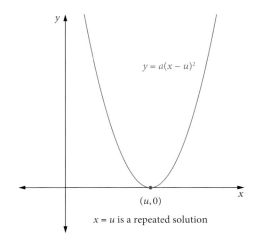

$x = u$ is a repeated solution

WORKED EXAMPLE 41 — Determining the equation – intercept form

A parabola has x-intercepts at $x = 1$ and $x = -8$. It passes through the point $(2, 30)$. What is the equation of this parabola?

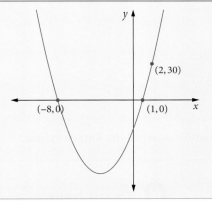

Steps	Working
1 Decide which quadratic form to use.	Use intercept form because the x-intercepts are provided. $y = a(x - u)(x - v)$
2 Fill in the known information.	The x-intercepts are 1 and -8. Considering the null factor law, the equation becomes: $y = a(x - 1)(x + 8)$
3 Use the remaining information to solve for the unknown(s).	There is one unknown, a. One new piece of information, the point $(2, 30)$ can be used to find it. When $x = 2$, $y = 30$. $30 = a(2 - 1)(2 + 8)$ $30 = 10a$ $a = 3$
4 Write the equation using all the known values.	$y = 3(x - 1)(x + 8)$

Chapter 3 | Linear and quadratic relationships

WORKED EXAMPLE 42 — Determining the equation – turning point form

Determine the equation of this parabola.

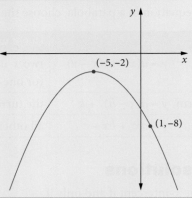

Steps	Working
1 Identify the key features provided.	The parabola has a turning point of $(-5, -2)$ and passes through the point $(1, -8)$.
2 Decide which quadratic form to use.	Use turning point form because the turning point is provided. $y = a(x - h)^2 + k$
3 Fill in the known information.	turning point $= (-5, -2)$ $h = -5$ $k = -2$ $y = a(x - (-5))^2 - 2$ $y = a(x + 5)^2 - 2$
4 Use the remaining information to solve for the unknown(s).	There is one unknown, a. One new piece of information, the point $(1, -8)$, can be used to find it. When $x = 1$, $y = -8$. $-8 = a(1 + 5)^2 - 2$ $-6 = 36a$ $a = -\dfrac{1}{6}$
5 Write the equation using all the known values.	$y = -\dfrac{1}{6}(x + 5)^2 - 2$

 Exam hack

If the graph is reflected in the x-axis, you don't need to write $-a$. When a is found, it will turn out to be negative.

WORKED EXAMPLE 43 | Determining the equation – general form

A parabola passes through the points $(1, 7)$, $(0, 6)$ and $(-2, 22)$. Find its equation.

Steps	Working
1 Decide which quadratic form to use.	Use the general form because neither the turning point nor the x-intercepts were provided. $y = ax^2 + bx + c$
2 Fill in the known information to create three simultaneous equations.	When $x = 1$, $y = 7$. $7 = a(1)^2 + b(1) + c$ $7 = a + b + c$ \[1\] When $x = 0$, $y = 6$. $6 = a(0)^2 + b(0) + c$ $c = 6$ \[2\] When $x = -2$, $y = 22$. $22 = a(-2)^2 + b(-2) + c$ $22 = 4a - 2b + c$ \[3\]
3 Solve the simultaneous equations.	$7 = a + b + c$ \[1\] $c = 6$ \[2\] $22 = 4a - 2b + c$ \[3\] Use \[2\] to create simpler versions of \[1\] and \[3\]: $a + b = 1$ \[1a\] $4a - 2b = 16$ \[3a\] Solve by elimination. \[1a\] × 2: $2a + 2b = 2$ \[1b\] \[1b\] + \[3a\] $6a = 18$ $a = 3$ Substitute a into one of the equations to find b. $a \rightarrow$ \[1a\] $3 + b = 1$ $b = -2$ $a = 3$, $b = -2$, $c = 6$
4 Write the equation using all the known values.	$y = 3x^2 - 2x + 6$

 Exam hack

The y-intercept can be used like any other point. For example, if the y-intercept is 5, then when $x = 0$, $y = 5$.

EXERCISE 3.11 Determining the equation of a parabola

ANSWERS p. 462

Recap

1 Find the coordinates of the turning point of $y = (x - 5)(x + 9)$.

2 Write the equation $y = x^2 - 3x + 2$ in the form $y = a(x - h)^2 + k$ by completing the square.

Mastery

3 WORKED EXAMPLE 40 For each quadratic function below
   i determine how many pieces of information are needed to find the unknown parameters
   ii give an example of information that could be used to find each unknown.

a $y = -(x + 10)^2 + k$
b $y = a(x - 10)(x + 4)$
c $y = ax^2 + bx + 5$
d $y = a(x - u)(x - v)$
e $y = ax^2 + bx + c$
f $y = a(x - h)^2 + k$

4 WORKED EXAMPLE 41 Find the equation of each parabola.

a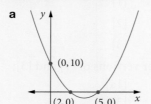
b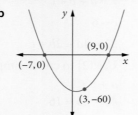
c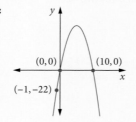
d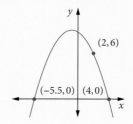

5 WORKED EXAMPLE 42 Find the equation of each parabola.

a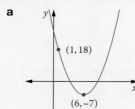
b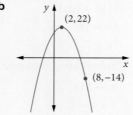
c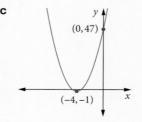
d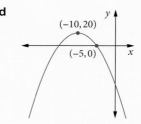

6 WORKED EXAMPLE 43 Find the equation of each parabola.

a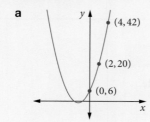
b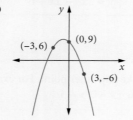
c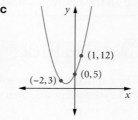
d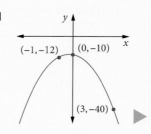

7 WORKED EXAMPLES 40–43 For each parabola described
   i state which form is most useful in determining the equation: $y = a(x - u)(x - v)$, $y = a(x - h)^2 + k$, or $y = ax^2 + bx + c$
   ii find the equation.

   a A parabola which passes through $(1, -5)$ with turning point $(5, 11)$.
   b A parabola which passes through $(2, -9)$ with x-intercepts -4 and -1.
   c A parabola which passes through $(0, 5)$, $(1, 9)$ and $(2, 21)$.

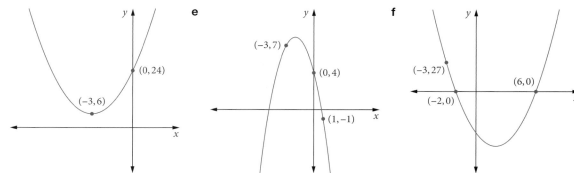

Calculator-free

8 (4 marks)

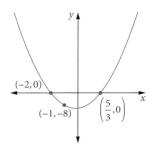

   a The equation of this parabola can be expressed in the form $y = a(sx - u)(tx - v)$. Find the values of a, s, t, u and v if they are all integers and $t > s$. (3 marks)
   b Hence, write the equation in the form $y = ax^2 + bx + c$. (1 mark)

9 (3 marks) A parabola passes through the point $(4, 1)$ and has its turning point on the x-axis at $(m, 0)$.
   a Find the equation of the parabola in terms of m. (2 marks)
   b How many x-intercepts does the parabola have? (1 mark)

Calculator-assumed

10 (4 marks)
   a Find the equation of the parabola passing through $(-1, 6)$, $(2, 3)$ and $(3, 10)$. (2 marks)
   b Find the equation of the parabola passing through $(1, 50)$, $(-2, 110)$ and $(3, 80)$. (2 marks)

11 (7 marks) A parabola passes through the points $(-1, 32)$ and $(2, 98)$ and has one x-intercept.
   a Considering the form $y = ax^2 + bx + c$, write three simultaneous equations in terms of a, b and c. (3 marks)
   b Considering the form $y = a(x - h)^2 + k$:
      i state the value of k. (1 mark)
      ii write two simultaneous equations in terms of a and h. (2 marks)
   c By solving either system of equations, find the equation of the parabola. (1 mark)

3.12 Applications of quadratic relationships

Applications of quadratic relationships

Quadratic relationships exist in the world around us. They can model visible scenarios, for example, the position of an object under constant acceleration $\left(s = ut + \dfrac{1}{2}at^2\right)$. They also model the underlying nature of the universe, for example, an electrostatic force between two charged particles ($Fr^2 = kq_1q_2$) or a gravitational force between two objects ($Fr^2 = Gm_1m_2$).

> **Applications of quadratic relationships**
>
> When applying quadratics skills to practical scenarios, consider the meaning of key features of the parabola:
> - The x-intercepts occur when the y value is zero, e.g. zero height.
> - The y-intercept occurs when the x value is zero, e.g. zero time.
> - The turning point occurs when the y value is a maximum or minimum, e.g. maximum area.

Finding the maximum or minimum value of a function is a common application. Here is a reminder of how to find the x value of the turning point from each quadratic form:

Quadratic form	How to find the x value of the turning point
$y = a(x - u)(x - v)$	halfway between the x-intercepts
$y = a(x - h)^2 + k$	h
$y = ax^2 + bx + c$	$-\dfrac{b}{2a}$ or halfway between the x-intercepts

The maximum or minimum refers to the output of the function; that is, the y value. To find it, substitute the x value of the turning point into the original equation.

WORKED EXAMPLE 44 — Calculator-free quadratic applications

A rectangular pool has perimeter 60 m and a length of x m.
a Find an equation for the area of the pool in terms of x.
b What value(s) of x give an area of 200 m²?
c Find the maximum area of the pool.
d Sketch the graph of A vs x for logical values of x.

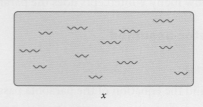

Steps	Working
a 1 Label the width of the rectangle, y.	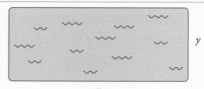
2 Write an equation to describe what you want to find.	$A = xy$
3 Identify the variable to remove. The goal is to find an expression for A in terms of x.	y

4 Find an equation for the perimeter. This extra information can be used to remove y.		$P = 2x + 2y$ $60 = 2x + 2y$ $30 = x + y$
5 Use this equation to find an expression for y in terms of x.		$y = 30 - x$
6 Substitute this into the original equation.		$A = xy$ $A = x(30 - x)$
b 1 Substitute the desired area into the equation. **2** Solve for x.		$200 = x(30 - x)$ $200 = 30x - x^2$ $x^2 - 30x + 200 = 0$ $(x - 20)(x - 10) = 0$ $x = 10\,\text{m},\ 20\,\text{m}$
3 If there is more than one solution, check whether both are logical.		If x is 10 m or 20 m, the perimeter of the pool can still be 60 m. Both are positive. Both are viable solutions. $x = 10\,\text{m},\ 20\,\text{m}$
c 1 The maximum/minimum value of a quadratic expression occurs at the turning point.		This parabola is reflected in the x-axis, shown in the $-x^2$. The turning point will be a maximum. The turning point is halfway between the x-intercepts. Halfway between 10 and 20 is 15.
2 Find the y value for the given x value.		When $x = 15$, $A = x(30 - x) = 15(15) = 225\,\text{m}^2$. The maximum area is 225 m².
d 1 Sketch the parabola, labelling all key features with their coordinates.		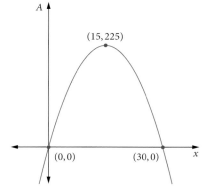
2 Consider which values of x are logical.		The A-axis is area. Area cannot be zero or negative, so $x \leq 0$ and $x \geq 30$ should not be part of the graph. Sketch for $0 < x < 30$.
3 Sketch the graph over the logical values of x. Label any endpoints with their coordinates.		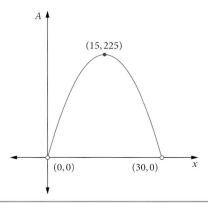

WORKED EXAMPLE 45 — Calculator-assumed quadratic applications

An AFL player kicks the footy 55 m out from the goal line. The ball follows a parabolic path represented by the equation $h = -0.0165d^2 + 0.985d + 0.3$. h represents the height, in metres, of the ball above the ground. d measures the distance from the player, on the grass, along the line to the goals, in metres.

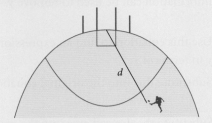

a At what starting height, in cm, is the ball kicked?

b What is the maximum height of the ball to one decimal place?

c Does the ball clear the goal line?

d A defender runs and jumps at the goal line and reaches a height of 305 cm with their hand. Do they touch the ball?

Steps	Working
a The starting position of the ball is when $d = 0$.	When $d = 0$, $h = 0.3$. The starting height is 30 cm.
b 1 The maximum occurs when $d = -\dfrac{b}{2a}$. Find this value.	$h = -0.0165d^2 + 0.985d + 0.3$ $d = -\dfrac{b}{2a} = -\dfrac{0.985}{2 \times (-0.0165)} = 29.85$ m
2 Substitute the x value to find the desired maximum/minimum y value.	When $d = 29.8484…$ (use all decimal places) $h = -0.0165(29.8484…)^2 + 0.985(29.8484…) + 0.3$ $= 15.0$ m
c 1 The ball hits the ground when $h = 0$. Find the value of d for which this occurs. **ClassPad** solve(0=−0.0165·d²+0.98·d+0.3|d≥0, d) {d=59.69850011} **TI-Nspire** solve(0=−0.0165·d²+0.98·d+0.3,d)|d≥0 d=59.6985 $d = -0.303$, 60, $d \geq 0$ $d = 60$ m	$0 = -0.0165d^2 + 0.985d + 0.3$
2 Compare this value to the information in the question.	The player started 55 m out. The distance along the grass that the ball flies over is 60 m. Yes, the ball clears the goal line.
d 1 Calculate the height of the ball as it crosses the goal line.	The player kicked from 55 m out. The ball crosses the line when $d = 55$. $-0.0165(55)^2 + 0.985(55) + 0.3 = 4.56$ m
2 Compare this to the height the defender reaches.	$4.56 > 3.05$ The defender does not touch the ball.

EXAMINATION QUESTION ANALYSIS

Calculator-assumed (9 marks)

An object lighter than water is projected from the top of a seaside cliff. It travels in a parabolic path and hits the water 15 metres below its starting point as shown in the diagram.

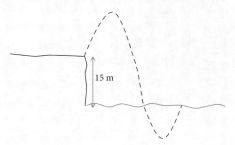

The equation of the parabolic path is $y = 5x(b - x)$, where x and y are in metres and b is a constant.

The starting point is $(0, 0)$. The maximum height of the object is 20 metres above the cliff. After hitting the water, the object's buoyancy makes it follow another parabolic path until it reaches the surface of the water. Its path under the water reaches a depth half the height it reached above the cliff.

a Find the value of b. (3 marks)

b Write an equation representing the water level (ignoring any waves). (1 mark)

c Find the horizontal distance from the cliff to where the object hits the water, correct to two decimal places. (2 marks)

d The equation of the object's parabolic path under the water is $y = (x - h)^2 + k$, where x and y are in metres and h and k are constants. Find the equation of this path, writing values of h and k correct to two decimal places. (3 marks)

Reading the question

- The two parabolas have separate equations.
- The measurements 15 m and 20 m are needed for this question.
- The point of projection at the top of the cliff is $(0, 0)$.

Thinking about the question

- The parabola under the water is not reflected, so x^2 has a positive coefficient.
- The water level is parallel to the x-axis (horizontal).
- The turning point of a parabola occurs halfway between the x-intercepts (or at $x = -\frac{b}{2a}$).
- The answers for parts **c** and **d** are to a number of decimal places, so a calculator should be used.

Worked solution (✓ = 1 mark)

a $y = 5x(b - x)$ has x-intercepts at $x = 0$ and $x = b$, so the x value of its turning point is halfway:

$x = \dfrac{0 + b}{2} = \dfrac{b}{2}$.

or

$y = 5x(b - x) = -5x^2 + 5bx$, so the x value of its turning point is

$x = -\dfrac{b}{2a}$

$= \dfrac{-5b}{2(-5)} = \dfrac{-5b}{-10} = \dfrac{b}{2}$

$y = 20$ at $x = \dfrac{b}{2}$:

$5\left(\dfrac{b}{2}\right)\left(b - \dfrac{b}{2}\right) = 20$

$\dfrac{5b^2}{4} = 20$

$b^2 = 16$

$b > 0$ because the x-intercept from $(b - x) = 0$ must be positive (to the right of the cliff).

$b = 4$

finds the x value of the turning point ✓

writes an equation that can be solved for b ✓

finds b ✓

b The water level is a horizontal line 15 m below the origin, so its equation is

$y = -15$. ✓

c The point where the object hits the water is the intersection of $y = -15$ and $y = 5x(4 - x)$.

$-15 = 5x(4 - x)$

ClassPad

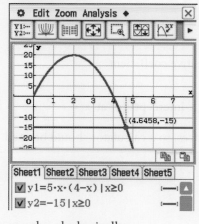

TI-Nspire

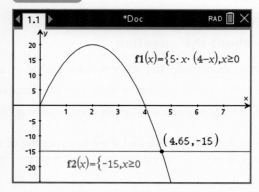

or solve algebraically:

$5x(4 - x) = -15$

$x(4 - x) = -3$

$-x^2 + 4x = -3$

$x^2 - 4x - 3 = 0$

Identify a, b and c and solve using the quadratic formula.

$a = 1$, $b = -4$, $c = -3$

$$x = \frac{-(-4) \pm \sqrt{(-4)^2 - 4(1)(-3)}}{2(1)}$$

$$= \frac{4 \pm \sqrt{28}}{2}$$

$$= \frac{4 \pm 2\sqrt{7}}{2}$$

$$= 2 \pm \sqrt{7}$$

But $x > 4$, so $x = 2 + \sqrt{7} \approx 4.65$.

It hits the water **4.65 m** horizontally from the cliff.

writes an equation whose solution is the intersection ✓

finds the horizontal distance to two decimal places ✓

d The object reaches a depth of $\frac{1}{2} \times 20 = 10$ m under the water.

The parabola passes through the point approximately $(4.64575, -15)$ and has minimum $-15 - 10 = -25$.

The equation is of the form $y = (x - h)^2 + k$.

The turning point is (h, k) so $k = -25$.

At $x \approx 4.64575$, $y = -15$:

$$-15 = (4.65 - h)^2 - 25$$
$$(4.65 - h)^2 = 10$$
$$h = 4.64575 \pm \sqrt{10}$$

The lowest point is at $x > 4.64575$, so

$h = 4.64575 + \sqrt{10}$
≈ 7.81

The equation is $y = (x - 7.81)^2 - 25$.

finds the value for k ✓

writes an equation which can be solved for h ✓

writes the equation in the correct form with the correct values for h and k ✓

EXERCISE 3.12 Applications of quadratic relationships   ANSWERS p. 463

Recap

1 Determine the equation of the parabola that has x-intercepts of -5 and $-\frac{4}{3}$ and passes through the point $(2, -420)$.

2 Find the point(s) of intersection between $y = -2x + 9$ and $5x + 12y = -6$.

Mastery

3 WORKED EXAMPLE 44 The length of a rectangular plot of land is 28 m longer than the width.

  a Write an equation for the area, A, of the plot of land in terms of its width, x.

  b If the area is 480 m², find the dimensions of the plot.

4. **WORKED EXAMPLE 44** The volume, $V\,\text{m}^3$, of gas in a tank is given by the equation $V = -\dfrac{2t}{5}(t-11)$, where t is the time in minutes since the tank started being filled. Find the maximum volume of gas in the tank.

5. **WORKED EXAMPLE 44** A park ranger is conducting a survey of the wildlife in her forest. To do this, she needs to lay a perimeter of non-lethal traps and record which animals are caught, before releasing them again. She uses a stone boundary as one edge of the rectangular enclosure. Traps form the other three sides.

    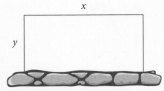

   a. Write an equation for the area of the enclosure, A, and an equation for the perimeter of traps, P, each in terms of x and y.
   b. If the park ranger has 28 m of trapping, find an equation for the area in terms of x.
   c. Find the maximum area the ranger can trap and the value of x for which this occurs.

6. **WORKED EXAMPLE 44** A segment of a theme park ride follows the path of a parabola. The height of the car, h metres, above the ground is modelled by $h = t^2 - 9t + 18$, $0 \le t \le 9$, where t is the time in seconds after the ride begins.

   a. What is the lowest height the car reaches and when does this occur?
   b. Sketch the graph, labelling all key features with their coordinates, including endpoints.
   c. How many seconds of the ride are below ground?

7. **WORKED EXAMPLE 45** A ball is projected vertically upwards. At time t seconds, its height, s metres, is modelled by the equation $s = -5t^2 + 11t$.

   a. Sketch the graph of s vs t for non-negative values of s.
   b. What is the maximum height of the ball to two decimal places?
   c. At what time does the ball reach its maximum height to one decimal place?
   d. The ball passes a height of 3 m on its way up. How many seconds between that moment and when the ball passes a height of 3 m on its way down? Answer to one decimal place.

8. **WORKED EXAMPLE 45** A circle is inscribed within a square.

    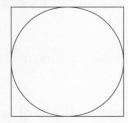

   a. If the area of the square is $5\,\text{cm}^2$ larger than the area of the circle. Write an equation, in terms of the circle's radius, r, to represent this idea.
   b. Hence, find the radius of the circle to two decimal places.

9. **WORKED EXAMPLE 45** The value, $\$V$, of an investment after two months is given by $V = 3000\left(1 + \dfrac{r}{100}\right)^2$, where r is the monthly interest rate.

   a. Find the values of a, h and k, where V is written in the form: $V = a(r-h)^2 + k$.
   b. Find the monthly interest rate that produces a value of $\$3060.30$ after two months.

10 Newton's law of gravitation can be used to calculate the gravitational force, in N (newtons), between two objects. The law is $F = G\dfrac{m_1 m_2}{r^2}$, where m_1 and m_2 are the masses, in kg, of each object, r is the distance, in m, between the two objects and G is the constant 6.67×10^{-11}. Earth weighs 5.97×10^{24} kg and its moon weighs 7.35×10^{22} kg. If the gravitational force of the Moon on Earth is 1.98×10^{20} N, find the distance between Earth and its moon to the nearest thousand kilometres.

Calculator-free

11 (6 marks) A rectangular sheet of cardboard is to be made into a box. The dimensions of the sheet are 10 cm × 20 cm. To fold up the sides, square corners of width x cm are cut out. The external faces of the box, including the base, will then be painted.

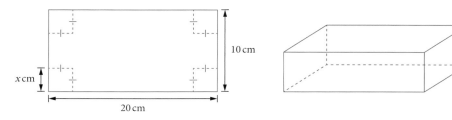

a Find an expression for the total surface area, A m², to be painted. (1 mark)
b This expression is only valid for $0 < x < 5$. Why? (1 mark)
c Sketch the graph of A vs x for $0 < x < 5$. Label the endpoints with their coordinates. (2 marks)
d If the cost of paint is $1.50 per square metre and $246 was spent painting the box, what is the width of the square corners? (2 marks)

Calculator-assumed

12 (6 marks) Tim crosses a soccer ball to the team's striker, Sam. Let x m represent the distance of the ball from the corner flag post, along the grass to Sam. Let h m represent the height of the ball.

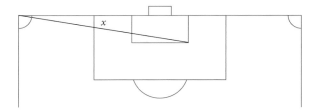

The ball's path can be modelled by the equation $h = -0.005x^2 + 0.31x - 0.3$.

a Sketch the graph for all positive h values. Label the x-axis intercepts and the turning point accurate to two decimal places. (2 marks)
b How far from the flag post, to two decimal places, does Tim take the initial kick? (1 mark)
c What is the maximum height, to two decimal places, that the ball reaches? (1 mark)
d Sam stands 51 m from the corner flag. Calculate the height of the ball at this point. If Sam's jump height is 2.56 m, will Sam succeed in heading the ball? (2 marks)

Chapter summary

Linear graphs
- The **gradient–intercept form** is $y = mx + c$. m is the **gradient** and c is the **y-intercept**.
- gradient $= m = \dfrac{\text{rise}}{\text{run}} = \dfrac{y_2 - y_1}{x_2 - x_1}$
- The x-intercept is found by letting $y = 0$ and solving for x.
- The y-intercept is found by letting $x = 0$ and solving for y.
- To sketch a linear graph, find the axes intercepts, draw a line through them and label the intercepts.
- A horizontal line has zero gradient and is represented by $y = c$.
- A vertical line has undefined gradient and is represented by $x = d$.

Determining the equation of a linear graph
- The gradient and a point are required. The gradient can be calculated first from other information.
- To determine the equation using $y = mx + c$ (gradient–intercept form):
  1. Find the gradient and substitute it for m.
  2. Substitute the coordinates of a point for x and y. Solve for c.
  3. Rewrite the full $y = mx + c$ equation using the known values for m and c.
- To determine the equation using $y - y_1 = m(x - x_1)$ (**point–gradient form**):
  1. Find the gradient and substitute it for m.
  2. Substitute the coordinates of a point for x_1 and y_1.
  3. Rearrange for y.
- Two **parallel lines** have the same gradient: $m_1 = m_2$.
- The gradients of two **perpendicular lines** are negative reciprocals of each other:
  $m_1 m_2 = -1$ or, rearranged, $m_2 = -\dfrac{1}{m_1}$.

Expanding
- First, Outer, Inner, Last

$$(a + b)(c + d) = ac + ad + bc + bd$$

with F = ac, O = ad, I = bc, L = bd.

- **Perfect squares**

$$(a + b)^2 = a^2 + 2ab + b^2 \qquad (a - b)^2 = a^2 - 2ab + b^2$$

- **Difference of perfect squares**

$$(a - b)(a + b) = a^2 - b^2$$

Factorising
- Look first for a **highest common factor**, then a difference of perfect squares, then a **quadratic trinomial**.
- Perfect squares

$$a^2 + 2ab + b^2 = (a + b)^2 \qquad a^2 - 2ab + b^2 = (a - b)^2$$

- Difference of perfect squares
  $a^2 - b^2 = (a - b)(a + b)$
- Factorising by grouping:
  1. Separate the terms into groups with their own common factor and factorise each group.
     After factorising, the expression in brackets should be the same across all groups.
  2. Use the expression in brackets as a new common factor and factorise again.

Factorising quadratic trinomials

- Cross method

  To factorise any quadratic trinomial $ax^2 + bx + c$ by the cross method:
  1. Draw a cross and write a factor pair of the ax^2 term on the left-hand side. If $a = 1$, use x and x.
  2. Write a factor pair of c on the right-hand side.
  3. Check if the choice of factor pairs is correct by multiplying along each diagonal and adding the resultant terms together. The sum should equal bx.
  4. Once the correct cross is found, write each linear factor by reading horizontally.

- Grouping method

  To factorise any quadratic trinomial $ax^2 + bx + c$ by grouping, split bx into two terms whose coefficients multiply to give the product ac.

- Quadratic trinomials cannot always be factorised.

Solving quadratic equations

- **Null factor law**: if $mn = 0$, then $m = 0$ or $n = 0$.
- **Completing the square**
  - takes $ax^2 + bx + c = 0$ and transposes it to the form $a(x - h)^2 + k = 0$
  - the equation can then be solved by rearranging for x
  - remember $\pm$ in front of the $\sqrt{}$.

Steps for completing the square on $x^2 + bx + c$	
1 Halve b and square the result.	$\left(\dfrac{b}{2}\right)^2$
2 Add this value after the x term. Subtract as well to keep the expression unchanged.	$x^2 + bx + \left(\dfrac{b}{2}\right)^2 - \left(\dfrac{b}{2}\right)^2 + c$
3 Place brackets around the first three terms and simplify the terms outside the brackets.	$\left(x^2 + bx + \left(\dfrac{b}{2}\right)^2\right) - d$
4 Factorise to a perfect square. The second term in the linear factor is half of b.	$\left(x + \dfrac{b}{2}\right)^2 - d$

Note: To complete the square on $ax^2 + bx + c$, where $a \neq 1$, take out a factor of a from all three terms then complete the square within the brackets.

- The **quadratic formula**

  For a quadratic equation $ax^2 + bx + c = 0$, the solutions are
  $$x = \dfrac{-b \pm \sqrt{b^2 - 4ac}}{2a}$$

- The **discriminant**, Δ

  For any quadratic equation $ax^2 + bx + c = 0$, the discriminant is $\Delta = b^2 - 4ac$.
  - If $\Delta > 0$, the quadratic equation has two distinct real solutions.
  - If $\Delta < 0$, the quadratic equation has no real solutions.
  - If $\Delta = 0$, the quadratic equation has one real solution.
- The discriminant can be used to determine which value(s) of a parameter lead to 0, 1 or 2 solutions, such as in $5x^2 + 2x + k = 0$.

Sketching parabolas

- To sketch a **parabola**, find and label:
  - the y-intercept
  - any x-intercepts, if they exist
  - the **turning point**.
- Three forms of a **quadratic function**:
  - **Intercept form**: $y = a(x - u)(x - v)$
  - **Turning point form**: $y = a(x - h)^2 + k$
  - **General form**: $y = ax^2 + bx + c$

  For all three forms:
  - the turning point is a minimum if $a > 0$
  - the turning point is a maximum if $a < 0$.

Quadratic form	Turning point	x-intercepts	y-intercept
Turning point form $y = a(x - h)^2 + k$	(h, k)	Let $y = 0$ and rearrange for x.	Let $x = 0$, simplify for y.
Intercept form $y = a(x - u)(x - v)$	x value: halfway between the x-intercepts y value: substitute the x value into the equation	Let $y = 0$ and use the null factor law.	
General form $y = ax^2 + bx + c$	x value: $x = -\dfrac{b}{2a}$ or halfway between the x-intercepts y value: substitute the x value into the equation	Let $y = 0$ and • factorise then null factor law; or • quadratic formula; or • complete the square and rearrange for x.	

- The solutions to '**quadratic expression** = 0' are the x-intercepts of 'y = quadratic expression'. They are also called '**roots**' or 'zeros'.

Simultaneous equations

- The **elimination method** adds or subtracts equations to eliminate x or y. One or more equations may first need to be multiplied by a constant to make elimination possible.
- The **substitution method** substitutes an expression from one equation into the other to create an equation with just one variable.
- Use elimination or substitution for linear **simultaneous equations**.
- Use substitution for mixed quadratic and linear simultaneous equations.
- The number of solutions to simultaneous linear equations can be:
  - one unique solution (two lines with different gradients)

  - no solutions (parallel lines: same gradient, different y-intercepts)

  - infinite solutions (same line: same gradient, same y-intercepts).

- The number of intersection points between a parabola and a line (0, 1 or 2) can be determined using the discriminant.

Determining the equation of a parabola

- To find n unknowns, n pieces of information are required. Use one of these forms depending on the information provided.

Form	Use it when provided with:
Intercept form $y = a(x - u)(x - v)$	One or two x-intercepts and other information.
Turning point form $y = a(x - h)^2 + k$	The turning point and other information.
General form $y = ax^2 + bx + c$	All other scenarios, e.g. three points on the parabola.

Applications of quadratic relationships

- Consider the meaning of key features of the parabola:
  - The x-intercepts occur when the y value is zero, e.g. zero height.
  - The y-intercept occurs when the x value is zero, e.g. zero time.
  - The turning point occurs when the y value is a maximum or minimum, e.g. maximum area.

Cumulative examination: Calculator-free

Total number of marks: 31   Reading time: 4 minutes   Working time: 31 minutes

1 (8 marks) For two events A and B, it is known that $P(A) = x + 0.1$, $P(B) = 0.6$ and $P(A|B) = x - 0.2$.

    a If A and B are mutually exclusive, determine the value of

       i x (3 marks)

       ii $P(\overline{A \cup B})$ (1 mark)

    b Determine the value of x such that A and B are complementary events.   (2 marks)

    c Show that it is not possible for A and B to be independent events.   (2 marks)

2 (5 marks) Consider this system of linear equations:

$$-3x + y = -1$$
$$2x + 6y = 9$$

    a Solve the system of linear equations.   (2 marks)

    b Are the two graphs parallel or perpendicular? Why?   (1 mark)

    c Sketch the graph of both lines on the same axes, labelling all key features.   (2 marks)

3 (2 marks)

    a Factorise $x^2 - 14x - 32$.   (1 mark)

    b Hence, solve $x^2 - 14x - 32 = 0$.   (1 mark)

4 (2 marks) Solve $4x^2 - 44x - 41 = 0$ by completing the square.

5 (2 marks) Find the value of k if $-2x^2 + 10x + k = 0$ has one solution.

6 (6 marks) Sketch each graph, labelling key features with their coordinates.

    a $y = -3(x - 5)^2 + 12$ (3 marks)

    b $y = -4(x + 2)(x - 15)$ (3 marks)

7 (4 marks) Find the equation of the parabola

    a with x-intercepts -4 and 8 and a y-intercept of 7 (2 marks)

    b passing through the points $(3, 10)$, $(0, -14)$ and $(-1, -10)$.   (2 marks)

8 (2 marks) Find the point(s) of intersection between $y = x^2 - 5x + 3$ and $y = -\frac{1}{5}x + 4$.

Cumulative examination: Calculator-assumed

Total number of marks: 43   Reading time: 5 minutes   Working time: 43 minutes

1 (11 marks) A mobile service provider recorded some observations over the course of 6 months regarding what brand of mobile phone customers of various ages purchased. Some of the results are shown in the table below.

	Apple	Other	Total
≤ 25 years old	250		370
> 25 years old		130	
Total	430		

  a Copy and complete the two-way table. (2 marks)

  b Estimate the probability, correct to four decimal places where appropriate, that a randomly selected new customer will purchase

    i an Apple phone (2 marks)

    ii an Apple phone, given that they are 25 years old or younger (2 marks)

    iii a non-Apple branded phone, given that they are older than 25. (2 marks)

  c The data analyst of the service provider claims that 'the brand of mobile phone purchased does not seem to be significantly influenced by the customers' age.' Use the appropriate relative frequencies to comment on the validity of the analyst's claim. Justify your answer. (3 marks)

2 (7 marks) David runs a small business making and selling small statues of well-known Western Australian landmarks. The statues are made in a mould, then finished (smoothed and then hand-painted using a special gold paint) by David himself. David sends the statues **in order of completion** to an inspector, who classifies them as either 'Superior' or 'Regular', depending on the quality of their finish.

If a statue is Superior, then the probability that the next statue completed is Superior is p.

If a statue is Regular, then the probability that the next statue completed is Superior is $p - 0.2$. On a particular day, David knows that $p = 0.9$.

On that day

  a if the **first statue inspected is Superior**, determine the probability that the third statue is Regular (2 marks)

  b if the **first statue inspected is Superior**, determine the probability that the next three statues are Superior. (2 marks)

On another day, David finds that if the **first statue inspected is Superior** then the probability that the third statue is Superior is 0.7.

  c Show that the value of p on this day is 0.75. (3 marks)

3 (2 marks) The relationship between Celsius, c, and Fahrenheit, f, is linear. If 5°C equals 41°F and 20°C equals 68°F, find the rule for c in terms of f.

4 (3 marks) Sketch the graph of $y = x^2 - 6x - 1$, labelling all key features with their coordinates.

5 (2 marks) For what value(s) of b will the equation $y = 3x^2 - bx - b - 1$ have no x-intercepts?

6 (5 marks) A rectangle is inscribed in the parabola $y = -x^2 + 25$ such that its base sits on the x-axis and its top corners just touch the parabola.

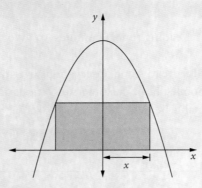

    a Find an expression for the perimeter, P, of the rectangle in terms of x. (1 mark)

    b If the perimeter of the rectangle is 44 units^2, what is the value of x? (2 marks)

    c Find the maximum possible perimeter of the rectangle and the value of x for which this occurs. (2 marks)

7 (3 marks) Consider the parabola $y = ax^2 + bx + c$ passing through the points $(-2,-32)$, $(1,-5)$ and $(4,-32)$.

    a Write three simultaneous equations in terms of a, b and c. (1 mark)

    b Hence, find the equation of the parabola. (2 marks)

8 (5 marks) Javelin Frankston receives the football 70 m out from the goal line and kicks towards goal. The ball follows a parabolic path represented by the equation $h = -0.011d^2 + 0.74d + 0.32$. h m represents the height of the ball above the ground. d m measures the distance from Frankston, on the grass, along the line to the goals.

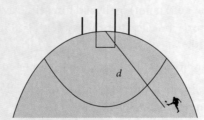

    a At what starting height, in cm, is the ball kicked? (1 mark)

    b What is the maximum height of the ball to one decimal place? (1 mark)

    c How far before the goal line does the ball hit the ground to one decimal place? (1 mark)

    d A lone defender runs into the forward 50 and jumps on the line between the player and the goals, 6 metres out from goal. They reach a height of 252 cm. What is the gap between the player's hand and the ball to the nearest centimetre? (2 marks)

9 (5 marks) Consider the parabola $y = x^2 + px - 7$, which has two x-intercepts.

    a Rewrite the equation in the form $y = a(x - h)^2 + k$. (2 marks)

    b Find the x-intercepts of the graph, in terms of p, in the form $x = \dfrac{-p \pm \sqrt{p^2 + s}}{t}$, where s and t are positive integers. (2 marks)

    c Hence, rewrite the equation in the form $y = (x + u)(x + v)$ where u and v are in terms of p. (1 mark)

CHAPTER 4
FUNCTIONS AND RELATIONS

Syllabus coverage
Nelson MindTap chapter resources

4.1 Polynomials and power functions
  Polynomials
  Equating coefficients
  Using CAS 1: Defining and evaluating polynomials
  Power functions
  The hyperbola
  Square root functions

4.2 Solving cubic equations
  Using CAS 2: Finding factors of polynomials
  Using CAS 3: Solving polynomial equations

4.3 Graphing cubic functions

4.4 Functions and relations
  Relations
  Function notation
  Domain and range
  Using CAS 4: Finding the range

4.5 Transformations of functions
  Translations
  Dilations from the x- and y-axes
  Reflections in the x- and y-axes
  Combined transformations
  Using CAS 5: Transformations of power functions

4.6 Graphs of relations

4.7 Inverse proportion

Examination question analysis
Chapter summary
Cumulative examination: Calculator-free
Cumulative examination: Calculator-assumed

Syllabus coverage

Inverse proportion
1.2.9   examine examples of inverse proportion
1.2.10   recognise features and determine equations of the graphs of $y = \dfrac{1}{x}$ and $y = \dfrac{a}{x-b}$, including their hyperbolic shapes and their asymptotes

Powers and polynomials
1.2.11   recognise features of the graphs of $y = x^n$ for $n \in \mathbb{N}$, $n = -1$ and $n = \dfrac{1}{2}$, including shape, and behaviour as $x \to \infty$ and $x \to -\infty$
1.2.12   identify the coefficients and the degree of a polynomial
1.2.13   expand quadratic and cubic polynomials from factors
1.2.14   recognise features and determine equations of the graphs of $y = x^3$, $y = a(x - b)^3 + c$ and $y = k(x - a)(x - b)(x - c)$, including shape, intercepts and behaviour as $x \to \infty$ and $x \to -\infty$
1.2.15   factorise cubic polynomials in cases where all roots are given or easily obtained from the graph
1.2.16   solve cubic equations using technology, and algebraically in cases where all roots are given or easily obtained from the graph

Graphs of relations
1.2.17   recognise features and determine equations of the graphs of $x^2 + y^2 = r^2$ and $(x - a)^2 + (y - b)^2 = r^2$, including their circular shapes, their centres and their radii
1.2.18   recognise features of the graph of $y^2 = x$, including its parabolic shape and its axis of symmetry

Functions
1.2.19   understand the concept of a function as a mapping between sets and as a rule or a formula that defines one variable quantity in terms of another
1.2.20   use function notation; determine domain and range; recognise independent and dependent variables
1.2.21   understand the concept of the graph of a function
1.2.22   examine translations and the graphs of $y = f(x) + a$ and $y = f(x - b)$
1.2.23   examine dilations and the graphs of $y = cf(x)$ and $y = f(dx)$
1.2.24   recognise the distinction between functions and relations and apply the vertical line test

Mathematics Methods ATAR Course Year 11 syllabus p. 10 © SCSA

Video playlists (8):
4.1   Polynomials and power functions
4.2   Solving cubic equations
4.3   Graphing cubic functions
4.4   Functions and relations
4.5   Transformations of functions
4.6   Graphs of relations
4.7   Inverse proportion
Examination question analysis Functions and relations

Worksheets (8):
4.2   Factorising polynomials
4.3   Cubic functions • Graphing cubics
4.4   Functions and relations
4.5   Translations of functions • Dilations of functions • Graphing transformed functions
4.7   Variation problems

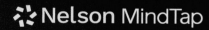

To access resources above, visit
cengage.com.au/nelsonmindtap

4.1 Polynomials and power functions

Polynomials

A **polynomial** is an expression involving a sum of terms.

Polynomial comes from the Greek words poly- (meaning 'many') and -nomial (meaning 'term'), so it translates to 'many terms'.

A polynomial in x is of the form:

$$P(x) = a_n x^n + a_{n-1} x^{n-1} + a_{n-2} x^{n-2} + \ldots + a_1 x^1 + a_0$$

where n is a non-negative whole number, $a_0, a_1, a_2 \ldots a_{n-1}, a_n$ are real numbers and $a_n \neq 0$.

- The numbers multiplying the different powers of x are called the coefficients. The coefficients are $a_0, a_1, a_2 \ldots a_{n-1}, a_n$.
- The constant is a_0 and is also known as the **constant** term.
- The highest power of x is called the **degree of a polynomial**. The degree of $P(x) = a_n x^n + a_{n-1} x^{n-1} + a_{n-2} x^{n-2} + \ldots + a_1 x^1 + a_0$ is n.
- The term that contains the highest power is called the **leading term**, $a_n x^n$, and the coefficient of the leading term, a_n, is called the **leading coefficient**.
- When the leading coefficient, a_n, is 1, the polynomial is called a **monic polynomial**.

Some examples of polynomials are:

5	A polynomial of degree 0 (it could be written as $5x^0$)
$2x - 3$	A polynomial of degree 1, called a **linear polynomial**
$4x^2 + 3\sqrt{2}x + 9$	A polynomial of degree 2, called a **quadratic polynomial**
$x^3 - 4x^2 + x$	A polynomial of degree 3, called a **cubic polynomial**
$\frac{2}{3}x^4 + 2x^2 - 5x + 0.28$	A polynomial of degree 4, called a **quartic polynomial**

WORKED EXAMPLE 1 Polynomial expressions and their features

Determine whether each expression is a polynomial, giving reasons if it is not. State the degree, leading term and coefficients of any polynomial found.

a $4x - 3x^2 + 2$

b $x^4 + \dfrac{x^2}{4} - \dfrac{3}{x} + 7x^3$

Steps	Working
a 1 Write the expression with descending powers of x. All of the powers are positive whole numbers, so it is a polynomial.	$-3x^2 + 4x + 2$ This is a polynomial.
2 Identify the degree, leading term and coefficients.	degree = 2 leading term = $-3x^2$ coefficients = $-3, 4, 2$
b Write the expression with descending powers. Not all of the powers are positive whole numbers, so it is not a polynomial.	$x^4 + 7x^3 + \dfrac{1}{4}x^2 - 3x^{-1}$ This is not a polynomial as $3x^{-1}$ has a negative power.

Equating coefficients

If two polynomials are equal for all values of x, then they are identical and the coefficients of the same powers of x are equal. This allows us to **equate the coefficients** of each corresponding term on either side of the '=' sign.

> **Equating coefficients of identical polynomials**
>
> If $P(x) = a_n x^n + a_{n-1} x^{n-1} + a_{n-2} x^{n-2} + \ldots + a_1 x^1 + a_0$ and
> $Q(x) = b_n x^n + b_{n-1} x^{n-1} + b_{n-2} x^{n-2} + \ldots + b_1 x^1 + b_0$
> are equal for all values of x, then $a_n = b_n$, $a_{n-1} = b_{n-1}$, $a_{n-2} = b_{n-2}$ and so on up to $a_0 = b_0$.

WORKED EXAMPLE 2 — Equating coefficients

If $x^2 + 4x + 2 = a(x + 2)^2 + b$ for all x, find the values of a and b.

Steps	Working
1 Expand the RHS.	$a(x + 2)^2 + b = a(x^2 + 4x + 4) + b$ $= ax^2 + 4ax + 4a + b$
2 Let LHS = RHS. LHS = left-hand side RHS = right-hand side	So $x^2 + 4x + 2 = ax^2 + 4ax + 4a + b$.
3 Equate coefficients.	coefficients of x^2: $\quad 1 = a$ coefficients of x: $\quad 4 = 4a \Rightarrow a = 1$ constant term: $\quad 2 = 4a + b$ Substituting $a = 1$: $2 = 4 + b$ $b = -2$ So $a = 1$, $b = -2$. (That is, $x^2 + 4x + 2 = (x + 2)^2 - 2$.)

WORKED EXAMPLE 3 — Substituting values into polynomials

a Given that $P(x) = 5x^2 + 4x - 2$, find

  i $P(-3)$

  ii $P(2a - 1)$

b x when $P(x) = 7$.

Steps	Working
a i Substitute $x = -3$ into $P(x)$ and evaluate.	$P(-3) = 5(-3)^2 + 4(-3) - 2$ $= 31$
ii Substitute $x = 2a - 1$ into $P(x)$ and simplify.	$P(2a - 1) = 5(2a - 1)^2 + 4(2a - 1) - 2$ $= 5(4a^2 - 4a + 1) + 8a - 4 - 2$ $= 20a^2 - 20a + 5 + 8a - 4 - 2$ $= 20a^2 - 12a - 1$
b Substitute $P(x) = 7$ and solve for x.	$5x^2 + 4x - 2 = 7$ $5x^2 + 4x - 9 = 0$ $(5x + 9)(x - 1) = 0$ $x = -\dfrac{9}{5}$ or $x = 1$

USING CAS 1 | Defining and evaluating polynomials

Given $P(x) = -2x^4 + 5x^3 + 3x^2 + 11$, find $P(-2)$.

ClassPad

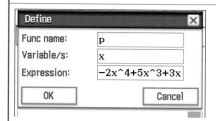
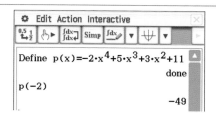

1. Tap **Main** and clear all calculations.
2. Enter and highlight the polynomial.
3. Tap **Interactive** > **Define**.
4. Change the **Func name:** to **p** (use the letter **p**, not the variable *p*).
5. The polynomial will appear in the **Expression:** field, as shown above.
6. Tap **OK**.

7. The defined function will be displayed.
8. Enter **p(–2)**.
9. Press **EXE**. The solution will be displayed.

TI-Nspire

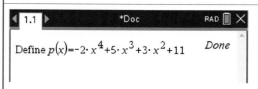

1. Add a **Calculator** page.
2. Press **menu** > **Actions** > **Define**.
3. Enter the function as shown above.
4. Press **enter**.

5. Enter **p(–2)**.
6. Press **enter**. The solution will be displayed.

$P(-2) = -49$

Power functions

As studied in Chapter 3, $y = x$ is a linear function (straight line) and $y = x^2$ is a quadratic function (parabola). The graphs of $y = x^3, x^4, x^5$, etc. become progressively steeper as the power increases, and the curvature near $(0,0)$ becomes more pronounced.

The shapes of even powers of $y = x^n$ are similar to $y = x^2$, but the higher the power the steeper the curve. For all even powers, as $x \to \pm\infty$, $y \to \infty$.

The shapes of odd powers of $y = x^n$ are similar to the shape of $y = x^3$, but the higher the power the steeper the curve. For all odd powers, as $x \to \pm\infty$, $y \to \pm\infty$.

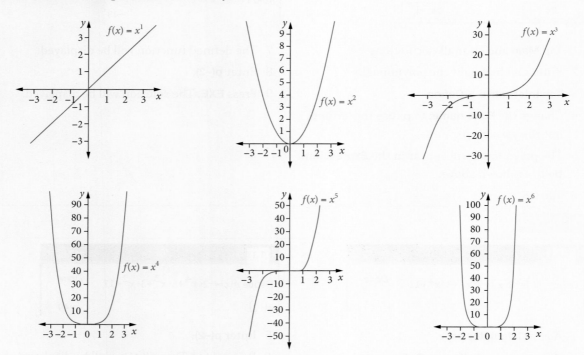

At $(0,0)$, graphs of *even powers* have a turning point and graphs of *odd powers* ≥ 3 have a point of inflection.

As the power gets greater, the curve gets steeper at both ends, but is *flatter* near the turning point.

The hyperbola

The simple **hyperbola** is the graph of the function $f(x) = x^{-1} = \dfrac{1}{x}$.

As $x \to \pm\infty$, $\dfrac{1}{x} \to 0$.

The function is undefined at $x = 0$.

We can also plot points or use CAS to find the general shape of $f(x)$.

The hyperbola

There is a **discontinuity** at $x = 0$ (there is a break in the graph at $x = 0$). So the graph of $y = \dfrac{1}{x}$ is separated into 2 branches.

The x-axis ($y = 0$) is a **horizontal asymptote**: as $x \to \pm\infty$ (as x approaches positive or negative infinity), the curve approaches $y = 0$, but never reaches it.

The y-axis ($x = 0$) is a **vertical asymptote**: as $x \to 0$, $y \to \pm\infty$ (y approaches positive or negative infinity) so the curve approaches $x = 0$, but never reaches it.

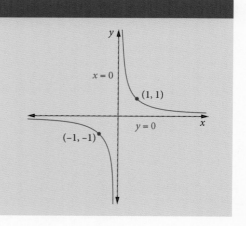

The families of hyperbolas given by $f(x) = ax^{-1} = \dfrac{a}{x}$ and $f(x) = (x + b)^{-1} = \dfrac{1}{x + b}$ have the same relationship to the basic hyperbola as the families of other functions. The graph is moved closer or further away from the origin by the parameter a. The vertical asymptote is moved left or right by the parameter b.

WORKED EXAMPLE 4 — Sketching $f(x) = ax^{-1}$

Sketch the graph of $f(x) = \dfrac{3}{x}$.

Steps	Working
1 Write the comparison with $y = x^{-1}$.	$f(x) = \dfrac{3}{x} = 3x^{-1}$ is $y = x^{-1}$ dilated by a factor 3 from the x-axis.
2 State the asymptotes.	The asymptotes are still $x = 0$ and $y = 0$.
3 Sketch the graph.	

WORKED EXAMPLE 5 — Sketching $f(x) = (x + b)^{-1}$

Sketch the graph of $f(x) = \dfrac{1}{x - 1}$.

Steps	Working
1 Write the comparison with $y = x^{-1}$.	$f(x) = \dfrac{1}{x - 1}$ is moved 1 unit to the right from $\dfrac{1}{x}$.
2 State the asymptotes.	The asymptotes are $x = 1$ and $y = 0$.
3 State the y-intercept.	$f(0) = -1$
4 Sketch the graph.	

Square root functions

The **square root function**, $f(x) = x^{\frac{1}{2}} = \sqrt{x}$ is shown on the right.

As $x \to \infty$, $y \to \infty$.

The graph starts at $(0,0)$ and does not exist for negative x values.

$y = \sqrt{x-2}$ is the graph of $y = \sqrt{x}$ translated 2 units to the right.

$y = \sqrt{x} - 2$ is the graph of $y = \sqrt{x}$ translated 2 units down.

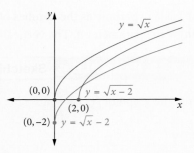

EXERCISE 4.1 Polynomials and power functions   ANSWERS p. 464

Mastery

1. **WORKED EXAMPLE 1** State whether or not each expression is a polynomial.

   a $5x^4 - 3x^2 + x + \dfrac{1}{x}$ b $x^2 + 3^x$ c $x^2 + 3x - 7$

   d $3x + 5$ e $6x^3 + 4\sqrt{x} - 8$ f $3x^2 - \dfrac{1}{2}x + 1$

2. For the polynomial $P(x) = x^5 - 3x^4 - 5x + 4$, find the

   a degree of $P(x)$ b constant term

   c coefficient of x^4 d coefficient of x^2.

3. **WORKED EXAMPLE 2** Find the values of a and b such that $a(x+2) + b(x+3) = 18x + 8$ for all values of x.

4. **WORKED EXAMPLE 3**

   a Given $f(x) = x + 3$, find

      i $f(1)$ ii $f(-3)$

   b If $h(x) = x^2 - 2$, find

      i $h(0)$ ii $h(2)$ iii $h(-4)$

   c If $f(x) = 2x - 5$, find x when $f(x) = 13$.

   d Given $P(x) = x^2 + 3$, find any values of x for which $P(x) = 28$.

   e If $f(x) = 2x - 9$, find $f(x + h)$.

   f Find $g(x - 1)$ when $g(x) = x^2 + 2x + 3$.

5. **Using CAS 1**

   a Given $f(x) = x^2 - 5x + 2$, find

      i $f(0)$ ii $f(-1)$

   b If $h(x) = x^3 - 2x^2 - x + 6$, find

      i $h(4)$ ii $h(-2)$

6. **WORKED EXAMPLE 4** Sketch $y = -\dfrac{2}{x}$.

7. **WORKED EXAMPLE 5** Sketch $y = \dfrac{1}{x+2}$.

194   Nelson WAmaths Mathematics Methods 11

Calculator-free

8 (2 marks) Compare the graphs of $y = x^3$ and $y = x^5$, commenting on any differences and similarities.

9 (5 marks) $P(x) = (a + 1)x^3 + (b - 7)x^2 + c + 5$. Find values for a, b or c if

    a $P(x)$ is monic   (1 mark)

    b   the coefficient of x^2 is 3   (1 mark)

    c   the constant is -1 (1 mark)

    d $P(x)$ has degree 2   (1 mark)

    e   the leading term has a coefficient of 5.   (1 mark)

10 (4 marks) Find the value of A, B and C if $6x^2 - 24x + 14 = A(x + B)^2 + C$.

Calculator-assumed

11 (2 marks) If $f(x) = 4x^3 + bx^2 - 3x$ and $f(-2) = 3f(1)$, find b.

12 (3 marks) Find the value of m, n and p if $3x^3 - 18x^2 + 36x - 11 = m(x + n)^3 + p$.

13 (2 marks) Explain why the graph of $y = \dfrac{1}{x}$ has asymptotes at $y = 0$ and $x = 0$.

4.2 Solving cubic equations

Consider the polynomial $P(x) = x^3 + 4x^2 + x - 6$, which is written in expanded form.

This can also be written in factored form, which in this case is $P(x) = (x + 2)(x + 3)(x - 1)$.

Important points from this are

- -2, -3 and 1 are the x-intercepts (or roots)
- $P(-2) = 0$, $P(-3) = 0$ and $P(1) = 0$, which means that $(x + 2)$, $(x + 3)$ and $(x - 1)$ are factors of the polynomial
- -2, -3 and 1 are factors of -6, the constant term of the polynomial.

We can use this information to solve cubic equations.

Properties of factors of polynomials

- For a polynomial $P(x)$, if $P(a) = 0$, then $(x - a)$ is a factor of the polynomial.
- a is called a **zero** of the polynomial $P(x)$ or a **root** of the polynomial equation $P(x) = 0$.
- If $(x - a)$ is a factor of $P(x)$, then $P(a) = 0$.
- If $P(x) = (x - a)(x - b)(x - c) = 0$, then the solutions are $x = a$, $x = b$, $x = c$.

Video playlist Solving cubic equations

Worksheet Factorising polynomials

WORKED EXAMPLE 6 — Solving cubic equations

Consider the cubic polynomial $P(x) = x^3 - 4x^2 - 7x + 10$.

a Show that $(x - 1)$ is a factor of $P(x)$.

b Hence, solve the equation $P(x) = 0$.

Steps	Working
a 1 Substitute $x = 1$ into $P(x)$.	Replace x with 1. $P(1) = 1^3 - 4 \times 1^2 - 7 \times 1 + 10 = 0$, therefore, $(x - 1)$ is a factor.
b 1 Write the cubic in partially factored form (if $(x - 1)$ is a factor, then a quadratic factor remains).	$(x - 1)(ax^2 + bx + c) = 0$
2 Expand and equate coefficients to determine a, b and c.	$(x - 1)(ax^2 + bx + c) = ax^3 + (b - a)x^2 + (c - b)x - c$ $\therefore ax^3 + (b - a)x^2 + (c - b)x - c = x^3 - 4x^2 - 7x + 10$ Equating coefficients, $a = 1$, $c = -10$, $b - a = -4$ so $b = -3$.
3 Factorise the quadratic and write down the fully factored cubic.	$x^2 - 3x - 10 = (x - 5)(x + 2)$ Therefore, $(x - 1)(x - 5)(x + 2) = 0$
4 State the solutions to the cubic equation.	$x = 1$, $x = 5$ and $x = -2$

Note: In some cases, all the factors can also be found by 'trial and error' by trying different factors of the constant term to see which make $P(x) = 0$.

USING CAS 2 — Finding factors of polynomials

Factorise $P(x) = x^3 - 8x^2 + 19x - 12$.

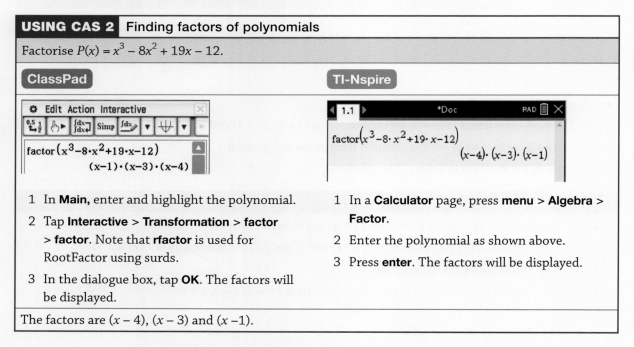

ClassPad

1. In **Main**, enter and highlight the polynomial.
2. Tap **Interactive** > **Transformation** > **factor** > **factor**. Note that **rfactor** is used for RootFactor using surds.
3. In the dialogue box, tap **OK**. The factors will be displayed.

TI-Nspire

1. In a **Calculator** page, press **menu** > **Algebra** > **Factor**.
2. Enter the polynomial as shown above.
3. Press **enter**. The factors will be displayed.

The factors are $(x - 4)$, $(x - 3)$ and $(x - 1)$.

When solving a cubic equation, it should be written in **standard form**, $ax^3 + bx^2 + cx + d = 0$. The equation $x^2 + 4x - 1 = \dfrac{6}{x}$ is a cubic, though it is not written in the standard form. We need to multiply through by x and then subtract 6 from both sides, giving us $x^3 + 4x^2 - x - 6 = 0$. This is now in standard form. However, cubics of the form $a(x - b)^3 + c$ can be solved as shown in Worked example 7.

WORKED EXAMPLE 7 | Solving equations involving perfect cubes

Solve $2(x - 1)^3 - 250 = 0$.

Steps	Working
1 Add 250 to both sides.	$2(x - 1)^3 = 250$
2 Divide both sides by 2.	$(x - 1)^3 = 125$
3 Take the cube root of both sides.	$x - 1 = 5$
4 Add 1 to both sides.	$x = 6$

USING CAS 3 | Solving polynomial equations

Solve $2x^3 - 3x^2 - 39x + 20 = 0$.

ClassPad

1. In **Main**, set the mode to **Standard**.
2. Enter and highlight the equation.
3. Tap **Interactive** > **Equation/Inequality** > **solve**.
4. In the dialogue box, tap **OK**. The solutions will be displayed.

TI-Nspire

1. In a **Calculator** page, press **menu** > **Algebra** > **Solve**.
2. Enter the equation followed by **,x** as shown above.
3. Press **enter**. The solutions will be displayed.

The solutions are $x = -4$, $x = \dfrac{1}{2}$ and $x = 5$.

EXERCISE 4.2 Solving cubic equations

ANSWERS p. 464

Recap

1. Given that $P(x) = 3x^3 - 2x^2 + x - 2$, determine $P(-2)$.

2. Which of the following expressions are polynomials?

    a $3x^2 - 2x^3 + \sqrt{7x^2}$ b $x^5 + \dfrac{3}{x^2} + 7x^2$ c $2^x + x^2$ d $(x^4)^{\frac{1}{2}} - \dfrac{x^2}{3x}$

Mastery

3. **WORKED EXAMPLE 6**

    a Consider the cubic polynomial $P(x) = x^3 + 3x^2 - 10x - 24$.
       i Show that $(x + 2)$ is a factor of $P(x)$.
       ii Hence, solve the equation $P(x) = 0$.
    b Consider the cubic polynomial $P(x) = x^3 + x^2 - 9x - 9$.
       i Show that $(x + 1)$ is a factor of $P(x)$.
       ii Hence, solve the equation $P(x) = 0$.

4. **Using CAS 2** Factorise $x^3 - 11x^2 + 4x + 60$.

5 WORKED EXAMPLE 7 Solve each cubic equation.

a $3(x-4)^3 - 24 = 0$
b $3(x+5)^3 + 81 = 0$
c $-2(x-2)^3 - 2 = 0$
d $(3x-2)^3 - 8 = 0$

6 Using CAS 3 Solve each cubic equation.

a $x^3 - 13x + 12 = 0$
b $-x^3 + 2x^2 + 9x - 18 = 0$
c $x^3 + 2x^2 - 4x - 8 = 0$
d $x^3 - 5x^2 + 8x - 4 = 0$

Calculator-free

7 (8 marks)

a Given that $(x+5)$ and $(x+1)$ are factors of $P(x) = x^3 + 4x^2 - 7x - 10$, determine the third factor of the cubic polynomial. (2 marks)

b Given that $(x+7)$ and $(x-6)$ are factors of $P(x) = x^3 + 2x^2 - 41x - 42$, determine the third factor of the cubic polynomial. (2 marks)

c Given that $(x+2)^2$ are factors of $P(x) = x^3 + 2x^2 - 4x - 8$, determine the third factor of the cubic polynomial. (2 marks)

d Given that $(2x+3)$ and $(x+4)$ are factors of $P(x) = 2x^3 + 7x^2 - 10x - 24$, determine the third factor of the cubic polynomial. (2 marks)

8 (2 marks) If $x + a$ is a factor of $4x^3 - 13x^2 - ax$, determine a.

9 (3 marks) Part of the graph of the curve with equation $y = x^3 - 2x^2 - 5x + 6$ is shown.

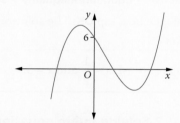

a Write the equation in the form $y = (x+2)(x^2 + bx + c)$. (1 mark)
b Hence, by factorising, find the exact values of the x-intercepts. (2 marks)

Calculator-assumed

10 (4 marks) Factorise each polynomial.

a $x^3 - 7x^2 + 4x + 12$ (1 mark)
b $x^3 + 2x^2 - 9x - 18$ (1 mark)
c $2x^3 - 5x^2 - 23x - 10$ (1 mark)
d $2x^3 + 3x^2 - 11x - 6$ (1 mark)

11 (2 marks) If $x + a$ is a factor of $8x^3 - 14x^2 - a^2x$, determine a.

12 (3 marks) A part of the track for Tim's model train follows the curve passing through A, B, C, D, E and F as shown. Tim has designed it by putting axes on the drawing as shown. The track is made up of two curves, one to the left of the y-axis and the other to the right.

B is the point (0, 7).

The curve from B to F is part of the graph of $f(x) = px^3 + qx^2 + rx + s$ where p, q, r and s are constants.

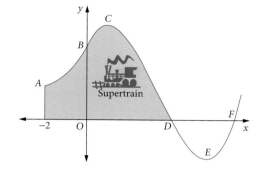

a Show that s = 7. (1 mark)

b Assume $f(x) = 0.25x^3 - 2.5x^2 + 4.25x + 7$.
   Find the exact coordinates of D and F. (2 marks)

4.3 Graphing cubic functions

Video playlist
Graphing cubic functions

A **cubic function** is a polynomial of degree 3, whose equation has the general form $y = ax^3 + bx^2 + cx + d$. The simplest cubic function is $y = x^3$, shown below. The centre of the graph at (0, 0) is flat and is a special turning point called a **point of inflection**, at which the graph changes from concave up to concave down, or from concave down to concave up. (Inflection means 'bend'.) As $x \to \pm\infty$, $y \to \pm\infty$.

Worksheets
Cubic functions

Graphing cubics

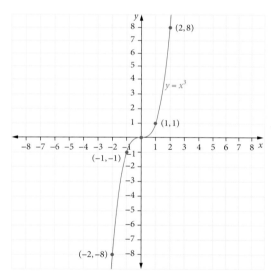

The graph of $y = ax^3$

The graphs of $y = ax^3$ are shown. As can be seen the value of the coefficient a makes the graph more steep or flat. If a is negative, such as in $y = -x^3$, then the graph is reflected in the x-axis, pointing in the opposite directions.

Compared to the graph of $y = x^3$, the effect of the value of a is to dilate ('stretch' or 'squash') $y = ax^3$ in the y direction.

If a is large ($y = 5x^3$, for example), the cubic is **steeper**.

If a is small ($y = \frac{1}{5}x^3$, for example), the cubic is **flatter**.

If a is negative ($y = -x^3$, for example), the cubic is **reflected in the x-axis** (flipped upside-down).

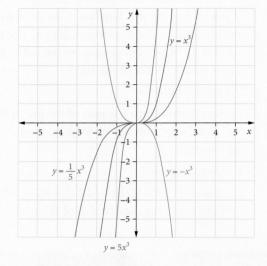

The graph of $y = x^3 + c$

The graphs of $y = x^3 + c$ are shown. As can be seen the value of the constant c moves the graph of $y = x^3$, c units up (or down if c is negative), so that c is the y-intercept and $(0, c)$ is the point of inflection.

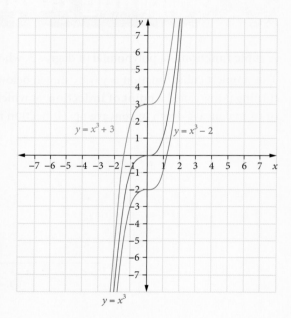

The graph of $y = (x - b)^3$

The graphs of $y = (x - b)^3$ are shown. As can be seen the value of the constant b moves the graph of $y = x^3$, b units right (or left if b is negative), so that b is the x-intercept and $(b, 0)$ is the point of inflection.

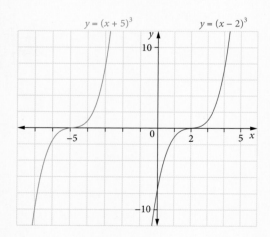

The graph of $y = a(x - b)^3 + c$

Previously we learnt that compared to the graph $y = x^2$, the graph of $y = a(x - b)^2 + c$ is a parabola that is stretched by a factor a from the x-axis and the turning point is at (b, c).

The same is true for the cubic curve $y = a(x - b)^3 + c$, which is the graph of $y = x^3$ stretched by a factor a from the x-axis and has point of inflection at (b, c).

WORKED EXAMPLE 8 — Graphing $y = (x - b)^3 + c$

Sketch the graph of $y = (x + 2)^3 - 1$.

Steps	Working
1 Compare $y = (x + 2)^3 - 1$ with $y = x^3$.	$y = (x + 2)^3 - 1$ is $y = x^3$ translated 2 units left and 1 unit down. Its new centre (point of inflection) is $(-2, -1)$.
2 Find its x-intercept(s).	Substitute $y = 0$: $$0 = (x + 2)^3 - 1$$ $$1 = (x + 2)^3$$ $$\sqrt[3]{1} = x + 2$$ $$x + 2 = 1$$ $$x = -1$$
3 Find its y-intercept.	Substitute $x = 0$: $$y = (0 + 2)^3 - 1$$ $$= 7$$
4 Sketch the graph. It will have the same shape as $y = x^3$.	(graph showing $y = (x+2)^3 - 1$ with points 7 on y-axis, -1 on x-axis, and inflection point $(-2, -1)$)

The graph of $y = ax^3 + bx^2 + cx + d$

The graph of the general cubic function $y = ax^3 + bx^2 + cx + d$ does not have a flat point of inflection. Instead, it has 2 turning points or vertices: a maximum point and a minimum point. The shape of the cubic depends on the sign of the leading coefficient of the function, a:

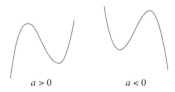

$a > 0$ $a < 0$

- If a is positive, the cubic curve starts at the bottom and finishes at the top (see above diagram); that is, $f(x) \to -\infty$ as $x \to -\infty$, and $f(x) \to +\infty$ as $x \to +\infty$.
- If a is negative, the cubic curve starts at the top and finishes at the bottom (see above diagram); that is, $f(x) \to +\infty$ as $x \to -\infty$, and $f(x) \to -\infty$ as $x \to +\infty$.

The graph of $y = a(x - p)(x - q)(x - r)$

If a cubic function can be factorised using the factor theorem so that it is of the form $y = a(x - p)(x - q)(x - r)$, then p, q and r are its x-intercepts. For example, $y = (x + 3)(x + 1)(x - 2)$ is shown, with x-intercepts at -3, -1 and 2.

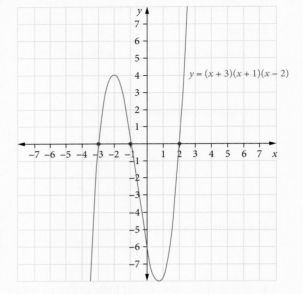

If one of the factors is squared, then the x-intercept 'touches' (rather than crosses) the x-axis For example, $y = 3(x - 1)^2(x + 5)$ 'touches' the x-axis at $x = 1$. $(x - 1)^2$ is called a repeated factor or a double factor.

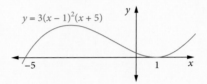

WORKED EXAMPLE 9 — Graphing $y = ax^3 + bx^2 + cx + d$

Sketch the graph of $y = -x^3 + x^2 + 2x$, showing all intercepts.

Steps	Working
1 Factorise $y = -x^3 + x^2 + 2x$.	$y = -x^3 + x^2 + 2x$ $= -x(x^2 - x - 2)$ $= -x(x - 2)(x + 1)$
2 Find the x-intercepts.	Substitute $y = 0$: $0 = -x(x - 2)(x + 1)$ $x = 0, 2, -1$
3 Find the y-intercept.	Substitute $x = 0$: $y = -0(0 - 2)(0 + 1)$ $= 0$
4 Determine the direction of the graph.	For $y = -x^3 + x^2 + 2x$, the leading coefficient $a = -1 < 0$, so the graph points down at the right.
5 Sketch the graph.	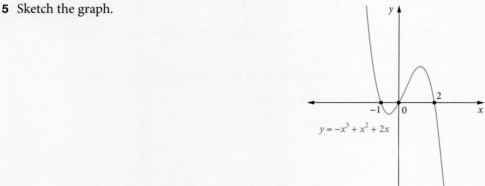

Graphing cubic functions

1 Determine the shape of the cubic curve from its equation and leading coefficient, a: it can have up to one maximum point and one minimum point.
2 Factorise the function if possible and note any repeated factors.
3 Find the x- and y-intercepts: it can have up to three x-intercepts but only one y-intercept.
The y-intercept of $y = ax^3 + bx^2 + cx + d$ is d.

WORKED EXAMPLE 10 Finding the equation from the graph

Find the rule for the function with the graph shown.

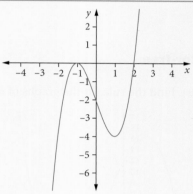

Steps	Working
1 Write down the x-intercepts.	$(2, 0), (-1, 0)$
2 Enter values into $y = a(x - p)(x - q)(x - r)$.	$y = a(x - 2)(x + 1)^2$ $(x + 1)$ is a repeated factor because curve touches x-axis at $x = -1$.
3 Substitute in any coordinates from the graph (except x-intercepts) to determine a.	Look at graph to see that when $x = 1$ then $y = -4$. $-4 = a(1 - 2)(1 + 1)^2$ $-4 = a(-1)(2)^2$ $-4 = a \times -4$ $\therefore a = 1$
4 Write down the equation.	$y = (x - 2)(x + 1)^2$

EXERCISE 4.3 Graphing cubic functions   ANSWERS p. 465

Recap

1 Factorise $P(x) = x^3 - 3x^2 - 4x + 12$.

2 Solve $2(x - 3)^3 = 16$.

Mastery

3 **WORKED EXAMPLE 8** Graph each cubic function.
  a $y = x^3 - 3$
  b $y = -x^3 + 2$
  c $y = -x^3 - 1$
  d $y = x^3 + 1$
  e $y = (x - 1)^3 + 3$
  f $y = 4 - (x + 2)^3$
  g $y = 3(x - 2)^3 + 1$
  h $y = (2x - 1)^3 + 3$
  i $y = (3x + 5)^3 - 2$
  j $y = 4 - 2(3x - 6)^3$

4 **WORKED EXAMPLE 9** Graph each cubic function.
  a $y = (x + 1)(x - 2)(x - 3)$
  b $y = x(x + 4)(x - 2)$
  c $y = x(x + 2)^2$
  d $y = (5 - x)(x + 2)(x + 5)$
  e $y = -x^3 - 4x^2 + 5x$
  f $y = 2x^3 + x^2 - 15x$

5 **WORKED EXAMPLE 10** Find the rule for the function with the graph shown.

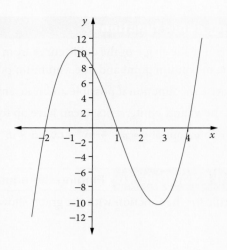

Calculator-free

6 (6 marks) Find the rule for the graphs of each of the following functions.

a (2 marks)

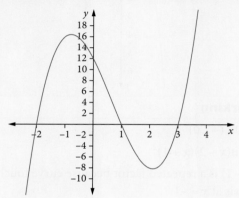

b (2 marks)

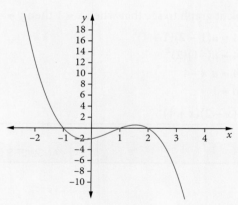

c (2 marks)

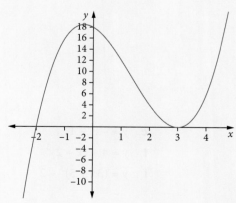

7 (2 marks) Determine the values of a and b if $(x - b)(ax + 2)(x - 1) = 3x^3 - 10x^2 + x + 6$.

8 (5 marks) A polynomial function p has degree 3. A part of its graph, near the point on the graph with coordinates $(2, 0)$, is shown.

State whether each of the following could be the rule for polynomial p.
  a $p(x) = -x(x - 2)^2$ (1 mark)
  b $p(x) = (x - 1)(x - 2)^2$ (1 mark)
  c $p(x) = (x - 2)^3$ (1 mark)
  d $p(x) = x^2(x - 2)$ (1 mark)
  e $p(x) = x(x - 2)^2$ (1 mark)

9 (5 marks) Let f be a polynomial function of degree 3. The graph of the curve with rule $y = f(x)$ either intersects or touches the x-axis at exactly two points, $(a, 0)$ and $(b, 0)$ where $a > 0$ and $b > 0$. State which equations are possible rules for f.
  a $f(x) = (x - a)(x - b)$ (1 mark)
  b $f(x) = (x - a)(x + b)^2$ (1 mark)
  c $f(x) = (x - a)(x - b)^2$ (1 mark)
  d $f(x) = (x - a)^2(x - b)$ (1 mark)
  e $f(x) = (x + a)^2(x + b)$ (1 mark)

10 (4 marks) Kim wants to determine the length and height of her bookcase. The function $f(x) = x^3 + 14x^2 + 57x + 72$ represents the volume of the bookcase, where the width is $x + 3$ units. What are the possible dimensions for the length and height of the bookcase?

11 (9 marks) A specimen of bacteria is smeared on a round agar plate of diameter 12 cm and left for testing. The area (in cm^2) of the bacterial colony is given by the function $A(t) = -t^3 + 6t^2 + 16t + 0.5$ where t is the number of weeks after the culture is started.
  a What is the initial size of the colony? (1 mark)
  b Sketch the graph for the first 10 weeks. (2 marks)
  c When does the model of growth become inadequate? (2 marks)
  d What percentage of the plate is covered after 4 weeks? (2 marks)
  e What is the maximum percentage of the plate that is covered? (2 marks)

12 (2 marks) Given the y-intercept is $(0, 24)$, determine the equation of the graph in the form $y = ax^3 + bx^2 + cx + d$.

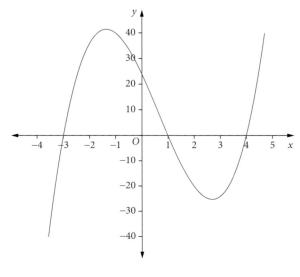

4.4 Functions and relations

Relations

A **relation** R is a set of ordered pairs (x, y), where x and y are usually both in the set of real numbers. The set of values of the **independent variable** x is called the **domain** and is often the set X. The set Y is called the **co-domain** and the set of values of the **dependent variable** y is called the **range**.

We can show a relation as a rule (formula), a list of ordered pairs, a table, a mapping or a graph.

A **mapping** shows ordered pairs connected by arrows from the x values to the y values.

WORKED EXAMPLE 11 — Mappings and ordered pairs

The mapping shows the relation M.

a Write M as a set of ordered pairs.
b Show M as a table.
c State the domain and range of M.

Steps	Working
a Write the ordered pairs.	$M = \{(3, 2), (3, 12), (5, 5), (7, 4), (7, 5), (9, 4), (9, 12)\}$
b Write the ordered pairs in a table.	<table><tr><td>x</td><td>3</td><td>3</td><td>5</td><td>7</td><td>7</td><td>9</td><td>9</td></tr><tr><td>y</td><td>2</td><td>12</td><td>5</td><td>4</td><td>5</td><td>4</td><td>12</td></tr></table>
c 1 The domain is the set of first numbers.	Write the domain and range in order. domain = {3, 5, 7, 9}
2 The range is the set of second numbers.	range = {2, 4, 5, 12}

Function notation

> **Functions**
>
> A **function** $y = f(x)$ is a relation such that each x value in the domain has only one ordered pair (x, y). In other words, each x value matches with only one y value.

A function may be shown by a rule like $f(x) = 5x - 1$.

$f(x)$, pronounced 'f of x', is called **function notation**, and means the rule for operating on the independent variable, x.

The value of $f(x)$ at $x = c$ is written as $f(c)$.

For example, $f(2)$ is the value of $f(x) = 5x - 1$ when $x = 2$.

$f(2) = 5 \times 2 - 1 = 9$

WORKED EXAMPLE 12 | Determining if a relation is also a function

State whether each relation is a function.

a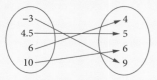

b $x^2 = y^2$

c $y = \sqrt{x - 2}$

d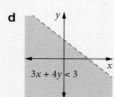

$3x + 4y < 3$

Steps	Working
a 1 Write the ordered pairs.	Relation is $(-3, 9), (4.5, 5), (6, 4), (10, 6)$.
2 Write the conclusion.	It is a function because each x value occurs only once.
b 1 Do any ordered pairs have the same x value?	$(2, -2)$ and $(2, 2)$ have the same x value. $x = 2$ has 2 matching y values.
2 Write the conclusion.	It is not a function as there are 2 ordered pairs with the same x value.
c 1 Do any ordered pairs have the same x value?	$\sqrt{x - 2}$ means the positive value so each ordered pair (x, y) is unique.
2 Write the conclusion.	It is a function because each x value has one y value.
d 1 Do any ordered pairs have the same x value?	$(0, 0.5), (0, 0), (0, -0.2), (0, -1) \ldots$ have the same x value.
2 Write the conclusion.	It is not a function as there is an infinite number of ordered pairs with the same x value.

We can usually determine from a graph if a relation is a function. A function will show any value on the x-axis with (at most) one value for y. A relation may have 2 or more y values for one x value: there will be 2 or more parts of the graph directly above each other.

If any **vertical line** intersects the graph of a relation at most once, then it is a function.

If a vertical line intersects the graph of a relation in more than one place, then it is not a function.

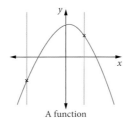

A function

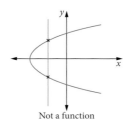

Not a function

WORKED EXAMPLE 13 — Using graphs to determine if a relation is also a function

State whether each of the following shows the graph of a function.

a

b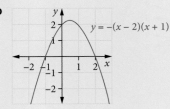

c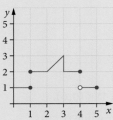

d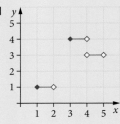

Steps	Working
a 1 Do any points have the same x value?	There are 2 points at $x = 2$: $(2, 5)$ and $(2, -1)$.
2 Write the conclusion.	$(x - 2)^2 + (y - 2)^2 = 9$ is not a function.
b 1 Do any points have the same x value?	There are no points with the same x value.
2 Write the conclusion.	$y = -(x - 2)(x + 1)$ is a function.
c 1 Do any points have the same x value?	$(3, 2), (3, 2.1) \ldots (3, 3)$ have the same x value. $(1, 1)$ and $(1, 2)$ have the same x value.
2 Write the conclusion.	This graph does not show a function.
d 1 Do any points have the same x value?	There are no points with the same x value. At $x = 4$, there is no value, so the relation is undefined there.
2 Write the conclusion.	This graph does show a function.

Most relations with rules like 'y = algebraic rule involving x' are functions.

For example, $y = \dfrac{\sqrt{x - 3}}{x - 5} + \dfrac{1}{\sqrt{7 - x}}$ is a function. Each value of x in its domain has only one corresponding value of y.

Example 1: One-to-one function

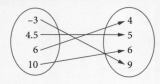

Example 2: Many-to-one function

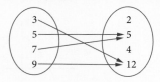

Consider the above function mapping.

- In example 1, each x value in the domain is matched with a unique y value. In the pairings, there aren't any repeated x or y values. This is called a **one-to-one function**.
- In example 2, each x value in the domain is matched with one y value. But notice that for both $x = 5$ and $x = 7$, $y = 5$. And for both $x = 3$ and $x = 9$, $y = 12$. In the pairings, there are repeated y values. This is called a **many-to-one function**.

Domain and range

The range of a function is the set of y values that the domain produces. The y values are the images of the x values, so the range is the image of the domain.

The **implied domain** of a function such as $f(x) = \sqrt{x-3}$ is the set of all real values of x for which the function is defined. This is also called the **natural domain**.

For example, for $f(x) = x^2$ the domain is all real numbers, and the range is $f(x) \geq 0$. For $y = \sqrt{x+1}$, the domain is $x \geq -1$, and the range is $y \geq 0$.

WORKED EXAMPLE 14	Finding the domain and range of a function
\multicolumn{2}{l	}{What is the domain and range of each of these functions?}
\multicolumn{2}{l	}{**a** $y = \sqrt{x-3}$ **b** $h(x) = \dfrac{1}{x-4}$}
Steps	**Working**
a 1 When is $\sqrt{x-3}$ defined?	$x-3$ cannot be negative. $x - 3 \geq 0$ $x \geq 3$
2 Consider the possible y values for the range.	The function can only be positive (because of the square root sign), so $y \geq 0$.
3 Write the answer.	The domain of $y = \sqrt{x-3}$ is $x \geq 3$ and the range is $y \geq 0$.
b 1 When is $\dfrac{1}{x-4}$ defined?	x cannot be 4, but can be any other value.
2 Consider the possible y values for the range.	As function is undefined when $x = 4$, $h(x)$ cannot be zero, but can be any other values.
3 Write the answer.	The domain is all real numbers except $x = 4$, and the range is all real numbers except $h(x) = 0$.

WORKED EXAMPLE 15	Finding the range of a quadratic function
\multicolumn{2}{l	}{Find the range of the function given by $f(x) = x^2 + 6x + 11$.}
Steps	**Working**
1 Complete the square.	$f(x) = x^2 + 6x + 11$ $= x^2 + 6x + 9 + 2$ $= (x+3)^2 + 2$
2 State the turning point.	There is a minimum at $(-3, 2)$.
3 Sketch the graph.	*(graph showing parabola with vertex at $(-3, 2)$ and y-intercept at $(0, 11)$)*
4 Write the range.	The range of f is $y \geq 2$.

USING CAS 4 — Finding the range

Find the range of $f(x) = x^2 - 4x + 3$ for $-5 \le x \le 6$.

ClassPad

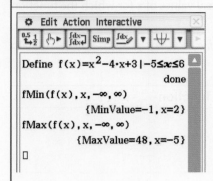
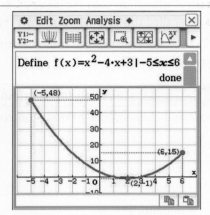

1. Enter and highlight the expression $x^2 - 4x + 3$.
2. Tap **Interactive > Define** to get $f(x) = x^2 - 4x + 3$.
3. Open the **Keyboard > Math3** and access the | and ≤ symbol to include the domain.
4. Copy and paste $f(x)$ to the next line, then tap **Interactive > Calculation > fMin/fMax > fMin**.
5. The minimum y value and corresponding x value will be displayed.
6. Repeat by selecting **fMax** to determine the maximum y and x values.
7. Open the **Graph** window and drag down from the **left side of screen** with the stated domain.
8. Tap **View Window** and adjust the window settings for the minimum and maximum values.
9. Tap **Analysis > G-Solve > Min** to find the local minimum.
10. Tap **Analysis > G-Solve > x-Cal/y-Cal** or **Analysis > G-Solve > fMax** to identify endpoints at $x = -5$ and $x = 6$.
11. Press **EXE** to label the coordinates of the points on the graph.

TI-Nspire

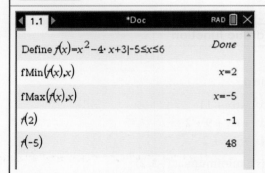
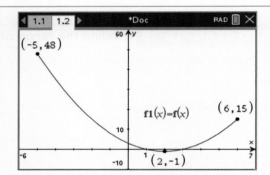

1. Press **menu > Actions > Define**.
2. Define the function as shown. Press **ctrl + =** to access the mini-palette for the | and ≤ symbols.
3. Press **menu > Calculus > Function Minimum**.
4. Enter **f(x),x** to determine the x value when the function is a minimum.
5. Repeat by selecting **Function Maximum** to determine the x value when the function is a maximum.
6. Substitute these values into the function to determine the corresponding y values.
7. Add a **Graphs** page.
8. Press **menu > Window/Zoom > Window Settings** and adjust the window settings for these minimum and maximum values.
9. Press **menu > Trace**.
10. When you enter an x value, the cursor will jump to that point.
11. Press **enter** to label the coordinates of the point.
12. Use **Trace** to confirm the minimum and maximum values of the function.

The range of the function is $-1 \le f(x) \le 48$.

We can work backwards to find the domain of a function when we are given the range.

WORKED EXAMPLE 16 Finding domain when given range

What is the domain of $h(x) = 5 - 4x$ if the range is $3 \leq h(x) < 29$?

Steps	Working
1 Find the endpoint on the line when $y = 3$.	$3 = 5 - 4x$ $x = 0.5$ endpoint $(0.5, 3)$
2 Find the endpoint on the line when $y = 29$ (but excluded from domain).	$29 = 5 - 4x$ $x = -6$ other endpoint $(-6, 29)$
3 Write the domain.	Line goes from points $(-6, 29)$ excluded, to $(0.5, 3)$ included, so the domain is $-6 < x \leq 0.5$.

EXERCISE 4.4 Functions and relations

ANSWERS p. 466

Recap

1 Sketch the graph of $y = -2x^3 - 8$.

2 Determine the equation of the below graph.

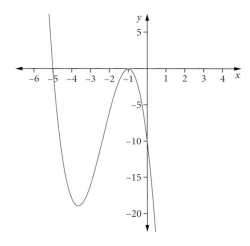

Mastery

3 WORKED EXAMPLE 11 Write each mapping as ordered pairs. State the domain and range and whether the mapping is a function or not. If it is a function, state whether it is one-to-one or many-to-one.

a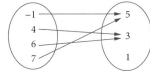

b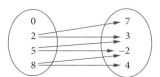

c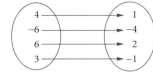

d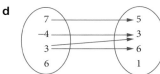

4 **WORKED EXAMPLE 12** State whether each relation is a function.

a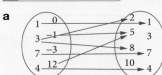

b $y^2 = x + 1$

c $xy = 3$

d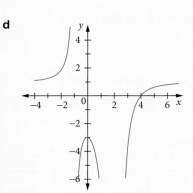

5 **WORKED EXAMPLE 13** State whether each graph represents a function.

a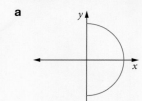
b
c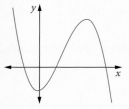
d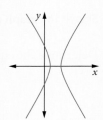

e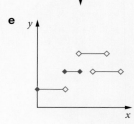
f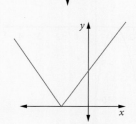
g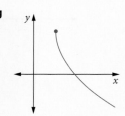
h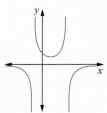

6 **WORKED EXAMPLE 14** What is the domain and range of each function?

a $y = \dfrac{1}{x + 5}$

b $f(x) = \sqrt[3]{x + 1}$

c $f(x) = \dfrac{1}{x^2 - 4}$

d $g(x) = \dfrac{1}{4x^2 + 1}$

e $y = \sqrt{x^2 - 16}$

f $g(x) = x^2\sqrt{x}$

g $h(x) = \sqrt{(x + 2)(x - 3)}$

h $y = \sqrt{(3 + x)(5 - x)}$

i $f(x) = \dfrac{1}{\sqrt{7 - x}}$

7 **WORKED EXAMPLE 15** What is the range of $f(x) = x^2 + 2x - 15$.

8 **Using CAS 4** Find the range of each function.

a $y = 6 - x - x^2$ for $-4 \le x \le 4$

b $x^2 - 7x - 8$ for $3 < x < 10$

c $y = 2^x + 5$ for $-2 \le x \le 4$

d $\dfrac{4}{x^2 - 36}$ for $-4 < x \le 5$

9 **WORKED EXAMPLE 16** Given the range of each function, find its domain.

a $f(x) = 5x - 7$, range: $-17 < f(x) \le 3$

b $g(x) = 7 - 2x$, range: $-5 \le g(x) < 5$

c $h(x) = 4 - 3x$, range: $3 < h(x) \le 8$

d $k(x) = 4x - 5$, range: $-3 \le k(x) < 6$

Calculator-free

10 (4 marks) State whether each set of ordered pairs is a function or not. If it is a function, state whether it is one-to-one or many-to-one.

    a $\{(3, 5), (4, 7), (5, 7), (6, 7), (8, 9), (9, 5), (12, 5)\}$ (1 mark)

    b $\{(4, 7), (5, 6), (5, 4), (7, 3), (8, 2), (9, 1), (10, 0)\}$ (1 mark)

    c $\{(7, 3), (3, 7), (9, 2), (2, 9), (4, 8), (8, 4), (5, 7), (7, 5)\}$ (1 mark)

    d $\{(0.1, 3), (0.3, 9), (0.5, 15), (0.7, 21)\}$ (1 mark)

11 (4 marks) State whether each relation is a function.

    a $\dfrac{x-3}{5} - \dfrac{y+2}{3} = 1$ (1 mark)

    b $\dfrac{(x-3)^2}{25} - \dfrac{(y+2)^2}{9} = 1$ (1 mark)

    c $12x - 3y^2 = 5$ (1 mark)

    d $15x - 3y^3 = 7$ (1 mark)

12 (2 marks) What is the natural domain of $y = \sqrt{x+3}$?

13 (2 marks) What is the natural domain of $f(x) = \dfrac{3}{\sqrt{3-x}}$?

14 (2 marks) If the linear function $f(x) = 5 - x$ has range $-4 \leq f(x) < 5$, determine the domain.

Calculator-assumed

15 (10 marks) State whether each of the following are functions or not.

    a $y = x^2 - 5$ (1 mark)

    b $y^2 + 5 = x$ (1 mark)

    c $y^2 = x + 5$ (1 mark)

    d $\dfrac{1}{x} = \dfrac{1}{y^2 + 1}$ (1 mark)

    e $x^2 - y^2 = 5$ (1 mark)

    f $y = \dfrac{1}{x+3}$ (1 mark)

    g $y = x^2 - 5^2$ (1 mark)

    h $\dfrac{1}{x} = \dfrac{1}{y}$ (1 mark)

    i $x^2 = y^2 + 9$ (1 mark)

    j $\sqrt{y} = x^2 + 4$ (1 mark)

16 (20 marks) What is the range of each function?

    a $y = 25 - x^2$ (2 marks)

    b $y = x^2 - 4x + 3$ (2 marks)

    c $y = 8 + 2x - x^2$ (2 marks)

    d $y = \dfrac{1}{x+2}$ (2 marks)

e $y = \dfrac{1}{(x-4)^2}$ (2 marks)

   f $y = 2x - 10$ for $3 \le x \le 7$ (2 marks)

   g $y = 5x + 3$ for $-4 < x < 4$ (2 marks)

   h $y = 6 - 3x$ for $-4 \le x < 9$ (2 marks)

   i $y = x^3 + 4$ for $x \ge 0$ (2 marks)

   j $y = (x-6)(x+2)$ for $-4 \le x < 4$ (2 marks)

17 (2 marks) State the natural domain and range of $f(x) = \sqrt{x^2 - 4}$.

18 (2 marks) The linear function $f(x) = 4 - x$ has range $-2 \le f(x) < 6$. Determine the domain.

4.5 Transformations of functions

Earlier in the chapter, we sketched families of power functions changed by simple parameters. Changes to the appearance of a graph (or geometric figure) can be regarded as **transformations**.

Translations

A **translation** in the direction of the y-axis (or from the x-axis) changes the appearance by moving the shape in the y direction. Every point is changed by adding the same amount to the y-coordinate.

A translation in the direction of the x-axis (or from the y-axis) changes the appearance by moving the shape in the x direction. Every point is changed by adding the same amount to the x-coordinate.

In general, under a translation of a units in the y direction, $(x, y) \to (x, y + a)$, and $f(x) \to g(x)$ is given by $g(x) = f(x) + a$. Under a translation of b units in the positive x direction, $(x, y) \to (x + b, y)$, and $f(x) \to g(x)$ is given by $g(x) = f(x - b)$. Under a translation of b units in the negative x direction, $(x, y) \to (x - b, y)$, and $f(x) \to g(x)$ is given by $g(x) = f(x + b)$.

WORKED EXAMPLE 17 Finding and graphing functions of translations

The function $f(x) = x^3$ is translated 2 units to the right and 1 unit up to give the function $g(x)$. State the rule for $g(x)$ and sketch f and g on the same axes.

Steps	Working
1 Use $g(x) = f(x - b) + a$.	$g(x) = (x - 2)^3 + 1$
2 State the points of inflections.	The points of inflection of $f(x)$ and $g(x)$ are $(0, 0)$ and $(2, 1)$, respectively.
3 State other points.	$f(1) = 1$ and $g(3) = 2$. The y-intercepts of $f(x)$ and $g(x)$ are 0 and -7 respectively. The zero of $g(x)$ is $(1, 0)$.
4 Sketch the functions on the same axes.	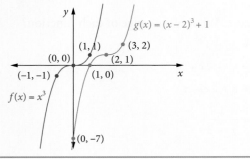

Dilations from the *x*- and *y*-axes

A **dilation** from the *x*-axis (or parallel to the *y*-axis) changes the appearance by expansion or compression ('vertically') from the *x*-axis. Every *y*-coordinate is multiplied by the same factor *c*. The dilation is a stretch if $c > 1$ or $c < -1$ and a compression (shrink) if $-1 < c < 1$.

A dilation from the *y*-axis (or parallel to the *x*-axis) changes the appearance by expansion or compression ('horizontally') from the *y*-axis. Every *x*-coordinate is multiplied by the same factor *d*.

In general, under a dilation of factor *c* from the *x*-axis, $(x, y) \rightarrow (x, cy)$, and $f(x) \rightarrow g(x)$ is given by $g(x) = cf(x)$. Under a dilation of factor *d* from the *y*-axis, $(x, y) \rightarrow (dx, y)$ and $f(x) \rightarrow g(x)$ is given by $g(x) = f(dx)$.

The family of the function $f(x)$ given by $f(dx)$ is dilated from the *y*-axis by the factor $\frac{1}{d}$. The dilation is a stretch if $-1 < d < 1$ and a compression (shrink) if $d > 1$ or $d < -1$.

WORKED EXAMPLE 18 — Finding and graphing functions of dilations

a The function $f(x) = x^3$ is dilated from the *x*-axis by the factor 2 to give the function $g(x)$. State the rule for $g(x)$ and sketch *f* and *g* on the same axes.

b The function $f(x) = \sqrt{x}$ is dilated from the *y*-axis by the factor 3 to give the function $g(x)$. State the rule for $g(x)$ and sketch *f* and *g* on the same axes.

Steps	Working
a 1 Use $g(x) = cf(x)$.	$g(x) = 2x^3$
2 State the points of inflection.	Both functions have points of inflection of $(0, 0)$.
3 State other points.	*f* has points $(-1, -1)$ and $(1, 1)$ and *g* has the corresponding points $(-1, -2)$ and $(1, 2)$.
4 Sketch the functions on the same axes. The graph is stretched parallel to the *y*-axis (vertically) by factor 2.	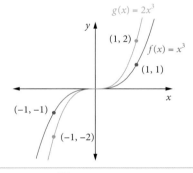
b 1 Use $g(x) = f(dx)$. $\frac{1}{d} = 3$ so $d = \frac{1}{3}$. $g(x) = f\left(\frac{1}{3}x\right)$.	$g(x) = \sqrt{\frac{x}{3}}$
2 State the starting points.	Both functions start at $(0, 0)$.
3 State other points.	*f* has the point $(1, 1)$ and $g(x)$ has the point $(3, 1)$.
4 Sketch the functions on the same axes.	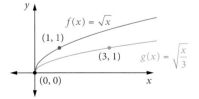

Reflections in the *x*- and *y*-axes

A **reflection** in the *x*-axis (or parallel to the *y*-axis) changes the appearance by inverting the shape (flipping it upside-down). The *y*-coordinate of every point is changed to its negative. The negative of the function $f(x)$ given by $g(x) = -f(x)$ is inverted or reflected in the *x*-axis, compared to $f(x)$.

A reflection in the *y*-axis (or parallel to the *x*-axis) changes the appearance by swapping the left and right sides (flipping it back-to-front). The *x*-coordinate of every point is changed to its negative. So the function given by $g(x) = f(-x)$ is swapped from left-to-right or reflected in the *y*-axis.

WORKED EXAMPLE 19	Finding and graphing reflections of functions

a The function $f(x) = x^3$ is reflected in the *x*-axis to give the function $g(x)$.

State the rule for $g(x)$ and sketch f and g on the same axes.

b The function $f(x) = \dfrac{1}{x+2} - 1$ is reflected in the *y*-axis to give the function $g(x)$.

State the rule for $g(x)$ and sketch f and g on the same axes.

Steps	Working
a 1 Use $g(x) = -f(x)$.	$g(x) = -x^3$
2 State the points of inflection.	Both functions have the points of inflection at $(0, 0)$.
3 State other points.	f and g both have $(0,0)$; f has points $(-1, -1)$ and $(1, 1)$; g has points $(-1, 1)$ and $(1, -1)$.
4 Sketch the functions on the same axes.	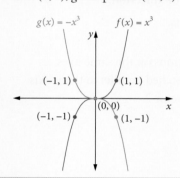
b 1 Use $g(x) = f(-x)$.	$g(x) = \dfrac{1}{-x+2} - 1 = \dfrac{1}{2-x} - 1$
2 State the asymptotes.	The asymptotes of $f(x)$ are $x = -2$ and $y = -1$; those of $g(x)$ are $x = 2$ and $y = -1$.
3 State other points.	The intercepts of f are $\left(0, -\dfrac{1}{2}\right)$ and $(-1, 0)$; for g they are $\left(0, -\dfrac{1}{2}\right)$ and $(1, 0)$.
4 Sketch the functions on the same axes.	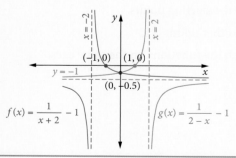

Combined transformations

Combined transformations

Transformations of the form $T: (x, y) \to (dx - b, cy + a)$ or $f(x) \to cf(d(x - b)) + a$ are performed in the order **dilation, reflection, translation**. Remember this as DRT or Dr T ('Doctor T').

Transformations in the (horizontal) direction of the x-axis:

- The dilation has a factor of $\frac{1}{d}$.
- If $d < 0$, there is a reflection in the y-axis.
- The translation is b units left (if $b < 0$, then the translation is to the right).

Transformations in the (vertical) direction of the y-axis:

- The dilation has a factor of c.
- If $c < 0$, there is a reflection in the x-axis.
- The translation is a units up (if $a < 0$, then the translation is downwards).

With the DRT order, it doesn't matter if horizontal or vertical transformations are done first, as long as those for each direction are done in the correct order.

Transformations of the form $T: (x, y) \to (d(x - b), c(y + a))$ or $f(x) \to c(f(dx - b) + a)$ follow the reverse order TRD: the translations are first.

The transformation $T: (x, y) \to (dx + b, cy + a)$ transforms the function $f(x) \to cf\left(\frac{1}{d}(x - b)\right) + a$.

WORKED EXAMPLE 20 — Combined transformations of points

The points $A(3, 7)$, $B(-2, 9)$ and $C(5, -3)$ are on the function $f(x)$.

a Find the new points after a dilation by factor 2 in the x direction, and translations 2 units in the positive y direction and 3 units in the negative x direction.

b Find the positions of A, B and C after translations of 4 units down and 2 units to the left, and then dilation by factor 0.5 from the x-axis and reflection in the y-axis.

Steps	Working
a 1 Write the transformation.	$(x, y) \to (2x - 3, y + 2)$
2 Find the points.	$A(3, 7) \to (2 \times 3 - 3, 7 + 2) = (3, 9)$
	$B(-2, 9) \to (2 \times (-2) - 3, 9 + 2) = (-7, 11)$
	$C(5, -3) \to (2 \times 5 - 3, -3 + 2) = (7, -1)$
3 Write the answer.	The new points are $A(3, 9)$, $B(-7, 11)$ and $C(7, -1)$.
b 1 Write the transformation.	$(x, y) \to (-(x - 2), 0.5(y - 4))$
2 Find the points.	$A(3, 7) \to (-(3 - 2), 0.5(7 - 4)) = (-1, 1.5)$
	$B(-2, 9) \to (-(-2 - 2), 0.5(9 - 4)) = (4, 2.5)$
	$C(5, -3) \to (-(5 - 2), 0.5(-3 - 4)) = (-3, -3.5)$
3 Write the answer.	The new points are $A(-1, 1.5)$, $B(4, 2.5)$ and $C(-3, -3.5)$.

WORKED EXAMPLE 21 — Combined transformations of functions

Sketch the graph of

a $f(x) = 1 - \dfrac{2}{x+3}$

b $g(x) = \sqrt{3 - 2x} - 1$

Steps	Working
a 1 State the basic function.	The basic function is $y = x^{-1}$.
2 State the transformations. *(Transformations parallel to the x-axis first.)*	Translation 3 units to the left (negative direction of the x-axis), reflection in the x-axis, dilation from the x-axis by the factor 2 and translation 1 unit up (positive direction of the y-axis).
3 State the asymptotes.	The asymptotes are $y = 1$ and $x = -3$.
4 State the y-intercept.	$f(0) = 1 - \dfrac{2}{3} = \dfrac{1}{3}$
5 Find zeros if possible.	$1 - \dfrac{2}{x+3} = 0$ $x + 3 = 2$ $x = -1$, the zero is $(-1, 0)$.
6 Sketch the graph, including coordinates of important points and asymptote equations. *(You might want to include the coordinates of vertices.)*	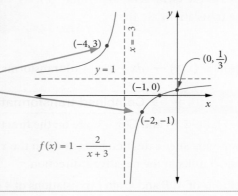
b 1 State the basic function.	The basic function is $y = \sqrt{x}$.
2 Write the function in the form $af(x+b) + c$.	$g(x) = \sqrt{3 - 2x} - 1 = \sqrt{-2\left(x - \dfrac{3}{2}\right)} - 1$
3 State the transformations.	Dilation by the factor $\dfrac{1}{2}$ parallel to the x-axis, reflection in the y-axis and translation 1.5 units to the right and 1 unit down.
4 State the starting point.	The starting point is $\left(\dfrac{3}{2}, -1\right)$.
5 State other points.	The y-intercept is $(0, \sqrt{3} - 1)$. The zero is at $(1, 0)$ and $(-3, 2)$ is on the curve.
6 Sketch the graph.	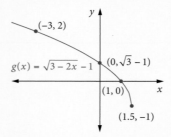

We can find the equation of a new function obtained by transformation of a given function.

WORKED EXAMPLE 22 — Finding the equation of a transformed function

Each function $f(x)$ is transformed to $g(x)$ by the operations given. Find each function $g(x)$.

a $f(x) = \sqrt{x}$: dilation parallel to the x-axis by the factor 0.5, reflection in the x-axis and translation 3 units up and 1 unit to the left.

b $f(x) = \dfrac{2}{x-3}$: reflection in the y-axis, translation 5 units in the negative direction of the x-axis and 3 units in the negative direction of the y-axis, and dilation by the factor $\dfrac{1}{4}$ from the x-axis. *(Note that this is not in DRT order.)*

Steps	Working
a 1 Do the dilation using $\dfrac{1}{0.5} = 2$.	$\sqrt{x} \rightarrow \sqrt{2x}$
2 Do the reflection in the x-axis.	$\sqrt{2x} \rightarrow -\sqrt{2x}$
3 Do the 2 translations and write $g(x)$.	$-\sqrt{2x} \rightarrow -\sqrt{2(x+1)} + 3$ $g(x) = 3 - \sqrt{2(x+1)}$
b 1 Do the reflection in the y-axis.	$\dfrac{2}{x-3} \rightarrow \dfrac{2}{-x-3}$
2 Do the translation to the left.	$\dfrac{2}{-x-3} \rightarrow \dfrac{2}{(-x-3)+5} = \dfrac{2}{2-x}$
3 Do the translation 3 units down.	$\dfrac{2}{2-x} \rightarrow \dfrac{2}{2-x} - 3$
4 Do the dilation by a factor of $\dfrac{1}{4}$ from the x-axis.	$\dfrac{2}{2-x} - 3 \rightarrow \dfrac{1}{4}\left(\dfrac{2}{2-x} - 3\right) = \dfrac{1}{2(2-x)} - \dfrac{3}{4}$
5 Write $g(x)$.	$g(x) = \dfrac{1}{2(2-x)} - \dfrac{3}{4} \quad \left(\text{or } \dfrac{1}{4-2x} - \dfrac{3}{4}\right)$

When we need to find the transformations that change one function to another, we work backwards. To find the transformations of a basic function that gives a graph, we also need to work backwards.

WORKED EXAMPLE 23 — Finding transformations from a graph

This graph has been obtained by applying transformations to the graph of $y = x^4$.

What transformations have been used and what is the equation of the graph?

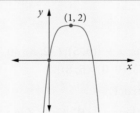

Steps	Working
1 Use the turning point to find the translations.	The graph has been flipped upside-down and translated 1 unit to the right and 2 units up.
2 State the form of the equation.	The equation is of the form $y = a(x-1)^4 + 2$, where a is negative.
3 Substitute $(0,0)$ to find the value of a and the equation.	$0 = a(0-1)^4 + 2$ $0 = a + 2$, so $a = -2$ Equation is $y = -2(x-1)^4 + 2$.
4 State the answer.	The graph has a dilation from the x-axis by a factor of 2, reflection in the x-axis, and translation 1 unit right and 2 units up. Its equation is $y = -2(x-1)^4 + 2$.

Use CAS to check the answer when finding an equation of a graph.

USING CAS 5 Transformations of power functions

Draw the graph of $y = -2(x - 1)^4 + 2$ and find the turning point.

ClassPad

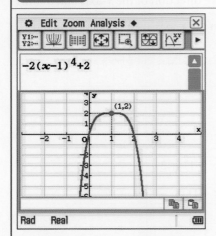

TI-Nspire

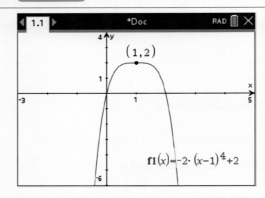

1. In **Main**, enter and highlight the expression $-2(x - 1)^4 + 2$.
2. Tap **Graph** and drag the expression down into the graph window.
3. Adjust the window settings to suit.
4. Tap **Analysis > G-Solve > Max**.
   Tap **EXE** to display the label. The coordinates of the turning point will be displayed on the graph.

1. Add a **Graphs** page.
2. Enter $f1(x) = -2(x - 1)^4 + 2$ and press **enter**.
3. Adjust the window settings to suit.
4. Press **menu > Analyze Graph > Maximum**.
5. When prompted for the **lower bound?**, click to the left of the turning point.
6. When prompted for the **upper bound?**, click to the right of the turning point. The coordinates of the turning point will be displayed on the graph.

The turning point is (1, 2).

EXERCISE 4.5 Transformations of functions

ANSWERS p. 467

Recap

1 Consider the three relations: $X = \{(3, 5), (-2, 3), (0, 4), (2, 5), (5, 3)\}$, $Y = \{(6, 2), (2, 5), (3, 7), (6, 3), (7, 3)\}$ and $Z = \{(2, 2), (1, 4), (3, 7), (0, 0), (5, 8)\}$.

Which of the following is true?

A None of them are functions. **B** Only Z is a function. **C** Y and Z are functions.
D X and Z are functions. **E** All of them functions.

2 Consider the three relations

$f: 3y^3 - 4x^2 = 1$ $g: 5x^3 - 4y^2 = 1$ $h: \dfrac{y - 5}{x + 4} = 3$

Which of the following is true?

A Only h is a function. **B** f and h are functions. **C** They are all functions.
D g and h are functions. **E** None of them are functions.

▶ **Mastery**

3 **WORKED EXAMPLES 17–19** For each of the below, state the rule for $g(x)$ and sketch f and g on the same axes.

a The function $f(x) = x^2 + 1$ is translated 2 units to the right and 1 unit up to give the function $g(x)$.
b The function $f(x) = x^2 + 1$ is dilated from the x-axis by the factor 2 to give the function $g(x)$.
c The function $f(x) = x^2 + 1$ is reflected in the x-axis to give the function $g(x)$.

4 **WORKED EXAMPLE 20** The points $A(3, -2)$, $B(-4, 6)$ and $C(2, 5)$ are on the function $f(x)$. Find their positions after each set of transformations.

a Dilation from the x-axis by the factor 3, from the y-axis by the factor 2, reflection in the y-axis and translation 2 units left and 4 units up.
b Dilation from the y-axis by the factor 4, from the x-axis by the factor 0.5, reflection in both axes and translation 2 units to the right and 1 unit up.
c Translation 4 units in the positive x direction, reflection in the x-axis and dilation from the y-axis by the factor 3.
d Translation 2 units down and 4 units to the left, then reflection in both axes and dilation by the factor 3 parallel to the y-axis.
e Dilation by the factor $\frac{1}{3}$ from the x-axis, reflection in the y-axis and translation 4 units to the right and 3 units down.

5 **WORKED EXAMPLE 21** Sketch the graph of each function.

a $f(x) = 2(x + 1)^3 - 4$
b $y = 3(-2 - x)^{-1} - 1$
c $f(x) = -(2x - 1)^4 + 3$
d $y = (1 - x)^3 + 2$
e $f(x) = 4\sqrt{x + 3} - 2$
f $y = 5 - 4(0.5x - 1)^3$
g $f(x) = (3(x - 1))^{-1} - 2$
h $y = \dfrac{\sqrt{x + 1}}{3} + 2$
i $y = \dfrac{2}{1 - 0.25x} + 3$
j $f(x) = -0.5\sqrt{3x - 2} - 1$
k $y = 2 - \dfrac{1}{(0.5x + 3)}$
l $f(x) = 1 + \dfrac{\sqrt{2(2 - x)}}{3}$

6 **WORKED EXAMPLE 22** Find the equation of $g(x)$ from the transformations on $f(x)$.

a $f(x) = x^2$: reflection in the x-axis, dilation by a factor of 3 from the x-axis, translation 1 unit down and 4 units to the right.
b $f(x) = x^4$: dilation by a factor of 4 parallel to the y-axis, reflection in the x-axis, translation 1 unit up and translation 3 units to the left.
c $f(x) = \dfrac{1}{x^2}$: reflection in the x-axis, dilation by a factor of $\dfrac{1}{3}$ from the y-axis, translation 3 units in the positive direction of the y-axis and 2 units in the positive direction of the x-axis.
d $f(x) = -2(x + 4)^2 - 3$: translation 2 units up 3 units to the right and dilation by a factor of $\dfrac{1}{4}$ parallel to the x-axis.
e $f(x) = \sqrt[3]{x + 3} - 2$: dilation by a factor of 3 from the x-axis, translation 1 unit in the negative x direction, 3 units in the negative y direction and reflection in the y-axis.
f $f(x) = -3(x + 4)^{-1} + 4$: translation 3 units to the right, reflection in the x-axis, translation 2 units down and dilation by a factor of $\dfrac{1}{2}$ from the y-axis.
g $f(x) = 6(x - 2)^3 - 3$: translation 6 units in the negative x direction, dilation by a factor of $\dfrac{1}{2}$ from the x-axis, reflection in the y-axis and translation 4 units in the positive y direction.
h $f(x) = 0.25\sqrt{2 - x} + 3$: reflection in the y-axis, dilation by a factor of 4 parallel to the y-axis, translation 3 units down and 2 units to the right.

7. **WORKED EXAMPLE 23** These graphs have been obtained by applying transformations to the given function. What are the transformations and equation of each graph?

a $f(x) = \sqrt{x}$

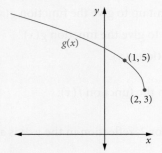

b $f(x) = x^3$

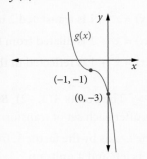

c $f(x) = x^3$

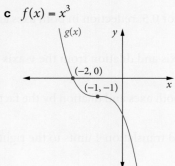

d $f(x) = \sqrt{x}$

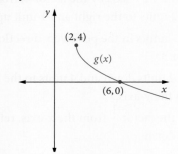

e $f(x) = -4x^2 + 2$

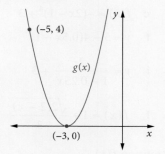

f $f(x) = 2(x + 3)^4$

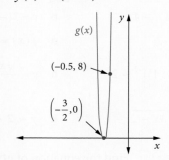

g $f(x) = \dfrac{2}{x - 1} - 3$

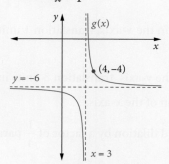

h $f(x) = \sqrt{2 - x} + 4$

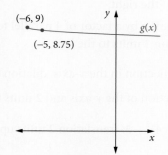

i $f(x) = 4\sqrt{2x + 3} - 1$

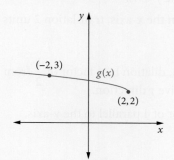

j $f(x) = 3 - \dfrac{2}{0.25x - 1}$

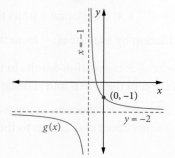

8 **Using CAS 5** Graph each function.

a $f(x) = 0.4(x+3)^4 - 2$
b $y = 3 - \dfrac{1}{2x-4}$
c $f(x) = \sqrt{x+4} - 3$

d $y = \dfrac{2}{x+1} + 3$
e $f(x) = 1 - \sqrt{x-3}$
f $y = 2(x-1)^3 + 4$

Calculator-free

9 (24 marks) Sketch the graph of each function, showing the coordinates of any intercepts and the equations of any asymptotes.

a $f(x) = 4x^3$ (2 marks)

b $y = -\dfrac{1}{x}$ (2 marks)

c $y = \sqrt{0.5x}$ (2 marks)

d $f(x) = (x+2)^4 - 8$ (2 marks)

e $f(x) = \dfrac{2}{x}$ (2 marks)

f $f(x) = -\dfrac{\sqrt{x}}{2}$ (2 marks)

g $f(x) = \dfrac{1}{3x}$ (2 marks)

h $y = 4 + \sqrt{x-1}$ (2 marks)

i $y = -x^3$ (2 marks)

j $y = \dfrac{1}{x-2} + 3$ (2 marks)

k $f(x) = \sqrt{-x}$ (2 marks)

l $y = \sqrt{1-x}$ (2 marks)

10 (3 marks) Let $f(x) = 1 + \dfrac{2}{x-3}$. Sketch the graph of f. Label the axes intercepts with their coordinates and label any asymptotes with the appropriate equation.

11 (3 marks) Let $f(x) = \dfrac{1}{2x-4} + 3$. Sketch the graph of $y = f(x)$. Label the axes intercepts with their coordinates and label each of the asymptotes with its equation.

12 (2 marks) The point $P(4, -3)$ lies on the graph of a function f. The graph of f is reflected in the x-axis and translated four units vertically up. What are the coordinates of the final image of P?

▶ **Calculator-assumed**

13 (3 marks) The graph of a function f is obtained from the graph of the function g with rule $g(x) = 2\sqrt{x-5}$ by a reflection in the x-axis followed by a dilation from the x-axis by a factor of $\frac{1}{2}$. Determine the rule for the function f.

14 (3 marks) Part of the graph of the function with the rule $y = \dfrac{a}{x+b} + c$ is shown below. Determine the values of a, b and c.

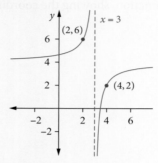

15 (2 marks) Determine the transformation that maps the graph of $y = \sqrt{4x-4}$ onto the graph of $y = \sqrt{x+1}$.

16 (3 marks) If $f(x-1) = x^2 - 2x + 3$, then determine $f(x)$.

4.6 Graphs of relations

Video playlist
Graphs of relations

There are two special relations that we can consider.

A circle is a relation that has the general equation of $(x-a)^2 + (y-b)^2 = r^2$, with centre (a, b) and radius r. If the centre of the circle is $(0, 0)$, then the equation becomes $x^2 + y^2 = r^2$.

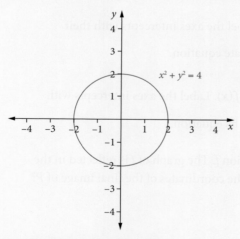

Centre $(0, 0)$ and radius of 2 units

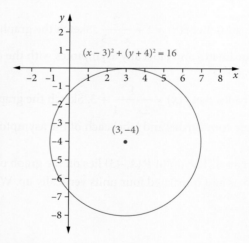

Centre $(3, -4)$ and radius of 4 units

WORKED EXAMPLE 24 — Graphing using centre and radius of a circle

Graph $x^2 + y^2 - 4x + 6y = 12$.

Steps	Working
1 Group the x and y terms.	$x^2 - 4x + y^2 + 6y = 12$
2 Complete the squares.	$(x - 2)^2 - 4 + (y + 3)^2 - 9 = 12$
3 Write in $(x - a)^2 + (y - b)^2 = r^2$ form.	$(x - 2)^2 + (y + 3)^2 = 5^2$
4 State centre and radius.	Centre is at $(2, -3)$ and radius is 5 units.
5 Graph the equation.	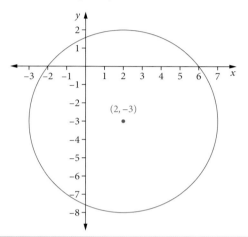

Another special type of relation is the graph of $y^2 = x$.

The graph has a parabolic shape (like a horizontal parabola), and a line of symmetry at the x-axis.

Note: The graphs of both of these relations fail the vertical line test, which means that they are not functions.

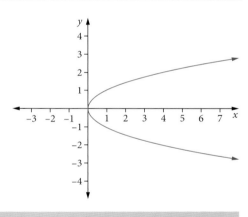

EXERCISE 4.6 Graphs of relations

ANSWERS p. 471

Recap

1 What transformations have been performed on $f(x) = \sqrt{x}$ to obtain $g(x) = -\sqrt{x+1}$?

2 Determine the equation of this graph.

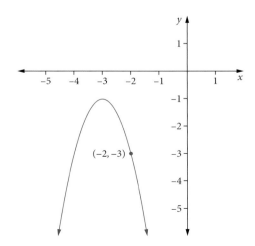

Mastery

3 [WORKED EXAMPLE 24] Graph $x^2 + y^2 - x + 4y = 2$.

4 State the centre and radius of $2x^2 + 2y^2 - 4x + 8y = 16$.

5 Sketch the relation $y^2 = x$.

Calculator-free

6 (11 marks)

    a Find the centre and radius of the circle with equation $x^2 + y^2 - 6x + 8y = 56$. (3 marks)

    b Find the centre and radius of the circle with equation $4x^2 + 28x + 4y^2 - 40y = 43$. (3 marks)

    c Write the equation of a circle with centre $(5, -3)$ and radius 7. (1 mark)

    d Write the equation of a circle with centre $(-2, 4)$ and radius 4. (1 mark)

    e Find the equation of a circle that just fits in the square $A(3, 5)$, $B(8, 9)$, $C(12, 4)$, $D(7, 0)$. (3 marks)

7 (2 marks) By substituting in various values for x, show that the following are not functions.

    a $y^2 = x$ (1 mark)

    b $x^2 + y^2 = 36$ (1 mark)

Calculator-assumed

8 (3 marks) A circle with centre $(a, -2)$ and radius 5 units has the equation $x^2 - 6x + y^2 + 4y = b$, where a and b are constants.

What are the values of a and b?

9 (3 marks) What is the rule for this graph?

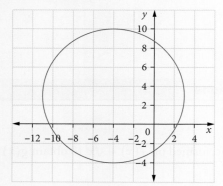

4.7 Inverse proportion

The graph of $y = \dfrac{1}{x}$ is an example of **inverse proportion**. As x increases in value, y decreases proportionally, or vice-versa. That is, if x is multiplied by a certain value, y is divided by the same value.

x	1	2	3	4	6
y	12	6	4	3	2

In this table, $y = \dfrac{12}{x}$ or $xy = 12$.

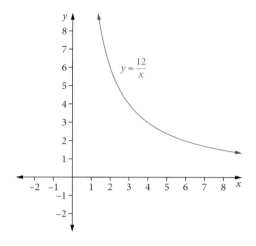

Inverse proportion

$$y = \dfrac{k}{x} \quad \text{or} \quad xy = k \quad \text{where } x \neq 0.$$

WORKED EXAMPLE 25 — Using inverse proportion

Two people are placing flyers into envelopes for a local politician to deliver in the local area. They calculate that if only the two of them are doing the task, it will take them 36 hours. They want to employ more people to help out. (Assume that everyone is working at the same pace).

a Given that this situation is an inverse proportion, write a rule for this situation, if x is number of people working and y is the number of hours taken.
b How long will it take if they have 4 people working?
c If the job only ended up taking 3 hours, how many people were working?

Steps	Working
a 1 Substitute in values for x and y, and calculate k.	x is 2 and y is 36. Therefore, k equals 72.
2 Write the rule.	$xy = 72$
b 1 Substitute into the rule.	$4y = 72$ $\therefore y = 18$
2 Answer the question.	The job will take 18 hours.
c 1 Substitute into the rule.	$3x = 72$ $\therefore x = 24$
2 Answer the question.	They had 24 people working.

Video playlist
Inverse proportion

Worksheet
Variation problems

Video
Examination question analysis: Functions and relations

EXAMINATION QUESTION ANALYSIS

Calculator-free (9 marks)

Let $f(x) = \dfrac{1}{3x - 1}$.

a For what value(s) of x is the function undefined? (1 mark)

b State the domain and range of f. (2 marks)

c Let g be the function obtained by applying the transformation T to the function f, where $T: (x, y) \to (x + c, y + d)$.

Find the values of c and d given that $g(x) = \dfrac{1}{3x + 1} + 2$. (2 marks)

d Write the transformation that changes $f(x)$ to its basic function $\dfrac{1}{x}$ and express in the form $T: (x, y) \to (ax + b, cy + d)$. (2 marks)

e Write the transformation that changes $g(x)$ to its basic function $\dfrac{1}{x}$ and express in the form $T: (x, y) \to (ax + b, cy + d)$. (2 marks)

Reading the question

- f is a hyperbola.
- The domain is the x values, and the range is the corresponding y values.
- Transformations are translations, dilations and reflections.
- The expression in part **c** is for a translation.

Thinking about the question

- In part **c**, c is the translation in the x direction and d is the translation in the y direction.
- The transformation in parts **d** and **e** undoes the transformations of $\dfrac{1}{x}$.

Worked solution (✓ = 1 mark)

a $3x - 1 \neq 0$, so the function is undefined for $x = \dfrac{1}{3}$ ✓

b domain: all real values excluding $x = \dfrac{1}{3}$ ✓

range: all real values excluding $y = 0$ ✓

c As $\dfrac{1}{3(x + c) - 1} = \dfrac{1}{3x + 1}$, solving for c gives $c = \dfrac{2}{3}$. This is a horizontal translation. $d = 2$ as the function has a vertical translation of $+2$.

$c = \dfrac{2}{3}$ ✓ and $d = 2$ ✓

d $\dfrac{1}{3x-1} \to \dfrac{1}{x} = 3\left[\dfrac{1}{3\left(x+\dfrac{1}{3}\right)-1}\right] = 3f\left(x+\dfrac{1}{3}\right).$

The transformation is dilation from the x-axis by the factor 3 and translation left by $\dfrac{1}{3}$ units. ✓

The required expression is $T: (x, y) \to \left(x - \dfrac{1}{3}, 3y\right).$ ✓

OR $\dfrac{1}{3x-1} \to \dfrac{1}{x} = \dfrac{1}{3\left(\dfrac{x}{3}+\dfrac{1}{3}\right)-1} = f\left(\dfrac{x}{3}+\dfrac{1}{3}\right) = f\left(\dfrac{1}{3}(x+1)\right).$

The transformation is dilation from the y-axis by the factor of 3 and translation left by 1 unit. ✓

The required expression is $T: (x, y) \to (3x - 1, y).$ ✓

e $\dfrac{1}{3x+1} + 2 \to \dfrac{1}{x} = 3\left[\dfrac{1}{\left(3\left(x-\dfrac{1}{3}\right)+1\right)} + 2 - 2\right] = 3\left[g\left(x-\dfrac{1}{3}\right)-2\right]$

The transformation is translation down by 2 units, right by $\dfrac{1}{3}$ unit, and dilation from the x-axis by the factor 3. ✓

The required expression is $T: (x, y) \to \left(x + \dfrac{1}{3}, 3(y-2)\right).$ ✓

OR $\dfrac{1}{3x+1} + 2 \to \dfrac{1}{x} = \dfrac{1}{3\left(\dfrac{x}{3}-\dfrac{1}{3}\right)+1} + 2 - 2 = g\left(\dfrac{x}{3}-\dfrac{1}{3}\right) - 2 = g\left(\dfrac{1}{3(x-1)}\right) - 2.$

The transformation is dilation from the y-axis by the factor 3 and translation right by 1 unit and translation down by $\dfrac{21}{3}$ units. ✓

The required expression is $T: (x, y) \to (3x + 1, y - 2).$ ✓

EXERCISE 4.7 Inverse proportion ANSWERS p. 471

Recap

1 Write the equation for a circle with centre $(-4, 2)$ and a radius of 10 units.

2 Draw a graph of $x^2 + y^2 + 6x - 8y = 12$.

Mastery

3 **WORKED EXAMPLE 25** A receptionist at a local dentist needs to purchase some diaries for the next calendar year for staff. From the budget her boss has given her, she knows that she can purchase 2 items at $15 each. She looks at other diaries to compare costs.

  a Given this situation is an inverse proportion, write a rule for this situation if x is the number of diaries purchased and y is the cost of the diaries.

  b How many diaries can she purchase if they are $10 each?

  c If she ends up purchasing diaries for 4 people, how much does each diary cost?

Calculator-free

4 (7 marks) A car is travelling at an average speed of 60 km/h and will take 3 hours to reach its destination.

   a Explain why this situation is an inverse proportion. (2 marks)

   b What is the distance to the destination? (1 mark)

   c How long will it take if the car travels at 100 km/h?
   (Give your answer in hours and minutes). (2 marks)

   d If the journey takes 6 hours, at what speed is the car travelling? What could account for this answer? (2 marks)

5 (4 marks) Calculate the inverse proportion rule for each of the following tables of values, and calculate the value of a and b.

   a (2 marks)

x	6	a	3	4	8
y	4	12	8	6	b

   b (2 marks)

x	2	3	6	−9	0.5
y	−9	−6	−3	a	b

Calculator-assumed

6 (2 marks) A cyclone is coming, and the local shops need to be sand bagged within the next 12 hours. The emergency services know that 2 people would take 60 hours to complete the job. How many people do they need sand bagging to complete the task in time?

7 (3 marks) It takes 5 hoses 420 minutes to fill the local swimming pool.

   a How long will it take if 8 hoses are used? (Give your answer in hours and minutes.) (2 marks)

   b What assumptions have you made in your answer for part **a**? (1 mark)

Chapter summary

Polynomials

- A **polynomial** is an expression involving a sum of terms of the form:
  $P(x) = a_n x^n + a_{n-1} x^{n-1} + a_{n-2} x^{n-2} + \ldots + a_1 x^1 + a_0$, where n is a non-negative whole number, $a_0, a_1, a_2 \ldots a_{n-1}, a_n$ are real numbers and $a_n \neq 0$.
- The **degree** of a polynomial is the highest power of x.
- The **leading term** is the term that contains the highest power of x.
- The **constant** term is a_0.
- The **leading coefficient** is a_n.
- A **monic polynomial** is a polynomial whose leading coefficient is 1.

Equating coefficients

- If $P(x) = a_n x^n + a_{n-1} x^{n-1} + a_{n-2} x^{n-2} + \ldots + a_1 x^1 + a_0$ and $Q(x) = b_n x^n + b_{n-1} x^{n-1} + b_{n-2} x^{n-2} + \ldots + b_1 x^1 + b_0$ are equal for all values of x, then $a_n = b_n$, $a_{n-1} = b_{n-1}$, $a_{n-2} = b_{n-2}$ and so on up to $a_0 = b_0$.

The hyperbola

- The simple **hyperbola** is the graph of the function $f(x) = x^{-1} = \dfrac{1}{x}$.
- There is a **discontinuity**, a break in the graph, at $x = 0$ that separates the graph into 2 branches.
- The x-axis is the **horizontal asymptote** $y = 0$: as $x \to \pm\infty$, $y \to 0$.
- The y-axis is the **vertical asymptote** $x = 0$: as $x \to 0$, $y \to \pm\infty$.
- The families of hyperbolas given by $f(x) = ax^{-1} = \dfrac{a}{x}$ and $f(x) = (x + b)^{-1} = \dfrac{1}{x + b}$ have the same relationship to the basic hyperbola as the families of other functions.

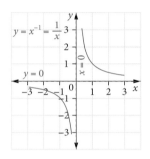

Square root functions

The basic **square root function**, $f(x) = x^{\frac{1}{2}} = \sqrt{x}$ is the inverse of $y = x^2$ for $x > 0$, whose graph is the graph of $y = x^2$ reflected in the line $y = x$. It has the domain and range $0 \leq x \leq \infty$, $0 \leq y \leq \infty$ respectively, since the square root of a negative number is not real and $\sqrt{x}$ is positive or 0. The shape is a half-parabola turned on its side.

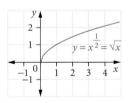

Solving polynomial equations

- For a polynomial $P(x)$, if $P(a) = 0$, then $(x - a)$ is a factor of the polynomial.
- a is called a zero of the polynomial $P(x)$ or a **root** of the polynomial equation $P(x) = 0$.
- If $(x - a)$ is a factor of $P(x)$, then $P(a) = 0$.
- If $P(x) = (x - a)(x - b)(x - c) = 0$, then the solutions are $x = a$, $x = b$, $x = c$.

Graphing cubic functions

- A **cubic** function is a polynomial of degree 3, whose equation has the general form $y = ax^3 + bx^2 + cx + d$.
- Compared to the graph of $y = x^3$, the effect of the value of a is to dilate ('stretch' or 'squash') $y = ax^3$ in the y direction.
- If a is large ($y = 5x^3$, for example), the cubic is steeper.
- If a is small ($y = \frac{1}{5}x^3$, for example), the cubic is flatter.
- If a is negative ($y = -x^3$, for example), the cubic is reflected in the x-axis (**flipped upside-down**).
- The cubic curve $y = a(x - b)^3 + c$ is the graph of $y = x^3$ that is stretched by a factor a from the x-axis and has point of inflection at (b, c).

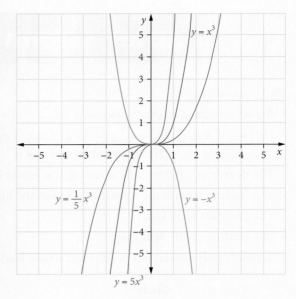

- The graph of the general cubic function $y = ax^3 + bx^2 + cx + d$ has 2 turning points: a maximum point and a minimum point. The shape of the cubic depends on the sign of the leading coefficient of the function, a.

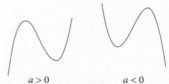

- If a cubic function can be factorised to the form $y = a(x - b)(x - c)(x - d)$, then b, c and d are its x-intercepts. For example, $y = (x + 3)(x + 1)(x - 2)$ has x-intercepts at -3, -1 and 2.
- If one of the factors is squared, then the x-intercept 'touches' the x-axis. For example, $y = 3(x - 1)^2(x + 5)$ 'touches' the x-axis at $x = 1$. $(x - 1)^2$ is called a double factor.

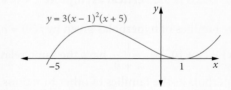

To graph a cubic function:
1. Determine the shape of the cubic curve from its equation and leading coefficient, a: it can have up to one maximum point and one minimum point.
2. Factorise the function if possible and note any repeated factors.
3. Find the x- and y-intercepts: it can have up to three x-intercepts but only one y-intercept. The y-intercept of $y = ax^3 + bx^2 + cx + d$ is d.

Functions and relations

- A **relation** R is a set of ordered pairs (x, y).
- The set of values of the independent variable x is called the **domain** and is often the set X. The set Y is called the co-domain and the set of values of the dependent variable y is called the **range**.
- A relation can be shown as a rule, a list of ordered pairs, a table, a mapping or a graph.
- A **function** f is a relation $f(x)$ such that each x value in the domain matches with only one y value.
- The value of a function at $x = c$ is written as $f(c)$.
- A function is either **one-to-one** or many-to-one.
- In a one-to-one function, each x value in the domain is matched with a unique y value. In the pairings, there aren't any repeated x or y values. For example, $y = 2x + 1$ is one-to-one.

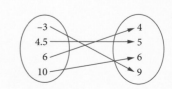

- In a many-to-one function, 2 or more x values in the domain may be matched with the same y value. In the pairings, there are repeated y values. For example, the function $y = x^2$ is many-to-one.

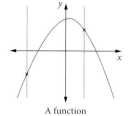

- If any vertical line intersects the graph of a relation at most once, then it is a function.

  If a vertical line intersects the graph of a relation in more than one place, then it is not a function.

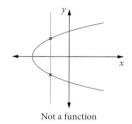

A function   Not a function

Domain and range

- The range is the set of y values that the domain produces.
- The **natural domain** of a function is the set of all real values of x for which the function is defined.

Transformations of power functions

Translations

- A **translation** in the direction of the y-axis (or from the x-axis) changes the appearance by moving the shape in the y direction. Every point is changed by adding the same amount to the y-coordinate:
  $f(x) \rightarrow f(x) + a$.
- A translation in the direction of the x-axis (or from the y-axis) changes the appearance by moving the shape in the x direction. Every point is changed by adding the same amount to the x-coordinate:
  $f(x) \rightarrow f(x - b)$.

Dilations from the x- and y-axes

- A **dilation** from the x-axis (or parallel to the y-axis) changes the appearance by expansion or compression ('vertically') from the x-axis. Every y-coordinate is multiplied by the same factor c: $f(x) \rightarrow cf(x)$.
- The family of the function $f(x)$ given by $cf(x)$ is dilated from the x-axis by the factor c. The dilation is a stretch if $c > 1$ or $c < -1$ and a shrink if $-1 < c < 1$.
- A dilation from the y-axis (or parallel to the x-axis) changes the appearance by expansion or compression ('horizontally') from the y-axis.
- The family of the function $f(x)$ given by $f(dx)$ is dilated from the y-axis by the factor $\frac{1}{d}$. The dilation is a stretch if $-1 < d < 1$ and a shrink if $d > 1$ or $d < -1$.

Reflections in the x- and y-axes

- A **reflection** in the x-axis (or parallel to the y-axis) changes the appearance by inverting the shape (flipping it upside-down). The y-coordinate of every point is changed to its negative: $f(x) \rightarrow -f(x)$.
- A reflection in the y-axis (or parallel to the x-axis) changes the appearance by swapping the left and right sides (flipping it back-to-front). The x-coordinate of every point is changed to its negative: $f(x) \rightarrow f(-x)$.

Relations

A circle is a relation that has the general equation of $(x - a)^2 + (y - b)^2 = r^2$, with centre (a, b) and radius r.
If the centre of the circle is $(0, 0)$ then the equation becomes $x^2 + y^2 = r^2$.

Inverse proportion

Inverse proportion is where as x increases in value, y decreases proportionally, or vice-versa.

$$y = \frac{k}{x} \quad \text{or} \quad xy = k \quad \text{where } x \neq 0.$$

Cumulative examination: Calculator-free

Total number of marks: 23   Reading time: 3 minutes   Working time: 23 minutes

1 (4 marks) Show the use of combinations to solve the following problems.

  a William is choosing a meal from a Chinese restaurant. He has to select one option from each of the following:
- Meat: pork, chicken, duck or beef
- Sauce: satay, sweet and sour, vegetable or black bean
- Carbohydrates: rice or noodles.

Calculate the number of possible meals that William has to choose from. (2 marks)

  b Gretta is travelling from Perth to Joondalup. She can
- drive directly to Joondalup, or
- take the train directly to Joondalup, or
- take the train part way and then one of five buses.

Calculate the total number of ways Gretta can make this journey. (2 marks)

2 (6 marks) A biased coin is tossed three times. Let the probability of obtaining heads from a toss of this coin be p.

  a Find, in terms of p, the probability of obtaining

    i three heads from the three tosses (1 mark)

    ii two heads and a tail, in any order, from the three tosses. (2 marks)

  b If the probability of obtaining three heads equals the probability of obtaining two heads and a tail in any order, find p, where p is not equal to zero. (3 marks)

3 (4 marks) Let $f(x) = (x + 2)^2(x - 1)$.

  a Show that $(x + 2)^2(x - 1) = x^3 + 3x^2 - 4$. (1 mark)

  b Copy the axes below and on them sketch the graph of f for the interval $-3 \le x \le 0$. Label the axes intercepts and any turning points with their coordinates. (3 marks)

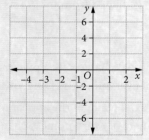

4 (2 marks) State the natural domain of $f(x) = \dfrac{1}{\sqrt{x^2 - 16}}$.

5 (3 marks) Sketch the graph of $f(x) = 4x - 4x^2 + 1$, labelling the coordinates of all intercepts and other significant points.

6 (4 marks) Determine the coordinates of the intersection of the circle $x^2 + y^2 = 16$ and the line $y - x = 4$.

Cumulative examination: Calculator-assumed

Total number of marks: 16   Reading time: 2 minutes   Working time: 16 minutes

1 (4 marks)

  a Find the equation of the linear graph which is parallel to the line $y = -2x + 7$ and passes through the point $(1, 8)$. (2 marks)

  b Find the equation of the linear graph which is perpendicular to the line $y = 3x$ and passes through the point $(-2, 1)$. (2 marks)

2 (2 marks) The graph with equation $y = x^2$ is translated 3 units down and 2 units to the right. Determine the equation of the resulting graph.

3 (2 marks) If $f(x + 1) = x^2 - 2x + 3$, then determine $f(x)$.

4 (2 marks) Determine the gradient of the line perpendicular to the line which passes through $(-2, 0)$ and $(0, -4)$.

5 (2 marks) Given that the range of $f(x) = 3 - 2x$ is $f(x) \leq -9$ or $f(x) \geq 5$, determine the domain.

6 (2 marks) The graph of the function f passes through the point $(-2, 7)$.

If $h(x) = f\left(\dfrac{x}{2}\right) + 5$, determine a point the graph of the function h must pass through.

7 (2 marks) Determine the rule for the graph.

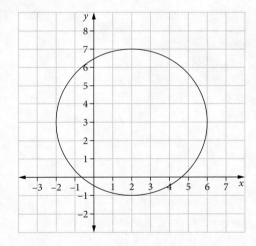

CHAPTER 5
TRIGONOMETRIC FUNCTIONS

Syllabus coverage
Nelson MindTap chapter resources

5.1 Right triangles and the angle of inclination
Right triangles
The angle of inclination

5.2 Non-right triangles
Sine rule
Ambiguous case
Using CAS 1: The sine rule
Cosine rule
Using CAS 2: The cosine rule
Area of a triangle

5.3 Radian measure, arc lengths, sectors and segments
Radian measure
Arc length
Sectors and segments

5.4 The trigonometric functions
The sine and cosine functions
The tangent function
Angles in the 4 quadrants

5.5 The exact values
$\dfrac{\pi}{6}$ and $\dfrac{\pi}{4}$ exact values
Points on the axes
Angle sum and difference identities

5.6 Graphs of trigonometric functions
The sine curve
The cosine curve
Using CAS 3: Graphing cosine
The tangent curve

5.7 Applying trigonometric functions
Using CAS 4: Applying trigonometric functions

Examination question analysis
Chapter summary
Cumulative examination: Calculator-free
Cumulative examination: Calculator-assumed

Syllabus coverage

TOPIC 1.3: TRIGONOMETRIC FUNCTIONS

Cosine and sine rules

1.3.1 review sine, cosine and tangent as ratios of side lengths in right-angled triangles
1.3.2 understand the unit circle definition of $\cos\theta$, $\sin\theta$ and $\tan\theta$ and periodicity using degrees
1.3.3 examine the relationship between the angle of inclination of a line and the gradient of that line
1.3.4 establish and use the cosine and sine rules, including consideration of the ambiguous case and the formula Area $= \frac{1}{2}bc\sin A$ for the area of a triangle

Circular measure and radian measure

1.3.5 define and use radian measure and understand its relationship with degree measure
1.3.6 use radian measure to calculate lengths of arcs and areas of sectors and segments in a circle

Trigonometric functions

1.3.7 understand the unit circle definition of $\sin\theta$, $\cos\theta$ and $\tan\theta$ and periodicity using radians
1.3.8 recognise the exact values of $\sin\theta$, $\cos\theta$ and $\tan\theta$ at integer multiples of $\frac{\pi}{6}$ and $\frac{\pi}{4}$
1.3.9 recognise the graphs of $y = \sin x$, $y = \cos x$, and $y = \tan x$ on extended domains
1.3.10 examine amplitude changes and the graphs of $y = a\sin x$ and $y = a\cos x$
1.3.11 examine period changes and the graphs of $y = \sin bx$, $y = \cos bx$ and $y = \tan bx$
1.3.12 examine phase changes and the graphs of $y = \sin(x - c)$, $y = \cos(x - c)$, and $y = \tan(x - c)$
1.3.13 examine the relationships $\sin\left(x + \frac{\pi}{2}\right) = \cos x$ and $\cos\left(x - \frac{\pi}{2}\right) = \sin x$
1.3.14 prove and apply the angle sum and difference identities
1.3.15 identify contexts suitable for modelling by trigonometric functions and use them to solve practical problems
1.3.16 solve equations involving trigonometric functions using technology, and algebraically in simple cases

Mathematics Methods ATAR Course Year 11 syllabus pp. 10–11 © SCSA

Video playlists (8):

5.1 Right triangles and the angle of inclination
5.2 Non-right triangles
5.3 Radian measure, arc lengths, sectors and segments
5.4 The trigonometric functions
5.5 The exact values
5.6 Graphs of trigonometric functions
5.7 Applying trigonometric functions
Examination question analysis Trigonometric functions

Worksheets (4):

5.2 Trigonometric equations
5.5 Angles of any magnitude • Radians of any magnitude
5.6 The sine and cosine curves

Puzzle (1):

5.6 Trigonometric graphs match-up

Nelson MindTap

To access resources above, visit
cengage.com.au/nelsonmindtap

Video playlist
Right triangles and the angle of inclination

5.1 Right triangles and the angle of inclination

Right triangles

For a right-angled triangle, the trigonometric ratios **sine**, **cosine** and **tangent** (abbreviated as sin, cos and tan respectively) are defined in terms of one of the acute angles and two sides related to the angle. The angle, θ, is measured in degrees.

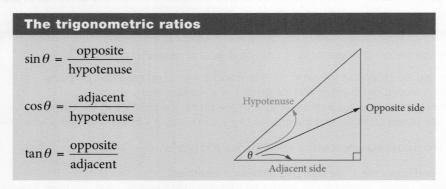

The trigonometric ratios

$$\sin \theta = \frac{\text{opposite}}{\text{hypotenuse}}$$

$$\cos \theta = \frac{\text{adjacent}}{\text{hypotenuse}}$$

$$\tan \theta = \frac{\text{opposite}}{\text{adjacent}}$$

Trigonometric ratios can be used to find the size of an unknown side or angle in a triangle.

WORKED EXAMPLE 1 Finding unknown sides in right-angled triangles

Find, correct to three decimal places, the value of the pronumeral in each triangle.

a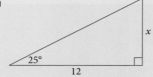

b

Steps	Working
a 1 Set the calculator to degree mode.	
2 Decide which trigonometric ratio is required. For the 25° angle, x is the opposite side and 12 is the adjacent side, so use tan.	$\tan \theta = \frac{\text{opposite}}{\text{adjacent}}$ $\tan(25°) = \frac{x}{12}$
3 Multiply both sides by 12 to solve the equation.	$12 \tan(25°) = x$
4 Evaluate and round the answer.	$x = 5.59569\ldots$ ≈ 5.596

From the diagram, an answer of $x \approx 5.596$ looks reasonable.

238 Nelson WAmaths Mathematics Methods 11

b 1 Set the calculator to degree mode.

2 Decide which trigonometric ratio is required. For the 36° angle, y is the hypotenuse and 32 is the opposite side, so use sin.

$$\sin\theta = \frac{\text{opposite side}}{\text{hypotenuse}}$$

$$\sin(36°) = \frac{32}{y}$$

3 Multiply both sides by y.

$$y\sin(36°) = 32$$

4 Divide both sides by $\sin(36°)$ to solve the equation.

$$y = \frac{32}{\sin(36°)}$$
$$= 54.44165\ldots$$
$$\approx 54.442$$

From the diagram, an answer of $y \approx 54.442$ looks reasonable.

WORKED EXAMPLE 2 — Finding an unknown angle in a right-angled triangle

a Find, to the nearest degree, the size of angle θ.

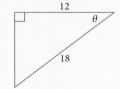

b Find, correct to one decimal place, angle θ.

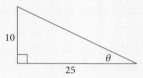

Steps	Working
a 1 Set the calculator to degree mode.	
2 Decide which trigonometric ratio is required. For θ, 12 is the adjacent side and 18 is the hypotenuse, so use cos.	$\cos\theta = \dfrac{12}{18}$
3 The angle is the inverse of the **trigonometric function**.	Use $\cos^{-1}$. $\theta = \cos^{-1}\left(\dfrac{12}{18}\right)$
4 Evaluate and round the answer.	$= 48.189\ldots°$ $\approx 48°$

From the diagram, an answer of $x \approx 48°$ looks reasonable.

b 1 Set the calculator to degree mode.	
2 Decide which trigonometric ratio is required. For θ, 10 is the opposite side and 25 is the adjacent side, so use tan.	$\tan\theta = \dfrac{10}{25}$
3 The angle is the inverse of the trigonometric function.	Use $\tan^{-1}$. $\theta = \tan^{-1}\left(\dfrac{10}{25}\right)$
4 Evaluate and round the answer.	$= 21.8014$ $\approx 21.8°$

The angle of inclination

The diagram shows a straight line making an angle of θ with the positive x-axis. This angle is called the **angle of inclination** of a straight line.

The gradient of the straight line (m) is given by

$m = \dfrac{y_2 - y_1}{x_2 - x_1}$, which is equal to $\tan \theta$; hence, $m = \tan \theta$.

The equation of the straight line is $y = mx + c$
or $y = x \tan \theta + c$.

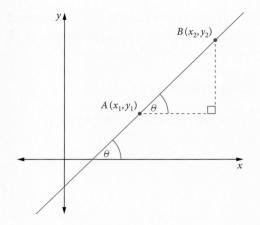

WORKED EXAMPLE 3 Finding the equation of a line using the angle of inclination

Determine the equation of a straight line that makes an angle of $45°$ with the x-axis and passes through the point $(1, 3)$.

Steps	Working
1 Determine the gradient of the line using the angle of inclination.	$m = \tan \theta = \tan 45° = 1$
2 Substitute $m = 1$, $x = 1$ and $y = 3$ into the equation $y = mx + c$ to determine the value of c.	$y = mx + c$ $3 = 1 \times 1 + c$ $\therefore c = 2$
3 Write down the required equation.	$y = x + 2$

EXERCISE 5.1 Right triangles and the angle of inclination   ANSWERS p. 472

Mastery

1 WORKED EXAMPLE 1   Find the value of each variable correct to three decimal places.

a 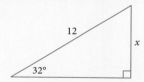 b 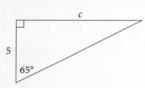 c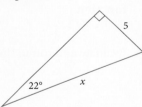

2 WORKED EXAMPLE 2   Find the size of each angle correct to the nearest degree.

a 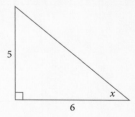 b 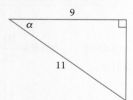 c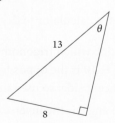

3 WORKED EXAMPLE 3   Determine the equation of the straight line that passes through the point $(-1, 2)$ and makes an angle of $135°$ with the positive x-axis.

Calculator-free

4 (5 marks) In a right-angled triangle ABC, $\angle A = 90°$, $\angle B = \theta$, $AC = x$, $AB = y$ and $BC = z$.
State whether each of the following statements is true or false.

a $\sin \theta = \dfrac{z}{x}$ (1 mark)

b $\cos \theta = \dfrac{y}{z}$ (1 mark)

c $\tan \theta = \dfrac{y}{x}$ (1 mark)

d $\cos(90° - \theta) = \dfrac{z}{y}$ (1 mark)

e $\sin(90° - \theta) = \dfrac{y}{z}$ (1 mark)

5 (3 marks)

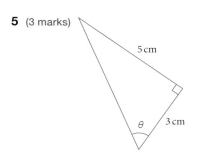

Determine exact expressions for

a $\tan \theta$ (1 mark)

b $\cos \theta$ (1 mark)

c $\sin \theta$ (1 mark)

Calculator-assumed

6 (3 marks) Tim is on level ground and measures the angle of elevation of the top of a building of height 25 metres to be 50°. How far from the base of the building is he standing? Give your answer correct to two decimal places.

7 (3 marks) A vertical flagpole is 8 m tall. There is a 20 m wire attached from the top of the flagpole to the ground, away from the base of the flagpole. What is the angle of elevation from the end of the wire to the top of the flagpole? Give your answer correct to two decimal places.

8 (2 marks) If $y = -\sqrt{3}x + 3$, determine the angle of inclination for this line.

9 (3 marks) A child lying on the ground looks up and sees a ladder of length 14 metres just reaching to the top of a brick wall at an angle of elevation of 49°. Find the height of the brick wall, to the nearest whole number.

10 (4 marks) A rectangular-based pyramid with a base measuring 16 cm by 12 cm and a height of 10 cm is to be constructed using perspex sheeting.

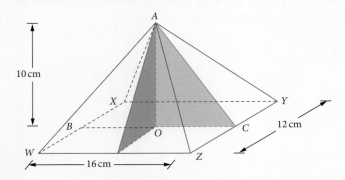

Calculate the angle that each of the triangular faces AYZ and AZW makes with the horizontal base.

11 (3 marks) From the top of a vertical building, the angle of depression of a sports car on the road is 30° and the angle of depression of a truck parked behind the sports car is 60°.

Show that the distance of the sports car from the base of the building is three times the truck's distance from the building.

12 (3 marks) A helicopter is sighted at the same time by two ground observers who are 2.4 km apart in the same direction from the helicopter. The angles of elevation from the observers to the helicopter (measured to the horizontal) are 18.4° and 31.6°. How high is the helicopter, correct to two decimal places?

5.2 Non-right triangles

Sine rule

In geometry, solving a triangle means finding the sizes of all unknown sides and angles. We already know how to solve a **right-angled triangle** for its sides and angles. The sine and cosine rules can be used to find the sides and angles in a **non-right-angled triangle**.

In any triangle ABC with angles labelled A, B, C, the opposite side lengths are labelled a, b, c respectively.

Consider the triangle ABC drawn below.

Diagram 1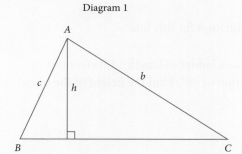

Diagram 2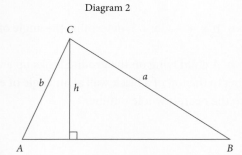

From diagram 1, $\sin C = \dfrac{h}{b} \Rightarrow h = b \sin C$ and $\sin B = \dfrac{h}{c} \Rightarrow h = c \sin B$.

Putting the two equations equal to each other gives $\dfrac{b}{\sin B} = \dfrac{c}{\sin C}$. [1]

Using diagram 2 (diagram 1 re-labelled), $\sin A = \dfrac{h}{b} \Rightarrow h = b \sin A$ and $\sin B = \dfrac{h}{a} \Rightarrow h = a \sin B$.

Putting the two equations equal to each other gives $\dfrac{b}{\sin B} = \dfrac{a}{\sin A}$, therefore using [1]:

$$\dfrac{a}{\sin A} = \dfrac{b}{\sin B} = \dfrac{c}{\sin C}$$

The **sine rule** relates each side of a triangle to its opposite angle. The longest side is opposite the largest angle, and the shortest side is opposite the smallest angle. The sine rule is used when the triangle problem gives two sides and one non-included angle, or two angles and one side.

The sine rule

In any triangle ABC:

$$\dfrac{a}{\sin A} = \dfrac{b}{\sin B} = \dfrac{c}{\sin C}$$

or

$$\dfrac{\sin A}{a} = \dfrac{\sin B}{b} = \dfrac{\sin C}{c}$$

WORKED EXAMPLE 4 | The sine rule

For this triangle, find the unknown side a correct to two decimal places.

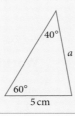

Steps	Working
1 Look for two pairs of sides and their opposite angles. Put the unknown variable in the numerator. $\dfrac{a}{\sin A} = \dfrac{b}{\sin B}$	$\dfrac{a}{\sin 60°} = \dfrac{5}{\sin 40°}$
2 Solve for a.	$a = \dfrac{5 \sin 60°}{\sin 40°}$ $\approx 6.74 \text{ cm}$

🔓 **Exam hack**

From the diagram, an answer of $a \approx 6.74$ cm looks reasonable. Side a is opposite the larger angle (60°), so it should be longer than 5 cm.

Ambiguous case

An angle θ in a triangle could be **acute**, **right** or **obtuse**, so it has the range $0° < \theta < 180°$.

This is important when finding the size of an unknown angle in a triangle using the sine rule, when it is possible to have two answers: one acute, one obtuse.

USING CAS 1 | The sine rule

Find the angle, C, for $\triangle ABC$, with $b = 6.4$ m, $c = 4.7$ m and $B = 68°$.

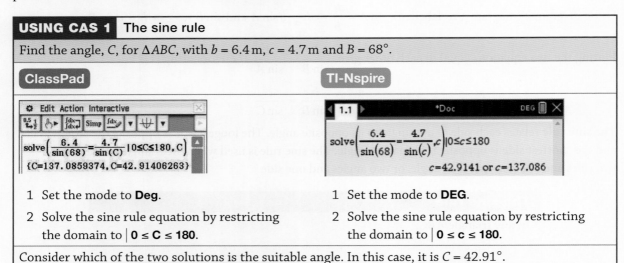

ClassPad	TI-Nspire
1 Set the mode to **Deg**.	1 Set the mode to **DEG**.
2 Solve the sine rule equation by restricting the domain to $\|0 \leq C \leq 180$.	2 Solve the sine rule equation by restricting the domain to $\|0 \leq c \leq 180$.

Consider which of the two solutions is the suitable angle. In this case, it is $C = 42.91°$.

WORKED EXAMPLE 5 | The sine rule for angles 1

For this triangle, find the possible values for B, correct to two decimal places.

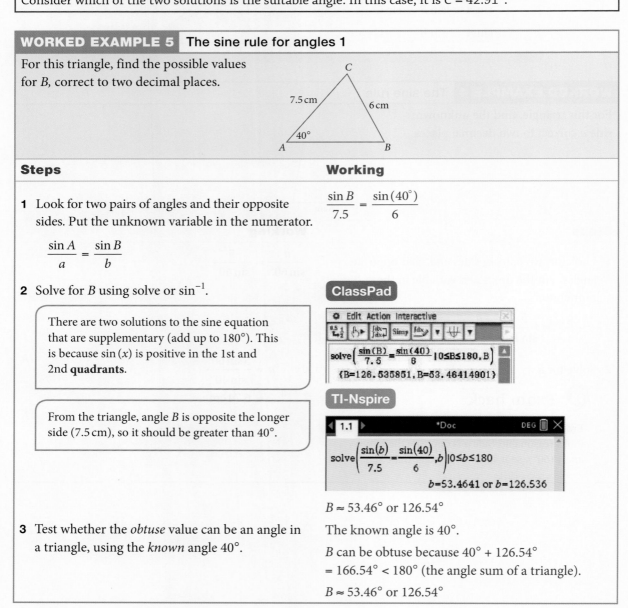

Steps	Working
1 Look for two pairs of angles and their opposite sides. Put the unknown variable in the numerator. $$\frac{\sin A}{a} = \frac{\sin B}{b}$$	$$\frac{\sin B}{7.5} = \frac{\sin(40°)}{6}$$
2 Solve for B using solve or $\sin^{-1}$. There are two solutions to the sine equation that are supplementary (add up to $180°$). This is because $\sin(x)$ is positive in the 1st and 2nd **quadrants**. From the triangle, angle B is opposite the longer side (7.5 cm), so it should be greater than $40°$.	$B \approx 53.46°$ or $126.54°$
3 Test whether the *obtuse* value can be an angle in a triangle, using the *known* angle $40°$.	The known angle is $40°$. B can be obtuse because $40° + 126.54° = 166.54° < 180°$ (the angle sum of a triangle). $B \approx 53.46°$ or $126.54°$

The two triangles show the two possible solutions to the previous example.

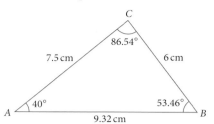 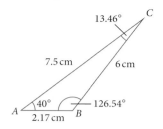

This situation is called the **ambiguous case**, where two solutions for an unknown angle are possible.

In some cases, both the acute and obtuse angles are correct, but in other cases, only one angle is correct for that triangle. It depends on the size of the given angle.

The ambiguous case only happens when we are finding an angle, not a side.

WORKED EXAMPLE 6 | The sine rule for angles 2

Find θ, correct to the nearest degree.

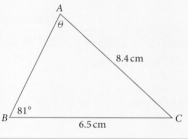

Steps	Working
1 Find two pairs of matching angles and opposite sides with unknown on top.	$\dfrac{\sin\theta}{6.5} = \dfrac{\sin(81°)}{8.4}$
2 Solve for $\sin\theta$.	$\sin\theta = \dfrac{6.5\sin(81°)}{8.4}$ $\sin\theta = 0.7642\ldots$
3 Use $\sin^{-1}$ to find the acute value of θ.	$\theta = 49.8432\ldots$
4 Find the possible obtuse angle.	obtuse $\theta \approx 180° - 49.8432\ldots°$ $= 130.1567\ldots°$
5 Check the obtuse angle using the known angle 81°.	sum of angles $= 81° + 130.1567\ldots° > 180°$
6 Sum $> 180°$ means triangle cannot exist for obtuse θ.	$\theta = 49.8432\ldots°$ only $\approx 50°$

Cosine rule

The **cosine rule** relates one side of a triangle to the other two sides and its opposite angle. It is an extension of Pythagoras' theorem, $c^2 = a^2 + b^2$. The cosine rule is used for triangle problems involving two sides and the included angle, or all three sides.

Consider the triangle ABC shown.

By Pythagoras, $b^2 = x^2 + h^2$ and $c^2 = (a-x)^2 + h^2$.

Therefore, $b^2 = x^2 + c^2 - (a-x)^2 = c^2 - a^2 + 2ax$ after expanding the bracket and collecting like terms. Now, $x = b\cos C$ from triangle ADC.

Hence, rearranging for c^2: $c^2 = a^2 + b^2 - 2ab\cos C$.

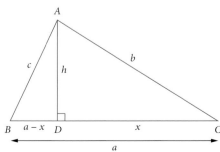

The cosine rule

In any triangle ABC,

$$c^2 = a^2 + b^2 - 2ab \cos C$$

or $\quad \cos C = \dfrac{a^2 + b^2 - c^2}{2ab}$

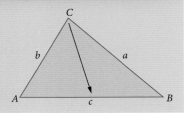

WORKED EXAMPLE 7 | **The cosine rule**

Find the length of XZ, correct to two decimal places.

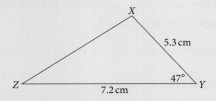

Steps	Working
1 Use the cosine rule of the form $c^2 = a^2 + b^2 - 2ab \cos C$, where c is the unknown length and C is the known opposite angle.	$XZ^2 = 5.3^2 + 7.2^2 - 2(5.3)(7.2)\cos(47°)$
2 Solve for XZ, and take the positive solution.	$XZ = 5.2801\ldots$
3 Round the answer.	$XZ \approx 5.28$ cm

> 🔓 **Exam hack**
>
> From the measurements given on the diagram, the answer looks reasonable.

WORKED EXAMPLE 8 | **The cosine rule for angles**

Find the angle M, to the nearest degree.

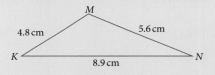

> Remember to start the cosine rule with the side opposite the angle we know or want to find out. Here, we start with side $m = 8.9$.

Steps	Working
1 Use the cosine rule of the form $\cos C = \dfrac{a^2 + b^2 - c^2}{2ab}$, where C is the unknown angle and c is the known opposite side.	$\cos M = \dfrac{4.8^2 + 5.6^2 - 8.9^2}{2 \times 4.8 \times 5.6}$
2 Solve for M using CAS by solving the cosine rule equation or using $\cos^{-1}$, in the range $0° < M < 180°$.	$M = 117.4836\ldots°$
3 Round the answer.	$M \approx 117°$

Note: The ambiguous case does not apply to the cosine rule.

> 🔓 **Exam hack**
>
> From the measurements given on the diagram, the answer looks reasonable.

USING CAS 2 | The cosine rule

Find the angle θ, to the nearest degree.

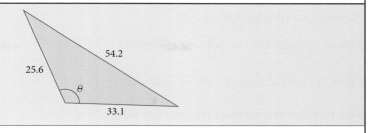

ClassPad

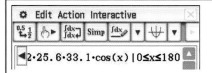   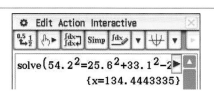

1. Set the mode to **Deg**.
2. Enter the equation using the cosine rule.
3. To restrict the domain, open the **Keyboard** > **Math3** for the | and the ≤ symbols.
4. Highlight the equation, including the restricted domain.
5. Tap **Interactive** > **Equation/Inequality** > **solve**.
6. In the dialogue box, keep the default variable x and tap **OK** to find the angle.

TI-Nspire

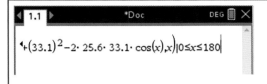   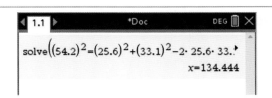

1. Set the mode to **DEG**.
2. Solve the equation for x using the cosine rule.
3. To restrict the domain, press **ctrl + =** to open the mini-palette for the | and the ≤ symbols.
4. At the end of the solve command, enter the restricted domain **| 0 ≤ x ≤ 180**.
5. Press **enter** to find the angle.

The angle is approximately 134°.

Solving problems with the sine and cosine rules:
- Use the **sine rule** for problems involving two sides and two angles.
- Use the **cosine rule** for problems involving two sides and the included angle or three sides.
- Use a combination of the **sine rule** and **cosine rule** in the one problem, if suitable, remembering the ambiguous case in the sine rule for angles.
- Consider larger/smaller side lengths that match larger/smaller opposite angles.

Area of a triangle

Consider the triangle ABC shown.

Area of triangle $ABC = \frac{1}{2}ah$, but $h = b \sin C$,

hence, the area of triangle $ABC = \frac{1}{2}ab\sin C$.

Note that this area formula requires two sides of the triangle and the included angle.

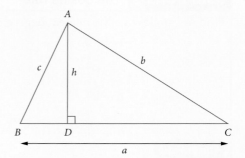

WORKED EXAMPLE 9 — Finding the area of a triangle

Determine the area of triangle ABC where $a = 10$ cm, $b = 12$ cm and $C = 32°$. Give your answer to two decimal places.

Steps	Working
1 Draw a sketch of the given triangle.	*A triangle with $b = 12$ cm, $a = 10$ cm and included angle $32°$ at C.*
2 Substitute into the formula and determine the area.	area $= \frac{1}{2} \times 10 \times 12 \times \sin(32°)$ area $= 31.80$ cm^2

EXERCISE 5.2 Non-right triangles

ANSWERS p. 472

Recap

1. In a right-angled triangle, the hypotenuse has length 5 cm and the adjacent side has length 2 cm. The angle between the adjacent side and the hypotenuse is closest to

   A 21.80° **B** 23.30° **C** 23.58° **D** 66.42° **E** 66.70°

2. Determine the equation of a straight line that passes through the point $(-3, -2)$ and has an angle of inclination of 60°. Give your answer correct to two decimal places.

Mastery

3. **WORKED EXAMPLE 4** Find the unknown lengths to one decimal place, in each triangle.

   a

   b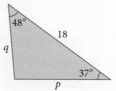

   c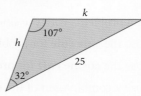

   d

   e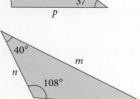

   f *Triangle with angle 27.2°, side 76, angle 115.5°, side x, side y.*

4 **Using CAS 1** Solve each triangle for all unknown angles and sides, to the nearest integer.
   a $\triangle XYZ$ where $Z = 65°$, $y = 11$ m and $z = 16$ m
   b $\triangle PQR$ where $P = 35°$, $Q = 53°$ and $q = 68$ cm

5 **WORKED EXAMPLE 5** Find the angle(s) θ, to the nearest degree, in each triangle.

a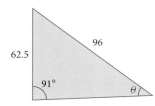
b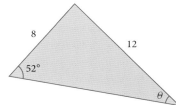
c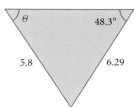

d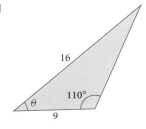
e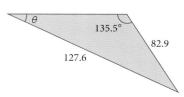
f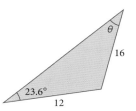

6 **WORKED EXAMPLE 6** Determine the value of θ in the triangle, and show that there is only one possible answer.

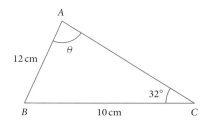

7 **WORKED EXAMPLE 7** Determine the length of the unknown side of the triangle.

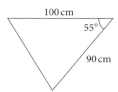

8 **WORKED EXAMPLE 8** **Using CAS 2** Find the angle θ, to the nearest degree, in each triangle.

a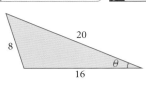
b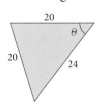
c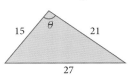

d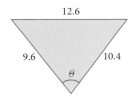
e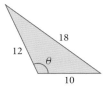
f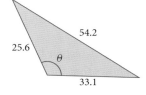

9 **WORKED EXAMPLE 9** Determine the area of triangle ABC where $a = 20$ cm, $b = 18$ cm and $C = 70°$. Give your answer correct to two decimal places.

Calculator-free

10 (5 marks) State whether each of the following, about this triangle, is true or false.

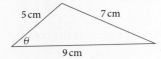

a $\cos\theta = \dfrac{5^2 + 7^2 - 9^2}{2 \times 7 \times 9}$ (1 mark)

b $\cos\theta = \dfrac{9^2 + 5^2 - 7^2}{2 \times 9 \times 5}$ (1 mark)

c $\cos\theta = \dfrac{5^2 + 7^2 - 9^2}{2 \times 7 \times 9}$ (1 mark)

d $\cos\theta = \dfrac{5^2 + 7^2 - 9^2}{2 \times 5 \times 7}$ (1 mark)

e $\cos\theta = \dfrac{5^2 + 9^2 - 7^2}{2 \times 5 \times 9}$ (1 mark)

11 (5 marks) For this triangle, state whether the following expressions are true or false for $\sin Q$.

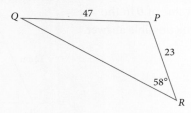

a $\dfrac{47}{23 \sin(58°)}$ (1 mark)

b $\dfrac{23}{47 \sin(58°)}$ (1 mark)

c $\dfrac{23}{47} \sin(58°)$ (1 mark)

d $\dfrac{23 \sin(58°)}{47}$ (1 mark)

e $\dfrac{58 \sin(23°)}{\sin(47°)}$ (1 mark)

Calculator-assumed

12 (3 marks) The area of a triangle is $64\,\text{cm}^2$, and two sides have lengths of 10 cm and 15 cm. Determine the size of the included angle between these two sides.

13 (3 marks) A farmer has a triangular paddock that he is going to use to grow potatoes. He knows that the three sides measure 30 m, 25 m and 42 m, and the largest angle is 100°. Determine how much land the farmer has for growing potatoes. Give your answer correct to two decimal places.

14 (4 marks) Find the value *x*, to the nearest whole number, in each diagram.

a (2 marks)

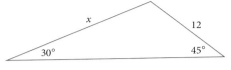

b (2 marks)

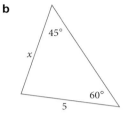

15 (2 marks) Find, to the nearest millimetre, the length of the longest side of $\triangle PQR$ if $P = 32°$, $Q = 29°$ and $PR = 42$ mm.

16 (2 marks) Calculate the length of the unknown side to the nearest metre.

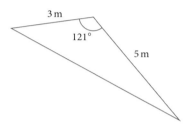

17 (2 marks) A weight is supported by cables attached to opposite sides of an overhead beam 18 m apart. One cable is 12 m long and the other is 9 m long.

What angles do each of the cables make with the ceiling? Give the answers correct to one decimal place.

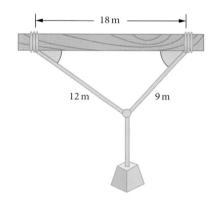

18 (2 marks) The folding chair shown has a seat that is 36 cm deep. The legs of the chair are inclined at an angle of 55° to the seat and join at an angle of 70°. The seat is 51 cm above the floor.

Find the distance, *d*, from the seat to where the legs are joined.

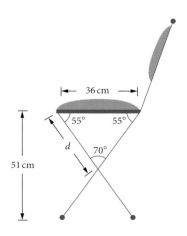

19 (4 marks) $\triangle ABC$ is an isosceles with $\angle BAC = 52°$ between the equal sides. BC is extended to D so that $AC = CD$. $AB = 12.3$ cm. Find, correct to one decimal place, the

   a length of AD (2 marks)

   b length of BD.   (2 marks)

20 (3 marks) Three gears are arranged as shown. The radii of the gears are 7.2 cm, 5.4 cm and 3.2 cm. Calculate θ, correct to one decimal place.

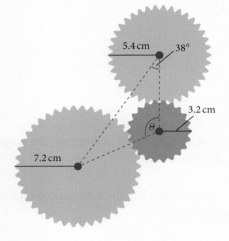

21 (6 marks) Sophie and Will set out from point A at the same time. Sophie travels at 30 km/h along a straight road in the direction 042°. Will travels at 24 km/h along another straight road in the direction 142°. After 4 hours, to one decimal place,

   a what is the distance between Will and Sophie   (3 marks)

   b what is the direction Sophie would have to go in a straight line to reach Will?   (3 marks)

22 (10 marks) A sculptor has removed a section from a copper block with dimensions as shown in the diagram. The base of the block is square.

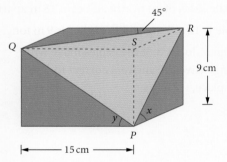

Calculate

   a QR (2 marks)

   b x (3 marks)

   c y (2 marks)

   d the angle that the triangular face PQR makes with the horizontal base.   (3 marks)

5.3 Radian measure, arc lengths, sectors and segments

Radian measure

We use degrees to measure the size of an angle, but we can also use **radians**.

One radian is the angle at the centre of a circle of **radius** r that is opposite an **arc** of length r. The length of the arc opposite an angle of 1 radian is equal to the radius of the circle.

One radian can be written 1 rad, but it is usually written simply as 1, without units. Radian measure is also called **circular measure**, because it is based on the circle.

1 rad corresponds to an arc length of r.

Multiplying both values by 2π gives the **circumference** of a circle:

2π rad corresponds to an arc length of $2\pi r$.

This means that 2π rad = 360° (a **revolution**).

So, π rad = 180°.

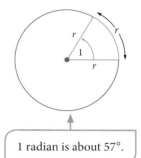

1 radian is about 57°.

Degrees and radians

The relationship between degrees and radians is

$$180° = \pi \text{ rad}$$

To convert degrees to radians, multiply the number of degrees by $\frac{\pi}{180°}$.

To convert radians to degrees, multiply the number of radians by $\frac{180°}{\pi}$.

Some common conversions are:

$90° = \frac{\pi}{2}$ $\quad 45° = \frac{\pi}{4}$ $\quad 60° = \frac{\pi}{3}$ $\quad 30° = \frac{\pi}{6}$ $\quad 270° = \frac{3\pi}{2}$ $\quad 360° = 2\pi$

WORKED EXAMPLE 10 — Converting between degrees and radians

Convert

a $\frac{3\pi}{4}$ to degrees **b** 225° to radians **c** 60° to radians.

Steps	Working
a Multiply by $\frac{180°}{\pi}$.	$\frac{3\pi}{4} = \frac{3\pi}{4} \times \frac{180°}{\pi} = 135°$
b Multiply by $\frac{\pi}{180°}$ and simplify.	$225° = 225° \times \frac{\pi}{180°} = \frac{225\pi}{180} = \frac{5\pi}{4}$
c Use same method as above: multiply by $\frac{\pi}{180°}$.	$60° = 60° \times \frac{\pi}{180°} = \frac{\pi}{3}$

> **Exam hack**
>
> Set your calculator to either degree mode or radian mode to satisfy the requirements of the question.

1 In the bottom right-hand corner of the screen, tap to toggle between **Rad** and **Deg** mode (ignore **Gra** mode, which stands for gradians).

2 Tap **Math1** or **Trig** to access the ° and ʳ symbols.

1 In the top right-hand corner of the screen, click to toggle between **RAD** and **DEG** mode.

2 Press **π** to access the ° and ʳ symbols.

Parts of a circle

A radius joins the centre of a circle to the circumference.
A **chord** joins two points on the circumference of a circle.
A diameter is a chord that passes through the centre of a circle.

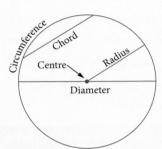

An arc is part of the circumference.
A **sector** is a region of a circle cut off by 2 radii.
A **segment** is a region of a circle cut off by a chord.
A **major arc**, **sector** or **segment** is more than half of the circle.
A **minor arc**, **sector** or **segment** is less than half of the circle.

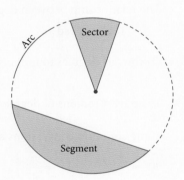

Arc length

When radians are used, finding the length (l) of an arc in a circle is simple.
l is a fraction of the circumference of the circle $C = 2\pi r$.
θ is a fraction of a revolution 2π.

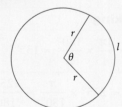

So $\dfrac{l}{2\pi r} = \dfrac{\theta}{2\pi}$.

$l = \dfrac{\theta}{2\pi} \times 2\pi r = r\theta$

> ### Arc length
> The arc length l that subtends an angle of θ radians at the centre of the circle of radius r is
> $l = r\theta$.

WORKED EXAMPLE 11 — Finding the arc length

Find the

a exact length of the arc formed by an angle of $\frac{\pi}{3}$ in a circle of radius 6 cm

b angle in radians subtended at the centre of a circle of radius 5.4 cm by an arc 3.6 cm long, correct to two decimal places

c exact length of the arc formed by an angle of 45° in a circle of radius 2 cm.

> **Exam hack**
> The formula uses radians and not degrees, so all angles must be in radians

Steps	Working
a Use $l = r\theta$ with $r = 6$ cm and $\theta = \frac{\pi}{3}$.	$l = r\theta$ $= 6 \times \frac{\pi}{3}$ $= 2\pi$ cm
b 1 Use $l = r\theta$ with $l = 3.6$ cm, $r = 5.4$ cm. 2 Divide both sides by 5.4. 3 Round the answer.	$3.6 = 5.4\theta$ $\theta = \frac{3.6}{5.4}$ $= 0.6666...$ ≈ 0.67 cm
c 1 Convert the angle into radians first. 2 Use $l = r\theta$ with $r = 2$ cm and $\theta = \frac{\pi}{4}$.	$45° = 45° \times \frac{\pi}{180°} = \frac{\pi}{4}$ $l = 2 \times \frac{\pi}{4}$ $= \frac{\pi}{2}$ cm

> **Exam hack**
> 'Exact length' means leave the answer in terms of π.

> **Exam hack**
> Subtended means 'sits opposite'.

WORKED EXAMPLE 12 — Finding the arc length in a practical context

The minute hand of a clock is 15 cm long. If the end of the hand travelled an arc of 10.5 cm, through what angle, to the nearest degree, did the hand rotate?

Steps	Working
1 Use $l = r\theta$ with $l = 10.5$ cm, $r = 15$ cm.	$10.5 = 15\theta$
2 Divide both sides by 15.	$\theta = \frac{10.5}{15}$ $= 0.7$
3 Convert this angle from radians to degrees.	$0.7 \text{ rad} = 0.7 \times \frac{180°}{\pi}$ $= 40.1070...$ $\approx 40°$

For the formula $l = r\theta$, θ must be in radians.

Sectors and segments

Using radians also makes finding the area of a sector of a circle easy.

> **Area of a sector**
>
> The **area of a sector** of a circle is
>
> $$A = \frac{1}{2}r^2\theta$$
>
> where
>
> r = radius of the circle
>
> θ = angle at the centre of the circle in radians.

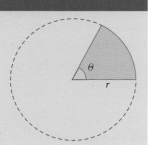

The area of a segment can be considered as the area of the sector less the triangle, as shown in the diagram.

area of segment = area of sector − area of triangle

$$\text{area of triangle} = \frac{1}{2} \times 2 \text{ sides} \times \sin(\text{included angle})$$

$$= \frac{1}{2}r^2 \sin\theta$$

$$\text{area of segment} = \frac{1}{2}r^2\theta - \frac{1}{2}r^2 \sin\theta$$

$$= \frac{1}{2}r^2(\theta - \sin\theta)$$

> **Area of a segment**
>
> The formula for the **area of a segment** of a circle is
>
> $$A = \frac{1}{2}r^2(\theta - \sin\theta)$$
>
> where
>
> r = radius of the circle
>
> θ = angle at the centre of the circle in radians.

WORKED EXAMPLE 13 | Arc length, sector and segment

For the sector of a circle with radius 10 cm forming an angle of $\frac{\pi}{3}$ radians at the centre of the circle, O, find correct to two decimal places, the

a arc length AB
b area of the sector AOB
c area of the segment.

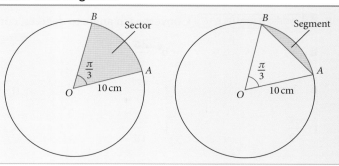

Steps	Working
a Use the formula $l = r\theta$.	$l = 10 \times \frac{\pi}{3}$ $= \frac{10\pi}{3}$ ≈ 10.47 units
b Use the formula $A = \frac{1}{2}r^2\theta$.	$A = \frac{1}{2} \times 10^2 \times \frac{\pi}{3}$ $= \frac{50\pi}{3}$ ≈ 52.36 units²
c Use the formula $A = \frac{1}{2}r^2(\theta - \sin\theta)$ An exact area is preferred, but a decimal approximation helps to compare values.	$A = \frac{1}{2} \times 10^2 \left(\frac{\pi}{3} - \sin\left(\frac{\pi}{3}\right)\right)$ $= 50\left(\frac{\pi}{3} - \frac{\sqrt{3}}{2}\right)$ ≈ 9.06 units²

Note: If an angle is given in degrees, we need to change that angle to radians to use these formulas.
CAS can convert radians to degrees and vice-versa.

EXERCISE 5.3 Radian measure, arc lengths, sectors and segments   ANSWERS p. 472

Recap

1 In the triangle, $B \approx$

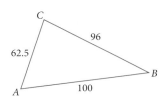

A 37.1° only
B 37.1° or 142.9°
C 38.9° only
D 38.9° or 141.1°
E 96.8° or 83.2°

2 In a triangle ABC, where side lengths are $a = 5$ m, $b = 7.1$ m and $c = 8.1$ m, angle B, correct to the nearest degree, is

A 38°   B 60°   C 82°   D 98°   E 120°

Mastery

3 WORKED EXAMPLE 10 Convert each angle to degrees, correct to one decimal place for parts **k** and **l**.

a $\dfrac{\pi}{5}$
b $\dfrac{2\pi}{3}$
c $\dfrac{5\pi}{4}$
d $\dfrac{7\pi}{6}$

e 3π
f $\dfrac{7\pi}{9}$
g $\dfrac{4\pi}{3}$
h $\dfrac{7\pi}{3}$

i $\dfrac{\pi}{9}$
j $\dfrac{5\pi}{18}$
k 1.09
l 0.768

4 WORKED EXAMPLE 10 Convert each angle to radians, correct to two decimal places.

a 56°
b 68°
c 127°
d 289°
e 312°

5 WORKED EXAMPLE 11 Find the arc length of the circle for each given radius and angle subtended at the centre. For parts **a** to **e**, write the answer in terms of π. For parts **f** to **h**, write the answer correct to two decimal places.

a 4 cm, π
b 3 m, $\dfrac{\pi}{3}$
c 10 cm, $\dfrac{5\pi}{6}$
d 3 cm, 30°

e 7 mm, 45°
f 1.5 m, 0.43
g 3.21 cm, 1.22
h 7.2 mm, 55°

6 WORKED EXAMPLE 12 The string of a pendulum is 50 cm in length and the bob at its end rotates through an angle of $\dfrac{1}{4}$.

a What arc length does the bob sweep out as it swings?

b If the angle of swing is now 20°, how much further will the bob travel? Give the answer correct to one decimal place.

7 WORKED EXAMPLE 13 The sector of a circle of radius 5 cm forms an angle of $\dfrac{2\pi}{3}$ radians at the centre of the circle. Determine, correct to two decimal places, the

a arc length

b area of the sector

c area of the segment.

Calculator-free

8 (10 marks) Convert each angle to radians in terms of π.

a 135° (1 mark)
b 30° (1 mark)
c 150° (1 mark)
d 240° (1 mark)
e 300° (1 mark)
f 63° (1 mark)
g 15° (1 mark)
h 450° (1 mark)
i 225° (1 mark)
j 120° (1 mark)

9 (2 marks) State the exact answer to $\dfrac{\dfrac{2\pi}{3} + 60°}{\dfrac{5\pi}{6} - \dfrac{\pi}{4} - 15°}$.

10 (4 marks) A major sector is bigger than a semicircle while a minor sector is a smaller than a semicircle. State whether each angle subtends a major or minor sector.

    a $\dfrac{2\pi}{3}$ (1 mark)

    b $\dfrac{11\pi}{6}$ (1 mark)

    c $\dfrac{8\pi}{9}$ (1 mark)

    d $\dfrac{7\pi}{6}$ (1 mark)

Calculator-assumed

11 (16 marks) Find the angle (in radians) subtended at the centre of a circle (correct to two decimal places where necessary) by an arc with

    a radius 16 m, length 28 m (2 marks)

    b radius 3 cm, length $\dfrac{2\pi}{7}$ cm (2 marks)

    c radius $\dfrac{\pi}{3}$ m, length $\dfrac{\pi}{2}$ m (2 marks)

    d radius 18 cm, length 28 cm (2 marks)

    e radius 194 mm, length 218 mm (2 marks)

    f radius 9 m, length 6 m (2 marks)

    g radius 19.2 cm, length 12 cm (2 marks)

    h radius π cm, length 2π cm. (2 marks)

12 (6 marks) Determine the value of each variable correct to three decimal places.

    a (2 marks)

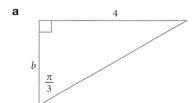

    b (2 marks)

    c (2 marks)

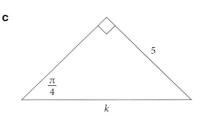

13 (2 marks) A circle of radius $\dfrac{9}{\pi}$ cm has arc length 8 cm. Determine the value of the angle the arc subtends at the centre of the circle in degrees.

14 (2 marks) This car's windscreen wiper sweeps through an angle of 110°. Calculate, correct to the nearest centimetre, the distance the red marker shown on the wiper covers in one sweep.

15 (2 marks) Determine the area of the segment for the sector of a circle with radius of 10 mm, subtended by the angle of 45°.

16 (2 marks) Determine the arc length, in cm, for the sector of a circle with radius of 10 cm, subtended by the angle of $\dfrac{9\pi}{8}$ radians.

17 (3 marks) The sector of a circle with radius of 10 cm is subtended by an angle of $\dfrac{9\pi}{8}$. What percentage of the circumference of the circle is the arc length of the sector?

5.4 The trigonometric functions

Video playlist
The trigonometric functions

The sine and cosine functions

Trigonometric functions are also called circular functions because they are based on the **unit circle**.

Right-angled triangles can be constructed within a quadrant in a unit circle to determine the trigonometric values of angles of any size.

Consider the unit circle below with equation $x^2 + y^2 = 1$, centre (0, 0) and radius 1. We can use it to define sine, cosine and tangent of any angle of size θ.

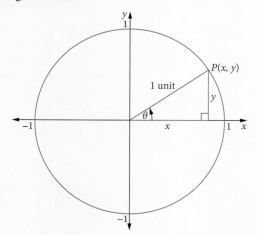

Let $P(x, y)$ be a point on the unit circle that makes an angle of θ with the positive direction of the x-axis.

Then $\sin \theta = \dfrac{\text{opposite}}{\text{hypotenuse}} = \dfrac{y}{1} = y$

and $\cos \theta = \dfrac{\text{adjacent}}{\text{hypotenuse}} = \dfrac{x}{1} = x$.

So, for any angle of size θ corresponding to a point $P(x, y)$ on the unit circle, $\sin \theta = y$-coordinate, $\cos \theta = x$-coordinate.

We can extend this definition of the sine and cosine functions to angles in the 2nd, 3rd and 4th quadrants, and therefore angles of any size.

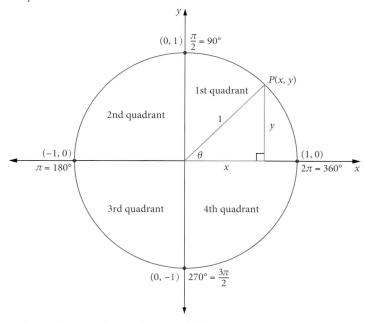

By examining the signs of x and y in each quadrant, it follows that:
- in the 1st quadrant, $\cos \theta$ is positive and $\sin \theta$ is positive
- in the 2nd quadrant, $\cos \theta$ is negative and $\sin \theta$ is positive
- in the 3rd quadrant, $\cos \theta$ is negative and $\sin \theta$ is negative
- in the 4th quadrant, $\cos \theta$ is positive and $\sin \theta$ is negative.

Using Pythagoras' theorem for the right-angled triangle in the diagram, we obtain $x^2 + y^2 = 1$, the equation of the circle.

Therefore, $(\cos \theta)^2 + (\sin \theta)^2 = 1$. This is usually written with $\sin \theta$ first as $\sin^2 \theta + \cos^2 \theta = 1$ and is known as the **Pythagorean identity**.

Note: the word 'identity' is used when the two sides of an equation are equal for any value of the variable.

Pythagorean identity
$$\cos^2 \theta + \sin^2 \theta = 1$$

The tangent function

In the diagram, OQ passes through P, QS is a vertical tangent to the circle at S and $\triangle QSO$ is right-angled.

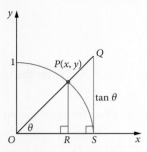

Then $\tan\theta = \dfrac{\text{opposite}}{\text{adjacent}} = \dfrac{QS}{OS} = \dfrac{y\text{-coordinate of }Q}{1} = y\text{-coordinate of }Q$.

But also, $\tan\theta = \dfrac{\text{opposite}}{\text{adjacent}} = \dfrac{PR}{OR} = \dfrac{y\text{-coordinate of }P}{x\text{-coordinate of }P} = \dfrac{\sin\theta}{\cos\theta}$.

> **The tangent identity**
>
> $$\tan\theta = \dfrac{\sin\theta}{\cos\theta}$$

By examining the signs of $\sin\theta$ and $\cos\theta$ in each quadrant, it follows that:

- in the 1st quadrant, $\tan\theta$ is positive
- in the 2nd quadrant, $\tan\theta$ is negative
- in the 3rd quadrant, $\tan\theta$ is positive
- in the 4th quadrant, $\tan\theta$ is negative.

1st quadrant	2nd quadrant	3rd quadrant	4th quadrant
$\sin\theta$ is positive	$\sin\theta$ is positive	$\sin\theta$ is negative	$\sin\theta$ is negative
$\cos\theta$ is positive	$\cos\theta$ is negative	$\cos\theta$ is negative	$\cos\theta$ is positive
$\tan\theta$ is positive	$\tan\theta$ is negative	$\tan\theta$ is positive	$\tan\theta$ is negative

Note that in the 1st quadrant, all the trigonometric functions are positive, but in the 2nd, 3rd and 4th quadrants, only one of the trigonometric functions is positive. This can be summarised by the **ASTC rule**.

> **The ASTC rule**
>
> The ASTC rule gives the signs of the trigonometric functions in each quadrant. We can remember it using a mnemonic such as 'All Stations To Claremont'.
>
>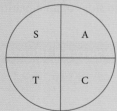
>
> **All** the trigonometric functions are positive in the 1st quadrant.
> **Sine** is positive and the others are negative in the 2nd quadrant.
> **Tangent** is positive and the others are negative in the 3rd quadrant.
> **Cosine** is positive and the others are negative in the 4th quadrant.
> Alternatively, we could remember the word CAST.

WORKED EXAMPLE 14 — Determining trigonometric values using the unit circle

Given $\cos\theta = -\dfrac{5}{13}$ and $\pi < \theta < \dfrac{3\pi}{2}$, without use of a calculator, find the exact value of

a $\sin\theta$ **b** $\tan\theta$

Steps	Working
a 1 Draw a right-angled triangle with adjacent side 5 and hypotenuse 13 and use Pythagoras' theorem to find the opposite side. 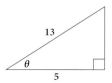	opposite side = $\sqrt{13^2 - 5^2} = 12$
2 Determine value of $\sin\theta$.	$\sin\theta = \dfrac{12}{13}$
3 Use the quadrant to determine if $\sin\theta$ is positive or negative. $\pi < \theta < \dfrac{3\pi}{2}$ means θ is in the 3rd quadrant.	θ is in the 3rd quadrant, so $\sin\theta$ is negative. $\sin\theta = -\dfrac{12}{13}$
b Use $\tan\theta = \dfrac{\sin\theta}{\cos\theta}$.	$\tan\theta = \dfrac{\sin\theta}{\cos\theta} = \dfrac{-\dfrac{12}{13}}{-\dfrac{5}{13}} = \dfrac{12}{5}$ $\cos\theta$ is negative in the 3rd quadrant, and $\tan\theta$ is positive.

Angles in the 4 quadrants

Trigonometric ratios involving the 2nd, 3rd and 4th quadrants can be expressed in terms of trigonometric ratios in the 1st quadrant.

Let the angle θ be formed using the point $P(a, b)$ on the unit circle, where θ is the angle that point P makes with the positive direction of the x-axis.

Consider a **negative angle**, $-\theta$, formed using the point $P'(a, -b)$.

P and P' have the same x-coordinate, but the y-coordinate of P' is the negative of the y-coordinate of P, so

$$\cos(-\theta) = \cos\theta$$

and $\sin(-\theta) = -\sin\theta$.

Also, $\tan(-\theta) = \dfrac{\sin(-\theta)}{\cos(-\theta)} = \dfrac{-\sin\theta}{\cos\theta} = -\tan\theta$.

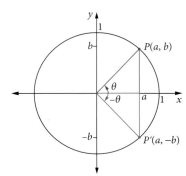

Consider an **angle in the 2nd quadrant**, $\pi - \theta$, formed using the point $P'(-a, b)$.

P and P' have the same y-coordinate, but the x-coordinate of P' is the negative of the x-coordinate of P, so

$$\cos(\pi - \theta) = -\cos\theta$$

and $\sin(\pi - \theta) = \sin\theta$.

Also, $\tan(\pi - \theta) = \dfrac{\sin(\pi - \theta)}{\cos(\pi - \theta)} = \dfrac{\sin\theta}{-\cos\theta} = -\tan\theta$.

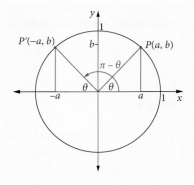

> In the 2nd quadrant, sin is positive.

Consider an **angle in the 3rd quadrant**, $\pi + \theta$, formed using the point $P'(-a, -b)$.

The x- and y-coordinates of P' are the negative of the x- and y-coordinates of P respectively, so

$$\cos(\pi + \theta) = -\cos\theta$$

and $\sin(\pi + \theta) = -\sin\theta$.

Also, $\tan(\pi + \theta) = \dfrac{\sin(\pi + \theta)}{\cos(\pi + \theta)} = \dfrac{-\sin\theta}{-\cos\theta} = \tan\theta$.

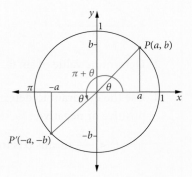

> In the 3rd quadrant, tan is positive.

Consider an **angle in the 4th quadrant**, $2\pi - \theta$, formed using the point $P'(a, -b)$. This is similar to the negative angles described earlier.

$$\cos(2\pi - \theta) = \cos\theta$$

and $\sin(2\pi - \theta) = -\sin\theta$.

Also, $\tan(2\pi - \theta) = \dfrac{\sin(2\pi - \theta)}{\cos(2\pi - \theta)} = \dfrac{-\sin\theta}{\cos\theta} = -\tan\theta$.

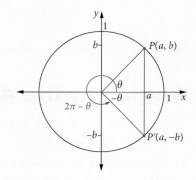

> In the 4th quadrant, cos is positive.

2nd quadrant	3rd quadrant	4th quadrant	Negative angles
$\sin(\pi - \theta) = \sin\theta$	$\sin(\pi + \theta) = -\sin\theta$	$\sin(2\pi - \theta) = -\sin\theta$	$\sin(-\theta) = -\sin\theta$
$\cos(\pi - \theta) = -\cos\theta$	$\cos(\pi + \theta) = -\cos\theta$	$\cos(2\pi - \theta) = \cos\theta$	$\cos(-\theta) = \cos\theta$
$\tan(\pi - \theta) = -\tan\theta$	$\tan(\pi + \theta) = \tan\theta$	$\tan(2\pi - \theta) = -\tan\theta$	$\tan(-\theta) = -\tan\theta$

Note that $2\pi + \theta$ is in the 1st quadrant.

Hence, $\sin(2\pi + \theta) = \sin\theta$, $\cos(2\pi + \theta) = \cos\theta$ and $\tan(2\pi + \theta) = \tan\theta$.

WORKED EXAMPLE 15 — Determining equivalent ratios

State the equivalent trigonometric function in quadrant 1 for each of the following.

a $\sin(360° - 30°)$ **b** $\tan(180° + 26°)$ **c** $-\cos\left(\dfrac{\pi}{8} - \pi\right)$

Steps	Working
a Use $\sin(360° - \theta°) = -\sin(\theta°)$.	$\sin(360° - 30°) = -\sin(30°)$
b Use $\tan(180° + \theta°) = \tan(\theta°)$.	$\tan(180° + 26°) = \tan(26°)$
c 1 Write in the form $\cos(\pi - \theta)$ using $-\cos\theta = \cos(\pi - \theta)$, where $\theta = \dfrac{\pi}{8} - \pi$.	$-\cos\left(\dfrac{\pi}{8} - \pi\right) = \cos\left(\pi - \left(\dfrac{\pi}{8} - \pi\right)\right)$ $= \cos\left(2\pi - \dfrac{\pi}{8}\right)$
2 Apply $\cos(2\pi - \theta) = \cos(\theta)$.	$= \cos\left(\dfrac{\pi}{8}\right)$

WORKED EXAMPLE 16 — Rewrite the angle to find equivalent ratios

State the equivalent trigonometric function in quadrant 1 for each of the following.

a $\cos\left(\dfrac{3\pi}{4}\right)$ **b** $\sin(240°)$ **c** $\tan\left(\dfrac{7\pi}{3}\right)$

Steps	Working
a 1 Write as $\cos(\pi - \theta)$.	$\cos\left(\dfrac{3\pi}{4}\right) = \cos\left(\pi - \dfrac{\pi}{4}\right)$
2 Use $\cos(\pi - \theta) = -\cos\theta$.	$\cos\left(\pi - \dfrac{\pi}{4}\right) = -\cos\left(\dfrac{\pi}{4}\right)$
b 1 Write as $\sin(180° + \theta)$.	$\sin(240°) = \sin(180° + 60°)$
2 Use $\sin(180° + \theta) = -\sin\theta$.	$\sin(180° + 60°) = -\sin(60°)$
c 1 Write in the form $\tan(2\pi + \theta)$.	$\tan\left(\dfrac{7\pi}{3}\right) = \tan\left(2\pi + \dfrac{\pi}{3}\right)$
2 Use $\tan(2\pi + \theta) = \tan\theta$.	$= \tan\left(\dfrac{\pi}{3}\right)$

EXERCISE 5.4 The trigonometric functions

ANSWERS p. 472

Recap

1 For a sector in a circle with radius 6 cm subtended by an angle of $\dfrac{\pi}{6}$, find the

a arc length **b** area of the sector **c** area of the minor segment.

2 For a sector of a circle with radius 1 cm and an angle of 120° at the centre, calculate correct to two decimal places

a the arc length **b** the area of the sector **c** the area of the segment.

Mastery

3 WORKED EXAMPLE 14 Given $\sin\theta = -\dfrac{4}{5}$ and $270° < \theta < 360°$, find the exact value of

a $\cos\theta$ **b** $\tan\theta$

4 **WORKED EXAMPLE 15** State the equivalent trigonometric ratio in the 1st quadrant for each expression.

a $\cos(180° + 80°)$ **b** $\tan\left(\pi - \dfrac{\pi}{7}\right)$ **c** $\sin(360° + 3°)$

5 **WORKED EXAMPLE 16** State the equivalent trigonometric ratio in the 1st quadrant for each expression.

a $\sin(100°)$ **b** $\cos\left(\dfrac{9\pi}{4}\right)$ **c** $\tan\left(\dfrac{11\pi}{6}\right)$

Calculator-free

6 (3 marks) Show that $\dfrac{\sin(2\pi - x) + \cos(\pi + x)}{\sin(x) + \cos(x)} = -1$.

7 (4 marks) Given $\tan\theta = \dfrac{12}{5}$ and $\pi < \theta < 2\pi$, find the exact value of

a $\sin\theta$ (2 marks)

b $\cos\theta$ (2 marks)

8 (2 marks) Given $\sin\theta = -\dfrac{24}{25}$ and θ is not in the 3rd quadrant, find the exact value of $\tan\theta$.

9 (5 marks) State whether each of the following statements are true or false.

a An angle of $\dfrac{3\pi}{2}$ is in the 2nd quadrant. (1 mark)

b An angle of $-\dfrac{5\pi}{6}$ is in the 4th quadrant. (1 mark)

c -3.25 revolutions of a circle produces an angle of $585°$. (1 mark)

d $\dfrac{2\pi}{9}$ and $-320°$ are equivalent angles. (1 mark)

e $\pi° = \dfrac{180}{\pi^2}$ (1 mark)

10 (2 marks) Write an equivalent 1st quadrant angle for $209°$.

11 (2 marks) Simplify $\dfrac{\cos(2\pi - \theta)}{\sin(\pi + \theta)}$.

Calculator-assumed

12 (2 marks) If $\sin\theta = -\dfrac{\sqrt{3}}{2}$ and $0 < \theta < 2\pi$, find the two possible values of θ.

> **Exam hack**
> An angle should be given in the same form as the domain. For example, if the domain is given in radians, the answer should be in radians.

13 (3 marks) If $\cos\theta = \dfrac{\sqrt{3}}{2}$, show that the exact value of $\cos(2\pi + \theta) + \sin^2\theta$ is $\dfrac{2\sqrt{3} + 1}{4}$.

5.5 The exact values

$\frac{\pi}{6}$ and $\frac{\pi}{4}$ exact values

We can find **exact values** of trigonometric functions for the angles $\frac{\pi}{6}$ (30°) and $\frac{\pi}{4}$ (45°).

We can then use these to find exact values of trigonometric functions for multiples of these angles, such as $\frac{2\pi}{6} = \frac{\pi}{3}$ (60°), $\frac{2\pi}{4} = \frac{\pi}{2}$ (90°) and $\frac{5\pi}{6}$ (150°).

The exact values for $\frac{\pi}{6}, \frac{\pi}{4}$ and $\frac{\pi}{3}$ can be found using special right-angled triangles that have angles of size 30°, 45° and 60°. These are called the **reference triangles**.

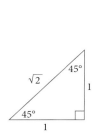
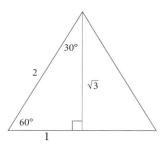

$\sin\left(\frac{\pi}{4}\right) = \sin(45°) = \frac{1}{\sqrt{2}}$ $\cos\left(\frac{\pi}{4}\right) = \cos(45°) = \frac{1}{\sqrt{2}}$ $\tan\left(\frac{\pi}{4}\right) = \tan(45°) = \frac{1}{1} = 1$

$\sin\left(\frac{\pi}{6}\right) = \sin(30°) = \frac{1}{2}$ $\cos\left(\frac{\pi}{6}\right) = \cos(30°) = \frac{\sqrt{3}}{2}$ $\tan\left(\frac{\pi}{6}\right) = \tan(30°) = \frac{1}{\sqrt{3}}$

$\sin\left(\frac{\pi}{3}\right) = \sin(60°) = \frac{\sqrt{3}}{2}$ $\cos\left(\frac{\pi}{3}\right) = \cos(60°) = \frac{1}{2}$ $\tan\left(\frac{\pi}{3}\right) = \tan(60°) = \sqrt{3}$

Points on the axes

The values for 0 and $\frac{\pi}{2}$ follow immediately from the unit circle definitions.

For the angle 0, P is at (1, 0) on the x-axis, so:

$\sin(0) = y = 0$, $\cos(0) = x = 1$, $\tan(0) = \frac{y}{x} = \frac{0}{1} = 0$.

For the angle $\frac{\pi}{2}$ (90°), P is at (0, 1) on the y-axis, so:

$\sin\left(\frac{\pi}{2}\right) = y = 1$, $\cos\left(\frac{\pi}{2}\right) = x = 0$, $\tan\left(\frac{\pi}{2}\right) = \frac{y}{x} = \frac{1}{0}$, which is not defined.

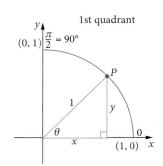

	0	$\frac{\pi}{6}$ (30°)	$\frac{\pi}{4}$ (45°)	$\frac{\pi}{3}$ (60°)	$\frac{\pi}{2}$ (90°)
sin θ	0	$\frac{1}{2}$	$\frac{1}{\sqrt{2}}$	$\frac{\sqrt{3}}{2}$	1
cos θ	1	$\frac{\sqrt{3}}{2}$	$\frac{1}{\sqrt{2}}$	$\frac{1}{2}$	0
tan θ	0	$\frac{1}{\sqrt{3}}$	1	$\sqrt{3}$	not defined

You can check these answers on your calculator.

The values of sin θ across the table are easily remembered as $\frac{\sqrt{0}}{2}, \frac{\sqrt{1}}{2}, \frac{\sqrt{2}}{2}, \frac{\sqrt{3}}{2}, \frac{\sqrt{4}}{2}$.

The values of cos θ are in the reverse order, and the values of tan θ are given by $\frac{\sin\theta}{\cos\theta}$.

As sine and cosine are complementary functions,

$$\sin\left(\frac{\pi}{2} - \theta\right) = \sin(90° - \theta) = \cos\theta$$

$$\cos\left(\frac{\pi}{2} - \theta\right) = \cos(90° - \theta) = \sin\theta$$

so, for example, $\sin\left(\frac{\pi}{6}\right) = \cos\left(\frac{\pi}{3}\right)$ and $\sin(45°) = \cos(45°)$.

WORKED EXAMPLE 17 Finding exact values

Find the exact value of each of the following.

a $\sin\left(\frac{3\pi}{4}\right)$ **b** $\cos(-240°)$ **c** $\tan\left(\frac{11\pi}{6}\right)$

Steps	Working
a 1 Identify the quadrant of the angle and use ASTC to establish the sign.	$\frac{3\pi}{4}$ is in the 2nd quadrant, so $\sin\left(\frac{3\pi}{4}\right)$ is positive.
2 For the 2nd quadrant, write the angle in the form $\pi - \theta$, using an acute angle θ.	$\sin\left(\frac{3\pi}{4}\right) = \sin\left(\pi - \frac{\pi}{4}\right)$ $= \sin\left(\frac{\pi}{4}\right)$ $= \frac{1}{\sqrt{2}}$

b 1 Use $\cos(-\theta) = \cos\theta$.

$\cos(-240°) = \cos(240°)$

> $\cos(-240°)$ on the unit circle is also equivalent to $\cos(360° - 240°) = \cos(120°)$.

2 Identify the quadrant of the angle and use ASTC to establish the sign.

$240°$ is in the 3rd quadrant, so $\cos(240°)$ is negative.

3 For the 3rd quadrant, write the angle in the form $180° + \theta$, using an acute angle for θ.

$\cos(240°) = \cos(180° + 60°)$
$= -\cos(60°)$
$= -\dfrac{1}{2}$

c 1 Identify the quadrant of the angle and use ASTC to establish the sign.

$\dfrac{11\pi}{6}$ is in the 4th quadrant, so $\tan\left(\dfrac{11\pi}{6}\right)$ is negative.

2 For the 4th quadrant, write the angle in the form $2\pi - \theta$ using an acute angle for θ.

$\tan\left(\dfrac{11\pi}{6}\right) = \tan\left(2\pi - \dfrac{\pi}{6}\right)$
$= -\tan\left(\dfrac{\pi}{6}\right)$
$= -\dfrac{1}{\sqrt{3}}$

Angle sum and difference identities

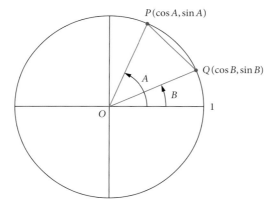

For the unit circle shown above, using the cosine rule in triangle POQ

$PQ^2 = 1^2 + 1^2 - 2 \times 1 \times 1 \times \cos(A - B)$
$= 2 - 2\cos(A - B)$

Also, $PQ^2 = (\cos B - \cos A)^2 + (\sin B - \sin A)^2$
$= \cos^2 B - 2\cos A\cos B + \cos^2 A + \sin^2 B - 2\sin A \sin B + \sin^2 A$

(Since $\sin^2 A + \cos^2 A = 1$ and $\sin^2 B + \cos^2 B = 1$)

$= 2 - 2\cos A \cos B - 2\sin A \sin B$

Therefore, $2 - 2\cos(A - B) = 2 - 2\cos A\cos B - 2\sin A\sin B$

Hence, $\cos(A - B) = \cos A\cos B + \sin A\sin B$ [1]

Putting $B = -B$ in [1], we get $\cos(A + B) = \cos A\cos B - \sin A\sin B$ since $\cos(-B) = \cos B$ and $\sin(-B) = -\sin B$.

Using complementary angles, i.e. $\cos(90° - \theta) = \sin\theta$

$\cos(90 - (A + B)) = \sin(A + B)$

i.e. $\cos((90 - A) - B)) = \sin(A + B)$

Using [1], $\sin(A + B) = \cos(90 - A)\cos B + \sin(90 - A)\sin B$

Therefore, $\sin(A + B) = \sin A \cos B + \cos A \sin B$ [2]

Putting $B = -B$ in [2], we get $\sin(A - B) = \sin A \cos B - \cos A \sin B$ since $\cos(-B) = \cos B$ and $\sin(-B) = -\sin B$.

It can be shown that $\tan(A \pm B) = \dfrac{\tan A \pm \tan B}{1 \mp \tan A \tan B}$.

Angle sum and difference identities

$\sin(A \pm B) = \sin A \cos B \pm \cos A \sin B$

$\cos(A \pm B) = \cos A \cos B \mp \sin A \sin B$

$\tan(A \pm B) = \dfrac{\tan A \pm \tan B}{1 \mp \tan A \tan B}$

WORKED EXAMPLE 18 Using sum and difference identities

Without the use of a calculator, determine the exact values of

a $\sin(105°)$ **b** $\cos\left(\dfrac{\pi}{12}\right)$

Steps	Working
a 1 Write 105° as the sum or difference of two angles for which exact values are known.	$105° = 60° + 45°$
2 Choose the correct identity and substitute in values.	$\sin(A \pm B) = \sin A \cos B \pm \cos A \sin B$ $\sin(60° + 45°) = \sin(60°)\cos(45°) + \cos(60°)\sin(45°)$
3 Evaluate the expression.	$\sin(105°) = \dfrac{\sqrt{3}}{2} \times \dfrac{1}{\sqrt{2}} + \dfrac{1}{2} \times \dfrac{1}{\sqrt{2}} = \dfrac{\sqrt{3}+1}{2\sqrt{2}}$
b 1 Write $\dfrac{\pi}{12}$ as the sum or difference of two angles for which exact values are known.	$\dfrac{\pi}{12} = \dfrac{\pi}{3} - \dfrac{\pi}{4}$
2 Choose the correct identity and substitute in values.	$\cos(A \pm B) = \cos A \cos B \mp \sin A \sin B$ $\cos\left(\dfrac{\pi}{3} - \dfrac{\pi}{4}\right) = \cos\dfrac{\pi}{3}\cos\dfrac{\pi}{4} + \sin\dfrac{\pi}{3}\sin\dfrac{\pi}{4}$
3 Evaluate the expression.	$\cos\left(\dfrac{\pi}{12}\right) = \dfrac{1}{2} \times \dfrac{1}{\sqrt{2}} + \dfrac{\sqrt{3}}{2} \times \dfrac{1}{\sqrt{2}} = \dfrac{1+\sqrt{3}}{2\sqrt{2}}$

WORKED EXAMPLE 19 Solving trigonometric equations

Solve $\cos x = \dfrac{\sqrt{3}}{2}$ over the domain $-\pi \leq x \leq \pi$.

Steps	Working
1 Determine the acute angle.	$x = \dfrac{\pi}{6}$
2 Use the ASTC rule to determine which other angles are positive and in the given domain.	$x = \dfrac{\pi}{6}, -\dfrac{\pi}{6}$

EXERCISE 5.5 The exact values

ANSWERS p. 473

Recap

1 For angle θ, if $\tan \theta = -5$ and $0 < \theta < \pi$, then $\cos \theta$ is

    A $-\dfrac{5}{\sqrt{26}}$ **B** $-\dfrac{1}{\sqrt{26}}$ **C** $\dfrac{1}{\sqrt{26}}$ **D** $\dfrac{1}{\sqrt{24}}$ **E** $\dfrac{\sqrt{24}}{5}$

2 If $\cos\left(\dfrac{\pi}{4}\right) = \dfrac{\sqrt{2}}{2}$, then $\cos\left(\dfrac{9\pi}{4}\right) - \cos\left(\pi + \dfrac{\pi}{4}\right) + \cos\left(2\pi - \dfrac{\pi}{4}\right)$ will simplify to

    A $-3\sqrt{2}$ **B** $-\dfrac{\sqrt{2}}{2}$ **C** $\sqrt{2}$ **D** $\dfrac{3\sqrt{2}}{2}$ **E** $2\sqrt{2}$

Mastery

3 **WORKED EXAMPLE 17** Find the exact value of each of the following.

    a $\sin\left(\dfrac{11\pi}{6}\right)$ **b** $\cos(-150°)$ **c** $\tan\left(\dfrac{5\pi}{6}\right)$

    d $\cos(315°)$ **e** $\sin(-135°)$ **f** $\cos\left(-\dfrac{7\pi}{6}\right)$

    g $\tan(270°)$ **h** $\sin(420°)$ **i** $\tan(4\pi)$

4 **WORKED EXAMPLE 18** Without the use of a calculator, evaluate $\cos(165°)$.

5 **WORKED EXAMPLE 19** Solve $\sin x = \dfrac{\sqrt{3}}{2}$ over the domain $-\pi \le x \le \pi$.

Calculator-free

6 (3 marks) Evaluate $\tan(15°)$.

7 (3 marks) Find the value of x in each case if $0 \le x \le \dfrac{\pi}{2}$.

    a $\sin(2\pi - x) = -\dfrac{1}{2}$ (1 mark)

    b $\tan(\pi - x) = -1$ (1 mark)

    c $\cos(x - 2\pi) = \dfrac{1}{2}$ (1 mark)

8 (3 marks) Evaluate $\dfrac{\sin\left(\dfrac{5\pi}{2}\right) \times \cos\left(\dfrac{3\pi}{4}\right) \times \sin\left(\dfrac{\pi}{4}\right)}{\sqrt{3} \times \cos\left(\dfrac{5\pi}{6}\right)}$.

9 (3 marks) Evaluate $\dfrac{\cos\left(\dfrac{\pi}{4}\right)}{\cos\left(\dfrac{\pi}{4}\right)\tan\left(-\dfrac{3\pi}{4}\right)}$.

10 (2 marks) Evaluate $\cos\left(\dfrac{2\pi}{3}\right) + \sin\left(\dfrac{7\pi}{6}\right) - \tan\left(\dfrac{3\pi}{4}\right)$.

11 (5 marks) If $\theta = \dfrac{7\pi}{2}$, state whether the following are true or false.

    a $\tan\theta = 0$ (1 mark)

    b $\sin\theta = -1$ (1 mark)

    c $\cos\theta$ is not defined   (1 mark)

    d $\sin\theta = 0$ (1 mark)

    e $\cos\theta = -1$ (1 mark)

12 (8 marks) Solve the following over the given domain.

    a $\sin x = \dfrac{1}{2}, -360° \le x \le 360°$ (2 marks)

    b $2\cos x - \sqrt{3} = 0, -\dfrac{\pi}{2} \le x \le \dfrac{\pi}{2}$ (3 marks)

    c $\tan x + \dfrac{3\sqrt{3}}{2} = \dfrac{\sqrt{3}}{2}, -\dfrac{\pi}{2} \le x \le \pi$ (3 marks)

> **Exam hack**
> First rearrange the equation so the trigonometric function is the subject.

Calculator-assumed

13 (3 marks) Solve the equation $2\sin x - 1 = 0$, $0° < x < 180°$.

14 (3 marks) If $\sin A = \dfrac{3}{5}$ and $\cos B = \dfrac{7}{25}$, evaluate each expression, showing use of the angle sum and difference identities. (Hint: Use right triangles and Pythagoras' theorem.)

    a $\sin(A + B)$ (2 marks)

    b $\cos(A - B)$ (1 mark)

15 (4 marks) Simplify each of the following to a single trigonometric expression.

    a $\cos(30°)\cos(45°) - \sin(30°)\sin(45°)$ (2 marks)

    b $\cos(3A + B)\sin(B) - \cos(B)\sin(3A + B)$ (2 marks)

5.6 Graphs of trigonometric functions

We defined the trigonometric functions according to the coordinates of a point P on the unit circle:

$\sin\theta = y$, $\cos\theta = x$, $\tan\theta = \dfrac{y}{x}$.

By examining the behaviour of the coordinates of $P(x, y)$ in different positions, the shape of the graphs of the trigonometric functions can be determined.

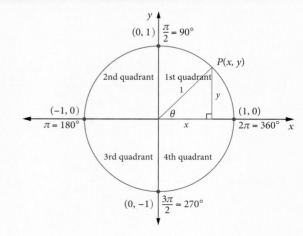

The sine curve

$\sin \theta = y$, which means that $\sin \theta$ is the 'signed' **height** of the right-angled triangle as P moves from $(1, 0)$ in an anticlockwise direction.

θ	0	$\frac{\pi}{2}$ (90°)	π (180°)	$\frac{3\pi}{2}$ (270°)	2π (360°)
$\sin \theta$	0	1	0	−1	0

- At $\theta = 0$, the height of the triangle, y, is 0, but as P moves through the 1st quadrant, $\sin \theta$ increases from 0 to 1.
- At $\theta = \frac{\pi}{2}$, $y = 1$, but as P moves through the 2nd quadrant, $\sin \theta$ decreases from 1 to 0.
- At $\theta = \pi$, $y = 0$, but as P moves through the 3rd quadrant, $\sin \theta$ decreases from 0 to −1.
- At $\theta = \frac{3\pi}{2}$, $y = -1$, but as P moves through the 4th quadrant, $\sin \theta$ increases from −1 back to 0.
- At $\theta = 2\pi$, $y = 0$ again, and the cycle repeats.

The graph of $y = \sin x$ looks like this:

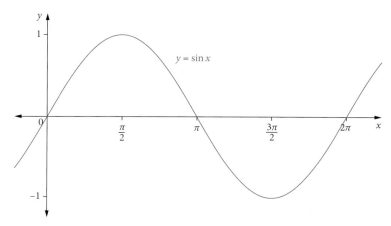

- Note that sine is positive (above the x-axis) in the 1st and 2nd quadrants, which is consistent with the ASTC rule.
- The domain of $y = \sin x$ is all real values of x, and the range is $-1 \le y \le 1$.
- The graph repeats itself after 2π, so we say that it has a **period** of 2π.
- The **amplitude** of a trigonometric function is the vertical distance from its centre to its turning points, so the amplitude of $y = \sin x$ is 1.
- A horizontal translation is called a **phase change**.

Note the use of x in place of variable θ. This is because we commonly use y as a function of x.

For $y = \sin(x - c)$ the graph translates in a horizontal direction. It moves 'c' units to the right if c is positive, and 'c' units to the left if c is negative.

Compare the graphs of $y = \sin x$ and $y = \sin\left(x - \dfrac{\pi}{2}\right)$. It can be seen that the second graph has moved $\dfrac{\pi}{2}$ units to the right. The period and amplitude has remained the same for both graphs.

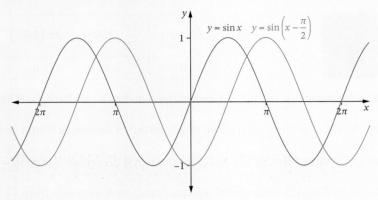

Cycle and period are not the same. Cycle refers to the repeating pattern and period refers to the horizontal length of the cycle.

Amplitude, period and phase change of a sine function

For $y = a \sin(bx)$, where $a > 0$ and $b > 0$, the amplitude is a, the period is $\dfrac{2\pi}{b}$.

For $y = \sin(x - c)$, the graph translates c units horizontally. It moves 'c' units to the right if c is positive, and to the left if c is negative.

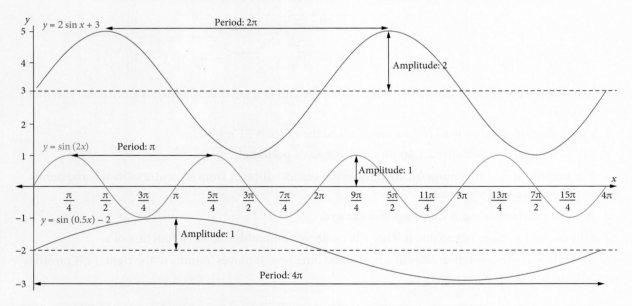

When technology is not used, the graph can be drawn by determining its general shape and properties, including amplitude, period and the mean value.

WORKED EXAMPLE 20 — Determining the amplitude and period

a State the amplitude and period of $y = 2\sin(3x)$.
b Sketch the graph of $y = 2\sin(3x)$ for $0 \le x \le \pi$.

Steps	Working
a $y = a\sin(bx)$ has amplitude a and the period is $\dfrac{2\pi}{b}$.	$y = 2\sin(3x)$ has amplitude 2 and period $\dfrac{2\pi}{3}$.
b 1 Decide on a scale for the x-axis.	The maximum occurs when $\sin(3x) = 1$. This is when $x = \dfrac{\pi}{6}$, so choose increments of $\dfrac{\pi}{6}$ from 0 to π.
2 Decide on a range for the y-axis.	The amplitude is 2, so the minimum is -2 and the maximum is 2.
3 Find some relevant points, and use the period to find other values.	$x = 0$, $y = 2\sin(0) = 0$. That is, $(0, 0)$. The period is $\dfrac{2\pi}{3}$, so another point is $\left(0 + \dfrac{2\pi}{3}, 0\right)$. $x = \dfrac{\pi}{6}$, $y = 2\sin\left(3 \times \dfrac{\pi}{6}\right) = 2$. That is, $\left(\dfrac{\pi}{6}, 2\right)$. Another point is $\left(\dfrac{\pi}{6} + \dfrac{2\pi}{3}, 2\right)$, or $\left(\dfrac{5\pi}{6}, 2\right)$. $x = \dfrac{\pi}{3}$, $y = 2\sin\left(3 \times \dfrac{\pi}{3}\right) = 0$. That is, $\left(\dfrac{\pi}{3}, 0\right)$. Another point is $\left(\dfrac{\pi}{3} + \dfrac{2\pi}{3}, 0\right)$ or $(\pi, 0)$.
4 Plot and connect the points smoothly, keeping in mind the general shape of a sine graph.	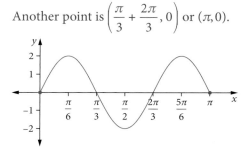

WORKED EXAMPLE 21 — Determining sine functions

a Determine the sine function that has an amplitude of 3 and a period of π.
b For the graph below, determine the equation of the function in the form $y = a\sin(x - c)$.

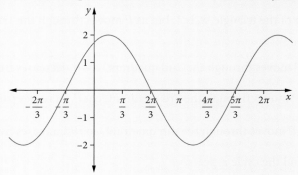

Steps	Working
a 1 $y = a\sin(bx)$ has amplitude a and the period is $\dfrac{2\pi}{b}$.	a will equal 3 as that is the amplitude. To determine b, let $\pi = \dfrac{2\pi}{b}$ and solve for b. $\therefore b = 2$
2 Write down the required function.	$y = 3\sin(2x)$

b
1. Determine the value of *a*.
   a will equal 2 as that is the amplitude.

2. Sketch the function $y = 2\sin x$ on the axes as a guide to determine the phase change.

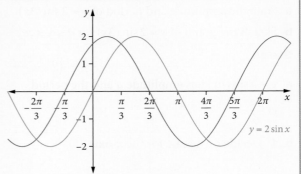

The graph has moved $\frac{\pi}{3}$ units to the left.

3. Write down the required function.
   $y = 2\sin\left(x + \frac{\pi}{3}\right)$

The cosine curve

On the unit circle, $\cos\theta = x$, the 'signed' base length of the right-angled triangle as point P moves from $(1, 0)$ around the unit circle.

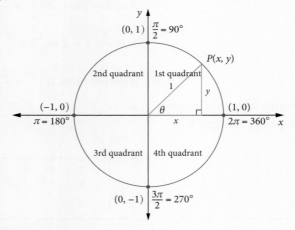

θ	0	$\frac{\pi}{2}$ (90°)	π (180°)	$\frac{3\pi}{2}$ (270°)	2π (360°)
$\cos\theta$	1	0	−1	0	1

- At $\theta = 0$, the base length of the triangle, x, is 1, but as P moves through the 1st quadrant, $\cos\theta$ decreases from 1 to 0.

- At $\theta = \frac{\pi}{2}$, $x = 0$, but as P moves through the 2nd quadrant, $\cos\theta$ decreases from 0 to −1.

- At $\theta = \pi$, $x = -1$, but as P moves through the 3rd quadrant, $\cos\theta$ increases from −1 to 0.

- At $\theta = \frac{3\pi}{2}$, $x = 0$, but as P moves through the 4th quadrant, $\cos\theta$ increases from 0 back to 1.

- At $\theta = 2\pi$, $x = 1$ again, and the cycle repeats.

The graph of $y = \cos x$ is as shown on the right.

- Note that cosine is positive in the 1st and 4th quadrants, consistent with the ASTC rule.
- The graph has a similar shape to the sine graph and also repeats itself after 2π, so it has a **period** of 2π.
- The domain of $y = \cos x$ is all real values of x, and the range is $-1 \leq y \leq 1$.
- It has an amplitude of 1.

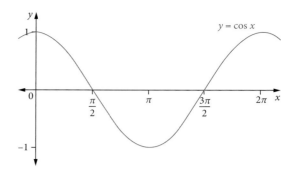

The amplitude, period and mean value of a cosine function are the same as those of a sine function.

> **Amplitude, period and phase change of a cosine function**
>
> For $y = a\cos(bx)$, where $a > 0$ and $b > 0$, the amplitude is a and the period is $\dfrac{2\pi}{b}$.
>
> For $y = \cos(x - c)$, the graph translates c units horizontally. It moves 'c' units to the right if c is positive, and to the left if c is negative.

USING CAS 3 — Graphing cosine

a Draw the graph of $3\cos(2x)$ for $0 \leq x \leq \pi$.

b State the period and amplitude.

ClassPad

TI-Nspire

a

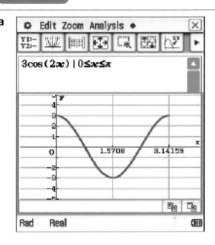
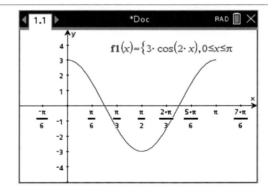

1 Using graph in **Main**, set the mode to **Rad**.

2 Enter the cosine function over the domain $0 \leq x \leq \pi$. Tap **Math3** for | and ≤.

3 Drag down into the **Graph** screen from the left-hand side.

4 Adjust the window settings to suit. It is useful to alter the x scale with fractions of radians.

1 Set the mode to **radian**.

2 Graph the cosine function over the domain $0 \leq x \leq \pi$. Press **ctrl + =** for | and ≤.

3 Adjust the window settings to suit.

Optional – To display the increments on the axes, move the cursor over one of the increments and press **ctrl + menu > Attributes**. Press the **down arrow** twice then the **left arrow** once to change '**Single Tick**' to '**Multiple Labels**'.

b The period is π and the amplitude is 3.

The tangent curve

$\tan\theta = \dfrac{y}{x} = \dfrac{\sin\theta}{\cos\theta}$, which means that $\tan\theta$ is the

gradient $\left(\dfrac{\text{rise}}{\text{run}}\right)$ of the hypotenuse of the right-angled

triangle as P moves around the unit circle.

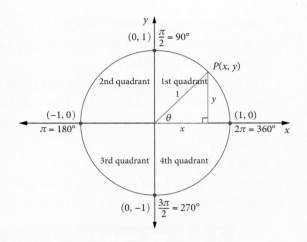

> $\tan\theta$ is also the height of QS, the vertical tangent to the circle shown on page 262.

θ	0	$\dfrac{\pi}{2}$ (90°)	π (180°)	$\dfrac{3\pi}{2}$ (270°)	2π (360°)
$\sin\theta$	0	1	0	−1	0
$\cos\theta$	1	0	−1	0	1
$\tan\theta$	0	not defined	0	not defined	0

- At $\theta = 0$, the gradient of the hypotenuse, or $\dfrac{y}{x}$, is 0, but as P moves through the 1st quadrant, $\tan\theta$ increases from 0 towards infinity (∞).

- At $\theta = \dfrac{\pi}{2}$, the gradient $\dfrac{y}{x}$ is not defined, but as P moves through the 2nd quadrant, $\tan\theta$ increases from $-\infty$ (negative infinity) back to 0.

- At $\theta = \pi$, the gradient $\dfrac{y}{x} = 0$, but as P moves through the 3rd quadrant, the cycle starts again and $\tan\theta$ increases from 0 towards ∞ again.

- At $\theta = \dfrac{3\pi}{2}$, the gradient $\dfrac{y}{x}$ is not defined, but as P moves through the 4th quadrant, $\tan\theta$ increases from $-\infty$ back to 0.

- At $\theta = 2\pi$, the gradient $\dfrac{y}{x} = 0$ again, and the cycle repeats.

The graph of $y = \tan x$ is shown on the right.

- tangent is positive in the 1st and 3rd quadrants, consistent with the ASTC rule.
- There are vertical asymptotes at $x = \pm\dfrac{\pi}{2}, \pm\dfrac{3\pi}{2}, \pm\dfrac{5\pi}{2}\ldots$
- The domain of $y = \tan x$ is all real values of x apart from odd integer multiples of $\dfrac{\pi}{2}$ and the range is **all real values of y**.
- The graph repeats itself after π, so it has a **period** of π.
- The **amplitude** of the graph is not defined.

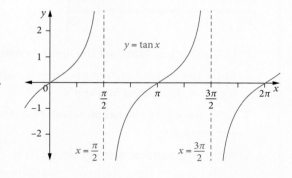

Period, amplitude and phase change of a tangent function

The period of $y = \tan(bx)$ is $\dfrac{\pi}{b}$, where $b > 0$, and the amplitude is not defined.

For $y = \tan(x - c)$, the graph translates c units horizontally. It moves 'c' units to the right if c is positive, and to the left if c is negative.

Function	Amplitude	Period	Phase change	Domain	Range
$y = \sin x$	1	2π	0	All real values	$-1 \le y \le 1$
$y = \cos x$	1	2π	0	All real values	$-1 \le y \le 1$
$y = \tan x$	Not defined	π	0	All real values apart from odd integer multiples of $\dfrac{\pi}{2}$	All real values
$y = a\sin(b(x-c))$	a	$\dfrac{2\pi}{b}$	c	All real values	$-a \le y \le a$
$y = a\cos(b(x-c))$	a	$\dfrac{2\pi}{b}$	c	All real values	$-a \le y \le a$
$y = a\tan(b(x-c))$	Not defined	$\dfrac{\pi}{b}$	c	All real values apart from $(2k+1)\left(\dfrac{\pi}{2b}\right) + c$, $k \in \mathbb{Z}$	All real values

EXERCISE 5.6 Graphs of trigonometric functions   ANSWERS p. 473

Recap

1 The exact value of $\sin\left(\dfrac{13\pi}{6}\right) + \cos\left(\dfrac{8\pi}{3}\right)$ is

  A -1 **B** $-\dfrac{1}{2}$ **C** 0 **D** $\dfrac{1}{\sqrt{2}}$ **E** $\dfrac{2}{\sqrt{3}}$

2 Evaluate $\sin(90°)$ showing use of the angle sum and difference identities.

Mastery

3 WORKED EXAMPLE 20

  a State the amplitude and period of $y = 4\sin x$.

  b Sketch the graph of $y = 4\sin x$ for $-\pi \le x \le \pi$.

4 WORKED EXAMPLE 21

  a Determine the sine function that has an amplitude of 2 and a period of 4π.

  b For the graph below, determine the equation of the function in the form $y = a\sin(x - c)$, where $a > 0$ and $c > 0$.

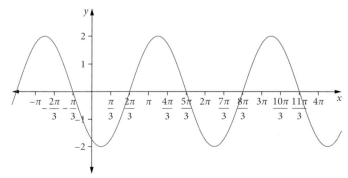

5 Using CAS 3

  a Draw the graph of $y = 2\cos(1.5x)$ for $0 \le x \le \pi$.

  b Find the coordinates of the maximum and the minimum values, where appropriate.

Calculator-free

6 (2 marks) State the range and period of the function h: $h(x) = 3\cos\left(\dfrac{\pi x}{3}\right)$.

7 (2 marks) Determine the value of p given that the period of $y = \sin\left(\dfrac{5\pi x}{2p+7}\right)$ is 6.

8 (2 marks) Determine the value(s) of p given that $y = \cos(p\pi x)$ has the same period as $y = \tan\left(\dfrac{\pi x}{p-1}\right)$ and $p > 0$.

9 (2 marks) Determine the period of the function $f(x) = -3\sin\left(\dfrac{\pi x}{5}\right)$.

10 (2 marks) Determine the sine function which has an amplitude of 3 and a period of $\dfrac{1}{10}$.

11 (3 marks) Determine the function of the cosine graph shown. Write your answer in the form $y = a\cos(x - c)$ where $a > 0$ and $c > 0$.

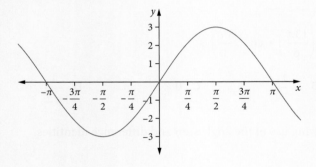

12 (2 marks) Determine the period for the function $f(x) = 4\tan\left(\dfrac{x}{3}\right)$.

Calculator-assumed

13 (2 marks) What changes in the period and amplitude are required for $y = \cos x$ to become $y = 2\cos(\pi x)$?

14 (2 marks) The diagram shows two cycles of the graph of a trigonometric function.

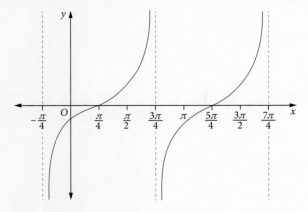

Determine the period of the function.

15 (3 marks) Sketch the graph of the function $f(x) = 3\tan\left(\dfrac{3x}{2}\right)$ for $-\pi \leq x \leq \pi$.

Clearly label any axes intercepts and asymptotes.

16 (2 marks) Sketch one cycle of the graph $y = -2\cos(4x)$ that starts at $x = 0$.

5.7 Applying trigonometric functions

Trigonometric functions, particularly sine and cosine, can be used to model real-life phenomena for which cycles and periodicity are observed, such as in the rise and fall of tides and the regular rotation of objects and motors.

WORKED EXAMPLE 22 | Applying trigonometric functions

The graph shows the vertical distance between a fixed point on the rim of a rotating bicycle wheel and the axle (centre) of the wheel, over time.

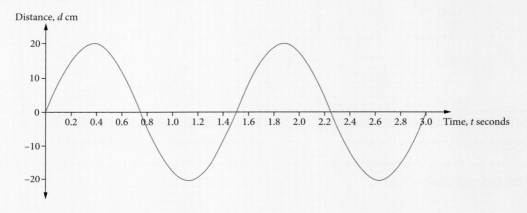

a What is the radius of the wheel?
b How long does it take the wheel to make one revolution?
c What is the minimum vertical distance reached by the point? Where on the wheel is this?
d What do the t-intercepts on the graph represent?

Steps	Working
a Find the amplitude.	The radius is the amplitude of the graph. From the graph, this is 20 cm.
b One revolution of the wheel corresponds to one period on the graph.	Two periods correspond to 3 s, so one period is 1.5 s. It takes the wheel 1.5 s to make a revolution.
c Find the lowest point on the graph.	Minimum distance is −20 cm, which represents when the point is at ground level, 20 cm below the centre of the wheel.
d The t-intercepts represent when $d = 0$, when the point is 0 cm vertically from the centre of the wheel.	The t-intercepts represent when the point is level with the centre of the wheel.

WORKED EXAMPLE 23 | Water level described by a trigonometric function

The water level (above sea level) of a lagoon changes periodically and is described by the function $h(t) = 0.5\cos\left(\dfrac{\pi t}{2}\right) + 0.5$, where h is in metres and t is the time in hours after midnight.

a Sketch the graph of the function in the interval $0 \le t \le 12$.

b At what times after midnight is the water level at its maximum?

c What is the water level at 3:45 am? Give your answer correct to three decimal places.

Steps	Working
a Identify the amplitude, period and transformations, and use them to sketch the graph. 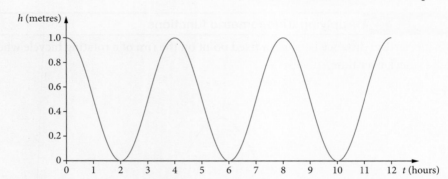	Amplitude is 0.5. Period is $2\pi \div \dfrac{\pi}{2} = 4$. Vertical translation is 0.5 units up.
b Find the t values after $t = 0$ corresponding to the maximum points on the graph.	The maximum water level occurs at $t = 4, 8, 12$. This corresponds to 4 am, 8 am and 12 pm.
c Determine the t value corresponding to 3:45 am and substitute to find h.	3:45 am is 3.75 hours after midnight, so $t = 3.75$. When $t = 3.75$, $h = 0.5\cos\left(\dfrac{\pi \times 3.75}{2}\right) + 0.5$ $= 0.9619\ldots$ $\approx 0.962\,\text{m}$

Exam hack

For trigonometric functions, make sure that your calculator is in radian mode, not degree mode.

USING CAS 4 — Applying trigonometric functions

The vertical displacement d cm at time t seconds of an old engine's piston is modelled by the equation $d = 5 \sin(9.5\pi t)$.

a   Sketch the graph of the function in the interval $0 \leq t \leq 1$.

b   How many times does the piston move away from its centre position in 1 second?

c   Calculate the distance the piston travels each $\frac{1}{5}$ of a second.

d   During the first 0.1 second, when was the piston at a distance of more than 2 cm away from its centre position? Give your answer correct to two decimal places.

ClassPad

a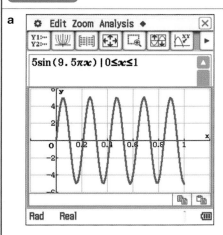

1   In **Main**, set the mode to **Rad**.
2   Enter the sine function over the domain [0, 1] using **Math3** for | and ≤.
3   Drag down into the **Graph** screen from the left-hand side.
4   Adjust the window settings to suit.

b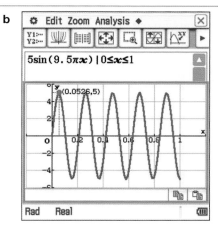

Count the number of maximums and minimums to get a total of 10.

c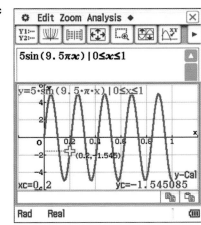

1   Tap **Analysis > G-Solve > x-Cal/y-Cal > y-Cal** and in the dialogue box enter **1/5**.
2   Tap **OK** and the graph will jump to the point **(0.2, −1.55)**.
3   The distance travelled is
    amplitude − 1.55 = 4 × 5 − 1.55
    = 18.45 cm

d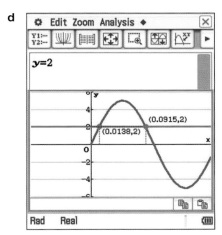

1   In the upper window, enter **y = 2**.
2   Drag down both graphs into the Graph window.
3   Tap **Analysis > G-Solve > Intersection**.
4   Press **EXE** to label the first point of intersection.
5   Press the **right arrow**, then **EXE** to label the second point of intersection.
6   The piston was over 2 cm away for $0.0915 - 0.0138 \approx 0.077 \approx 0.08$ s.

TI-Nspire

a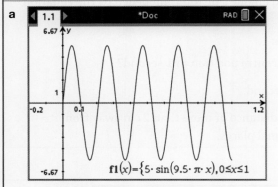

1. Graph the function over the domain [0, 1].
2. Adjust the window settings to suit.

b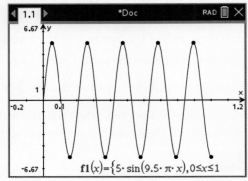

Count the number of maximums and minimums to get a total of 10

c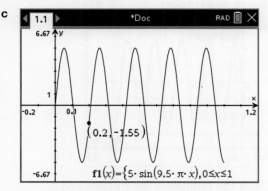

1. Press **menu** > **Trace** > **Graph Trace**.
2. Enter $\frac{1}{5}$ and the graph will jump to the point (0.2, −1.55).
3. The distance travelled is
   amplitude − 1.55 = 4 × 5 − 1.55
   = 18.45 cm

d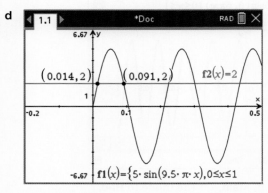

1. Graph $f(x) = 2$.
2. Press **menu** > **Analyze Graph** > **Intersection**.
3. Determine the coordinates of the first two points of intersection.
4. The piston was over 2 cm away for $0.091 − 0.014 ≈ 0.08$ s.

EXAMINATION QUESTION ANALYSIS

Calculator-assumed (8 marks)

Trigg the gardener is working in a temperature-controlled greenhouse. During a particular 24-hour time interval, the temperature ($T°C$) is given by $T(t) = 25 + 2\cos\left(\frac{\pi t}{8}\right), 0 \le t \le 24$, where t is the time in hours from the beginning of the 24-hour time interval.

a State the maximum temperature in the greenhouse and the values of t when this occurs. (2 marks)
b State the period of the function T. (1 mark)
c Find the smallest value of t for which $T = 26$. (3 marks)
d For how many hours during the 24-hour time interval is $T \ge 26$? (2 marks)

Video Examination question analysis: Trigonometric functions

Reading the question

- Read all parts of each question first to see where CAS will be used.
- Notice the key words and phrases. For example, 'the beginning of the 24-hour time interval' means $t = 0$, and 'how many hours' means the total number of hours in the 24-hour period.
- A cosine function means that the temperature will be cyclic.

Thinking about the question

- For questions worth more than 2 marks, sufficient working needs to be shown, or if CAS is used, clear explanations of how the solution was obtained.
- Decide when it will be more efficient to use algebra, particularly when exact answers are required.
- Since calculations using $T(t) = 25 + 2\cos\left(\dfrac{\pi t}{8}\right)$ involve radians, ensure that calculator settings are also in radian mode.

Worked solution (✓ = 1 mark)

a Maximum temperature occurs when $\cos\left(\dfrac{\pi t}{8}\right) = 1$.

Maximum temperature is $25 + 2 \times 1 = \mathbf{27\,°C}$. ✓

Maximum temperature when

$$\cos\left(\dfrac{\pi t}{8}\right) = 1$$

$$\dfrac{\pi t}{8} = 0, 2\pi, 4\pi \ldots$$

$t = 0, 16, 32 \ldots$

$t = 0$ (midnight) and $t = 16$ (4 pm) are in the interval $\mathbf{0 \le t \le 24}$. ✓

Alternatively, use CAS to find the maximum value.

ClassPad

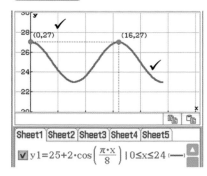

TI-Nspire

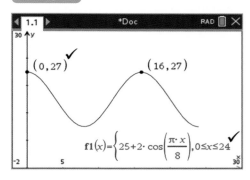

b The period of $\cos(bt)$ is $\dfrac{2\pi}{b}$, so the period of $\cos\left(\dfrac{\pi t}{8}\right)$ is $2\pi \div \dfrac{\pi}{8} = 16$.

Hence, the period of $T(t)$ is **16 hours**. ✓

c $T = 26$, so $26 = 25 + 2\cos\left(\dfrac{\pi t}{8}\right)$ ✓

$\cos\left(\dfrac{\pi t}{8}\right) = \dfrac{1}{2}$ ✓

Solve for t.

$\dfrac{\pi t}{8} = \dfrac{\pi}{3} \Rightarrow t = \dfrac{8}{3}$ hours. ✓

Confirm using CAS.

ClassPad

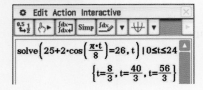

TI-Nspire

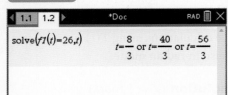

d Using CAS, find the times when the temperature is 26°C.

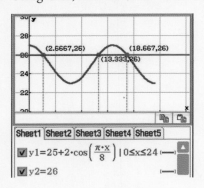

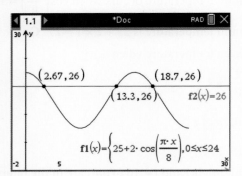

Determine the range of hours where the temperature is above 26°C.

The temperature is 26°C when $t = \dfrac{8}{3}, \dfrac{40}{3}, \dfrac{56}{3}$. ✓

From inspection of the graph, the total time when $T > 26$ is $\dfrac{8}{3} + \left(\dfrac{56}{3} - \dfrac{40}{3}\right) = 8$ hours. ✓

EXERCISE 5.7 Applying trigonometric functions   ANSWERS p. 473

Recap

1 State the period and amplitude of $y = -4\sin(3x)$.

2 Determine the equation of the function in the form $y = a\cos(bx)$.

Mastery

3 WORKED EXAMPLE 22 The function $y(t) = 50\sin\left(\dfrac{\pi t}{15}\right)$ shows the height of a horse above its rest position on a merry-go-round over 30 seconds.

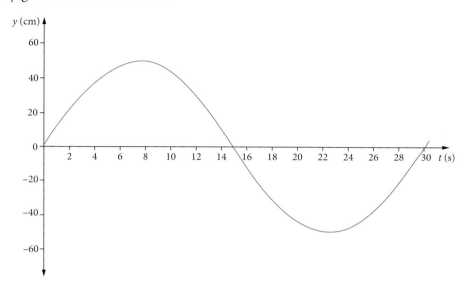

The merry-go-round itself takes 40 seconds to make one revolution.

a How many times does the horse go up and down during 10 revolutions of the merry-go-round?

b Find, correct to the nearest centimetre, the height of the horse after 1.25 revolutions of the merry-go-round.

c Calculate, correct to two decimal places, the amount of time during the first revolution of the merry-go-round that the horse is moving upwards at more than 40 cm away from its rest position.

4 WORKED EXAMPLE 23 The water level in a dam is controlled by adding and releasing water according to the function $H(t) = 3.5\sin\left(\dfrac{\pi t}{3}\right) + 6.5$. The depth of the water is H metres at time t hours after midday.

a Sketch the graph of the function in the interval $0 \le t \le 10$.

b What is the minimum and the maximum level of water in the dam at any time?

c At what time of day is the water level first at half its maximum capacity?

d By how much does the water level change between 2 pm and 5 pm? Give your answer correct to two decimal places.

5 Using CAS 4 When a mass suspended on the end of a spring is pulled down and released, it executes simple harmonic motion (moves up and down). The amount of extension and contraction of the spring is described by the equation $d = -4.5\cos(6t)$, where d is the position of the mass in centimetres and t is the time in seconds. A positive value for d means contraction of the spring (above the unstretched position).

a Sketch the graph of the function in the interval $0 \le t \le 1.5$.

b Describe the initial position of the mass.

c For how many seconds during each oscillation (that is, for one up-and-down cycle) is the spring stretched (below the unstretched position)? Give the answer correct to two decimal places.

d Calculate, correct to one decimal place, the position of the mass after 1 s.

Calculator-assumed

6 (2 marks) The temperature, $T\,°C$, in a greenhouse t hours after 6 am is described by the equation $T = 10\cos\left(\dfrac{\pi}{4}t\right) + 25$.

State the maximum and the minimum temperature, and the times of the day from 6 am to 6 pm when they occur.

7 (2 marks) The motion of a girl on a swing can be approximated by the equation $d = 1.5\sin\left(\dfrac{\pi}{2}t\right) + 2$, where d metres is her vertical height above the ground at time t seconds.

For how long during the first 30 seconds is the girl at least 2 metres above the ground?

8 (2 marks) A cork is bobbing periodically on the surface of the sea according to the equation $h = 8\sin(2\pi t)$, where h cm is the cork's height at time t seconds. A calm sea has zero height. How many cycles does the cork make in 2 seconds?

9 (2 marks) The water depth, D metres, of a river is described by $D = 1.2\sin\left(\dfrac{\pi}{3}t\right) + 1.8$, where t is the number of hours after midnight. What is the water level at 2:30 am?

10 (2 marks) The function $h(t) = 0.1\cos\left(\dfrac{1}{2}\pi t\right) + 0.5$ gives the vertical height h m of a bicycle pedal at time t seconds. What is the height of the pedal when it is at its lowest value?

11 (3 marks) The vertical position of a toy train above the centre of a circular track of radius 46 cm is given by the equation $P = 46\sin(at)$, where P cm is the position and t is time in seconds. The train takes 10 seconds to complete one half of the track. Determine the value of a.

12 (2 marks) The path of a rocket can be described by $h = 36\tan\left(\dfrac{\pi}{4}x\right)$, where x km is its horizontal distance from the take-off site and h is its height. What horizontal distance does the rocket approach as its height increases?

13 (5 marks) A pendulum's horizontal displacement (y cm) from its rest position is modelled by $y = 6.5\cos(3.5t)$, where t is the time in seconds.

Initially, the pendulum is released from its maximum possible height.

a How many complete swings does the pendulum make per minute? (1 mark)

Note: One swing means the pendulum moves from one side to the other side and back.

b Find the horizontal displacement after the first 0.2 seconds, correct to two decimal places. (2 marks)

c Determine for how long, during one swing, the pendulum's horizontal displacement is less than 3 cm. State your answer correct to two decimal places. (2 marks)

14 (5 marks) The vertical displacement, x cm, of an engine piston from its casing is described by $x = 4\sin\left(\dfrac{3\pi}{4}t\right)$, where t milliseconds is the time after the engine is switched on.

   a What is the maximum distance of the piston from the casing? (1 mark)

   b State the number of cycles (periods) the piston makes in 24 milliseconds. (1 mark)

   c How far does the piston travel in 4 milliseconds? (1 mark)

   d Determine the first time after the engine starts that the piston's position is $2\sqrt{2}$ cm from its casing. (2 marks)

15 (9 marks) P is a point on the wheel of a train that rotates as the train moves.

The vertical distance y cm of P in relation to point G (at ground level) is given by $y = 4\cos(at) + r$, where t is measured in seconds and r is the radius of the wheel in centimetres.

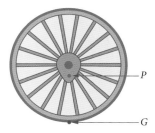

   a The circumference of the wheel is 40π cm and it rotates with a period of $\dfrac{\pi}{40}$ seconds. Find the value of r and a. (2 marks)

   b Sketch the graph of $y = 4\cos(at) + r$ in the interval $0 \leq t \leq 2$. (3 marks)

   c How many full rotations does the wheel make in the first 2 seconds? (2 marks)

   d In relation to point G, describe the position of P after 1.2 seconds. State your answer correct to one decimal place. (2 marks)

5 Chapter summary

Trigonometry of right-angled triangles

$\sin\theta = \dfrac{\text{opposite}}{\text{hypotenuse}}$

$\cos\theta = \dfrac{\text{adjacent}}{\text{hypotenuse}}$

$\tan\theta = \dfrac{\text{opposite}}{\text{adjacent}}$

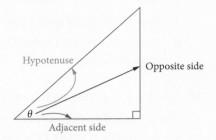

Angle of inclination

The diagram shows a straight line making an angle of θ with the positive x-axis. This angle is called the **angle of inclination** of a straight line. So for $y = mx + c$, $m = \tan\theta$.

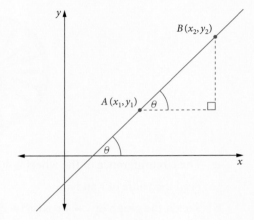

The sine rule

- In any triangle ABC with angles labelled A, B, C, the opposite side lengths are labelled a, b, c respectively.
- The **sine rule** relates each side of a triangle to its opposite angle.

$$\dfrac{a}{\sin A} = \dfrac{b}{\sin B} = \dfrac{c}{\sin C} \quad \text{or} \quad \dfrac{\sin A}{a} = \dfrac{\sin B}{b} = \dfrac{\sin C}{c}$$

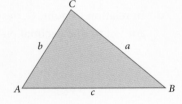

- The sine rule is used for triangle problems involving two sides and one non-included angle, or two angles and one side.
- To find an unknown side, look for two matching pairs of sides and opposite angles, and place the unknown side in the numerator.
- To find an unknown angle, look for two matching pairs of angles and opposite sides, and place the unknown angle in the numerator. There may be two supplementary solutions for the unknown angle (acute and obtuse, the ambiguous case), and you must test the obtuse angle using the given angle and the angle sum of a triangle.

The cosine rule

- The **cosine rule** relates one side of a triangle to the other two sides and its opposite angle. It is an extension of Pythagoras' theorem, $c^2 = a^2 + b^2$.

$$c^2 = a^2 + b^2 - 2ab\cos C$$

$$\text{or } \cos C = \dfrac{a^2 + b^2 - c^2}{2ab}$$

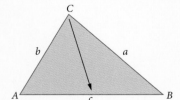

- The cosine rule is used for triangle problems involving two sides and the included angle, or all three sides.
- To find an unknown side, let c be the side and C be the opposite angle.
- To find an unknown angle, let C be the angle and c be the opposite side.

Area of a triangle

Consider the triangle ABC drawn below.

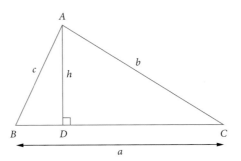

area of triangle $ABC = \dfrac{1}{2}ab\sin C$

Radian measure

- $180° = \pi$
- To convert degrees to radians, multiply by $\dfrac{\pi}{180}$.
- To convert radians to degrees, multiply by $\dfrac{180}{\pi}$.

Parts of a circle

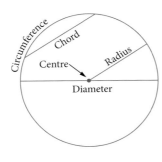 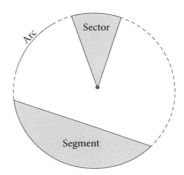

Arc length of a circle

$l = r\theta$

where r = radius of the circle

θ = angle at the centre of the circle in radians.

Area of a sector

$A = \dfrac{1}{2}r^2\theta$

where r = radius of the circle

θ = angle at the centre of the circle in radians.

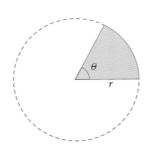

Area of a segment

$$A = \frac{1}{2}r^2(\theta - \sin\theta)$$

where r = radius of the circle

θ = angle at the centre of the circle in radians.

The trigonometric functions

The trigonometric functions **sine**, **cosine** and **tangent** are defined for any angle using the point $P(x, y)$ on the **unit circle**.

- $\sin\theta = y$
- $\cos\theta = x$
- $\tan\theta = \dfrac{y}{x}$

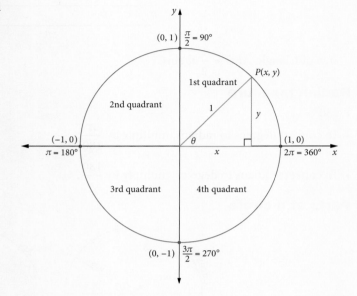

The ASTC rule

- **All** the trigonometric functions are positive in the 1st quadrant.
- **Sine** is positive and the others are negative in the 2nd quadrant.
- **Cosine** is positive and the others are negative in the 4th quadrant.
- **Tangent** is positive and the others are negative in the 3rd quadrant.

Trigonometric identities and properties

- $\tan\theta = \dfrac{\sin\theta}{\cos\theta}$
- $\sin^2\theta + \cos^2\theta = 1$
- $\sin\left(\dfrac{\pi}{2} - \theta\right) = \cos\theta$
- $\cos\left(\dfrac{\pi}{2} - \theta\right) = \sin\theta$

2nd quadrant	3rd quadrant	4th quadrant	Negative angles
$\sin(\pi - \theta) = \sin\theta$	$\sin(\pi + \theta) = -\sin\theta$	$\sin(2\pi - \theta) = -\sin\theta$	$\sin(-\theta) = -\sin\theta$
$\cos(\pi - \theta) = -\cos\theta$	$\cos(\pi + \theta) = -\cos\theta$	$\cos(2\pi - \theta) = \cos\theta$	$\cos(-\theta) = \cos\theta$
$\tan(\pi - \theta) = -\tan\theta$	$\tan(\pi + \theta) = \tan\theta$	$\tan(2\pi - \theta) = -\tan\theta$	$\tan(-\theta) = -\tan\theta$

The exact values

The reference triangles

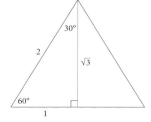

θ	0	$\frac{\pi}{6}$ (30°)	$\frac{\pi}{4}$ (45°)	$\frac{\pi}{3}$ (60°)	$\frac{\pi}{2}$ (90°)
$\sin \theta$	0	$\frac{1}{2}$	$\frac{1}{\sqrt{2}}$	$\frac{\sqrt{3}}{2}$	1
$\cos \theta$	1	$\frac{\sqrt{3}}{2}$	$\frac{1}{\sqrt{2}}$	$\frac{1}{2}$	0
$\tan \theta$	0	$\frac{1}{\sqrt{3}}$	1	$\sqrt{3}$	not defined

Points on the axes

θ	$\sin \theta$	$\cos \theta$	$\tan \theta$
0° (0)	0	1	0
90° $\left(\frac{\pi}{2}\right)$	1	0	not defined
180° (π)	0	−1	0
270° $\left(\frac{3\pi}{2}\right)$	−1	0	not defined
360° (2π)	0	1	0

Angle sum and difference identities

$\sin(A \pm B) = \sin A \cos B \pm \cos A \sin B$

$\cos(A \pm B) = \cos A \cos B \mp \sin A \sin B$

$\tan(A \pm B) = \dfrac{\tan A \pm \tan B}{1 \mp \tan A \tan B}$

Graphs of trigonometric functions

- $y = \sin x$ has period 2π and amplitude 1.
- $y = \tan x$ has period π and the amplitude is not defined.

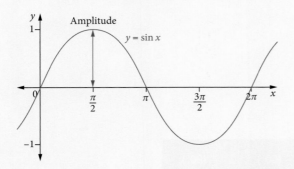
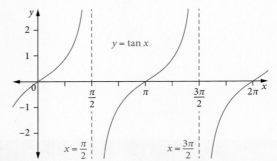

- $y = \cos x$ has period 2π and amplitude 1.

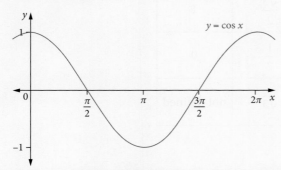

Function	Amplitude	Period	Average (central) value	Domain	Range
$y = \sin x$	1	2π	0	All real values	$-1 \le y \le 1$
$y = \cos x$	1	2π	0	All real values	$-1 \le y \le 1$
$y = \tan x$	not defined	π	0	All real values apart from odd integer multiples of $\dfrac{\pi}{2}$	All real values
$y = a \sin(b(x - c))$	a	$\dfrac{2\pi}{b}$	c	All real values	$-a \le y \le a$
$y = a \cos(b(x - c))$	a	$\dfrac{2\pi}{b}$	c	All real values	$-a \le y \le a$
$y = a \tan(b(x - c))$	not defined	$\dfrac{\pi}{b}$	c	All real values apart from $(2k + 1)\left(\dfrac{\pi}{2b}\right) + c$, $k \in \mathbb{Z}$	All real values

Applying trigonometric functions

- Real-life situations involving periodicity are modelled using trigonometric functions. This includes tides, weather, motors, light and soundwaves.

  For example, the function $h(t) = \dfrac{1}{2}\sin\left(\dfrac{\pi t}{6}\right) + 2$, where t represents the number of hours after midnight, models the water level, in relation to a fixed level, of daily tides. The period of each tide is 12 hours, the high tide level is 2.5 m at 3 am and 3 pm, and the low tide height is 1.5 m occurring at 9 am and at 9 pm.

Cumulative examination: Calculator-free

Total number of marks: 28   Reading time: 3 minutes   Working time: 28 minutes

1 (7 marks) Evaluate the following expressions.

   a $^6C_3$ (2 marks)

   b $^5C_2 \times {}^7C_2$ (3 marks)

   c $\binom{4}{4} + \binom{4}{3} + \binom{4}{2} + \binom{4}{1} + \binom{4}{0}$ (2 marks)

2 (4 marks) Consider the linear graph passing through $(1, 3)$ and $(4, -24)$.

   a Find the equation of the graph. (2 marks)

   b Sketch the graph, labelling x- and y-intercepts. (2 marks)

3 (2 marks) Consider the function $f(x) = \dfrac{1}{2}(x-1)(x+2)^2$ for $-3 \leq x \leq 2$.

Copy the axes below and on them sketch the graph of f, clearly indicating axes intercepts and turning points. Label the endpoints with their coordinates.

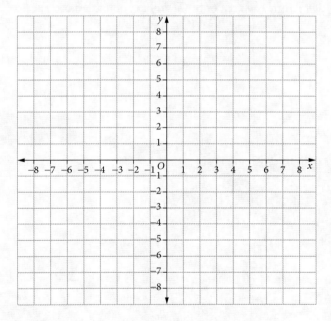

4 (2 marks) Simplify $\dfrac{\sin\theta[\sin(2\pi + \theta) - \sin(\pi - \theta) + 1]}{\cos(-\theta)}$.

5 (3 marks) On any given day, the depth of water in a river is modelled by the function

$$h(t) = 14 + 8\sin\left(\frac{\pi t}{12}\right), 0 \le t \le 24$$

where h is the depth of water, in metres, and t is the time, in hours, after 6 am.

    a Find the minimum depth of the water in the river. (1 mark)

    b Calculate as an exact value the depth of water at 10 am. (2 marks)

6 (6 marks) State whether each of the statements below is true or false.

    a The amplitude of $2\tan x$ is 2. (1 mark)

    b The amplitude of $y = -4\sin(5x)$ is -4. (1 mark)

    c $\dfrac{7\pi}{6}$ radians is $210°$. (1 mark)

    d The function $y(x) = 3\cos(x - \pi)$ is a phase shift of π units to the right of $y(x) = 3\cos x$. (1 mark)

    e The length of the arc formed by an angle of $30°$ in a circle of radius 12 cm is π. (1 mark)

    f $\sin\left(-\dfrac{4\pi}{3}\right) = -\dfrac{\sqrt{3}}{2}$ (1 mark)

7 (4 marks) An arc of a circle of radius 2 cm subtends an angle θ radians at the centre. If the arc length is $\dfrac{\pi}{3}$ cm, determine the area of the segment subtended by the same angle.

Cumulative examination: Calculator-assumed

Total number of marks: 29   Reading time: 3 minutes   Working time: 29 minutes

1 (7 marks) Danni is a tennis player. If she wins a game, the probability that she will win the next game is 0.8. If she loses a game, the probability that she will lose the next game is 0.9. Danni has just lost a game.

   a Construct a suitable display to represent the possible outcomes of Danni's next two games.   (3 marks)

   b Determine the probability that Danni

      i wins both of her next two games   (1 mark)

      ii wins exactly one of her next two games   (2 marks)

      iii wins the second of the two next games, given that she lost the first.   (1 mark)

2 (2 marks) Determine the period of $f(x) = -3\tan(2\pi x)$.

3 (4 marks) For the rectangle drawn below, find the length of the diagonal BD and the angle BDC to two decimal places.

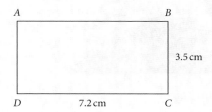

4 (5 marks) The population of wombats in a particular location varies according to the rule $n(t) = 1200 + 400\cos\left(\dfrac{\pi t}{3}\right)$, where n is the number of wombats and t is the number of months after 1 March 2023.

   a Find the period and amplitude of the function n.   (2 marks)

   b Find the maximum and minimum populations of wombats in this location.   (2 marks)

   c Find $n(10)$.   (1 mark)

5 (4 marks) Sammy enters a capsule on a giant Ferris wheel from a platform above the ground. The Ferris wheel is rotating anticlockwise. The capsule is attached to the Ferris wheel at point P. The height of P above the ground, h, is modelled by $h(t) = 63 - 55\cos\left(\dfrac{\pi t}{15}\right)$, where t is the time in minutes after Sammy enters the capsule and h is measured in metres. Sammy exits the capsule after one complete rotation of the Ferris wheel.

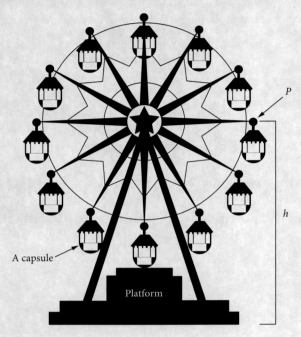

a State the minimum and maximum heights of P above the ground. (1 mark)

b For how much time is Sammy in the capsule? (1 mark)

c If the radius of the Ferris wheel is 55 metres, what is the arc length between any two consecutive capsules? Give the answer to the nearest centimetre. (2 marks)

6 (7 marks) John is marking out an orienteering course. From his starting point he walks on a bearing of 040° to the first checkpoint. He then turns and walks on a bearing of 110° to the second checkpoint, which is directly due east of the starting position. He then walks 200 m back to the starting point.

a Calculate the total length of the orienteering course correct to two decimal places. (5 marks)

b Calculate the area of the region bounded by the three legs of the course correct to two decimal places. (2 marks)

CHAPTER 6

EXPONENTIAL FUNCTIONS

Syllabus coverage

Nelson MindTap chapter resources

6.1 Index laws
Exponential terms with an integer base
Negative powers

6.2 Fractional powers

6.3 Exponential equations
Solving exponential equations using the null factor law

6.4 The exponential function $y = a^x$
Exponential functions
Using CAS 1: Graphing exponential functions
Finding equations of exponential functions

6.5 Exponential growth and decay
Exponential growth and decay
Significant figures and scientific notation

Examination question analysis

Chapter summary

Cumulative examination: Calculator-free

Cumulative examination: Calculator-assumed

Syllabus coverage

TOPIC 2.1: EXPONENTIAL FUNCTIONS

Indices and the index laws

2.1.1 review indices (including fractional and negative indices) and the index laws
2.1.2 use radicals and convert to and from fractional indices
2.1.3 understand and use scientific notation and significant figures

Exponential functions

2.1.4 establish and use the algebraic properties of exponential functions
2.1.5 recognise the qualitative features of the graph of $y = a^x$ ($a > 0$), including asymptotes, and of its translations ($y = a^x + b$ and $y = a^{x-c}$)
2.1.6 identify contexts suitable for modelling by exponential functions and use them to solve practical problems
2.1.7 solve equations involving exponential functions using technology, and algebraically in simple cases

Mathematics Methods ATAR Course Year 11 syllabus p. 12 © SCSA

Video playlists (6):
6.1 Index laws
6.2 Fractional powers
6.3 Exponential equations
6.4 The exponential function $y = a^x$
6.5 Exponential growth and decay
Examination question analysis Exponential functions

Worksheets (6):
6.1 Review of index laws
6.2 Fractional indices and radicals
6.3 Solving exponential equations
6.4 Exponential functions • Graphing exponentials
6.5 Using exponential models

Nelson MindTap

To access resources above, visit
cengage.com.au/nelsonmindtap

6.1 Index laws

In the term a^n, a is the base and n is called the **index** or the **exponent**. Index laws for simplifying index expressions are listed below. These laws are easy to understand if we use the definition

$a^n = a \times a \times a \times a \times \ldots$ (to n factors) where the base, a, is a positive real number.

For example, $2^5 = 2 \times 2 \times 2 \times 2 \times 2 = 32$.

The index laws

First index law: Multiply exponential terms with same base $\quad a^m \times a^n = a^{m+n}$

Second index law: Divide exponential terms with same base $\quad a^m \div a^n = a^{m-n}$

Third index law: Raise an exponential term to a power $\quad (a^m)^n = a^{mn}$

Power of zero $\quad a^0 = 1$

where $a > 0$, $a \neq 1$.

WORKED EXAMPLE 1 Simplifying using a single index law

Simplify these expressions.

a $\quad a^6 \times a^8 \times b^8 \times b^{-7}$

b $\quad \dfrac{12a^6 b^3}{3a^2 b}$

c $\quad (3m^3 n^5)^2$

Steps	Working
a 1 Use the first index law: $a^m \times a^n = a^{m+n}$	$a^6 \times a^8 \times b^8 \times b^{-7}$ $= a^{6+8} \times b^{8-7}$
2 Simplify indices.	$= a^{14} b$
b 1 Divide the numbers.	$\dfrac{12a^6 b^3}{3a^2 b}$ $= \dfrac{4a^6 b^3}{a^2 b}$
2 Use the second index law: $a^m \div a^n = a^{m-n}$	$= 4a^{6-2} b^{3-1}$
3 Simplify.	$= 4a^4 b^2$
c 1 Use the third index law: $(a^m)^n = a^{mn}$	$(3m^3 n^5)^2$ $= 3^2 m^{3 \times 2} n^{5 \times 2}$
2 Simplify.	$= 9m^6 n^{10}$

Video playlist
Index laws

Worksheet
Review of index laws

In some problems we need to use more than one index law.

Use the order: brackets, multiply then divide.

WORKED EXAMPLE 2 Simplifying using multiple index laws

Simplify the expression $\dfrac{(2a^2b^3)^3 \times 18a^7b^{11}}{(3a^2b)^2}$.

Steps	Working
1 Use the third index law: $(a^m)^n = a^{mn}$ Expand the brackets.	$\dfrac{(2a^2b^3)^3 \times 18a^7b^{11}}{(3a^2b)^2}$ $= \dfrac{2^3 a^6 b^9 \times 18a^7b^{11}}{3^2 a^4 b^2}$
2 Evaluate the integers in the expression and simplify.	$= \dfrac{8a^6b^9 \times 18a^7b^{11}}{9a^4b^2}$ $= \dfrac{8a^6b^9 \times 2a^7b^{11}}{a^4b^2}$ $= \dfrac{16a^6b^9 \times a^7b^{11}}{a^4b^2}$
3 Use the first index law: $a^m \times a^n = a^{m+n}$	$= \dfrac{16a^{13}b^{20}}{a^4b^2}$
4 Use the second index law: $a^m \div a^n = a^{m-n}$	$= 16a^9b^{18}$

Exponential terms with an integer base

Some problems have expressions involving exponential terms with different integer bases. The first step is to convert each base into its prime factors. Once this has been done, we can use the index laws to simplify terms with the same base.

Powers of prime numbers

Powers of 2		Powers of 3		Powers of 5	
2^0	1	3^0	1	5^0	1
2^1	2	3^1	3	5^1	5
2^2	4	3^2	9	5^2	25
2^3	8	3^3	27	5^3	125
2^4	16	3^4	81	5^4	625
2^5	32	3^5	243	5^5	3125
2^6	64	3^6	729		

> **WORKED EXAMPLE 3** Simplifying indices with an integer base

Simplify the expressions.

a $8^6 \times 16^4 \times 5^{12} \div 25^3$

b $\dfrac{12^3 \times 27}{6^2 \times 8}$

Steps	Working
a 1 Write the expression and express the bases in terms of their prime factors. 16 and 8 are powers of 2 and 25 is a power of 5.	$8^6 \times 16^4 \times 5^{12} \div 25^3$ $= (2^3)^6 \times (2^4)^4 \times 5^{12} \div (5^2)^3$
2 Use the third index law to expand the brackets.	$= 2^{18} \times 2^{16} \times 5^{12} \div 5^6$
3 Use the first and second index laws to simplify.	$= 2^{18+16} \times 5^{12-6}$ $= 2^{34} \times 5^6$
b 1 Express the bases as powers of 2 and 3 by writing each base in terms of their prime factors. For example, $12 = 4 \times 3 = 2^2 \times 3$.	$\dfrac{12^3 \times 27}{6^2 \times 8}$ $= \dfrac{(2^2 \times 3)^3 \times 3^3}{(2 \times 3)^2 \times 2^3}$
2 Use the third index law to expand the brackets.	$= \dfrac{2^6 \times 3^3 \times 3^3}{2^2 \times 3^2 \times 2^3}$
3 Use the first index law to multiply terms with the same base.	$= \dfrac{2^6 \times 3^6}{2^5 \times 3^2}$
4 Use the second index law to divide terms with the same base.	$= 2 \times 3^4$

Negative powers

Consider the exponential expression $\dfrac{a^3}{a^5}$.

Using the second index law: $\dfrac{a^3}{a^5} = a^{3-5} = a^{-2}$

In factor form: $\dfrac{a^3}{a^5} = \dfrac{a \times a \times a}{a \times a \times a \times a \times a}$

Now, cancel a factor of a^3: $\dfrac{\cancel{a} \times \cancel{a} \times \cancel{a}}{a \times a \times \cancel{a} \times \cancel{a} \times \cancel{a}} = \dfrac{1}{a \times a} = \dfrac{1}{a^2}$

It can be seen that $\dfrac{1}{a^2} = a^{-2}$.

> **Negative powers**
>
> $a^{-n} = \dfrac{1}{a^n}$

WORKED EXAMPLE 4 — Simplifying indices with negative powers

Simplify each of the following expressions, expressing your answers with positive powers.

a $\dfrac{4mn^3}{6m^8n}$

b $\dfrac{(a^2b^{-5})^{-2}}{a^{-7}b^{14}}$

Steps	Working
a 1 Cancel the common factor of 2.	$\dfrac{4mn^3}{6m^8n} = \dfrac{2mn^3}{3m^8n}$
2 Divide terms with the same base using the second index law to simplify.	$= \dfrac{2m^{1-8}n^{3-1}}{3}$
	$= \dfrac{2m^{-7}n^2}{3}$
3 Write the answer with positive powers. Use $a^{-n} = \dfrac{1}{a^n}$.	$= \dfrac{2n^2}{3m^7}$
b 1 Expand brackets and multiply powers.	$\dfrac{(a^2b^{-5})^{-2}}{a^{-7}b^{14}} = \dfrac{a^{-4}b^{10}}{a^{-7}b^{14}}$
2 Divide and subtract powers.	$= a^{-4+7}b^{10-14}$
	$= a^3 b^{-4}$
3 Express with positive powers.	$= \dfrac{a^3}{b^4}$

> m^{-7} moves from the numerator to the denominator and changes to m^7.

WORKED EXAMPLE 5 — Simplifying indices with a variable in the exponent

Simplify the following expressions.

a $\dfrac{4^{n-2} \times 32^{n+3}}{8^{n+1}}$

b $24^{3n-2} \times 9^{2n-3}$

Steps	Working
a 1 Express all bases as powers of 2.	$\dfrac{4^{n-2} \times 32^{n+3}}{8^{n+1}} = \dfrac{(2^2)^{n-2} \times (2^5)^{n+3}}{(2^3)^{n+1}}$
2 Expand the brackets using the third index law.	$= \dfrac{2^{2n-4} \times 2^{5n+15}}{2^{3n+3}}$
3 Multiply the terms in the numerator by adding the powers.	$= \dfrac{2^{7n+11}}{2^{3n+3}}$
4 Divide the terms in the numerator by subtracting the powers.	$= 2^{4n+8}$
b 1 Express all bases as powers of 2 and 3. $24 = 8 \times 3 = 2^3 \times 3$	$24^{3n-2} \times 9^{2n-3} = (2^3 \times 3)^{3n-2} \times (3^2)^{2n-3}$
2 Expand the brackets and multiply the powers, then use the first index law to simplify.	$= 2^{9n-6} \times 3^{3n-2} \times 3^{4n-6}$
	$= 2^{9n-6} \times 3^{7n-8}$

EXERCISE 6.1 Index laws

ANSWERS p. 475

Mastery

1 WORKED EXAMPLE 1   Simplify the following expressions.

a $a^6 \div a^3$

b $x^5 \times x^4 \times x^{11}$

c $(a^4)^6$

d $5a^6b^4 \times 8a^3b$

e $\dfrac{3m^7n^2}{81m^2n}$

f $(2a^6b^4)^3$

2 WORKED EXAMPLE 2   Simplify the following expressions.

a $(5x^7y^5)^2 \times (2x^4y^5)^3$

b $\dfrac{(10x^6y^7)^4}{500x^{12}y^{11}}$

c $\dfrac{(3a^3b^4)^2 \times 16a^4b^{13}}{(6ab^5)^2}$

3 WORKED EXAMPLE 3   Simplify each expression.

a $32^3 \times 16^5 \times 8^2$

b $27^2 \times 9^3 \div 3^5$

c $15^6 \times 25^3 \times 81^2$

d $\dfrac{144^7}{2^4 \times 3^3}$

4 WORKED EXAMPLE 4   Simplify each of the following expressions, expressing your answers with positive powers.

a $m^7 \div m^{14}$

b $\dfrac{50xy}{15x^6y^5}$

c $\dfrac{(3a^7b^{-2})^{-3}}{a^{-3}b^{10}}$

d $\dfrac{(5x^3y)^2}{20x^{-7}y^{-3}} \times \dfrac{6x^{-6}y^{-4}}{9(xy)^5}$

5 WORKED EXAMPLE 5   Simplify each of the following expressions, expressing your answers in terms of n.

a $6^{5n-2} \times 8^{3n+2}$

b $\dfrac{10^{1-n} \times 4^{4n-1}}{25^{3-n}}$

c $18^{4-3n} \div 16^{n+5}$

Calculator-free

6 (3 marks) Simplify each of the following expressions, expressing your answers in exponential form, with a prime base.

a $32^{2x} \times 2^x \times 16^x$ (1 mark)

b $3^{2x} \div 27^{3x}$ (1 mark)

c $(64^{3x})^5$ (1 mark)

7 (5 marks) Simplify the following expressions, expressing your answers with positive indices.

a $5x^3y^{-4} \times 4x^{-8}y^{-2}$ (2 marks)

b $\dfrac{24x^{-1}y^5 \times (2x^{-3}y^{-3})^{-2}}{12x^4y^4}$ (3 marks)

Calculator-assumed

8 (2 marks) Simplify

a $ab \times (4ab)^3 \div (ab)^2$ (1 mark)

b $b^6 \times (2b)^6 \div b^2$ (1 mark)

9 (6 marks)

a Find the prime factors of

  i 108   (1 mark)

  ii 54   (1 mark)

  iii 48   (1 mark)

b Use the prime factors to simplify $\dfrac{108^{n+2}}{54^{2n-1} \times 48^n}$.   (3 marks)

10 (6 marks) The number $36^n \times 72^{n-2} \div 20^{4-3n}$ can be written in the form $2^a \times 3^b \times 5^c$.

Find a, b and c, each a linear expression, in terms of n.

6.2 Fractional powers

Video playlist
Fractional powers

Worksheet
Fractional indices and radicals

The exponent of a number tells us how many times to multiply the number.

For example: $8^2 = 64$ can be written in factor form as $8 \times 8 = 64$.

Consider the equation: $64^n \times 64^n = 64$, or $64^{2n} = 64^1$.

The value of n can be found by solving $2n = 1$, therefore $n = \dfrac{1}{2}$.

It follows that $64^{\frac{1}{2}} = 8$; however, $\sqrt{64} = 8$, therefore $64^{\frac{1}{2}} = \sqrt{64}$.

Let's try another fraction.

$$27^{\frac{1}{3}} \times 27^{\frac{1}{3}} \times 27^{\frac{1}{3}} = 27$$

This tells us $27^{\frac{1}{3}} = \sqrt[3]{27}$.

The equation $27^{\frac{1}{3}} = \sqrt[3]{27}$ has a number written in two different forms:

$27^{\frac{1}{3}}$ is the exponential form and $\sqrt[3]{27}$ is the radical form.

Radical form	Exponential form
$\sqrt[n]{a}$	$a^{\frac{1}{n}}$

Fraction powers

Square root: $\sqrt{a} = a^{\frac{1}{2}}$

Cube root: $\sqrt[3]{a} = a^{\frac{1}{3}}$

nth root: $\sqrt[n]{a} = a^{\frac{1}{n}}$

$\sqrt[n]{a^x} = a^{\frac{x}{n}}$

WORKED EXAMPLE 6 — Expressing terms in radical and exponential form

a Write as a radical.

 i $x^{\frac{1}{4}}$ **ii** $(x^7)^{\frac{1}{3}}$

b Write in exponential form.

 i $\sqrt[8]{x}$ **ii** $\left(\sqrt[3]{x}\right)^4$

Steps	Working
a Write the term as a radical. **i** Use $\sqrt[n]{a} = a^{\frac{1}{n}}$. **ii** Use $\sqrt[n]{a^x} = a^{\frac{x}{n}}$.	$x^{\frac{1}{4}} = \sqrt[4]{x}$ $(x^7)^{\frac{1}{3}} = \sqrt[3]{x^7}$ This answer can also be written as $(x^7)^{\frac{1}{3}} = \left(\sqrt[3]{x}\right)^7$.
b Write the term in exponential form. **i** Use $a^{\frac{1}{n}} = \sqrt[n]{a}$. **ii** Use $a^{\frac{x}{n}} = \sqrt[n]{a^x}$.	$\sqrt[8]{x} = x^{\frac{1}{8}}$ $\left(\sqrt[3]{x}\right)^4 = \left(x^{\frac{1}{3}}\right)^4$ $= x^{\frac{4}{3}}$

WORKED EXAMPLE 7 — Evaluating fractional powers

Find the value of each term.

a $(81)^{\frac{1}{4}}$ **b** $(27)^{\frac{2}{3}}$ **c** $(125)^{-\frac{1}{3}}$

Steps	Working
a 1 Write the number as a radical. 2 Think: What number multiplied by itself 4 times equals 81? $3^4 = 81$	$(81)^{\frac{1}{4}} = \sqrt[4]{81}$ $= 3$
b 1 Write the number in radical form. $a^{\frac{1}{3}} = \sqrt[3]{a}$ 2 Evaluate the cube root. 3 Write the answer.	$(27)^{\frac{2}{3}} = \left(27^{\frac{1}{3}}\right)^2$ $= \left(\sqrt[3]{27}\right)^2$ $= (3)^2$ $= 9$
c 1 Write the number in radical form. 2 Evaluate the cube root. 3 Write the answer.	$(125)^{-\frac{1}{3}} = \left((125)^{\frac{1}{3}}\right)^{-1}$ $= 5^{-1}$ $= \dfrac{1}{5}$

WORKED EXAMPLE 8 — Simplifying expressions with fractional powers

Simplify the following.

a $x^{\frac{1}{2}} \times x^{\frac{1}{3}}$ **b** $\sqrt[5]{x^2} \times \sqrt[3]{x^5}$ **c** $\dfrac{\sqrt{a^3 b}}{\sqrt[3]{ab^3}}$

Steps	Working
a Use the first index law. Obtain a common denominator to add the fractions. $\dfrac{1}{2} + \dfrac{1}{3} = \dfrac{3}{6} + \dfrac{2}{6} = \dfrac{5}{6}$	$x^{\frac{1}{2}} \times x^{\frac{1}{3}} = x^{\frac{1}{2} + \frac{1}{3}}$ $= x^{\frac{5}{6}}$
b 1 Express the radicals in exponential form. $a^{\frac{1}{n}} = \sqrt[n]{a}$ 2 Use the third index law to expand the brackets. 3 Obtain a common denominator and add the fractions. $\dfrac{2}{5} + \dfrac{5}{3} = \dfrac{6}{15} + \dfrac{25}{15} = \dfrac{31}{15}$	$\sqrt[5]{x^2} \times \sqrt[3]{x^5} = (x^2)^{\frac{1}{5}} \times (x^5)^{\frac{1}{3}}$ $= x^{\frac{2}{5}} \times x^{\frac{5}{3}}$ $= x^{\frac{2}{5} + \frac{5}{3}}$ $= x^{\frac{31}{15}}$

> This answer can also be written in radical form as $\sqrt[15]{x^{31}}$.

c **1** Express the radicals in exponential form.
$$\frac{\sqrt{a^3 b}}{\sqrt[3]{ab^3}} = \frac{(a^3 b)^{\frac{1}{2}}}{(ab^3)^{\frac{1}{3}}}$$

2 Use the third index law.
$$= \frac{a^{\frac{3}{2}} b^{\frac{1}{2}}}{a^{\frac{1}{3}} b}$$

3 Obtain a common denominator and subtract the powers.
$$= a^{\frac{7}{6}} b^{-\frac{1}{2}}$$

$$\frac{3}{2} - \frac{1}{3} = \frac{9}{6} - \frac{2}{6} = \frac{7}{6}$$

This answer can also be written as $\dfrac{a^{\frac{7}{6}}}{b^{\frac{1}{2}}}$.

WORKED EXAMPLE 9 — Simplifying exponential terms with fractional powers

Find the value of n if $8^{\frac{3}{2}} \times 32^{\frac{3}{4}} \times 16^{\frac{1}{3}} = 2^n$.

Steps	Working
1 Express each term as a power of 2.	$8^{\frac{3}{2}} \times 32^{\frac{3}{4}} \times 16^{\frac{1}{3}} = 2^n$ $(2^3)^{\frac{3}{2}} \times (2^5)^{\frac{3}{4}} \times (2^4)^{\frac{1}{3}} = 2^n$
2 Use the third index law to expand the brackets.	$2^{\frac{9}{2}} \times 2^{\frac{15}{4}} \times 2^{\frac{4}{3}} = 2^n$
3 Add the powers. $\dfrac{9}{2} + \dfrac{15}{4} + \dfrac{4}{3} = \dfrac{54}{12} + \dfrac{45}{12} + \dfrac{16}{12}$ $= \dfrac{115}{12}$	$2^{\frac{115}{12}} = 2^n$ $n = \dfrac{115}{12}$

EXERCISE 6.2 Fractional powers

ANSWERS p. 475

Recap

1 Simplify.
  a $m^9 \div m^3$
  b $x^{11} \times x^{24} \times x^{10}$
  c $(2a^7)^4$

2 Simplify each of the following, expressing your answers in exponential form with a prime base.
  a $9^{2x} \times 27^x \times 81^x$
  b $128^{2x} \div 2^{3x}$
  c $(25^{3x})^2$

Mastery

3 WORKED EXAMPLE 6

  a Write the following exponent terms as radicals.
   i $x^{\frac{1}{2}}$
   ii $m^{\frac{1}{3}}$
   iii $x^{\frac{6}{5}}$

  b Write the following radicals in exponential form.
   i $\sqrt[6]{x}$
   ii $\sqrt[4]{a^2 b^3}$
   iii $\left(\sqrt[7]{p}\right)^2$

4 WORKED EXAMPLE 7 Find the value of each of the following.

a $(125)^{\frac{1}{3}}$
b $(16)^{\frac{1}{4}}$
c $(16)^{\frac{3}{4}}$
d $(16)^{-\frac{3}{4}}$

e $(32)^{\frac{3}{5}}$
f $(9)^{-\frac{3}{2}}$
g $(128)^{\frac{5}{7}}$
h $\left(\dfrac{25}{4}\right)^{-\frac{3}{2}}$

5 WORKED EXAMPLE 8 Simplify the following expressions.

a $x^{\frac{1}{5}} \times x^{\frac{1}{3}}$
b $\sqrt{x^5} \times \sqrt[3]{x^4}$
c $\dfrac{\sqrt[3]{x^2 y^3}}{\sqrt{x^3 y}}$

d $\sqrt{49x^3 y^5} \times \sqrt[4]{16xy}$
e $(x^4 y^{-2})^{\frac{1}{3}} \times (xy^{-3})^{-\frac{1}{2}}$
f $\dfrac{(81x^{-4} y)^{\frac{1}{4}} \times 15 x^{\frac{2}{3}} y^{\frac{1}{4}}}{(25x^3 y^6)^{\frac{1}{2}}}$

6 WORKED EXAMPLE 9

a Find the value of n if $(32)^{\frac{3}{5}} \times (8)^{\frac{4}{3}} \times (16)^{\frac{5}{4}} = 2^n$.

b Find the value of n if $(243)^{\frac{3}{5}} \times (27)^{\frac{2}{3}} \div (9)^{\frac{5}{2}} = 3^n$.

c Find the values of a and b if $(15)^{\frac{1}{4}} \times (25)^{\frac{1}{3}} \times (75)^{-\frac{1}{2}} = 5^a \times 3^b$.

Calculator-free

7 (4 marks)

a Evaluate

  i $64^{\frac{1}{2}}$ (1 mark)

  ii $27^{\frac{1}{3}}$ (1 mark)

b Hence find the value of $64^{\frac{1}{2}} \times 27^{\frac{2}{3}}$. (2 marks)

Calculator-assumed

8 (4 marks)

a Find the values of a and b if $\sqrt[4]{x^7} = x^a$ and $\sqrt{x^{-5}} = x^b$. (2 marks)

b Hence, simplify the expression $\sqrt[4]{x^7} \times \sqrt{x^{-5}}$. (2 marks)

9 (4 marks)

a Show that $\left(\sqrt[5]{x^3}\right)^{\frac{1}{2}} = x^{\frac{3}{10}}$. (2 marks)

b Hence, simplify $\dfrac{\left(\sqrt[5]{x^3}\right)^{\frac{1}{2}}}{\left(\sqrt{x^3}\right)^{\frac{1}{5}}}$. (2 marks)

6.3 Exponential equations

Video playlist
Exponential equations

Worksheet
Solving exponential equations

An **exponential equation** is an equation in which the variable is in the exponent or power. An example of an exponential equation is $8^x = 128$.

We can solve exponential equations by expressing all terms with a common base, usually a prime number, such as 2, 3 or 5. Some knowledge of the powers of prime numbers will help when solving exponential equations.

WORKED EXAMPLE 10 Solving simple exponential equations

Solve $8^x = 128$ for x.

Steps	Working
1 Write 8 and 128 as powers of 2.	$8^x = 128$ $(2^3)^x = 2^7$ $2^{3x} = 2^7$
2 Equate the powers.	$3x = 7$
3 Solve the linear equation.	$x = \dfrac{7}{3}$

WORKED EXAMPLE 11 Solving exponential equations

Solve $4^{x-2} = 32^{1-x}$ for x.

Steps	Working
1 Write 4 and 32 as powers of 2.	$4^{x-2} = 32^{1-x}$ $(2^2)^{x-2} = (2^5)^{1-x}$
2 Expand the brackets and multiply the powers.	$2^{2x-4} = 2^{5-5x}$
3 Equate the powers.	$2x - 4 = 5 - 5x$
4 Solve the linear equation.	$7x - 4 = 5$ $7x = 9$ $x = \dfrac{9}{7}$

WORKED EXAMPLE 12 Solving exponential equations using index laws

Solve $9^x \times 27^{x+1} = 81^{1-x}$ for x.

Steps	Working
1 Write 9, 27 and 81 as powers of 3.	$9^x \times 27^{x+1} = 81^{1-x}$ $(3^2)^x \times (3^3)^{x+1} = (3^4)^{1-x}$
2 Expand the brackets, multiply the powers then simplify.	$3^{2x} \times 3^{3x+3} = 3^{4-4x}$ $3^{2x+3x+3} = 3^{4-4x}$ $3^{5x+3} = 3^{4-4x}$
3 Equate the powers and solve the linear equation.	$5x + 3 = 4 - 4x$ $9x + 3 = 4$ $9x = 1$ $x = \dfrac{1}{9}$

4 Confirm using CAS.

ClassPad	TI-Nspire
solve$(9^x \cdot 27^{x+1} = 81^{1-x}, x)$ $\{x = \frac{1}{9}\}$	solve$(9^x \cdot 27^{x+1} = 81^{1-x}, x)$ $x = \frac{1}{9}$

Solving exponential equations using the null factor law

An exponential equation of the form $(2^x - 8)(2^x - 2) = 0$ can be solved using the null factor law. Either $2^x - 8 = 0$ or $2^x - 2 = 0$ will give two possible solutions for the equation.

WORKED EXAMPLE 13 Solving factorised exponential equations 1

Solve $(2^x - 8)(2^x - 2) = 0$ for x.

Steps	Working		
1 Use the null factor law to write two equations that solve for x.	either	$2^x - 8 = 0$ or	$2^x - 2 = 0$
2 Solve each equation.		$2^x = 8$	$2^x = 2$
		$2^x = 2^3$	$2^x = 2^1$
		$x = 3$	$x = 1$

WORKED EXAMPLE 14 Solving factorised exponential equations 2

Solve $(4(2^x) - 1)(2^x + 3) = 0$ for x.

Steps	Working	
1 Use the null factor law to write two equations that solve for x.	$4(2^x) - 1 = 0$	$2^x + 3 = 0$
2 Solve each exponential equation. The two factors only produce one solution because 2^x is always positive.	$2^x = \frac{1}{4}$ $2^x = 2^{-2}$ $x = -2$	$2^x = -3$ no solution, as $2^x > 0$

EXERCISE 6.3 Exponential equations   ANSWERS p.475

Recap

1 a Write $n^{\frac{5}{2}}$ as a radical.

   b Write $\sqrt[3]{a^7}$ in exponential form.

2 Simplify $\sqrt[3]{x^7} \times \sqrt[4]{x^3}$.

Mastery

3 WORKED EXAMPLE 10   Solve each equation for x.

    a $81^x = 243$ **b** $16^x = 512$

    c $125^x = 25$ **d** $8^x = \dfrac{1}{4}$

4 WORKED EXAMPLE 11   Solve each equation for x.

    a $27^{x-1} = 81^{3-2x}$ **b** $64^{x+1} = 32^{x+2}$

    c $4^{5x+3} = 8^{1-3x}$ **d** $125^{x-2} - 25^x = 0$

5 WORKED EXAMPLE 12   Solve each equation for x.

    a $9 \times 27^{x-1} = 81$ **b** $64^{x+1} \times 32^{x+2} = 4 \times 16^x$

    c $16 = 64^{x+1} \times 8^{x-1}$ **d** $125^{x-2} \times 25^x = 625$

6 WORKED EXAMPLE 13   Solve each equation for x.

    a $(2^x - 8)(2^x - 4) = 0$ **b** $(3^x - 81)(3^x - 1) = 0$

    c $(5^x - 5)(5^x - 125) = 0$ **d** $(2^x - 4)(2^x + 2) = 0$

7 WORKED EXAMPLE 14   Solve each equation for x.

    a $(9(3^x) - 1)(3^x - 3) = 0$ **b** $(4(2^x) - 1)(2^x - 1) = 0$

    c $(2(2^x) - 1)(2^x - 4) = 0$ **d** $(8(2^x) - 1)(2^x + 1) = 0$

Calculator-free

8 (8 marks)   Solve the following equations for x.

    a $2^{3x-3} = 8^{2-x}$ (2 marks)

    b $3^{-4x} = 9^{6-x}$ (2 marks)

    c $8^{x+3} \times 32^x = 64$ (2 marks)

    d $\dfrac{8^{3x+2}}{4^{3-x}} = 256$ (2 marks)

9 (8 marks)   Find the solution of each equation.

    a $128^x = 8$ (2 marks)

    b $25^{2x} = 125^{x+2}$ (2 marks)

    c $16 \times 2^{2x} = 128^{2-x}$ (2 marks)

    d $\dfrac{81^{2x+3}}{9} = 27^x$ (2 marks)

Calculator-assumed

10 (3 marks)   Solve each equation for x, correct to two decimal places.

    a $8^x = 5$ (1 mark)

    b $3^{x+1} = 21$ (1 mark)

    c $14^{x-2} = 7^{x+2}$ (1 mark)

6.4 The exponential function $y = a^x$

Exponential functions

Most plants and animals begin as a single cell. Consider a cell that divides into 2 every hour. The subsequent cells divide causing the total number of cells to double each hour. The number of cells is shown in this table.

Number of hours, t	0	1	2	3	4
Number of cells, n	1	2	4	8	16

The rule for the number of cells is $n = 2^t$, an **exponential function**.

It can be seen in the graph shown that the value of n doubles every hour.

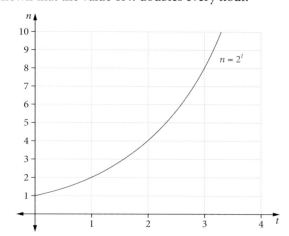

Video playlist
The exponential function $y = a^x$

Worksheets
Exponential functions

Graphing exponentials

Properties of the exponential function, $y = 2^x$

The table of values for $y = 2^x$ is shown below.

x	$y = 2^x$
−3	$2^{-3} = \dfrac{1}{8}$
−2	$2^{-2} = \dfrac{1}{4}$
−1	$2^{-1} = \dfrac{1}{2}$
0	$2^0 = 1$
1	$2^1 = 2$
2	$2^2 = 4$
3	$2^3 = 8$

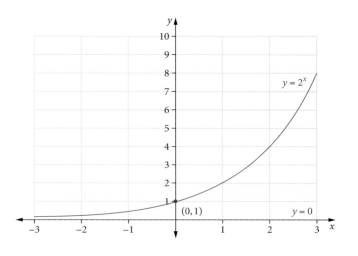

As x decreases ($x \to -\infty$), the value of y gets closer to zero ($y \to 0$).

The graph of the function gets closer and closer to the x-axis but will never touch it.

The line that the graph approaches as $x \to \pm\infty$ is called an **asymptote**.

$y = 2^x$ has a horizontal asymptote with the equation $y = 0$ (the x-axis).

Properties of the exponential function $y = a^x$ where $a > 1$

- The function is always positive.
- The gradient of the graph is always positive.
- As x increases the gradient also increases.
- The y-intercept is at $(0, 1)$ (because $a^0 = 1$).
- The horizontal asymptote is $y = 0$ (the x-axis).

USING CAS 1 Graphing exponential functions

Graph $y = 3^x$ for $-2 \leq x \leq 2$ and comment on the features of the graph.

ClassPad

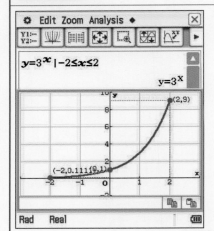

TI-Nspire

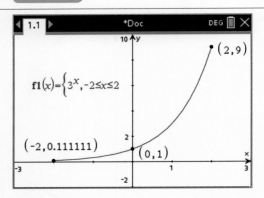

1. In **Main**, enter the expression using the **Math3** symbols | and ≤ for the domain $[-2, 2]$.
2. Highlight and drag the expression into the **Graph** window.
3. Adjust the window settings to suit.

1. Add a **Graphs** page and graph the function over the domain $[-2, 2]$.
2. Adjust the window settings to suit.

The graph has a y-intercept of 1. The graph is always positive and increases as x increases. The horizontal asymptote is the x-axis, $y = 0$.

The coordinates of the endpoints are $\left(-2, \dfrac{1}{9}\right)$ and $(2, 9)$.

Transformations of $y = a^x$

Changing the base of the graph of $y = a^x$

The graphs of $y = 2^x$ and $y = 5^x$ are shown below. Notice the effect of changing the base of an exponential function.

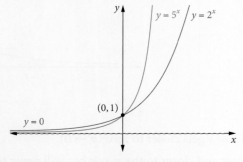

All graphs of $y = a^x$ have the same y-intercept $(0, 1)$, the same horizontal asymptote, $y = 0$, and the same basic shape.

When the base a increases, the graph becomes steeper for $x > 0$ and less steep for $x < 0$.

Translations of $y = a^x$

Graphs of the form $y = a^x + b$

The graphs of $y = 2^x$ and $y = 2^x + 3$ are shown below. Notice the effect of a vertical translation on the graph of an exponential function.

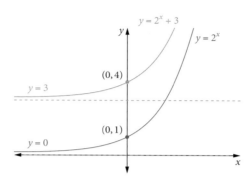

The graph of $y = 2^x$ has been translated up by 3 units to produce the graph of $y = 2^x + 3$. Notice the asymptote has moved from $y = 0$ to $y = 3$ and the y-intercept has moved from $(0, 1)$ to $(0, 4)$.

> **Graphs of the form $y = a^x + b$ where $b > 0$**
>
> The constant b produces a vertical translation of $y = a^x$ in the direction of the y-axis.
> - The graph of $y = a^x + b$ translates the graph of $y = a^x$ b units up.
>   The new horizontal asymptote is $y = b$ and the new y-intercept is $(1 + b, 0)$.
> - The graph of $y = a^x - b$ translates the graph of $y = a^x$ b units down.
>   The new horizontal asymptote is $y = -b$ and the new y-intercept is $(1 - b, 0)$.

Graphs of the form $y = a^{(x-c)}$

The graphs of $y = 2^x$ and $y = 2^{x-1}$ below show the effect of a horizontal translation on the graph of an exponential function.

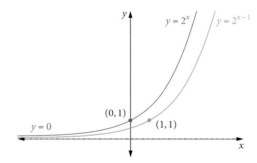

The graph of $y = 2^x$ has been translated right by 1 unit to produce the graph of $y = 2^{x-1}$.

> **Graphs of the form $y = a^{(x-c)}$ where $c > 0$**
>
> The constant c produces a horizontal translation in the direction of the x-axis.
> - The graph of $y = a^{(x-c)}$ translates the graph of $y = a^x$ right by c units (in the positive x direction).
> - The graph of $y = a^{(x+c)}$ translates the graph of $y = a^x$ left by c units (in the negative x direction).
>
> The horizontal translation does not change the horizontal asymptote.

WORKED EXAMPLE 15 — Using translations to sketch an exponential function

Sketch the graph of $y = 2^{(x-2)} + 1$.

Label the coordinates of the y-intercept and the asymptote with its equation.

Steps	Working
1 The graph of $y = 2^x$ has a horizontal asymptote at $y = 0$ and a y-intercept at $(0, 1)$. The horizontal translation is 2 units to the right and the vertical translation is 1 unit up.	The horizontal asymptote is $y = 1$. The point $(0, 1)$ on $y = 2^x$ is translated 2 units right and 1 unit up to $(2, 2)$.
2 Find the y-intercept by substituting $x = 0$.	The y-intercept $(x = 0)$: $y = 2^{(0-2)} + 1$ $y = 2^{-2} + 1$ $y = \dfrac{1}{4} + 1 = \dfrac{5}{4}$ y-intercept $= \left(0, \dfrac{5}{4}\right)$
3 Sketch the graph. **Exam hack** Include the guiding point $(2, 2)$ to improve the accuracy of your sketch graph.	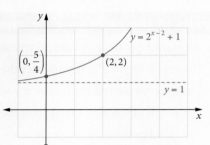

Finding equations of exponential functions

An exponential function of the form $y = a^{(x-c)} + b$ has an asymptote with the equation $y = b$.

Finding the equation of a particular exponential function is often a multi-step process, and every problem is different depending on the information we are given.

WORKED EXAMPLE 16 — Finding the rule of an exponential function from a graph

The rule for the function with the graph shown is of the form $y = 3^{(x-c)} + b$.

Find the values of b and c and state the equation of the function.

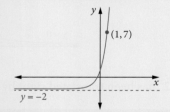

Steps	Working
1 Using the horizontal asymptote find the value of b.	$y = -2$ is the horizontal asymptote. Therefore, $b = -2$.
2 Substitute $(1, 7)$ into the equation to find c.	$y = 3^{(x-c)} - 2$ $7 = 3^{(1-c)} - 2$ $3^{(1-c)} = 9$ $3^{(1-c)} = 3^2$ $1 - c = 2$ $c = -1$ $c = -1, b = 2$ $y = 3^{(x+1)} - 2$

WORKED EXAMPLE 17 — Finding the rule of an exponential function given two points

The exponential function $y = 2^{(x-c)} + b$ passes through the points $(6, 5)$ and $(8, 11)$.
Find the values of b and c.

Steps	Working
1 Use the coordinates $(6, 5)$ and $(8, 11)$ to write two simultaneous equations in terms of b and c.	$(6, 5)$ $5 = 2^{(6-c)} + b$ [1] $(8, 11)$ $11 = 2^{(8-c)} + b$ [2]
2 Solve the simultaneous equations using CAS.	$b = 3, c = 5$

ClassPad

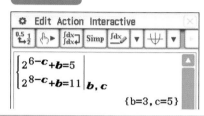

TI-Nspire

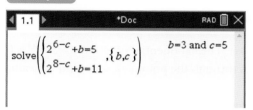

EXERCISE 6.4 The exponential function $y = a^x$

ANSWERS p. 475

Recap

1 Solve for x.

  a $25^x = 125$ **b** $8^x = \dfrac{1}{4}$

2 Solve $8^{x+1} \times 4^{x+2} = 32 \times 16^x$.

Mastery

3 **Using CAS 1**   Use CAS to graph each function. Comment on the features of the graph.

  a $y = 8.5 \times 2.5^x$ for $-3 \le x \le 3$ **b** $y = 16 \times 2^x$ for $-5 \le x \le 5$

4 **WORKED EXAMPLE 15**   For each function, find the coordinates of the y-intercept and the equation of the asymptote, then sketch the graph of the function.

  a $y = 4^x - 3$ **b** $y = 8^x - 1$ **c** $y = 5^{x-2} + 1$

5 Match each function to one of the graphs in the options below.

  a $y = 2^x + 3$ **b** $y = 2^x$ **c** $y = 2^x - 3$

A
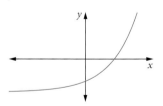

B
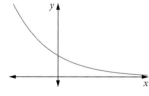

C
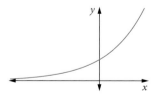

D
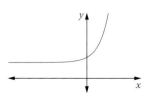

E
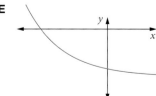

6 **WORKED EXAMPLE 16** The rule for the function with the graph shown is of the form $y = 2^{(x-c)} + b$.
Find the values of b and c and state the equation of the function.

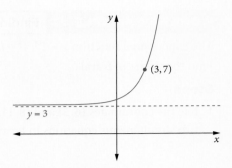

7 **WORKED EXAMPLE 17** The exponential function $y = 2^{(x-c)} + b$ passes through the points $(-3, -8)$ and $(1, 22)$. Find the values of b and c.

Calculator-free

8 (4 marks) The exponential function $y = 5^{(x-c)} + b$ has a horizontal asymptote of $y = -4$ and passes through the point $(-2, -3)$. Find the values of b and c.

9 (2 marks) The function $f(x) = 4^x$ is translated horizontally 5 units to the right and vertically 3 units down to produce a new function, $g(x)$. Write the rule for $g(x)$.

10 (4 marks) Let $f(x) = 2^{x+1} - 2$. Part of the graph of f is shown below.

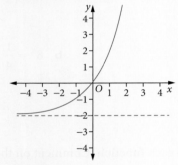

    a   **i** Find the equation of the asymptote. (1 mark)

        ii Find the coordinates of the y-intercept. (1 mark)

    b State the translation that maps the graph of $y = 2^x$ onto the graph of $f(x)$. (2 marks)

Calculator-assumed

11 (5 marks) The exponential function $y = 3^{(x-c)} + b$ passes through the points $(-4, -3)$ and $(-5, -9)$.

    a Find two equations in terms of b and c. (2 marks)

    b Find the values of b and c. (2 marks)

    c Find the equation of the asymptote. (1 mark)

12 (5 marks) Let $f(x) = a^x$ where $a > 0$. The graph of f passes through the point $(5, 32)$.

    a Show that $a = 2$. (1 mark)

A horizontal translation 3 units left, followed by a vertical translation 4 units up, maps $f(x)$ onto $g(x)$.

    b Find the rule for $g(x)$. (2 marks)

    c If $f(7) = 128$, find $g(4)$. (2 marks)

6.5 Exponential growth and decay

An epidemic occurs when new cases of a disease appear at a much greater rate than normally expected. Famous examples of epidemics include the bubonic plague (Black Death) in 14th century Europe, the influenza (with misnomer of Spanish flu) in 1918–19, the sudden acute respiratory syndrome (SARS) in Asia in 2003 and the coronavirus disease (COVID-19) in 2019.

One of the reasons influenza spreads so quickly is that the flu virus multiplies rapidly. Suppose a culture containing 20 mg of flu virus increases by 30% every 10 minutes. Growth at this rate for an hour is shown in the following table.

10-minute periods	Mass of virus (mg)
0	20
1	20×1.3
2	$20 \times 1.3 \times 1.3 = 20 \times (1.3)^2$
3	$20 \times (1.3)^2 \times 1.3 = 20 \times (1.3)^3$
4	$20 \times (1.3)^3 \times 1.3 = 20 \times (1.3)^4$
5	$20 \times (1.3)^4 \times 1.3 = 20 \times (1.3)^5$
6	$20 \times (1.3)^5 \times 1.3 = 20 \times (1.3)^6$

Video playlist
Exponential growth and decay

Adding an increase of 30% is the same as multiplying by 130% or 1.3.

After an hour, the virus culture has grown from 20 mg to $20 \times (1.3)^6 \approx 97$ mg, almost 5 times its original amount.

From the table, it can be seen that the mass in mg of the virus can be modelled as

$$M(t) = 20 \times (1.3)^t$$

where t is the number of 10-minute periods.

$M(t) = 20 \times (1.3)^t$ is an example of an exponential function, where the variable is a power. Recall that the **exponent** is another name for the power or index.

Exponential growth and decay

> ### Exponential growth and decay
> The general form of an exponential growth or decay function is
> $$A(t) = A_0(a^t)$$
> where the y-intercept, A_0, is the initial value when $t = 0$ and a is the growth factor.
> For a growth function, $a > 1$.
> For a decay function, $0 < a < 1$.

Exponential growth

$A(t) = A_0(a^t), a > 1$

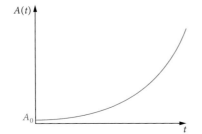

Exponential decay

$A(t) = A_0(a^t), 0 < a < 1$

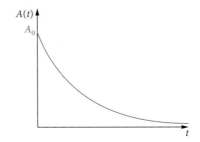

Any quantity that **increases** according to the exponential function $y = A(a^x)$, where $a > 1$, is showing **exponential growth**. Something that grows exponentially increases slowly at first, then more quickly. Examples of exponential growth are population size (people, animals, bacteria), an investment attracting compound interest, the size of a bushfire and a flu or computer virus.

Any quantity that **decreases** according to the exponential function $y = A(a^x)$, where a is between 0 and 1, is showing **exponential decay**. Something that decays exponentially decreases quickly at first, then more slowly. Examples of exponential decay are radioactive decay, the cooling of substances, the intensity of light in water and the dampening of vibrations.

Significant figures and scientific notation

Exponential growth calculations can often produce very large numbers, while exponential decay calculations can produce very small numbers. Consider the situation where the number of bacteria N, after t hours is modelled by the function $N(t) = 200 (2.5)^t$. If we calculate the number of bacteria after 30 hours we get $N(30) = 173\,472\,347\,597\,681$.

There are 15 digits in the answer, which can be cumbersome and time consuming to write and we cannot claim that amount of accuracy in the answer for this calculation. Scientific notation and significant figures show us ways of addressing this issue.

Rounding to significant figures

Rounding is an important skill when dealing with real-life data. In practical situations, it often makes sense to round an answer to a number of **significant figures** rather than to a number of decimal places, particularly when a combination of large whole numbers and decimals are involved. We need to be able to do both methods of rounding.

In the following examples, the significant figures are in red.

Example	No. of significant figures	What type of digits are significant?	What type of digits are *not* significant?
367.268	6	non-zero digits	
190.0043	7	zeros between non-zero digits	
3.540	4	trailing zeros in decimals	
0.0043	2		leading zeros in decimals
1390	3		trailing zeros in whole numbers

When rounding to significant figures, use the same rounding rules as for rounding to a number of decimal places:

- '0–4 round down' and '5–9 round up'

    e.g. 27 501 rounded to two significant figures is 28 000
- round a 9 to 0 and carry the rounding over to the next digit on the left

    e.g. 3.9722 rounded to two significant figures is 4.0

    497 rounded to two significant figures is 500.

> **Significant figures**
>
> Significant figures:
> - any non-zero digit
> - zeros between non-zero digits
> - trailing zeros in decimals.
>
> Not significant figures:
> - leading zeros in decimals
> - trailing zeros in whole numbers.
>
> When rounding to significant figures, use usual rounding rules:
> - '0–4 round down' and '5–9 round up'
> - round the 9 to 0 and carry the rounding over to the next digit on the left
> - include trailing zeros in decimals, if necessary.

WORKED EXAMPLE 19	Rounding a number to significant figures

Round the following numbers to the given number of significant figures:

a 1.333 to two significant figures

b 14 700 to two significant figures

c 12.8989 to four significant figures.

Steps	Working
a Focus on the first two significant figures.	1.333 rounded to two significant figures is 1.3.
b Focus on the first two significant figures.	14 700 rounded to two significant figures is 15 000.
c Focus on the first four significant figures.	12.8989 rounded to four significant figures is 12.90.

Scientific notation

Scientific notation is used to write very large or very small numbers. These occur regularly in calculations involving exponential growth and decay.

Scientific notation is a number between 1 and 10 multiplied by a power of 10.

For example:

    265 000, when written in scientific notation, is 2.65×10^5

    0.000 076, when written in scientific notation, is 7.6×10^{-5}.

The power tells us how many places and in which direction to move the decimal point.

WORKED EXAMPLE 20	Converting numbers to and from scientific notation

a Write 8 120 000 in scientific notation.

b Write 0.008 15 in scientific notation.

c Write 4.5×10^3 in decimal notation.

d Write 9.12×10^{-2} in decimal notation.

Steps	Working
a Write a number between 1 and 10 multiplied by a power of 10. The decimal point is moved left 6 places, so the power of 10 is 6.	$8\,120\,000 = 8.12 \times 10^6$
b Write a number between 1 and 10 multiplied by a power of 10. The decimal point is moved right 3 places, so the power of 10 is −3.	$0.008\,15 = 8.15 \times 10^{-3}$
c The power of 10 is 3 so move the decimal point 3 places to the right.	$4.5 \times 10^3 = 4500$
d The power of 10 is −2 so move the decimal point 2 places to the left.	$9.12 \times 10^{-2} = 0.0912$

Worksheet Using exponential models

WORKED EXAMPLE 21 — Exponential growth modelling

The number of bacteria N in a colony after t days is given by the function $N(t) = N_0(a^t)$.

A colony starts with 2000 bacteria and the number of bacteria triples every day. Assuming that growth continues in this way, find how many bacteria, correct to two significant figures, will be in the colony

a after 5 days **b** after 60 hours.

Steps	Working
This is an example of exponential growth.	
The number of bacteria triples every day, or the new number of bacteria is 300% of the previous number.	growth factor $a = 3$
The initial number is 2000.	initial number $N_0 = 2000$
Write the growth function for the number of bacteria after t days.	$N(t) = 2000 \times 3^t$
a For the number of bacteria after 5 days, substitute $t = 5$. Write the answer to two significant figures.	$N(5) = 2000(3^5)$ $= 486\,000$ There will be approximately 490 000 bacteria after 5 days.
b For the number of bacteria after 60 hours, convert to days first. 60 hours $= \dfrac{60}{24} = 2\dfrac{1}{2}$ days, so substitute $t = 2.5$.	$N(2.5) = 2000(3^{2.5})$ $= 31\,176.914\ldots$ There will be about 31 000 bacteria after 60 hours.

WORKED EXAMPLE 22 — Exponential decay modelling

The amount of air left (and thus pressure) P, in a leaking spaceship, after t minutes is given by the rule $P(t) = P_0(a^t)$. When the leak first occurred, the air pressure was 105 kPa (kilopascals) and the pressure is decreasing by 30% every minute. What will be the pressure (correct to one decimal place) after 6 minutes?

Steps	Working
1 This is an example of exponential decay. The pressure decreases by 30% every minute, or the new pressure is 70% of the previous value.	growth factor $a = 0.7$
The initial value, P_0, is 105 kPa.	$P_0 = 105$
2 Write the decay function for pressure after t minutes.	$P(t) = 105 \times (0.7)^t$
3 For the air pressure after 6 minutes, substitute $t = 6$.	$P(6) = 105(0.7)^6$ ≈ 12.4 After 6 minutes, the pressure will be about 12.4 kPa.

WORKED EXAMPLE 23 — Further exponential decay modelling

The difference between the temperature of a cup of coffee and its surroundings decreases by 4% every minute. A fresh cup has a temperature of 70°C, and the room temperature is 20°C. A cup of coffee is considered too cold to drink if it is below 40°C.

The temperature of the cup of coffee is given by the function $T(t) = T_s + (T_0 - T_s)(a^t)$, where T_s is the room temperature, T_0 is the initial temperature of the coffee and T is the temperature of the coffee cup after t minutes.

Find

a the function $T(t)$

b how long (to the nearest minute) it takes to for the coffee to become undrinkable.

322 Nelson WAmaths Mathematics Methods 11

Steps	Working
a The temperature of the coffee decreases by 4% each minute, so $a = 96\% = 0.96$. Use $T(t) = T_s + (T_0 - T_s)(a^t)$ where $T_s = 20$, $T_0 = 70$ and $a = 0.96$.	$T(t) = T_s + (T_0 - T_s)(a^t)$ $T(t) = 20 + (70 - 20)(0.96^t)$ $T(t) = 20 + 50(0.96^t)$
b Use CAS to solve $T(t) = 40$.	It takes about 22 minutes for the coffee to become undrinkable.

ClassPad

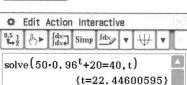

TI-Nspire

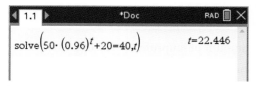

EXAMINATION QUESTION ANALYSIS

Calculator-assumed (10 marks)

An adult takes 400 mg of ibuprofen. After 1 hour, the amount of ibuprofen in the person's system is 300 mg. The number of mg of the drug D in the patient's system after t hours is modelled by the function $D(t) = D_0(a^t)$.

a Find the value of D_0. (1 mark)

b Find the percentage decrease in the amount of ibuprofen after 1 hour. (2 marks)

c Show that $a = 0.75$. (1 mark)

d Find the number of mg of the drug in the patient's system, correct to three significant figures, after 4 hours. (2 marks)

e Using the function $D(t)$, determine if there will ever be no ibuprofen in the person's blood stream. (1 mark)

f Find the amount of time, to the nearest minute, when the amount of ibuprofen in the person's blood stream has halved. (3 marks)

Reading the question

- The problem relates to exponential decay.
- Highlight the parts of the question that relate to the type of answer required. This may be a value, percentage or a function.
- Take note of which questions will require working to be shown, and the specified accuracy of the answers.
- Note, in the context of the question, the significance of keywords and phrases such as *using the function* and *show*.

Thinking about the question

- This question requires an understanding of exponential decay.
- You will need to know how each part of the formula relates to the problem.
- You will also need to pay particular attention to the unit and accuracy of answers required. Take into account instructions such as *answer to the nearest minute* or *significant figures* as full marks will only be given when all aspects of your answer are correct.
- Define any given function on CAS at the start of the question. You can use this defined function to calculate values, solve equations and sketch.

Worked solution (✓ = 1 mark)

a $D_0 = 400$ ✓

b decrease = $400 - 300 = \mathbf{100}$ ✓

percentage decrease = $\dfrac{100}{400} \times 100\% = \mathbf{25\%}$ ✓

c $a = 100\% - 25\% = \mathbf{75\%}$ ✓
$a = 0.75$

d $D(t) = D_0(a^t)$
$D(t) = 400(0.75^t)$
$D(4) = 400(0.75^t) = \mathbf{126.5625}$ ✓

The amount of ibuprofen is **127 mg**. ✓

e The function $D(t)$ has a **horizontal asymptote at $D = 0$** ✓
so there will never be zero mg of ibuprofen.

f Solve $D(t) = \dfrac{1}{2} \times 400 = \mathbf{200}$ ✓
$t = \mathbf{2.40942\ldots}$ ✓

It would take **2 hours 25 minutes** ✓ for the drug in the bloodstream to halve.

ClassPad

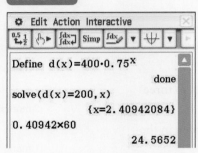

TI-Nspire

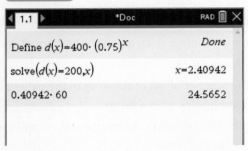

EXERCISE 6.5 Exponential growth and decay

ANSWERS p. 476

Recap

1 The exponential function $y = a^x$ passes through the point $(4, 81)$. Find the value of a.

2 Find the equation of the horizontal asymptote and the coordinates of the y-intercept of $y = 4^{x+2} - 8$.

Mastery

3 WORKED EXAMPLE 19 Round each number to two significant figures.

    a 7.421   **b** 12.919   **c** 0.363   **d** 1800

    e 20.666   **f** 72.0037   **g** 9.7962

4 WORKED EXAMPLE 20

    a Write the following numbers in scientific notation.

        i 436 000   **ii** 0.000 082   **iii** 13.5

    b Convert the following numbers from scientific notation to decimal notation.

        i 5.56×10^3 **ii** 8.6×10^{-4} **iii** 4.0×10^{-1}

5 For each exponential function, find the
   i initial amount
   ii growth factor.
   a $f(x) = 45 \times (1.15)^x$
   b $f(t) = 1100 \times (0.82)^t$

6 **WORKED EXAMPLE 21** An algal bloom in a polluted lake covers an area of $3\,m^2$. The bloom grows in size by 5% every day. The area of algal bloom A (m^2) after t days is given by $A(t) = A_0(a^t)$.
   a Find the function $A(t)$.
   b Find the area covered by the algal bloom, correct to three significant figures, after
      i 3 days   ii 10 days   iii 20 days.
   c Find when the algal bloom will cover $500\,m^2$.

7 **WORKED EXAMPLE 22** The value of a car, originally worth $24\,900, depreciates by 15% each year. The value of the car, V, after t years is modelled by the function $V(t) = V_0(a^t)$.
   a Find $V(t)$.
   b Find the value of the car after
      i 3 years   ii 5 years   iii 10 years.
   c Find the time taken for its value to reduce to $1000.

8 **WORKED EXAMPLE 23** A kiln is heated to a temperature of 500°C. The temperature of the room is 20°C. When the kiln is switched off, its temperature T decreases by 10% each minute. The temperature of the kiln is given by the function $T(t) = T_s + (T_0 - T_s)(a^t)$, where T_s is the room temperature, T_0 is the initial temperature of the kiln and T is the temperature of the kiln after t minutes.
   a Find the function $T(t)$.
   b Find the temperature of the kiln, correct to four significant figures, after
      i 5 minutes   ii 10 minutes   iii 25 minutes.
   c Find the time taken for the kiln to cool to a temperature of 400°C.

9 Given the following table of values,

m	0	1	2	3	4
p	3	6	12	24	48

   a find the rule for p as a function of m
   b find the value of p when $m = 8$.

Calculator-free

10 (3 marks) The population of Margaret River can be modelled by $P(t) = 6191(1.04)^t$, where t is the number of years since 1990.
   a What was the population in 1990? (1 mark)
   b By what percent did the population increase by each year? (2 marks)

11 (4 marks) The resale value, V, of a car t years after purchase is given $V(t) = 10\,000(0.8)^t$. Find the
   a purchase price of the car (1 mark)
   b percentage decrease in the value of the car each year (1 mark)
   c value of the car after one year. (2 marks)

Calculator-assumed

12 (8 marks) An investment of $3000 at compound interest grows by 8% every year. The value of the investment, V, after t years is modelled by the function $V(t) = V_0(a^t)$.

    a Find $V(t)$. (3 marks)

    b Find the value of the investment after

        i 3 years (1 mark)

        ii 10 years (1 mark)

        iii 25 years. (1 mark)

    c Find the time taken for the investment to double in value. (2 marks)

13 (2 marks) Find the bank account balance if the account starts with $100, has an annual compound interest rate of 4% and the money is left in the account for 12 years.

14 (2 marks) Bacteria can multiply at an alarming rate. When each bacteria splits into 2 new cells, the number of bacteria doubles. If we start with only one bacteria, which can double every hour, how many bacteria will there be by the end of one day? Write your answer in scientific notation, correct to three significant figures.

15 (2 marks) In 1985, there were 285 mobile phone subscribers in the small town of Centreville. The number of subscribers increased by 75% per year after 1985. How many mobile phone subscribers were in Centreville in 1994?

Chapter summary

The index laws

First index law: Multiply exponential terms with same base $a^m \times a^n = a^{m+n}$

Second index law: Divide exponential terms with same base $a^m \div a^n = a^{m-n}$

Third index law: Raise an exponential term to a power $(a^m)^n = a^{mn}$

                          Power of zero $a^0 = 1$

where $a > 0$, $a \neq 1$.

Powers of prime numbers

Powers of 2	
2^0	1
2^1	2
2^2	4
2^3	8
2^4	16
2^5	32
2^6	64

Powers of 3	
3^0	1
3^1	3
3^2	9
3^3	27
3^4	81
3^5	243
3^6	729

Powers of 5	
5^0	1
5^1	5
5^2	25
5^3	125
5^4	625
5^5	3125

Negative powers $a^{-n} = \dfrac{1}{a^n}$

Radical form: $\sqrt[n]{a}$ Exponential form: $a^{\frac{1}{n}}$

Square root: $\sqrt{a} = a^{\frac{1}{2}}$

Cube root: $\sqrt[3]{a} = a^{\frac{1}{3}}$

nth root: $\sqrt[n]{a} = a^{\frac{1}{n}}$ $\sqrt[n]{a^x} = a^{\frac{x}{n}}$

Properties of the graph of the exponential function $y = a^x$ where $a > 1$

- The function is always positive.
- The gradient of the graph is always positive.
- As x increases the gradient also increases.
- The y-intercept is at $(0, 1)$ (because $a^0 = 1$).
- The horizontal asymptote is $y = 0$ (the x-axis).

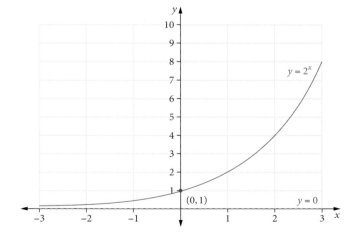

Graphs of the form $y = a^x + b$ where $b > 0$

The constant b produces a vertical translation of $y = a^x$ in the direction of the y-axis.
- The graph of $y = a^x + b$ translates the graph of $y = a^x$ b units up.
  The new horizontal asymptote is $y = b$ and the new y-intercept is $(1 + b, 0)$.
- The graph of $y = a^x - b$ translates the graph of $y = a^x$ b units down.
  The new horizontal asymptote is $y = -b$ and the new y-intercept is $(1 - b, 0)$.

Graphs of the form $y = a^{(x-c)}$ where $c > 0$

The constant c produces a horizontal translation in the direction of the x-axis.
- The graph of $y = a^{(x-c)}$ translates the graph of $y = a^x$ right by c units (in the positive x direction).
- The graph of $y = a^{(x+c)}$ translates the graph of $y = a^x$ left by c units (in the negative x direction).

Exponential growth and decay

The general form of an **exponential growth** or **decay** function is $A(t) = A_0(a^t)$, where the y-intercept, A_0, is the initial value when $t = 0$ and a is the growth factor.

For a growth function, $a > 1$.

For a decay function, $0 < a < 1$.

Significant figures and scientific notation

Significant figures	
Significant figures: • any non-zero digit • zeros between non-zero digits • trailing zeros in decimals.	Not significant figures: • leading zeros in decimals • trailing zeros in whole numbers.
When rounding to significant figures, use usual rounding rules: • '0–4 round down' and '5–9 round up' • round a 9 to 0 and carry the rounding over to the next digit on the left • include trailing zeros in decimals, if necessary.	

Scientific notation is used to write very large or very small numbers. These occur regularly in calculations involving exponential growth and decay.

Scientific notation is a number between 1 and 10 multiplied by a power of 10.

Cumulative examination: Calculator-free

Total number of marks: 32   Reading time: 4 minutes   Working time: 32 minutes

1 (7 marks) Consider the four sets defined below.

$A = \{1, 2, 3, 4\}$ $B = \{3, 4, 5, 6\}$ $C = \{4, 5, 6, 7\}$ $D = \{6, 7, 8\}$

  a Use set brackets to list the elements of the following sets.

    i $A \cap B$ (1 mark)

    ii $B \cup D$ (1 mark)

    iii $A \cap B \cap C$ (1 mark)

    iv $B \cap C \cap \bar{A}$ (1 mark)

  b State $n(B \cup C \cup D)$. (1 mark)

  c How many different sets containing two elements can be formed by selecting one element from set A and one element from set D? (2 marks)

2 (4 marks) Factorise the following expressions.

  a $x^2 - 8x - 33$ (1 mark)

  b $12x^2 - 60x + 75$ (2 marks)

  c $81x^2 - 10y^4$ (1 mark)

3 (3 marks) Sketch the graph of $f(x) = 2 - \dfrac{1}{x+3}$, showing all intercepts and the equations of any asymptotes.

4 (3 marks) Let $f(x) = -2\cos\left(\dfrac{\pi x}{2}\right)$. State the period, amplitude and range.

5 (3 marks) For the function $f(x) = 16(4)^{2x} - 1$, find

  a $f(0)$ (1 mark)

  b x when $f(x) = 0$. (2 marks)

6 (6 marks) Simplify the following expressions.

  a $\dfrac{(5a^4 b^5)^2 \times 15a^4 b^{13}}{(15ab^5)^2}$ (3 marks)

  b $\dfrac{12^{1-n} \times 9^{4n-1}}{8^{3-n}}$ (3 marks)

7 (6 marks)

  a For the function $y = 2^{x+2} + 3$, find the

    i coordinates of the y-intercept (2 marks)

    ii equation of the asymptote. (1 mark)

  b Sketch the graph, labelling important features. (3 marks)

Cumulative examination: Calculator-assumed

Total number of marks: 35   Reading time: 4 minutes   Working time: 35 minutes

1 (10 marks) On Michael's bookshelf, there are three different types of books: 6 reference books, 14 fiction books and 5 non-fiction books. Michael wants to randomly select three books to donate to the local high school.

   a Determine the total number of possible combinations of three books that Michael can donate. (2 marks)

   b Determine the number of possible combinations of three books that Michael can donate, if he selects one of each type. (2 marks)

   c Calculate the probability, to four decimal places, that Michael randomly selects

   i all three books as fiction books (2 marks)

   ii at least one non-fiction book (2 marks)

   iii one of each type, given there is at least one non-fiction book. (2 marks)

2 (3 marks) Sketch the parabola with the equation $y = -2x^2 + 7x - 4$, labelling all axes intercepts and the turning point with its coordinates.

3 (10 marks) A chef needs to use an oven to boil 100 mL of water in five minutes for a new experimental recipe. The temperature of the water must reach 100°C in order to boil. The temperature, T, of 100 mL of water t minutes after being placed in an oven set to T_0°C can be modelled by the equation

$$T(t) = T_0 - 175(0.93^t)$$

In a preliminary experiment, the chef placed a 100 mL bowl of water into an oven that had been heated to $T_0 = 200$°C.

   a What is the temperature of the water at the moment it is placed into the oven? (1 mark)

   b What is the temperature of the water five minutes after being placed in the oven? (2 marks)

   c How many minutes does it take, correct to three significant figures, for the water to heat to 50°C? (2 marks)

   d How long does it take, correct to three significant figures, for the water to reach the required temperature of 100°C? (2 marks)

   e Determine the temperature the oven must be preheated to, in order to boil 100 mL of water in five minutes. (3 marks)

4 (5 marks) The exponential function $y = 2^{(x-c)} + b$ passes through the points $(9, 1)$ and $(11, 13)$.

   a Find two equations in terms of b and c. (2 marks)

   b Find the values of b and c. (2 marks)

   c Find the equation of the asymptote. (1 mark)

5 (7 marks) Let $f(x) = a^x$, where $a > 0$. The graph of f passes through the point $(4, 81)$.

   a Find the value of a. (1 mark)

A horizontal translation of 4 units right, followed by a vertical translation of 3 units up, maps $f(x)$ onto $g(x)$.

   b Find the rule for $g(x)$. (2 marks)

   c If $f(3) = 27$, find $g(7)$. (2 marks)

   d If $f(m) = n$, find $g(m + 4)$ in terms of n. (2 marks)

CHAPTER 7
ARITHMETIC AND GEOMETRIC SEQUENCES AND SERIES

Syllabus coverage
Nelson MindTap chapter resources

7.1 Arithmetic sequences
Arithmetic sequences as numerical lists
The recursive rule for arithmetic sequences
The general (explicit) rule for arithmetic sequences
Using CAS 1: Determining the terms of an arithmetic sequence
Tabular and graphical forms of arithmetic sequences

7.2 Modelling linear growth and decay
Modelling linear growth
Modelling linear decay

7.3 Arithmetic series and their applications
The sum of arithmetic sequences
The arithmetic series formulas
Solving practical problems involving arithmetic series
Using CAS 2: Finding the sum of an arithmetic sequence

7.4 Geometric sequences
Geometric sequences as numerical lists
The recursive rule for geometric sequences
The general (explicit) rule for geometric sequences
Using CAS 3: Determining the nth term of a geometric sequence
Tabular and graphical forms of geometric sequences

7.5 Modelling exponential growth and decay
Modelling exponential growth, $r > 1$
Modelling exponential decay, $0 < r < 1$

7.6 Geometric series and their applications
Expressing the sum of geometric sequences
The finite geometric series formula
The infinite geometric series formula
Solving practical problems involving geometric series
Using CAS 4: Solving problems with geometric sequences and series

Examination question analysis
Chapter summary
Cumulative examination: Calculator-free
Cumulative examination: Calculator-assumed

Syllabus coverage

TOPIC 2.2: ARITHMETIC AND GEOMETRIC SEQUENCES AND SERIES

Arithmetic sequences

2.2.1 recognise and use the recursive definition of an arithmetic sequence: $t_{n+1} = t_n + d$

2.2.2 develop and use the formula $t_n = t_1 + (n-1)d$ for the general term of an arithmetic sequence and recognise its linear nature

2.2.3 use arithmetic sequences in contexts involving discrete linear growth or decay, such as simple interest

2.2.4 establish and use the formula for the sum of the first n terms of an arithmetic sequence

Geometric sequences

2.2.5 recognise and use the recursive definition of a geometric sequence: $t_{n+1} = t_n r$

2.2.6 develop and use the formula $t_n = t_1 r^{n-1}$ for the general term of a geometric sequence and recognise its exponential nature

2.2.7 understand the limiting behaviour as $n \to \infty$ of the terms t_n in a geometric sequence and its dependence on the value of the common ratio r

2.2.8 establish and use the formula $S_n = t_1 \dfrac{r^n - 1}{r - 1}$ for the sum of the first n terms of a geometric sequence

2.2.9 use geometric sequences in contexts involving geometric growth or decay, such as compound interest

Mathematics Methods ATAR Course Year 11 syllabus p. 13 © SCSA

Video playlists (7):

7.1 Arithmetic sequences
7.2 Modelling linear growth and decay
7.3 Arithmetic series and their applications
7.4 Geometric sequences
7.5 Modelling exponential growth and decay
7.6 Geometric series and their applications

Examination question analysis Arithmetic and geometric sequences and series

Worksheets (6):

7.1 Arithmetic sequences • Arithmetic progressions
7.2 Modelling with sequences
7.4 Geometric sequences • Geometric progressions
7.5 Modelling with sequences

Nelson MindTap

To access resources above, visit
cengage.com.au/nelsonmindtap

7.1 Arithmetic sequences

Arithmetic sequences as numerical lists

In your early Mathematics learning, you may have been asked to start at the number 1 and count by 2s, or to identify a pattern in the numbers 16, 12, 8, 4 … This very idea is the idea of a **sequence**. A mathematical sequence is a pattern of numbers and each number in the sequence is called a **term**. There are many types of sequences; however, in Year 11 Mathematics Methods, we only examine two types.

The sequences 1, 3, 5, 7 … and 16, 12, 8, 4 … are examples of **arithmetic sequences**. An arithmetic sequence is a sequence of numbers that have a **common difference** d; that is, a constant addition or subtraction pattern. The sequence 1, 3, 5, 7 … has a common difference of +2; 16, 12, 8, 4 … has a common difference of −4. When describing an arithmetic sequence, however, we cannot just rely on the common difference, we also need to know where to start. This is called the **first term** or **initial term**, a.

Video playlist
Arithmetic sequences

Worksheets
Arithmetic sequences

Arithmetic progressions

WORKED EXAMPLE 1 — Describing arithmetic sequences in words

Describe the following sequences:

a 5, 8, 11, 14 …

b 150, 100, 50, 0 …

Steps	Working
a 1 Identify the common difference and first term.	$d = 3$, $a = 5$
2 Write a statement describing the sequence.	It is an arithmetic sequence with a first term of 5 and a common difference of 3.
b 1 Identify the common difference and first term.	$d = -50$, $a = 150$
2 Write a statement describing the sequence.	It is an arithmetic sequence with a first term of 150 and a common difference of −50.

From these two examples, we can see that an arithmetic sequence can either be an **increasing sequence** or a **decreasing sequence**.

The recursive rule for arithmetic sequences

Describing sequences in words is time consuming, so we need an efficient way to name and represent the worded description using an appropriate notation. A sequence is often named by a letter, for example A or b, and subscript notation is used to represent the placement of a term in that sequence.

For example, let the sequence 1, 3, 5, 7 … be called A. The notation … A_{n-1}, A_n, A_{n+1} … can be used to represent consecutive terms of this sequence, where A_{n+1} is the 'next term' after A_n and A_n is the 'previous term' before A_{n+1}. For example,

$A_1 = 1, A_2 = 3, A_3 = 5, A_4 = 7$ …

To generalise a rule, we first want to think of a way to describe how to get from the previous term to the next term. In other words, we could say something like 'the next term is 2 more than the previous term, where the first term is 1'. This is called a recursive definition of a sequence; that is, a rule that describes how to get from one term to the next. As a result, the **recursive rule** for this sequence is $A_{n+1} = A_n + 2$, $A_1 = 1$.

The recursive rule – arithmetic sequences

Let T be an arithmetic sequence with a first term $T_1 = a$ and a common difference of d.

The recursive rule for such a sequence can be expressed as

$T_{n+1} = T_n + d, T_1 = a$

If $d > 0$, the sequence is increasing.

If $d < 0$, the sequence is decreasing.

Note that this is no different to

$T_n = T_{n-1} + d, T_1 = a$ or $T_{n+2} = T_{n+1} + d, T_1 = a$

as long as it is the 'next term' defined by the 'previous term'.

Exam hack

Remember, the first term is important in recursive rules, because if you do not know where to start, you cannot generate the sequence!

WORKED EXAMPLE 2 — Generating the terms of a sequence from a recursive rule

List the first five terms of the following sequences.

a $T_{n+1} = T_n + 5, T_1 = 4$

b $B_n = B_{n-1} - 10, B_1 = 105$

Steps	Working
a 1 Identify the first term and the common difference.	$a = 4, d = 5$
2 List the first five terms, starting at T_1 and adding 5 each time.	4, 9, 14, 19, 24
b 1 Identify the first term and the common difference.	$a = 105, d = -10$
2 List the first five terms, starting at B_1 and subtracting 10 each time.	105, 95, 85, 75, 65

WORKED EXAMPLE 3 — Writing a recursive rule from sufficient information

Use the following information to write a recursive rule for each of the sequences.

a The next term in sequence C is given by 6 less than the previous term, where the first term is 19.

b The first three terms of sequence X are 3.5, 3.9, 4.3.

Steps	Working
a 1 Identify the first term and the common difference.	$a = 19, d = -6$
2 Write a recursive rule using the correct letter for the name of the sequence.	$C_{n+1} = C_n - 6, C_1 = 19$
b 1 Identify the first term and the common difference.	$a = 3.5, d = 0.4$
2 Write a recursive rule using the correct letter for the name of the sequence.	$X_{n+1} = X_n + 0.4, X_1 = 3.5$

You may have noticed that an arithmetic sequence resembles a linear relationship due to the common difference between the terms, which represents the gradient of the linear relationship. As a result, we can use a similar approach to finding the gradient of a line given two points when finding the common difference of an arithmetic sequence given two terms.

> **Finding the common difference**
>
> Given two terms of an arithmetic sequence, $T_m = a$ and $T_n = b$, we can consider these as coordinates (m, a) and (n, b). Then, the common difference d can be found by
> $$d = \frac{b - a}{n - m} = \frac{T_n - T_m}{n - m}$$

WORKED EXAMPLE 4 — Writing a recursive rule from two non-consecutive terms

An arithmetic sequence has $T_1 = 42$ and $T_7 = 6$. Determine the recursive rule for the sequence.

Steps	Working
1 Find the common difference between the terms.	$d = \frac{6 - 42}{7 - 1}$ $d = -\frac{36}{6}$ $d = -6$
2 Identify the first term.	$T_1 = a = 42$
3 State the recursive rule of the sequence in the correct form.	$T_{n+1} = T_n - 6,\ T_1 = 42$

WORKED EXAMPLE 5 — Solving algebraic problems involving arithmetic sequences

An arithmetic sequence has the first three terms of $x + 3$, $2x - 1$, $x - 7$.
Determine the value of x and hence state the recursive rule for the sequence.

Steps	Working
1 Recognise that for an arithmetic sequence, $d = T_3 - T_2 = T_2 - T_1$	$2x - 1 - (x + 3) = x - 7 - (2x - 1)$
2 Solve the equation formed for x.	$x - 4 = -x - 6$ $2x = -2$ $x = -1$
3 Use x to deduce the terms of the sequence.	$2, -3, -8 \ldots$
4 State the recursive rule of the sequence in the correct form.	$T_{n+1} = T_n - 5,\ T_1 = 2$

The general (explicit) rule for arithmetic sequences

A recursive rule is efficient if we have one term and want to find the next term or a sequence of consecutive terms, but what if we had the first term and wanted to find the 50th term? Finding 49 consecutive terms using a recursive process is inefficient and we need for a rule for the nth term of an arithmetic sequence.

Recall the sequence A, with the first four terms of 1, 3, 5, 7 …

To go from $A_1 = 1$ to $A_4 = 7$, we can start at 1 and add 3 lots of the common difference of 2.

$A_4 = 1 + 3(2) = 7$

So, if we wanted to start from $A_1 = 1$ and find the 50th term, we would need to add 49 lots of the common difference of 2. That is,

$$A_{50} = 1 + 49(2) = 1 + 98 = 99$$

You should notice that the 'number of lots of the common difference' that needs to be added to the first term is always one less than the placement of the term being found. That is, to find the nth term, we add $n - 1$ lots of the common difference to the first term. This is called the **general rule** (or **explicit rule**) of an arithmetic sequence.

> **The general rule – arithmetic sequences**
>
> For an arithmetic sequence with a recursive rule
>
> $$T_{n+1} = T_n + d, \, T_1 = a$$
>
> the nth term of the sequence is found using the general rule
>
> $$T_n = a + (n - 1)d$$

The general rule can also be simplified such that

$$\begin{aligned} T_n &= a + (n - 1)d \\ &= a + dn - d \\ &= dn + (a - d) \end{aligned}$$

For example, with our previous sequence A, we would have the general rule

$$A_n = 1 + (n - 1)(2)$$

which can be simplified to

$$A_n = 1 + 2n - 2$$
$$A_n = 2n - 1$$

This form of a general rule should look very familiar to you as it resembles the equation of a straight line, $y = mx + c$, where y represents the value of the term in the sequence (A_n), x represents the term's placement in the sequence (n), m represents the common difference (d) and $+ c$ is the term before the first term ($a - d$), which can also be called the 0th term.

WORKED EXAMPLE 6 — Writing a rule for the nth term from different descriptions

Use the following information to write a simplified general rule for each of the sequences.

a $D_{n+1} = D_n - 6, \, D_1 = 19$

b $Y_4 = 3$ and $Y_9 = 28$

Steps	Working
a 1 Identify the first term and the common difference.	$a = 19, \, d = -6$
2 Use the general rule $D_n = a + (n - 1)d$ and simplify the result.	$D_n = 19 + (n - 1)(-6)$ $= 19 - 6n + 6$ $D_n = 25 - 6n$
b 1 Find the common difference using the two terms.	$d = \dfrac{28 - 3}{9 - 4} = \dfrac{25}{5} = 5$
2 Find a by working backwards from the closest term.	$a = Y_1 = Y_4 - 3d$ $a = 3 - 3(5)$ $a = -12$
3 State the simplified general rule.	$Y_n = -12 + (n - 1)(5)$ $Y_n = -12 + 5n - 5$ $Y_n = 5n - 17$

USING CAS 1 Determining the terms of an arithmetic sequence

Determine the first five terms of the following arithmetic sequences.

a $T_{n+1} = T_n + 4$, $T_1 = 11$

b $T_n = 5 - 9n$

ClassPad

a

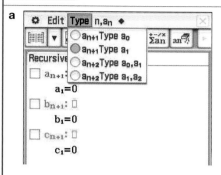

1. Open the **Sequence** application and tap on the **Recursive** tab.
2. Tap the **Type** menu and select a_{n+1}. **Type** a_1 to change the format.

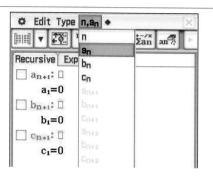

3. Tap the **n,a_n** menu and select a_n.

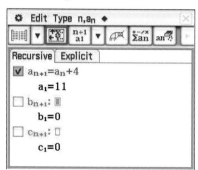

4. Enter **+4** and press **EXE**.
5. For a_1, enter **11** and press **EXE**.
6. Tap the **Sequence Table Input** tool.

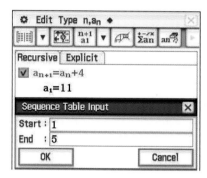

7. Set the **Start:** field to **1**.
8. Set the **End:** field to **5**.
9. Tap **OK**.

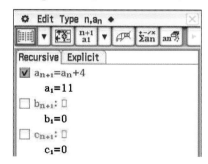

10. Tap the **Table** tool.

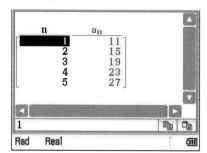

11. The first five terms of the sequence, 11, 15, 19, 23, 27, will be displayed in the lower window.

b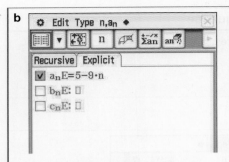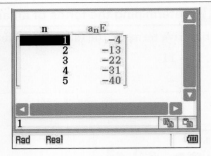

1. In the **Sequence** application, tap on the **Explicit** tab.
2. Enter the sequence **5−9n** and press **EXE**.
3. Tap the **Table** tool.
4. The first five terms of the sequence, −4, −13, −22, −31, −40, will be displayed in the lower window.

TI-Nspire

a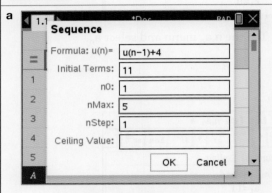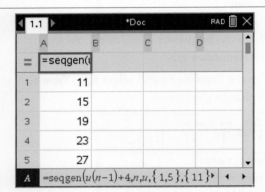

1. Add a **Lists & Spreadsheet** page.
2. Place the cursor in the cell immediately under the column **A** heading.
3. Press **menu > Data > Generate Sequence**.
4. Enter the expression and values into the fields as shown above and press **enter**.
5. The first five terms of the sequence, 11, 15, 19, 23, 27, will be displayed in column **A**.

Note the formula has to be input differently as $u(n-1)$ instead of $u(n)$.

b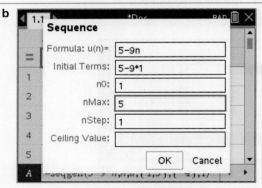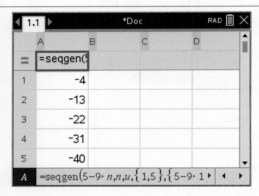

1. Add a **Lists & Spreadsheet** page.
2. Place the cursor in the cell immediately under the column **A** heading.
3. Press **menu > Data > Generate Sequence**.
4. Enter the expression and values into the fields as shown above and press **enter**.
5. The first five terms of the sequence, −4, −13, −22, −31, −40, will be displayed in column **A**.

a 11, 15, 19, 23, 27

b −4, −13, −22, −31, −40

Tabular and graphical forms of arithmetic sequences

Given that the terms of an arithmetic sequence can be considered discrete points on a linear relationship, we can also express the sequence in a tabular or graphical form. For example, the first 4 four terms of the sequence $A_n = 2n - 1$ can be represented as a table of values or as a plot.

A table of values

n	1	2	3	4
A_n	1	3	5	7

A plot

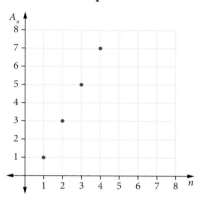

> 🔒 **Exam hack**
>
> It is important to note that you should not draw a straight line through the points, as the sequence is only defined for integer values of n.

WORKED EXAMPLE 7 — Plotting the terms of a sequence from a rule

Consider the sequence $T_{n+1} = T_n + 4$, $T_1 = 7$.

a Complete the table of values for T_n.

n	1	2	3	4	5
T_n					

b Hence, plot the first five terms of the sequence.

Steps	Working
a Use the first term and common difference to complete the table of values.	<table><tr><td>n</td><td>1</td><td>2</td><td>3</td><td>4</td><td>5</td></tr><tr><td>T_n</td><td>7</td><td>11</td><td>15</td><td>19</td><td>23</td></tr></table>
b Plot the set of points using the coordinates (n, T_n).	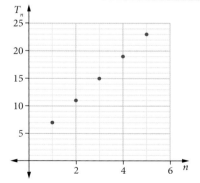

We can also use CAS to display a plot of a sequence after obtaining the table of values.

ClassPad

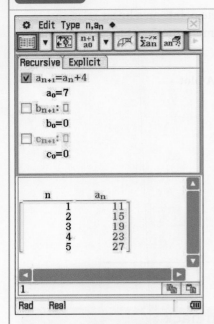 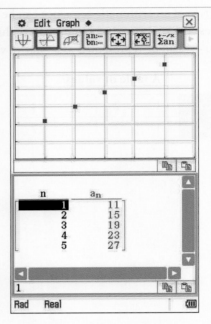

1. Open the **Sequence** application.
2. Enter the sequence $a_{n+1} = a_n + 4$, $a_0 = 7$ as shown above.
3. Tap the **Table** tool.
4. The sequence will be displayed in the lower window.
5. Tap in the lower window to highlight it.
6. Tap the **G-Plot** tool.
7. The points will be plotted in the upper window (adjust the window settings to suit).

TI-Nspire

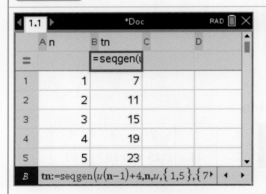 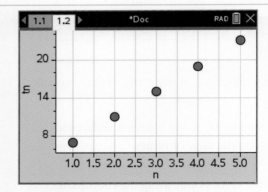

1. Add a **Lists & Spreadsheet** page.
2. Label column **A** as **n** and column **B** as **tn**.
3. In column **A**, enter the values 1 to 5 (or generate the sequence 1 to 5).
4. In column **B**, generate the sequence $u(n) = u(n-1)$, $u_0 = 7$ as shown in Using CAS 1.
5. Add a **Data & Statistics** page.
6. For the horizontal axis, select **n**.
7. For the vertical axis, select **tn**.
8. The points will be displayed on the page.

We can also reverse the process and deduce the rules of a sequence from a given plot.

WORKED EXAMPLE 8 — Deducing the rules of a sequence from a plot

The first three terms of an arithmetic sequence have been displayed on the axes provided.

a State the recursive rule for the sequence.
b State the simplified general rule for the nth term for the sequence.

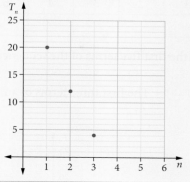

Steps	Working
a 1 Identify the first term and the common difference.	$a = 20$, $d = -8$
2 Write a recursive rule using the correct letter for the name of the sequence.	$T_{n+1} = T_n - 8$, $T_1 = 20$
b 1 Substitute the values of a and d into the general rule.	$T_n = a + (n-1)d$
2 Simplify the rule.	$T_n = 20 + (n-1)(-8)$ $T_n = 20 - 8n + 8$ $T_n = 28 - 8n$

EXERCISE 7.1 Arithmetic sequences

ANSWERS p. 477

Mastery

1 **WORKED EXAMPLE 1** Describe the following sequences.
   a 9, 12, 15, 18 …
   b 68, 64, 60, 56 …

2 **WORKED EXAMPLE 2** List the first five terms of the following sequences.
   a $V_{n+1} = V_n - 3$, $V_1 = 11$
   b $R_n = R_{n-1} + 9$, $R_1 = -4$

3 **WORKED EXAMPLE 3** Use the following information to state a recursive rule for each of the sequences.
   a The next term in sequence T is given by 2 more than the previous term, where the first term is -5.
   b The first three terms of sequence F are 6.1, 7, 7.9.

4 **WORKED EXAMPLE 4** An arithmetic sequence has $T_1 = -2$ and $T_9 = 22$. Determine the recursive rule for the sequence.

5 **WORKED EXAMPLE 5** An arithmetic sequence has the first three terms of $2x + 5$, $7 - x$, $4 + x$. Determine the value of x and hence state the recursive rule for the sequence.

6 **WORKED EXAMPLE 6** Use the following information to determine a simplified general rule for each of the sequences.
   a $A_{n+1} = A_n + 3$, $A_1 = 100$
   b $T_{10} = 14$ and $T_{15} = -6$

7 **Using CAS 1** Determine the first five terms of the following arithmetic sequences.
   a $T_{n+1} = T_n - 5$, $T_1 = 100$
   b $T_n = 11n + 6$

8 **WORKED EXAMPLE 7** Consider the sequence $T_{n+1} = T_n + 8$, $T_1 = -11$.

  a Complete the table of values for T_n.

n	1	2	3	4	5
T_n					

  b Hence, plot the first five terms of the sequence.

9 **WORKED EXAMPLE 8** The first three terms of an arithmetic sequence have been displayed on the axes provided.

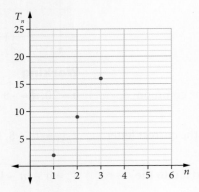

  a State the recursive rule for the sequence.

  b State the simplified general rule for the nth term for the sequence.

Calculator-free

10 (7 marks) Consider an arithmetic sequence with the first term $a = 3$ and a common difference $d = 4$.

  a State the first five terms of this arithmetic sequence. (2 marks)

  b Determine a simplified general rule for the nth term of the sequence. (2 marks)

  c Hence,

    i calculate the 50th term of the sequence (1 mark)

    ii determine the value of n that corresponds to a term of 27. (2 marks)

11 (9 marks) An arithmetic sequence is defined by the recursive rule $A_{n+1} = A_n + d$, $A_1 = a$.

  a Express the A_8 and A_{12} in terms of a and d. (2 marks)

  b If $A_8 = 25$ and $A_{12} = 41$, determine the values of a and d and hence state the recursive rule of the sequence. (4 marks)

  c Show that there are no values of n for which the value of nth term in the sequence is equal to the square of n. (3 marks)

Calculator-assumed

12 (3 marks) An arithmetic sequence is defined by the rule $t_n = 5 + (n-1)(-2)$.

  a State the 6th term of the sequence. (1 mark)

  b Determine the first term in the sequence that is less than -50, stating the corresponding value of n. (2 marks)

7.2 Modelling linear growth and decay

Much like we can model practical situations involving a constant rate of change using linear relationships, we can also use arithmetic sequences to model situations involving **linear growth** or **linear decay**.

> **Linear growth and decay**
>
> An arithmetic sequence with a common difference d can be used to model linear growth (when $d > 0$) or linear decay (when $d < 0$).

The only important difference that we should be aware of in practical situations is that an **initial value** that is given is often written in context as being the 0th term. For example, the initial population of a colony may need to be considered as P_0 or the initial amount of money invested into a bank account may need to be considered as A_0. This resetting of T_0 as the first term, a, has the following effect on the rules of an arithmetic sequence.

> **Rules for arithmetic sequences involving T_0**
>
> Recursive rule
>
> $T_{n+1} = T_n + d, \; T_0 = a$
>
> General rule
>
> $T_n = a + nd$
>
> The use of $a + nd$ instead of $a + (n-1)d$ is because 'n lots of the common difference' are required to go from the 0th term to the nth term.

 Exam hack

Depending on whether you are using $T_1 = a$ or $T_0 = a$, you may need to change the type of sequence on CAS and the starting value in the table.

Video playlist
Modelling linear growth and decay

Worksheet
Modelling with sequences

ClassPad

In the **Sequence** application, there are options to change the type of recursive sequence.

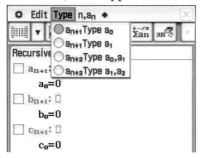

Tap on the **Type** menu to view the four options.

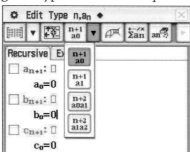

Tap on the **n+1** tool to view the four options.

> **TI-Nspire**
>
> In a **Lists & Spreadsheet > Generate Sequence** dialogue box, there is only the one recursive sequence option, **u(n) = u(n–1)**. Adjust the starting value(s) to suit.

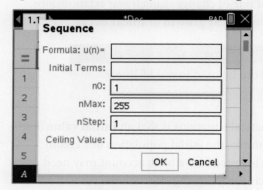 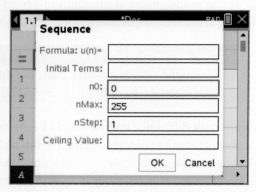

In the **Generate Sequence** dialogue box, keep the **n0** default field setting of 1 for relations in the form **u(n)** and **n–1**.

Change the **n0** field setting to **0** for relations in the form **u(n+1)** and **n**.

TI-Nspire has a **seq** function for generating explicit sequences and a **seqn** function for generating recursive sequences. These functions are available in the **Calculator, Lists & Spreadsheet** and **Notes** pages. Press **catalog** then **S** to jump to the functions starting with s, then scroll down to the functions.

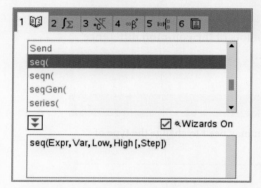 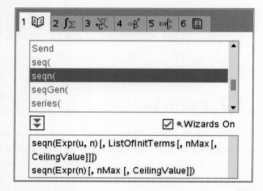

The **seq** function showing the parameters.

The **seqn** function showing the parameters.

Modelling linear growth

Problems involving linear growth can vary and tend to involve contexts from a variety of fields, such as finance or biology.

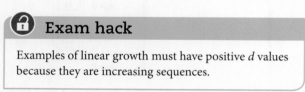

> **Exam hack**
>
> Examples of linear growth must have positive d values because they are increasing sequences.

WORKED EXAMPLE 9 **Modelling financial situations**

At the start of the year, Anthony has $25 in his piggy bank savings at home. His parents decide that they will give him $5 per week if he completes his weekly chores. Let A_n be the amount of money in Anthony's piggy bank after n weeks, assuming he completes his weekly chores.

a Write a rule for A_n, the amount of money Anthony would have in his piggy bank after n weeks of chores.

b Determine the amount of money in Anthony's piggy bank after 20 weeks.

c After how many weeks will Anthony have $200 in his piggy bank?

Steps	Working
a 1 Identify the first term a (A_0) and the common difference, d.	$a = 25$, $d = 5$
	$A_n = a + nd$
2 Substitute into the rule $A_n = a + nd$.	$A_n = 25 + 5n$
b 1 Substitute $n = 20$ and evaluate.	$A_{20} = 25 + 5(20)$
	$A_{20} = 125$
2 Answer in context of the question.	Therefore, Anthony has $125 after 20 weeks.
c 1 Substitute $A_n = 200$ and solve for n.	$200 = 25 + 5d$
2 Answer in context of the question.	$175 = 5d$
	$d = 35$
	Therefore, Anthony will have $200 after 35 weeks.

Modelling linear decay

The contexts of linear decay can also vary. Just be sure to note that these examples involve a negative d values and so are decreasing sequences.

WORKED EXAMPLE 10 — Modelling population decay

At the start of 2015, the population of birds in a local forest was estimated to be 210. Due to the loss of forest land each year, the population of birds after n years can be modelled by the recursive rule

$P_n = P_{n-1} - 14$, $P_0 = 210$

a Interpret the significance of the -14 in the context of the question.

b The population can be considered critically endangered when it falls below 20 birds. Predict at the start of which year this is expected to happen.

Steps	Working
a Describe the -14 as a constant decrease per unit of time in the question.	The -14 means that the population of birds decreases by 14 each year.
b 1 Use the general rule to solve for n when $P_n = 20$. Alternatively, use CAS to find the values either side of $P_n = 20$ and conclude in context.	$P_n = a + nd$
	$P_n = 210 - 14d$
	$20 = 210 - 14d$
	$d = 13.57 \ldots$
2 Interpret the answer in context of the question.	By the start of 2029, the population will be considered critically endangered.

ClassPad

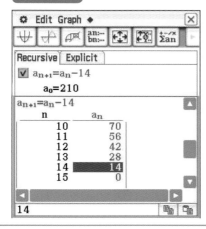

TI-Nspire

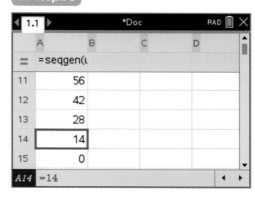

EXERCISE 7.2 Modelling linear growth and decay

ANSWERS p. 477

Recap

1. The fifth term of the sequence $T_n = T_{n-1} + 4$, $T_1 = 14$ is

   A 12   **B** 22   **C** 24   **D** 30   **E** 34

2. The common difference of the sequence $B_n = 21 - 3n$ is

   A −3   **B** 1   **C** 2   **D** 3   **E** 21

Mastery

3. **WORKED EXAMPLE 9** At the start of the year, Juliana has $4150 in her savings account. Each week, she deposits $30 into her account. Let S_n be the amount of money in Juliana's savings account after n weeks, assuming she does not withdraw any money and no other deposits are made.

   a. Write a rule for S_n.

   b. Determine the amount of money in Juliana's savings account after 26 weeks.

   c. After how many weeks will Juliana have $5000 in her savings account?

4. **WORKED EXAMPLE 10** A population of bacteria with 300 000 cells dies at a constant rate in unfavourable environmental conditions. The population after n hours can be modelled by the recursive rule $P_n = P_{n-1} - 5000$, $P_0 = 300\,000$.

   a. Interpret the significance of the −5000 in the context of the question.

   b. The population can be considered undetectable when it falls below 10 000 cells. Predict at the start of which hour this is expected to happen.

Calculator-free

5. (7 marks) A school athletics team is training for a 5-kilometre run. To prepare for the run, the coach plans a daily training schedule. On day 1, the team must complete a 400-metre run. Each day the coach increases the distance by 200 metres. Let R_n be the distance to be run on day n of the training schedule.

   a. Write a rule in the form $R_{n+1} = R_n + d$, $R_1 = a$ to represent the coach's training schedule. (2 marks)

   b. Write a simplified general rule for the distance to be run on day n of the training schedule. (2 marks)

   c. Calculate the distance to be run on day 10 of the training schedule. (1 mark)

   d. On which day of the training schedule will the team run five kilometres? (2 marks)

6. (4 marks) A boiled egg is removed from a pot of boiling water at 98°C and left to cool at room temperature. Each minute, the surface temperature of the egg decreases by 2°C. Let T_n be the surface temperature of the egg n minutes after being removed from the pot.

   a. State the surface temperature of the egg three minutes after being removed. (1 mark)

   b. Show the use of a general rule for T_n to calculate the number of minutes for the surface temperature of the egg to reach 26°C. (3 marks)

Calculator-assumed

7 (7 marks) The student council of a local high school is selling tickets for a fundraising event. After the first day of sales, they had sold 80 tickets. According to their predicted sales model they should have sold 300 tickets after five days. Assume that the council sells a constant number of tickets per day. Let T_n represent the number of tickets sold after n days.

  a Write a recursive rule for the number of tickets predicted to be sold over the fundraising campaign. (2 marks)

  b Predict the number of tickets sold after 10 days of campaigning. (2 marks)

  c The student council would like to sell 1500 tickets by the end of the campaign. According to the model, for how many days should they sell tickets? (3 marks)

8 (5 marks) The monthly sales, M, for each of the first four months since the release of a type of smartphone are shown in the table.

n	1	2	3	4
M_n	500	460	420	380

  a Use the table to write a simplified general rule for M_n, the monthly sales in the nth month since release. (2 marks)

  b According to the model for M_n,

    i calculate the number of sales in the 10th month (1 mark)

    ii determine the number of months for which this smartphone is expected to generate sales. (2 marks)

9 (5 marks) Steven borrows $8000 from his parents to buy his first car, who charge him simple interest at a rate of 1.75% per month. Let A_n represent the amount of money that Steven owes his parents after n months.

  a Write a rule for A_n. (2 marks)

  b Determine the amount of money that Steven owes his parents after 4 months. (1 mark)

Steven does not want to owe his parents any more than $12 000 by the end of the loan.

  c Within how many months should Steven aim to pay the entire loan? (2 marks)

7.3 Arithmetic series and their applications

Video playlist
Arithmetic series and their applications

The sum of arithmetic sequences

Recall the sequence $A_{n+1} = A_n + 2$, $A_1 = 1$, with the first five terms 1, 3, 5, 7, 9.

We can express the **finite sum** (also called a **partial sum**) of the first five terms as the numerical expression $1 + 3 + 5 + 7 + 9$. The finite sum of the terms of a sequence is the addition of consecutive terms up to a particular value of n. For example, $1 + 3 + 5 + 7 + 9$ is the finite sum for sequence A where $n = 5$. The notation for a finite sum of the first n terms of a sequence is S_n and so $S_5 = 25$ for this sequence.

> **Finite sums**
>
> The sum of the first n terms can be expressed in the form $S_n = T_1 + T_2 + \ldots + T_n$. Given that the finite sum of the first $(n + 1)$ terms is $S_{n+1} = T_1 + T_2 + \ldots + T_n + T_{n+1}$, then the finite sums form their own recursive sequence of the form
>
> $S_{n+1} = S_n + T_{n+1}$

WORKED EXAMPLE 11	Calculating sums for a small finite n

Let S_n be the finite sum of the first n terms of the sequence $T_{n+1} = T_n - 3$, $T_1 = 11$.

Determine

a S_1 b S_3 c $S_5 - S_4$

Steps	Working
a Recognise that $S_1 = T_1$.	$S_1 = T_1 = 11$
b 1 Express S_3 as a sum of terms.	$S_3 = T_1 + T_2 + T_3$
2 Evaluate the sum.	$= 11 + 8 + 5$
	$= 24$
c 1 Recognise that $S_5 - S_4 = T_5$.	$S_5 = T_1 + T_2 + T_3 + T_4 + T_5$
	$S_4 = T_1 + T_2 + T_3 + T_4$
2 Evaluate T_5 using the recursive rule.	$S_5 - S_4 = T_5 = -1$

The arithmetic series formulas

For the sequence $A_{n+1} = A_n + 2$, $A_1 = 1$, we can display the terms and corresponding values of S_n in a table of values.

n	1	2	3	4	5
A_n	1	3	5	7	9
S_n	1	4	9	16	25

We can see that the linear pattern of the arithmetic sequence becomes a quadratic pattern when considering its sum, S_n. That is, $S_n = n^2$. However, what about for a general arithmetic sequence $T_{n+1} = T_n + d$, $T_1 = a$? We can observe the generalisation of the terms with respect to a and d but we essentially need a rule in terms of n.

n	1	2	3	4	5
T_n	a	$a + d$	$a + 2d$	$a + 3d$	$a + 4d$
S_n	a	$2a + d$	$3a + 3d$	$4a + 6d$	$5a + 10d$

For the finite sum of the first n terms, we have the expression

$S_n = T_1 + T_2 + T_3 + T_4 + \ldots + T_n$

$S_n = a + a + d + a + 2d + a + 3d + \ldots + a + (n-1)d$

Consider the as in the expression. There are n lots of a; that is, an, and so

$S_n = an + d + 2d + 3d + \ldots + (n-1)d$

Consider the ds in the expression. We have the sum $d + 2d + 3d + \ldots + (n-1)d$.

Factoring out the d in this expression, we are left with $1 + 2 + 3 + \ldots + (n-1)$.

Consider the table showing the sum of the first five consecutive integers.

n	1	2	3	4	5
T_n	1	2	3	4	5
S_n	1 $\left(= \dfrac{1 \times 2}{2}\right)$	3 $\left(= \dfrac{2 \times 3}{2}\right)$	6 $\left(= \dfrac{3 \times 4}{2}\right)$	10 $\left(= \dfrac{4 \times 5}{2}\right)$	15 $\left(= \dfrac{5 \times 6}{2}\right)$

From this table, you can observe that the sum of the first n consecutive integers, that is $1 + 2 + 3 + \ldots + n$, is given by the rule

$$\frac{n(n + 1)}{2}$$

So, if we want $1 + 2 + 3 + \ldots + (n - 1)$, we replace n with $n - 1$ to obtain the rule

$$\frac{(n - 1)n}{2}$$

As a result, we have

$$S_n = an + d(1 + 2 + 3 + \ldots + (n - 1))$$

$$S_n = an + d\frac{(n - 1)n}{2}$$

$$= an + \frac{dn^2 - dn}{2}$$

$$= \frac{2an + dn^2 - dn}{2}$$

$$= \frac{n}{2}[2a + dn - d]$$

$$= \frac{n}{2}[2a + (n - 1)d]$$

Finite sum of an arithmetic sequence – an arithmetic series

The value of S_n for an arithmetic sequence with a first term a and a common difference d is given by

$$S_n = \frac{n}{2}[2a + (n - 1)d]$$

The formal name given to the sum of the terms in a sequence is a **series** and, hence, this is also referred to as the **arithmetic series** formula.

WORKED EXAMPLE 12 — Calculating finite sums using the first term and the common difference

Calculate the value of the following finite sums for each of the given sequences.

a S_{50} for $T_{n+1} = T_n + 3$, $T_1 = 5$

b S_{100} for $T_{n+1} = T_n - 9$, $T_1 = 1000$

Steps	Working
a 1 Identify the value of a and d.	$a = 5, d = 3$
2 Substitute the values of n, a and d into the arithmetic series formula.	$S_{50} = \frac{50}{2}[2(5) + (50 - 1)(3)]$
3 Evaluate, using CAS, as required.	$S_{50} = 25[10 + 147]$ $S_{50} = 25(157)$ $S_{50} = 3925$
b 1 Identify the value of a and d.	$a = 1000, d = -9$
2 Substitute the values of n, a and d into the arithmetic series formula.	$S_{100} = \frac{100}{2}[2(1000) + (100 - 1)(-9)]$
3 Evaluate, using CAS, as required.	$S_{100} = 50[2000 - 9(99)]$ $S_{100} = 55\,450$

Consider the graphical representation of the series
$S_5 = 1 + 3 + 5 + 7 + 9 = 25$.

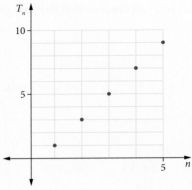

The value of 25 can be equivalently obtained by adding 5 lots of the average of the five terms. That is,

$$\frac{1+3+5+7+9}{5} = \frac{1+9}{2} = 5$$

$5 \times 5 = 25$

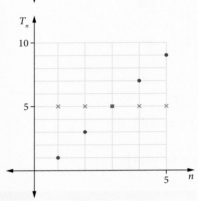

As a result, we can establish an alternative representation of the arithmetic series formula.

$$S_n = \frac{n}{2}[2a + (n-1)d]$$

Splitting the $2a$ into $a + a$,

$$S_n = \frac{n}{2}[a + a + (n-1)d]$$

Recognising that $T_1 = a$ and $T_n = a + (n-1)d$, then

$$S_n = \frac{n}{2}[T_1 + T_n]$$

Finite sum of an arithmetic sequence – an arithmetic series (alternative)

Given the 1st term and the nth term of a sequence, the value of S_n is given by n lots of the average of the first and last terms. That is,

$$S_n = n\left(\frac{T_1 + T_n}{2}\right)$$

WORKED EXAMPLE 13 — Calculating finite sums using the first and last terms

The first term of a sequence is 7 and the 40th term is 63. Determine S_{40}, the sum of the first 40 terms.

Steps	Working
1 Substitute into the formula $S_n = n\left(\frac{T_1 + T_n}{2}\right)$.	$S_{40} = 40\left(\frac{7 + 63}{2}\right)$
2 Evaluate the sum.	$S_{40} = 40(35)$ $S_{40} = 1400$

Solving practical problems involving arithmetic series

7.3

Much like arithmetic sequences can be used to model problems involving linear growth and decay, we can extend these problems to involve the concept of a series.

USING CAS 2 — Finding the sum of an arithmetic sequence

Sarah has started an investment plan to save money for her future financial goals. Every month, she plans to deposit a certain amount of money into the investment, with each month's deposit forming an arithmetic sequence. In the first month, she deposits $300 and in the 12th month, she deposits $700.

a Calculate the value of d, the increase in deposit per month.

b Determine Sarah's total deposits for the first two years, to the nearest whole dollar.

Steps	Working
a 1 Define the variable used to represent the sequence.	Let D_n represent the monthly deposit in the nth month. $D_1 = 300$, $D_{12} = 700$
2 Calculate the common difference.	$d = \dfrac{700 - 300}{12 - 1}$ $d = \dfrac{400}{11}$ $d = 36.36$
3 Interpret the result in context.	The increase is $36.36 per month.
b Establish the recursive rule for the monthly deposits.	$D_{n+1} = D_n + \dfrac{400}{11}$, $D_1 = 300$

ClassPad

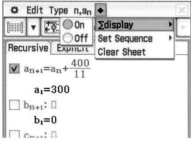

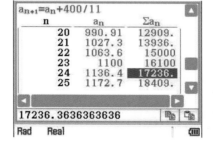

1 Generate the recursive function.
2 Tap ◆ > Σdisplay > **On**.

3 The sequence and the sum of sequence terms will be displayed in the lower window.
4 Scroll down to ***n* = 24** to find the total of the monthly deposits after two years.

The total of Sarah's deposits for the first two years is $17 236.

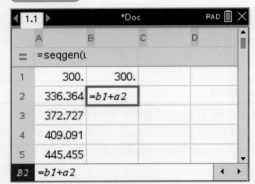

1. Generate the recursive sequence in column **A**.
2. In cell **b1**, enter the initial value of **300**.
3. In cell **b2**, enter the formula **=b1+a2** and press **enter**.
4. Place the cursor in cell **b2**.
5. Press **menu > Data > Fill**.
6. Press the **down arrow** to fill in column **B** and press **enter**.
7. Scroll to row **24** to find the total of the monthly deposits after two years.

The total of Sarah's deposits for the first two years is $17 236.

EXERCISE 7.3 Arithmetic series and their applications   ANSWERS p. 478

Recap

1. The value of an $1600 asset depreciates by $20 each month. Let V_n be the value of the asset after n months of depreciation. The general rule for V_n is
    A $V_n = 1600 + 20n$
    B $V_n = 1600 - 20n$
    C $V_n = 1620 - 20n$
    D $V_n = 20 - 1600n$
    E $V_n = 20 + 1600n$

2. The population of a colony of ants is modelled by the recursive rule $P_{n+1} = P_n + 40$, $P_0 = 150$, where n is measured in days since an initial estimation of the population. The number of ants estimated to be in the colony after one year is
    A 365   B 14 000   C 14 710   D 14 750   E 15 000

Mastery

3. **WORKED EXAMPLE 11** Let S_n be the finite sum of the first n terms of the sequence $T_{n+1} = T_n + 4$, $T_1 = -19$. Determine
    a S_1
    b S_5
    c $(S_6 - S_4)^2$

4. **WORKED EXAMPLE 12** Calculate the value of the finite sums for each of the following sequences.
    a S_{20} for $T_{n+1} = T_n - 1$, $T_1 = 19$
    b S_{45} for $T_{n+1} = T_n + 2$, $T_1 = -3$

5. **WORKED EXAMPLE 13** The first term of a sequence is 8 and the 25th term is 80. Determine S_{30}, the sum of the first 30 terms.

6 **Using CAS 2** Esther has started planning a road trip through different towns. She forms a budget plan for each day of the road trip, with the daily budget amounts forming an arithmetic sequence. On the first day, she budgets $80 and on the 15th day, she budgets $130.

   a Calculate the value of d, the increase in daily spending budget each day.

   b Determine Esther's total budget she has allocated if her road trip is planned for 15 days.

Calculator-free

7 (9 marks) An arithmetic sequence has $T_4 = 13$ and $T_9 = 28$.

   a State the recursive rule for this sequence. (3 marks)

   b Determine T_{15}. (2 marks)

   c Let S_n be the sum of the first n terms. Determine

      i S_4 (2 marks)

      ii S_{10} (2 marks)

8 (8 marks) An arithmetic sequence is defined by the general rule $B_n = 6n - 3$.

   a Determine the value of the first term, a, and the common difference, d. (2 marks)

   b Hence, write a simplified rule for S_n, the sum of the first n terms of the sequence. (2 marks)

   c Calculate S_{10}. (2 marks)

   d Calculate the value of n such that $S_n = 1200$. (2 marks)

9 (10 marks) An arithmetic series has the rule $S_n = 2n^2 - 3n$.

   a State the values of S_1, S_2 and S_3. (2 marks)

   b Hence, state the first three terms of the corresponding arithmetic sequence, T_n. (2 marks)

   c Determine the rule for the nth term of the arithmetic sequence and hence find T_{100}. (3 marks)

   d Show algebraically that there only exists one value of n for which $S_n = T_n$. (3 marks)

Calculator-assumed

10 (11 marks) A tower is constructed out of colourful blocks, such that the number of blocks used in the nth row, R_n, is determined by an arithmetic sequence. The number of blocks in the base of the tower, R_1, is 110 and each subsequent row has 5 fewer blocks than the previous.

   a Write a recursive rule for the number of blocks in each row and hence, state the number of blocks in the first five rows. (4 marks)

   b Write a rule for the total number of blocks used to construct the tower up to the nth row of the tower. (2 marks)

   c Hence, calculate the number of blocks in the tower if it has 15 rows. (2 marks)

   d The tower's construction will stop when no more blocks can be added. Determine the number of rows in the tower, and hence state the total number of blocks used to construct the tower. (3 marks)

7.4 Geometric sequences

Video playlist
Geometric sequences

Worksheets
Geometric sequences
Geometric progressions

Geometric sequences as numerical lists

When we observe numerical patterns, we need to be aware that there may be a multiplicative relationship between the terms rather than an additive one. For example, 2, 4, 8, 16 … is an increasing sequence in whch each term is twice the previous, or 270, 90, 30, 10 … is a decreasing sequence in which each term is a third of the previous. These are examples of **geometric sequences**. A geometric sequence is a sequence of numbers that have a **common ratio** r; that is, a constant multiplication or division factor. The sequence 2, 4, 8, 16 … has a common ratio of × 2, whereas 270, 90, 30, 10 … has a common ratio of ÷ 3. However, we need to always interpret the common ratio as a multiplication factor, so a ratio of ÷ 3 should be expressed as $\times \frac{1}{3}$.

WORKED EXAMPLE 14	Describing geometric sequences in words
Describe the following sequences. **a** 5, 15, 45, 135 … **b** 220, 110, 55, 27.5 …	
Steps	**Working**
a 1 Identify the common ratio and first term.	$r = 3, a = 5$
2 Write a statement describing the sequence.	It is a geometric sequence with a first term of 5 and a common ratio of 3.
b 1 Identify the common ratio and first term.	$r = \frac{1}{2}, a = 220$
2 Write a statement describing the sequence.	It is a geometric sequence with a first term of 220 and a common ratio of $\frac{1}{2}$.

The recursive rule for geometric sequences

Much like with arithmetic sequences, if we have the first term and common ratio of a geometric sequence, we can describe the sequence recursively.

> **The recursive rule – geometric sequences**
>
> Let T be a geometric sequence with a first term $T_1 = a$ and a common ratio of r, where $r \neq 0, 1$.
>
> The recursive rule for such a sequence can be expressed as
>
> $T_{n+1} = rT_n, T_1 = a$
>
> Technically, we will only examine situations for which $r > 0, r \neq 1$, but there is a special type of geometric sequence that exists with $r < 0$ called an **alternating sequence**.

WORKED EXAMPLE 15 — Generating the terms of a sequence from a recursive rule

List the first five terms of the following sequences.

a $T_{n+1} = 5T_n$, $T_1 = 4$

b $M_n = \dfrac{1}{10} M_{n-1}$, $M_1 = 1500$

Steps	Working
a 1 Identify the first term and the common ratio.	$a = 4$, $r = 5$
2 List the first five terms, starting at T_1 and multiplying by 5 each time.	4, 20, 100, 500, 2500
b 1 Identify the first term and the common ratio.	$a = 1500$, $r = \dfrac{1}{10}$
2 List the first five terms, starting at M_1 and dividing by 10 each time.	1500, 150, 15, 1.5, 0.15

WORKED EXAMPLE 16 — Writing a recursive rule from a list of terms

Write a recursive rule for the sequence with the first three terms 60, 12, and 2.4.

Steps	Working
1 Identify the common ratio using $\dfrac{T_2}{T_1}$.	$r = \dfrac{12}{60} = \dfrac{1}{5}$
2 State the recursive rule of the sequence in the correct form.	$T_{n+1} = \dfrac{1}{5} T_n$, $T_1 = 60$

Suppose for a geometric sequence, we know that $T_1 = 4$ and $T_5 = 2500$. To calculate the common ratio, we need to think multiplicatively. That is, 4 has been multiplied by r four times to go from 4 to 2500. Algebraically, this gives the equation

$$r^4 = \frac{2500}{4} = 625$$

To solve for r we need to take the fourth root of 625.

$$r = \sqrt[4]{625} = 5$$

Finding the common ratio

Given two terms of a geometric sequence, $T_m = a$ and $T_n = b$, we can consider these as coordinates (m, a) and (n, b). Then the common ratio r can be found by solving the equation

$$r^{n-m} = \frac{b}{a} = \frac{T_n}{T_m}$$

WORKED EXAMPLE 17 — Writing a recursive rule from two non-consecutive terms

A geometric sequence has $T_1 = 3$ and $T_7 = 192$. Determine the recursive rule for the sequence.

Steps	Working
1 Find the common ratio between the terms.	$r^{7-1} = \dfrac{192}{3}$ $r^6 = 64$ $r = \sqrt[6]{64} = (2^6)^{\frac{1}{6}} = 2$
2 Identify the first term.	$T_1 = a = 3$
3 State the recursive rule of the sequence in the correct form.	$T_{n+1} = 2T_n$, $T_1 = 3$

WORKED EXAMPLE 18 — Solving algebraic problems involving geometric sequences

A geometric sequence with $r > 0$ has the first three terms $x + 1$, $x - 1$, $2x + 1$.
Determine the value of x and hence state the recursive rule for the sequence.

Steps	Working
1 Recognise that for a geometric sequence, $r = \dfrac{T_3}{T_2} = \dfrac{T_2}{T_1}$.	$\dfrac{2x+1}{x-1} = \dfrac{x-1}{x+1}$
2 Solve the equation formed for x.	$(2x+1)(x+1) = (x-1)^2$ $2x^2 + 3x + 1 = x^2 - 2x + 1$ $x^2 + 5x = 0$ $x(x+5) = 0$ $x = 0$ or $x = -5$
3 Evaluate the sequence for each x and check which gives $r > 0$.	If $x = 0$, $r = \dfrac{1}{-1} = -1 < 0$. Reject $x = 0$. If $x = -5$, $r = \dfrac{-9}{-6} = \dfrac{3}{2}$. So, $x = -5$.
4 Use x to deduce the terms of the sequence.	$-4, -6, -9 \ldots$
5 State the recursive rule of the sequence in the correct form.	$T_{n+1} = \dfrac{3}{2} T_n,\ T_1 = -4$

The general (explicit) rule for geometric sequences

Once again, a recursive rule is only efficient if we have one term and want to find the next term, or a sequence of consecutive terms. To find a term with a greater value of n, we would need the general rule.

Recall the sequence, T, with the first five terms of 4, 20, 100, 500, 2500 …

To go from $T_1 = 4$ to $T_5 = 2500$, we can start at 4 and multiply by the common ratio of 5, four times.

$$T_5 = 4 \times 5 \times 5 \times 5 \times 5 = 4 \times 5^4$$

Similarly, if we wanted to start from $T_1 = 4$ and find the 50th term, we would need to multiply by 5, 49 times. That is,

$$T_{50} = 4 \times 5^{49}$$

which gives a significantly large number that we would need to express in scientific notation,

$$T_{50} = 7.105 \times 10^{34} \text{ (to four significant figures)}$$

You should notice that the 'number of times we multiply by the common ratio' is always one less than the placement of the term being found. That is, to find the nth term, we multiply the first term by r, $n - 1$ times. This is called the **general** (or **explicit rule**) of an geometric sequence.

> **The general rule – geometric sequences**
>
> For a geometric sequence with a recursive rule
>
> $$T_{n+1} = rT_n,\ T_1 = a$$
>
> the nth term of the sequence is found using the general rule
>
> $$T_n = ar^{n-1}$$

Using index laws, we can also simplify the rule such that

$T_n = ar^{n-1}$
$T_n = ar^n r^{-1}$
$T_n = \dfrac{a}{r} r^n$

For example, with our previous sequence T, we would have the general rule

$T_n = 4 \times 5^{n-1}$

or

$T_n = \dfrac{4}{5} 5^n$

where $\dfrac{4}{5}$ would represent the 0th term.

This form of a general rule should look very familiar to you as it resembles the equation of an exponential relationship, $y = a \times b^x$, where y represents the value of the term in the sequence (T_n), x represents the term's placement in the sequence (n), a represents the 0th term $\left(\dfrac{a}{r}\right)$ or the y-intercept, and the base b is the common ratio (r).

WORKED EXAMPLE 19 — Writing a rule for the nth term from different descriptions

Use the following information to write a general rule for each of the sequences.

a 7, 21, 63 …   **b** $X_{n+1} = \dfrac{1}{6} X_n$, $X_1 = 216$ **c** $Z_4 = \dfrac{1}{10}$ and $Z_7 = \dfrac{1}{80}$

Steps	Working
a 1 Identify the first term and common ratio.	$a = 7$, $r = 3$
2 Use the general rule $T_n = ar^{n-1}$.	$T_n = 7 \times 3^{n-1}$
b 1 Identify the first term and common ratio.	$a = 216$, $r = \dfrac{1}{6}$
2 Use the general rule $X_n = ar^{n-1}$.	$X_n = 216\left(\dfrac{1}{6}\right)^{n-1}$

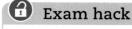 **Exam hack**

When r is fractional, be sure to put it in brackets, as $\dfrac{1^{n-1}}{6} \neq \left(\dfrac{1}{6}\right)^{n-1}$.

c 1 Find the common ratio using the two terms.	$r^{7-4} = \dfrac{\frac{1}{80}}{\frac{1}{10}}$
2 Find a by working backwards from the closest term.	$r^3 = \dfrac{1}{8}$ $r = \sqrt[3]{\dfrac{1}{8}} = \dfrac{1}{2}$
3 State the general rule.	$Z_1 = \dfrac{1}{10} 2^3 = \dfrac{8}{10} = \dfrac{4}{5}$ $Z_n = \dfrac{4}{5}\left(\dfrac{1}{2}\right)^{n-1}$

USING CAS 3 Determining the nth term of a geometric sequence

Determine the 20th term of the following geometric sequences, answering in scientific notation to four significant figures where appropriate.

a $T_{n+1} = 4T_n$, $T_1 = 1.5$

b $T_n = 500 \times (0.2)^{n-1}$

ClassPad

a

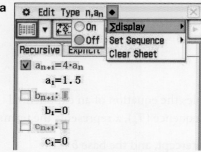 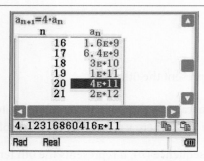

1. Open the **Sequence** application and tap the **Recursive** tab.
2. Tap ◆ > **display** > **Off** to turn off the sum column.
3. Enter the geometric recursive sequence as shown above and press **EXE**.
4. Tap the **Table** tool.
5. The sequence will be displayed in the lower window.
6. Scroll down to $n = 20$.

b

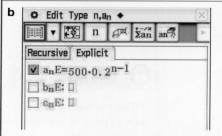 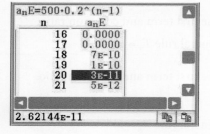

1. Tap the **Explicit** tab.
2. Enter the explicit equation as shown above and press **EXE**.
3. Tap the **Table** tool.
4. The sequence will be displayed in the lower window.
5. Scroll down to $n = 20$.

a $T_{20} = 412\,316\,860\,416 = 4.123 \times 10^{11}$

b $T_{20} = 2.621 \times 10^{-11}$

TI-Nspire

a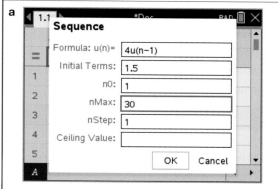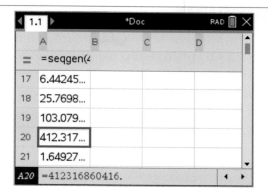

1. Add a **Lists & Spreadsheet** page.
2. Place the cursor in the cell immediately under the column **A** heading.
3. Press **menu > Data > Generate Sequence**.
4. Enter the expressions and values into the fields as shown above and press **enter**.
5. The sequence will be displayed in column **A**.
6. Scroll down to row **20**.

b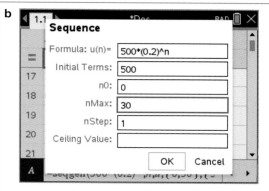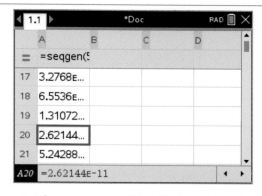

1. Add a **Lists & Spreadsheet** page.
2. Place the cursor in the cell immediately under the column **A** heading.
3. Press **menu > Data > Generate Sequence**.
4. Enter the expressions and values into the fields as shown above and press **enter**.
5. The sequence will be displayed in column **A**.
6. Scroll down to row 20.

a $T_{20} = 412\,316\,860\,416 = 4.123 \times 10^{11}$

b $T_{20} = 2.621 \times 10^{-11}$

Tabular and graphical forms of geometric sequences

The long-term behaviour of geometric sequences is not as simple as that of an arithmetic sequence, which was increasing if $d > 0$ and decreasing if $d < 0$. Instead, with geometric sequences, we need to consider the values of both a and r. To observe this, we will examine some tabular and graphical forms of geometric sequences. The graphs have been provided to display the shape of the sequence, rather than the plotted values.

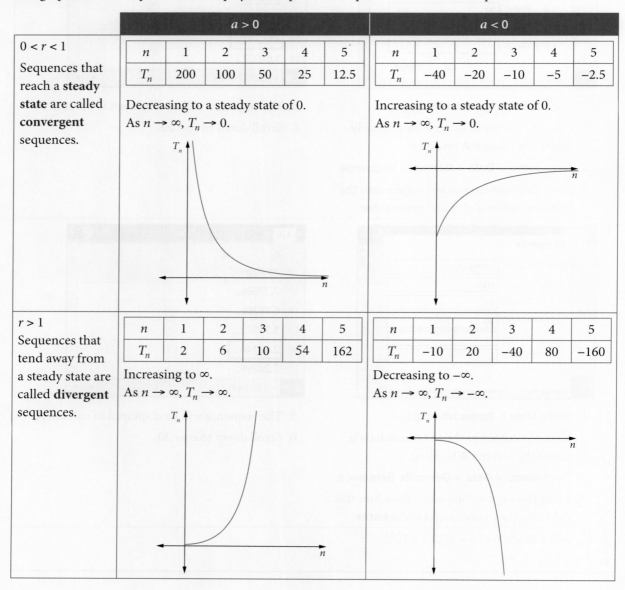

	$a > 0$	$a < 0$
$0 < r < 1$ Sequences that reach a **steady state** are called **convergent** sequences.	n: 1, 2, 3, 4, 5 T_n: 200, 100, 50, 25, 12.5 Decreasing to a steady state of 0. As $n \to \infty$, $T_n \to 0$.	n: 1, 2, 3, 4, 5 T_n: −40, −20, −10, −5, −2.5 Increasing to a steady state of 0. As $n \to \infty$, $T_n \to 0$.
$r > 1$ Sequences that tend away from a steady state are called **divergent** sequences.	n: 1, 2, 3, 4, 5 T_n: 2, 6, 10, 54, 162 Increasing to ∞. As $n \to \infty$, $T_n \to \infty$.	n: 1, 2, 3, 4, 5 T_n: −10, 20, −40, 80, −160 Decreasing to $-\infty$. As $n \to \infty$, $T_n \to -\infty$.

As mentioned, there is a special case of sequences when $r < 0$ called **alternating (oscillating) sequences**. Although these are not in our course, some examples for interest have been provided below.

$-1 < r < 0$						$r < -1$					
n	1	2	3	4	5	n	1	2	3	4	5
T_n	20	−10	5	−2.5	1.25	T_n	−10	20	−40	80	−160

Alternating sequence, converging to 0.
As $n \to \infty$, $T_n \to 0$.

Alternating sequence, diverging.
Alternating between $\pm \infty$.

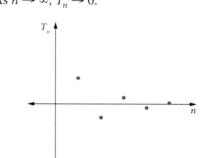
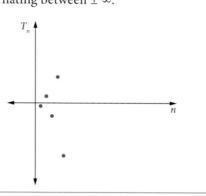

WORKED EXAMPLE 20 — Plotting the terms of a sequence from a rule

Consider the sequence $T_{n+1} = 2T_n$, $T_1 = \frac{1}{4}$.

a Complete the table of values for T_n.

n	1	2	3	4	5
T_n					

b Hence, plot the first five terms of the sequence.

Steps	Working
a Use the first term and common ratio to complete the table of values.	<table><tr><td>n</td><td>1</td><td>2</td><td>3</td><td>4</td><td>5</td></tr><tr><td>T_n</td><td>$\frac{1}{4}$</td><td>$\frac{1}{2}$</td><td>1</td><td>2</td><td>4</td></tr></table>
b Plot the set of points using the coordinates (n, T_n).	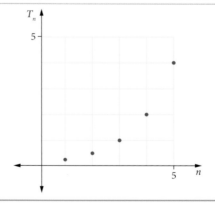

WORKED EXAMPLE 21 — Deducing the rules of a sequence from a plot

The first five terms of a geometric sequence have been displayed on the axes provided.

a State the recursive rule for the sequence.

b State the rule for the nth term for the sequence.

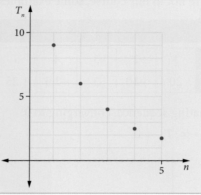

Steps	Working
a 1 Identify the first term and the common ratio.	$a = 9$ $r = \dfrac{6}{9} = \dfrac{2}{3}$
2 Write a recursive rule using the correct letter for the name of the sequence.	$T_{n+1} = \dfrac{2}{3} T_n,\ T_1 = 9$
b Substitute the values of a and r into the general rule.	$T_n = 9 \left(\dfrac{2}{3} \right)^{n-1}$

EXERCISE 7.4 Geometric sequences

ANSWERS p. 478

Recap

1 The sum of the first 10 terms of the sequence 5, 7, 9, 11 … is

A 23   B 77   C 96   D 117   E 140

2 An arithmetic sequence has $T_1 = 58$ and $T_{50} = -89$. The sum of the first 50 terms is

A −962   B −867   C −775   D −686   E −600

Mastery

3 WORKED EXAMPLE 14 Describe the following sequences.

a −2, −8, −32, −128 …

b 160, 40, 10, 2.5 …

4 WORKED EXAMPLE 15 List the first five terms of the following sequences.

a $A_{n+1} = \dfrac{1}{3} A_n,\ A_1 = 999$

b $B_n = 10 B_{n-1},\ B_1 = -0.2$

5 WORKED EXAMPLE 16 Write a recursive rule for the sequence with the first three terms 15, 75 and 375.

6 WORKED EXAMPLE 17 A geometric sequence has $T_1 = 540$ and $T_4 = 160$. Determine the recursive rule for the sequence.

7 WORKED EXAMPLE 18 A geometric sequence with $r > 0$ has the first two terms $T_1 = x + 4$ and $T_2 = 8x - 1$ and a common ratio of 2.

Determine the value of x and hence, state the recursive rule for the sequence.

8 WORKED EXAMPLE 19 Use the following information to write a general rule for each of the sequences.

a 9, 3, 1 …

b $B_{n+1} = 6 B_n,\ B_1 = -2$

c $T_4 = 4$ and $T_8 = 16$

9 **Using CAS 3** Determine the 30th term of the following geometric sequences, answering in scientific notation to four significant figures where appropriate.

  a $T_{n+1} = \frac{1}{2}T_n$, $T_1 = -700$

  b $T_n = 0.001 \times 3^{n-1}$

10 **WORKED EXAMPLE 20** Consider the sequence $T_{n+1} = 3T_n$, $T_1 = \frac{2}{3}$.

  a Complete the table of values for T_n.

n	1	2	3	4	5
T_n					

  b Hence, plot the first five terms of the sequence.

11 **WORKED EXAMPLE 21** The first five terms of a geometric sequence have been displayed on the axes provided.

  a State the recursive rule for the sequence.

  b State the rule for the nth term for the sequence.

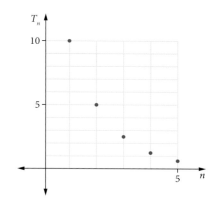

Calculator-free

12 (5 marks) Consider the first four terms of a geometric sequence $T_1 = a$, $T_2 = ar$, $T_3 = ar^2$ and $T_4 = ar^3$.

  a If $a = \frac{1}{2}$ and $r = 3$, determine T_5. (2 marks)

  b For some $a > 0$, if $T_3 = 216$ and $T_6 = 729$, determine the value of a and r and, hence, write a recursive rule for this sequence. (3 marks)

Calculator-assumed

13 (5 marks) A geometric sequence has the general rule $A_n = a\left(\frac{1}{3}\right)^{n-1}$.

  a Describe the behaviour of A_n as $n \to \infty$. (1 mark)

  b If $A_3 = \frac{2}{3}$, determine the value of a and, hence, state A_{20} in scientific notation to four significant figures. (4 marks)

14 (9 marks) Two sequences are defined by the following rules

  $X_{n+1} = X_n + d$, $X_1 = a$

  $Y_{n+1} = rY_n$, $Y_1 = b$

  a Suppose that $X_3 = Y_2$.

    i If $X_4 = 18$ and $Y_3 = 12$, show that $d = 18 - \frac{12}{r}$. (3 marks)

    ii Let $r = 2$. Hence, state the first three terms of each of the sequences. (4 marks)

  b Suppose $a = b = 2$ and $d = r = 3$. Justify whether the two sequences will ever have $X_n = Y_n$ other than for $n = 1$. If so, state the value of n. (2 marks)

Video playlist
Modelling exponential growth and decay

Worksheet
Modelling with sequences

7.5 Modelling exponential growth and decay

For any context that we can model using exponential growth and exponential decay, we can instead use geometric sequences, however, the situation must be discrete; that is, only defined for integer values of n.

> **Exponential growth and decay**
>
> A geometric sequence with a first term of a and common ratio r can be used to model exponential growth (when $a > 0, r > 0$) or exponential decay (when $a > 0, 0 < r < 1$).

Once again, be aware of the 'initial value' that is given and whether it represents T_0 or T_1.

> **Rules for geometric sequences involving T_0**
>
> Recursive rule
> $$T_{n+1} = rT_n, T_0 = a$$
> General rule
> $$T_n = ar^n$$
> The use of ar^n instead of ar^{n-1} is because we are required to multiply a by r, n times to go from the 0th term to the nth term.

Modelling exponential growth, $r > 1$

For examples involving exponential growth, you can consider a common ratio greater than 1 as the **growth rate** of the model. For example, if $r = 1.025$, then we can say there is a 2.5% increase per unit of time.

> **WORKED EXAMPLE 22** Modelling compound interest
>
> Jax took a $15 000 loan at an interest rate of 6% per annum, compounded annually. Let A_n be the value of the loan after n years.
>
> **a** Write a recursive rule for the value of the loan and, hence, state the value of the loan after one year.
>
> **b** Calculate the value of the loan after five years.
>
> Suppose Jax originally agreed upon a loan at an interest rate of 6% per annum, compounded monthly. Let B_n be the value of the loan after n months.
>
> **c** By first expressing the value of this loan scheme in the form $B_n = a \times r^n$, compare the value of the loan after five years to your answer in part **b**.

Steps	Working
a 1 Express a 6% increase per annum as a decimal.	$r = 1 + \dfrac{6}{100} = 1.06$
2 Write the recursive rule in the correct form using A_0.	$A_{n+1} = 1.06A_n, A_0 = 15\,000$
3 Find A_1, answering with the correct units.	$A_1 = 1.06(15\,000)$ $= \$15\,900$

b	1 Write the general rule in the correct form.	$A_n = 15\,000(1.06)^n$
	2 Find A_5, answering with the correct units to two decimal places.	$A_5 = 15\,000(1.06)^5$ $= \$20\,073.38$
c	1 Express a 6% increase per annum compounded monthly as a decimal.	$r = 1 + \dfrac{0.06}{12} = 1.005$
	2 Write the general rule in the correct form.	$B_n = 15\,000(1.005)^n$
	3 Find B_{60}, answering with the correct units to two decimal places.	5 years $\Rightarrow n = 60$ $B_{60} = 15\,000(1.005)^{60}$ $= \$20\,232.75$
	4 Compare the values of A_5 and B_{60}.	The value of the loan in scheme B is \$159.37 more than the loan in scheme A.

Modelling exponential decay, 0 < r < 1

When working with contexts involving exponential decay, be sure to interpret the rates of decay with respect to 100%. For example, if something decreases at a rate of 10% per unit of time, then the corresponding value of r is 0.9.

WORKED EXAMPLE 23 | Modelling depreciation

A \$1800 laptop loses 20% of its value at the start of every year after its purchase.

a Write a general rule for the value of the laptop n years after purchase, L_n.
b Hence, determine the value of the laptop after three years.
c Suppose the laptop is declared 'written-off', that is, no longer has marketable value, when it falls below \$500. Determine how many years after purchase this is expected to occur.

Steps	Working
a 1 Identify the values of a and r.	$a = 1800, r = 0.8$
2 Write the general rule in the correct form.	$L_n = 1800(0.8)^n$
b Evaluate the general rule for $n = 3$, answering with the correct units to two decimal places.	$L_3 = 1800(0.8)^3$ $= \$921.60$
c 1 Use CAS to find when the value of $L_n < 500$.	$L_5 = 589.82 > 500$ $L_6 = 471.86 < 500$
2 Answer in context of the question.	It is expected to be written-off after 6 years.

EXERCISE 7.5 Modelling exponential growth and decay   ANSWERS p. 478

Recap

1 The fifth term of the geometric sequence $T_{n+1} = \dfrac{1}{5}T_n$, $T_1 = 400$ is

    A $\dfrac{16}{25}$ **B** $\dfrac{16}{5}$ **C** 25   **D** 80   **E** 200

2 For a geometric sequence to be increasing, it must have a common ratio that is

    A $r > -1$ **B** $r > 1$ **C** $0 < r < 1$ **D** $r \neq 1$ **E** $r \neq 0$

▶ **Mastery**

3. **WORKED EXAMPLE 22** Faye took a $40 000 loan at an interest rate of 3.6% per annum, compounded annually. Let A_n be the value of the loan after n years.

   a Write a recursive rule for the value of the loan and, hence, state the value of the loan after one year.

   b Calculate the value of the loan after 10 years.

   Suppose Faye originally agreed upon a loan at 3.6% per annum, compounded every two months. Let B_n be the value of the loan after n months.

   c By first expressing the value of this loan scheme in the form $B_n = a \times r^n$, compare the value of the loan after 10 years to your answer in part **b**.

4. **WORKED EXAMPLE 23** A $1800 smartphone loses 40% of its value at the start of every year after its purchase.

   a Write a general rule for the value of the smartphone n years after purchase, P_n.

   b Hence, determine the value of the smartphone after three years.

   c Suppose the smartphone is declared 'written-off', that is, no longer has marketable value, when it falls below $300. Determine how many years after purchase this is expected to occur.

Calculator-free

5. (6 marks) Identify the value of a (the 0th term) and the value of r in each of the following contexts of exponential growth and decay.

   a A population of 3000 is growing at a rate of 15% per annum. Let P_n be the population after n years. (2 marks)

   b 30 micrograms of a radioactive substance is decaying at a rate of 2% per hour. Let R_n be the amount of substance (in micrograms) remaining after n hours. (2 marks)

   c The temperature of a pot of water placed on the stovetop at 26°C is increasing at a rate of 15% per minute. Let T_n be the temperature of the water after n minutes. (2 marks)

Calculator-assumed

6. (6 marks) The population of a certain city n years after January 2020 can be modelled by the geometric sequence $P_n = 120\,000 \times 1.02^n$.

   a Interpret the significance of 120 000 in context of the question. (1 mark)

   b Interpret the significance of 1.02 in context of the question. (1 mark)

   c Determine the predicted population in January 2035, rounded to the nearest whole. (2 marks)

   d According to this model, determine the number of years it would take for this city's population to double. (2 marks)

7. (7 marks) In a laboratory experiment, a colony of bacteria is observed such that the number of bacteria initially is $C_0 = a$ and the number of bacteria three hours later is $C_3 = 2a$.

   a Determine the value of r in the rule $C_{n+1} = rC_n$, $C_0 = a$ correct to four decimal places and interpret the result in context of the question. (3 marks)

   b Suppose there were 500 bacteria at the beginning of the experiment. Using the value of r from part **a**, determine the number of bacteria after 12 hours. (2 marks)

   c After what amount of time is the number of bacteria expected to exceed 10 times the original number of bacteria? (2 marks)

8 (4 marks) A sample of a radioactive isotope contains 800 grams initially. The isotope decays at a rate according to the geometric sequence $A_n = 800 \times 0.95^n$, where A_n is the mass of the isotope (in grams) remaining after n years.

a Calculate the mass of the isotope remaining after 10 years, correct to the nearest gram. (2 marks)

b The half-life of an isotope is the amount of time it takes for the substance to decay to half of its previous mass. Calculate the half-life of this isotope, correct to two decimal places. (2 marks)

9 (9 marks) Maria is considering two prospective employers. The salary increases offered by the two employers after each of the first three years of employment are shown in the table below, rounded to the nearest dollar.

Employer	Initial	After 1 year	After 2 years	After 3 years
A	$76 000	$76 760	$77 528	$78 303
B	$72 500	$73 950	$75 429	$76 938

a Assuming geometric sequences, write a recursive rule to represent the salary offered by each employer. (4 marks)

b Assuming this salary scheme is applicable for 10 years, compare the expected salary to be offered by each of the employers after 10 years. (3 marks)

c If Maria is planning to stay with her employer for no more than 5 years, justify which employer she should choose based on salary alone. (2 marks)

7.6 Geometric series and their applications

As with arithmetic series, a **geometric series** is the sum of the terms in a geometric sequence. These sums can be expressed numerically or algebraically.

Expressing the sum of geometric sequences

Consider the first four terms of the sequence 2, 4, 8, 16. The finite sum for first four terms, S_4, can be written numerically as $S_4 = 2 + 4 + 8 + 16 = 30$. When $0 < r < 1$, the calculations can become a little more tedious due to the addition of fractions – don't forget you need a common denominator when adding fractions!

For example, the finite sum $S_5 = 3 + 1 + \frac{1}{3} + \frac{1}{9} + \frac{1}{27}$ represents the sum of the first five terms of the sequence $T_{n+1} = \frac{1}{3}T_n$, $T_1 = 3$.

$$S_5 = 3 + 1 + \frac{1}{3} + \frac{1}{9} + \frac{1}{27}$$

$$= \frac{81}{27} + \frac{27}{27} + \frac{9}{27} + \frac{3}{27} + \frac{1}{27}$$

$$= \frac{121}{27}$$

WORKED EXAMPLE 24 — Calculating sums for a small finite n

Calculate the following finite sums in each of the following.

a S_3 for $T_n = 2(3)^{n-1}$

b S_4 for $T_{n+1} = 2T_n$, $T_1 = \dfrac{1}{5}$

c $S_5 = 20 + 10 + 5 + \dfrac{5}{2} + \dfrac{5}{4}$

Steps	Working
a 1 Identify the first term and the common ratio.	$a = 2$, $r = 3$
2 List the first three terms.	$T_1 = 2$, $T_2 = 6$, $T_3 = 18$
3 Evaluate the finite sum.	$S_3 = 2 + 6 + 18 = 26$
b 1 Use the recursive rule to write out the sum as a numerical expression.	$S_4 = \dfrac{1}{5} + \dfrac{2}{5} + \dfrac{4}{5} + \dfrac{8}{5}$
2 Evaluate the finite sum.	$S_4 = \dfrac{15}{5} = 3$
c 1 Obtain a common denominator.	$S_5 = 35 + \dfrac{10}{4} + \dfrac{5}{4}$ $= \dfrac{140}{4} + \dfrac{15}{4}$
2 Evaluate the finite sum.	$S_5 = \dfrac{155}{4}$

The finite geometric series formula

For the sequence $T_{n+1} = 2T_n$, $T_1 = 2$, we can display the terms and corresponding values of S_n in a table of values.

n	1	2	3	4	5
T_n	2	4	8	16	32
S_n	2	6	14	30	62

Observing the values in the row of S_n, it is not obvious what type of numerical pattern it is, but given the exponential nature of T_n, there is an element of an exponential relationship to the pattern. However, this exponential relationship is in the difference of sums terms.

S_n	2	6	14	30	62

+4   +8   +16   +32

The question is how do we express this algebraically?

Consider a general geometric sequence $T_{n+1} = rT_n$, $T_1 = a$. We can observe the generalisation of the terms with respect to a and r but we essentially need a rule in terms of n.

n	1	2	3	4	5
T_n	a	ar	ar^2	ar^3	ar^4

For the finite sum of the first n terms, we have the expression

$$S_n = T_1 + T_2 + T_3 + T_4 + \ldots + T_n$$
$$S_n = a + ar + ar^2 + ar^3 + \ldots + ar^{n-1}$$

There is not much we can do with this expression other than factoring out the a, which does not prove helpful. Instead, let's multiply S_n by r to obtain

$$rS_n = ar + ar^2 + ar^3 + ar^4 + \ldots + ar^{n-1} + ar^n$$

We can now see that S_n and rS_n have all terms in common except for a and ar^n. So, by subtracting the two expressions,

$$a + ar + ar^2 + ar^3 + \ldots + ar^{n-1}$$
$$-(ar + ar^2 + ar^3 + ar^4 + \ldots + ar^{n-1} + ar^n)$$
$$S_n - rS_n = a - ar^n$$

Remember, the objective here is to find a formula for S_n in terms of a, r and n.

Factoring out S_n from the left-hand side and a from the right-hand side, we have

$$S_n(1 - r) = a(1 - r^n)$$

Dividing both sides by $1 - r$, we obtain

$$S_n = \frac{a(1 - r^n)}{1 - r}$$

This formula shows that we cannot define a geometric sequence with $r \neq 1$, and additionally, if $r = 0$ then $S_n = a$ because the remainder of the terms in the sequence would be 0. These two cases ($r = 0$ and $r = 1$) can be ignored for geometric sequences.

Finite sum of a geometric sequence – a geometric series

The value of S_n for a geometric sequence with a first term a and a common ratio $r > 0$ is given by

$$S_n = \frac{a(1 - r^n)}{1 - r}$$

where $r \neq 0, 1$.

WORKED EXAMPLE 25 Calculating finite sums using the first term and the common ratio

Calculate the value of the following finite sums for each of the given sequences.

a S_{20} for $T_{n+1} = 2T_n$, $T_1 = \dfrac{1}{3}$

b S_{30} for $T_n = 10\left(\dfrac{1}{5}\right)^{n-1}$

Steps	Working
a 1 Identify the values of a, r and n from the sequence.	$a = \dfrac{1}{3}$, $r = 2$, $n = 20$
2 Substitute into the finite sum formula.	$S_{20} = \dfrac{\dfrac{1}{3}(1 - 2^{20})}{1 - 2}$ $= -\dfrac{1}{3}(1 - 2^{20})$
3 Evaluate, using CAS.	$= 349\,525$

b 1 Identify the values of a, r and n from the sequence.

$a = 10, r = \dfrac{1}{5}, n = 30$

2 Substitute into the finite sum formula.

$$S_{25} = \dfrac{10\left(1 - \left(\dfrac{1}{5}\right)^{30}\right)}{1 - \dfrac{1}{5}}$$

$$= \dfrac{10\left(1 - \left(\dfrac{1}{5}\right)^{30}\right)}{\dfrac{4}{5}}$$

$$= \dfrac{25}{2}\left(1 - \dfrac{1}{5^{30}}\right)$$

3 Evaluate, using CAS.

$= 12.5$

The infinite geometric series formula

In Worked example 25b, we saw that S_{30} for $T_n = 10\left(\dfrac{1}{5}\right)^{n-1}$ gave 12.5. However, if we use CAS to observe some other finite sums for this sequence, there is an interesting result!

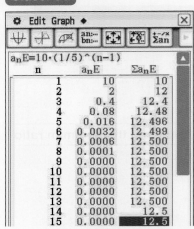

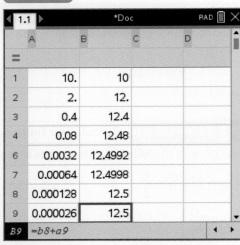

There are multiple finite sum values that give 12.5, getting more precise as n gets larger. If we review the line of working below, we can see why that is the case.

$$\dfrac{25}{2}\left(1 - \dfrac{1}{5^{30}}\right)$$

Replacing the value of 30 with n and letting $n \to \infty$

$$\dfrac{25}{2}\left(1 - \dfrac{1}{5^n}\right)$$

- 5^n will tend towards ∞, that is $5^n \to \infty$, and so
- $\dfrac{1}{5^n} \to 0$, and so
- $1 - \dfrac{1}{5^n} \to 1$.

Meaning that for a sufficiently large n value $S_n = \frac{25}{2}\left(1 - \frac{1}{5^n}\right) \to \frac{25}{2} = 12.5$.

This limiting value is called a **sum to infinity** or an **infinite geometric series** and will only occur when $0 < r < 1$, as the terms of the sequence converge to 0.

> **Sum to infinity of a geometric sequence – an infinite geometric series**
>
> For a convergent geometric sequence with $0 < r < 1$, as $n \to \infty$,
>
> $r^n \to 0$
>
> and so
>
> $1 - r^n \to 1$
>
> So, the value of S_n for a geometric sequence with a first term a and a common ratio $0 < r < 1$ as $n \to \infty$ is given by the formula
>
> $S_\infty = \dfrac{a}{1-r}$
>
> where $r \neq 0, 1$.

WORKED EXAMPLE 26 — Calculating the limiting value of an infinite series from a rule

For each of the following sequences, state whether a sum to infinity exists. If so, determine the value of S_∞. If not, determine S_{10}.

a $A_{n+1} = \dfrac{1}{2}A_n$, $A_1 = 40$ **b** $B_n = 200(0.4)^{n-1}$ **c** $C_{n+1} = 3C_n$, $C_1 = \dfrac{1}{9}$

Steps	Working
a 1 Check that $0 < r < 1$ for a sum to infinity to exist.	$r = \dfrac{1}{2} \Rightarrow 0 < r < 1$ Therefore, a sum to infinity exists.
2 Substitute a and r into the correct formula.	$S_\infty = \dfrac{a}{1-r}$ $= \dfrac{40}{1 - \dfrac{1}{2}}$ $= \dfrac{40}{\dfrac{1}{2}}$
3 Evaluate S_∞.	$S_\infty = 80$
b 1 Check that $0 < r < 1$ for a sum to infinity to exist.	$r = 0.4 \Rightarrow 0 < r < 1$ Therefore, a sum to infinity exists.
2 Substitute a and r into the correct formula.	$S_\infty = \dfrac{a}{1-r}$ $= \dfrac{200}{1 - 0.4}$ $= \dfrac{200}{0.6}$ $= \dfrac{2000}{6}$
3 Evaluate S_∞.	$S_\infty = \dfrac{1000}{3}$

c 1 Check whether $0 < r < 1$ for a sum to infinity to exist.

$r = 3 \Rightarrow r > 1$

Therefore, a sum to infinity does not exist.

2 Substitute a, r and n into the correct formula.

$$S_{10} = \frac{a(1 - r^n)}{1 - r}$$

$$= \frac{\frac{1}{9}(1 - 3^{10})}{1 - 3}$$

$$= -\frac{1}{18}(1 - 3^{10})$$

3 Evaluate S_{10}.

$$S_{10} = \frac{29524}{9}$$

WORKED EXAMPLE 27 — Calculating the limiting value of an infinite series from a sum

Consider the sum $5 + \frac{5}{3} + \frac{5}{9} + \frac{5}{27} + \ldots$

a Identify the values of a and r for the terms of the geometric sequence that produces this sum.
b Write a simplified rule for S_n, the finite sum of the first n terms.
c Justify why a sum to infinity exists and, hence, evaluate S_∞.

Steps	Working
a Identify the first term and the common ratio between terms in the sum.	$a = 5$, $r = \frac{1}{3}$
b 1 Use the finite geometric series formula and substitute in a and r.	$S_n = \frac{a(1 - r^n)}{1 - r}$ $= \frac{5\left(1 - \left(\frac{1}{3}\right)^n\right)}{1 - \frac{1}{3}}$ $= \frac{5\left(1 - \frac{1}{3^n}\right)}{\frac{2}{3}}$
2 Simplify the rule.	$S_n = \frac{15}{2}\left(1 - \frac{1}{3^n}\right)$
c 1 Use $0 < r < 1$ and $n \to \infty$ to explain why S_∞ exists.	As $n \to \infty$, $\frac{1}{3^n} \to 0$ given that $0 < r < 1$.
2 Use part **b** or $S_\infty = \frac{a}{1 - r}$ to evaluate the sum to infinity.	So, $S_\infty = \frac{15}{2} = 7.5$.

Solving practical problems involving geometric series

7.6

Practical problems can require you to combine your knowledge of both sequences and series in the same question.

> **Exam hack**
>
> Be sure to distinguish between a sequence and a series within the question. Series questions might often involve the idea of a 'total' quantity.

> **Exam hack**
>
> Be sure to define your variables if the sequences haven't already been defined for you in the question.

USING CAS 4 | **Solving problems with geometric sequences and series**

A car rental company is offering a loyalty program for extended car rentals over consecutive days. The program allows customers to rent a car at a decreasing daily rate, such that the first day's rate is $50 for the day, and then each subsequent day has a rate valued at 80% of that of the previous day. Let R_n be the daily rental rate for the nth day of the rental period.

a Calculate the daily rental rates for the first five days of consecutive rentals under this loyalty program.

b If a customer rents a car for 7 consecutive days, determine the total cost of the rental under this loyalty program.

c Suppose a customer needs to rent a car for a very long time. Describe the total cost for the car rental that the customer can expect to pay under this loyalty program.

ClassPad

a

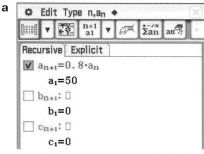

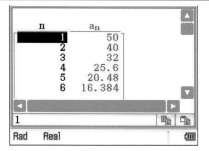

1 Write a recursive rule for the daily rental rates.

2 In the **Sequence** application, input the sequence using the **Recursive** tab.

3 Tap **Table** to generate the sequence.

4 The sequence will appear in the lower window.

5 Use the table to determine the first five rental rates.

$R_1 = \$50$

$R_2 = \$40$

$R_3 = \$32$

$R_4 = \$25.60$

$R_5 = \$20.48$

b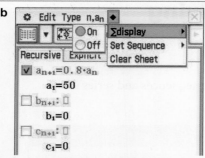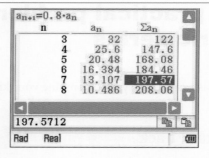

1. Tap ◆ > Σdisplay > On to turn on the sum column.
2. Tap **Table**.

3. In the lower window, scroll down to $n = 7$.

   $S_7 = \$197.57$

c Scroll down to a large value of n.

After approximately 50 consecutive days of rental, the customer can expect to pay a total of $250 for the car rental, regardless of the number of days.

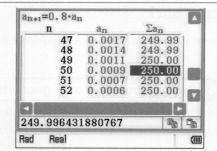

TI-Nspire

a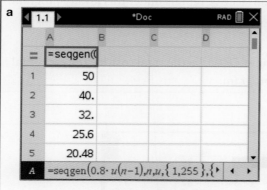

1. Write a recursive rule for the daily rental rates.
2. In a **List & Spreadsheet** page, use the **Generate Sequence** function to generate the sequence
3. Determine the first five rental rates.

b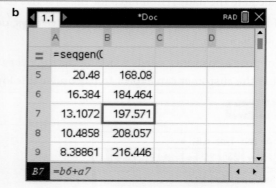

1. Create a second column which shows the cumulative sum of the values in the first column.
2. Scroll down to row **7**.

   $S_7 = \$197.57$

c Scroll down to a row with a large number.

After approximately 50 consecutive days of rental, the customer can expect to pay a total of $250 for the car rental, regardless of the number of days.

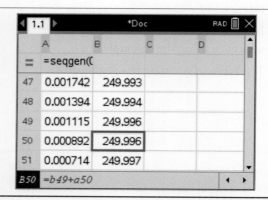

EXAMINATION QUESTION ANALYSIS

Calculator-assumed (13 marks)

Two balls, made of different materials, are dropped from the floor of a building 20 metres above the ground. The heights reached by ball A on subsequent bounces, A_n, follow the terms of an arithmetic sequence such that $A_{n+1} = A_n - 2.5$, $A_0 = 20$. Ball B on the other hand, reaches a height that is 75% of the height of its previous bounce. That is, when dropped from 20 metres above ground, $B_1 = 15$ metres, where B_n is the height reached on the nth bounce.

a Write a recursive rule for the height of the bounce of ball B. (2 marks)

b After which bounce will ball B reach a greater height than ball A? State the difference in the heights at this bounce. (2 marks)

c Based on the model for ball A, after how many bounces will ball A come to a stop? (2 marks)

d Ball B is said to come to a stop once the height of the bounce falls below 0.01 metres. After how many bounces will this occur? (2 marks)

e Compare the total vertical distances travelled by balls A and B. (5 marks)

Reading the question

- Highlight all key numerical information and any rules given in the question.
- Highlight the high-order commands of the question, e.g. *compare*.

Thinking about the question

- Distinguish between the arithmetic sequence of ball A and the geometric sequence of ball B.
- Which questions involve a sequence calculation compared to a series calculation?
- How can I use CAS efficiently to assist in the solving of this problem?

Worked solution (✓ = 1 mark)

a $B_{n+1} = 0.75 B_n$, $B_0 = 20$

expresses B_{n+1} in terms of B_n ✓

states the initial term ✓

b Inputting both sequences into CAS, we observe that ball A bounces higher than ball B until the 7th bounce.

$B_7 = 2.6697$, $A_7 = 2.5$

$B_7 - A_7 = 0.1697$ metres

identifies the 7th bounce ✓

calculates the difference between the heights when $n = 7$ ✓

ClassPad

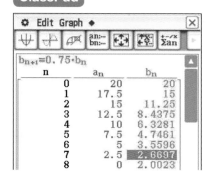

TI-Nspire

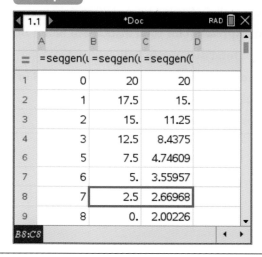

c $A_{n+1} = A_n - 2.5$, $A_0 = 20$

When $A_n = 0$, either using CAS or algebraically,

$A_n = 20 - 2.5n$
$0 = 20 - 2.5n$
$n = 8$

On the **8th bounce**, ball A will come to a stop.

recognises $A_8 = 0$ ✓

interprets answer in context ✓

d The question requires $B_n < 0.01$. Using CAS, $B_{26} = 0.0113$ and $B_{27} = 0.0085$.

Therefore, $B_{27} < 0.01$ and so, after approximately **27 bounces** ball B comes to a stop.

finds when $B_n < 0.01$ ✓

interprets answer in context ✓

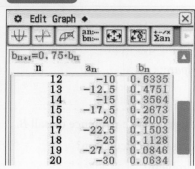

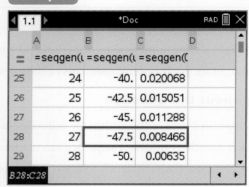

e The balls travel an initial 20 metres and then on each bounce, travel twice the bounce height (upwards and downwards).

So, the total distance, D, travelled by each of the balls can be expressed as such:

Ball A:

$D_A = A_0 + A_1 + A_1 + A_2 + A_2 + \ldots + A_7 + A_7$
$= A_0 + 2A_1 + 2A_2 + \ldots + 2A_7$
$= A_0 + 2(A_1 + A_2 + \ldots + A_7)$
$= A_0 + 2S_7$
$= 20 + 2\left[\frac{7}{2}(2(17.5) + (7-1)(-2.5))\right]$
$= 20 + 2(70)$
$= 160$ metres

Ball B:

$D_B = B_0 + B_1 + B_1 + B_2 + B_2 + \ldots$
$= B_0 + 2B_1 + 2B_2 + \ldots$
$= B_0 + 2(B_1 + B_2 + \ldots)$
$= B_0 + 2S_\infty$
$= 20 + 2\left[\frac{15}{1 - 0.75}\right]$
$= 20 + 2(60)$
$= 140$ metres

Therefore, ball A travels a total of **20 metres more** than ball B.

finds an expression for total distance of ball A in terms of a finite arithmetic series ✓

calculates total distance of ball A ✓

finds an expression for total distance of ball B in terms of an infinite geometric series ✓

calculates total distance of ball B ✓

compares distances travelled ✓

EXERCISE 7.6 Geometric series and their applications

ANSWERS p. 479

Recap

1 $12\,000$ is invested into an account earning 4.5% interest per annum compounded quarterly. Let A_n be the amount in the account after n quarters. The recursive rule to model this situation is

A $A_n = 12\,000(1.045)^n$

B $A_n = 12\,000(4.5)^n$

C $A_{n+1} = 1.011\,25 A_n,\ A_0 = 12\,000$

D $A_n = 1.045 A_{n-1},\ A_0 = 12\,000$

E $A_n = 12\,000(1.125)^n$

2 The temperature of an oven cools according to the general rule $T_n = 220 \times 0.9^n$, where T_n is the temperature (°C) of the oven n minutes after it is turned off. The amount of time it takes to cool down to 24°C is approximately

A 19 minutes   **B** 20 minutes   **C** 21 minutes   **D** 22 minutes   **E** 23 minutes

Mastery

3 WORKED EXAMPLE 24   Calculate the following finite sums for each of the following.

   a S_5 for $T_n = 40\left(\dfrac{1}{2}\right)^{n-1}$ **b** S_3 for $T_{n+1} = 6T_n,\ T_1 = -1$ **c** $S_4 = 300 + 60 + 12 + \dfrac{12}{5}$

4 WORKED EXAMPLE 25   Calculate the value of the following finite sums for each of the given sequences.

   a S_{10} for $T_{n+1} = 5T_n,\ T_1 = \dfrac{1}{1000}$ **b** S_{30} for $T_n = 2025\left(\dfrac{2}{5}\right)^{n-1}$

5 WORKED EXAMPLE 26   For each of the following sequences, state whether a sum to infinity exists. If so, determine the value of S_∞. If not, determine S_5.

   a $A_{n+1} = \dfrac{2}{3} A_n,\ A_1 = 60$ **b** $B_n = 50 \times (1.2)^{n-1}$ **c** $C_{n+1} = \dfrac{1}{2} C_n,\ C_1 = 360$

6 WORKED EXAMPLE 27   Consider the sum $9 + \dfrac{9}{2} + \dfrac{9}{4} + \dfrac{9}{8} + \ldots$

   a   Identify the values of a and r for the terms of the geometric sequence that produces this sum.

   b   Write a simplified rule for S_n, the finite sum of the first n terms.

   c   Justify why a sum to infinity exists and, hence, evaluate S_∞.

7 Using CAS 4   A sun lounge and beach umbrella rental company on a beach on the Amalfi Coast is offering a loyalty program for extended lounge and umbrella rentals over consecutive days. The program allows tourists to rent a spot on the beach at a decreasing daily rate, such that the first day's rate is $25 for the day, and then each subsequent day has a rate valued at 90% of that of the previous day. Let R_n be the daily rental rate for the nth day of the rental period.

   a   Calculate the daily rental rates for the first five days of consecutive rentals under this loyalty program.

   b   If a tourist rents a spot on the beach for 3 consecutive days, determine the total cost of the rental under this loyalty program.

   c   Suppose a tourist is staying on the Amalfi Coast for an extended period of time. Describe the total cost for the sun lounge and umbrella rental that the tourist can expect to pay under this loyalty program.

Calculator-free

8 (9 marks) The first two finite sums of a geometric series are $S_1 = 16$ and $S_2 = 20$.

a Determine the common ratio r of the corresponding geometric sequence, T_n. (2 marks)

b Hence, determine the first five terms of the corresponding geometric sequence. (2 marks)

c Prove that the sum of the first n terms is given by the rule $S_n = \dfrac{64 - 4^{3-n}}{3}$. (3 marks)

d Hence, or otherwise, determine S_3. (1 mark)

e Describe the behaviour of S_n as $n \to \infty$. (1 mark)

Calculator-assumed

9 (6 marks) A geometric sequence is defined such that $A_1 = 5$ and $A_4 = \dfrac{8}{25}$.

a Determine the value of r, the common ratio. (2 marks)

b Write a rule for the sum of the first n terms, S_n. (2 marks)

c Hence, determine S_8 to two decimal places. (2 marks)

10 (7 marks) A sales commission program is offered to salespersons such that the number of sales they make during the month determines the amount of commission they receive according to a geometric sequence. When a single sale is made, the commission earned is $50. For each subsequent sale in the same month, the commission increases by 10%. Let C_n be the value of the nth commission amount earned in a month.

a Determine C_4. (2 marks)

b Write a rule for S_n, the total commission earned by an employee who makes n sales within a month. (2 marks)

c Hence, determine S_8. (1 mark)

d Suppose an employee aims to earn a total commission of $1000 in a month. Determine the minimum number of sales required to reach that goal. (2 marks)

11 (6 marks) A pyramidal structure is being built out of blocks such that each subsequent row of the pyramid from the base to the apex has a quarter of the blocks in the row below it. The base row of the pyramid has 4096 blocks and the apex row has 1 block.

a Express the number of blocks in the nth row of the pyramidal structure as a rule, B_n. (2 marks)

b Determine the number of rows in this pyramidal structure. (2 marks)

c Use your understanding of geometric series to determine the total number of blocks used in this structure. (2 marks)

12 (6 marks) A ball is dropped from a height of 15 metres. After each bounce, the ball rebounds to a height that is 60% of the previous height. The rebound heights form a geometric sequence H_n, where H_n is the rebound height of the ball after the nth bounce. Show the use of geometric sequences and/or series to

a determine the height of the first four bounces of the ball (2 marks)

b calculate the total vertical distance travelled by the ball up until the fifth bounce (2 marks)

c describe the total vertical distance travelled up until the ball comes to rest. (2 marks)

Chapter 7: Chapter summary

Arithmetic sequences

Let T be an **arithmetic sequence** with a first term $T_1 = a$ and a **common difference** of d.

- The **recursive rule** for such a sequence can be expressed as
$$T_{n+1} = T_n + d, \; T_1 = a$$
- The **general rule** (or explicit rule) for such a sequence can be expressed as
$$T_n = a + (n-1)d$$
- If $d > 0$, the sequence is increasing.
- If $d < 0$, the sequence is decreasing.
- Given two terms of an arithmetic sequence, $T_m = a$ and $T_n = b$, we can consider these as coordinates (m, a) and (n, b). Then the common difference d can be found by
$$d = \frac{b-a}{n-m} = \frac{T_n - T_m}{n-m}$$
- Arithmetic sequences can be displayed in table and graphical forms as discrete points along a linear relationship.

Modelling linear growth and decay

- An arithmetic sequence with a common difference d can be used to model
  - **linear growth** (when $d > 0$), or
  - **linear decay** (when $d < 0$).
- If $T_0 = a$ instead of $T_1 = a$, then:
  - Recursive rule
  $$T_{n+1} = T_n + d, \; T_0 = a$$
  - General rule
  $$T_n = a + nd$$

Arithmetic series and their applications

- A **series** is the sum of the first n terms of a sequence and can be expressed in the form $S_n = T_1 + T_2 + \ldots + T_n$.
- Given that the finite sum of the first $(n+1)$ terms is $S_{n+1} = T_1 + T_2 + \ldots + T_n + T_{n+1}$, then the finite sums form their own recursive sequence of the form
$$S_{n+1} = S_n + T_{n+1}$$
- The **arithmetic series** formula has two forms:
  - When given a and d,
  $$S_n = \frac{n}{2}\left[2a + (n-1)d\right]$$
  - When given T_1 and T_n,
  $$S_n = n\left(\frac{T_1 + T_n}{2}\right)$$

Geometric sequences

Let T be a **geometric sequence** with a first term $T_1 = a$ and a **common ratio** of $r > 0$, where $r \neq 0, 1$.

- The recursive rule for such a sequence can be expressed as
  $$T_{n+1} = rT_n, T_1 = a$$
- The general rule (or explicit rule) for such a sequence can be expressed as
  $$T_n = ar^{n-1}$$
- Given two terms of a geometric sequence, $T_m = a$ and $T_n = b$, we can consider these as coordinates (m, a) and (n, b). Then the common ratio r can be found by solving the equation
  $$r^{n-m} = \frac{b}{a} = \frac{T_n}{T_m}$$
- When $r < 0$, it is an **alternating sequence**.

Modelling exponential growth and decay

- A geometric sequence with a first term of a and common ratio r can be used to model
  - exponential growth (when $a > 0, r > 0$), or
  - exponential decay (when $a > 0, 0 < r < 1$).
- If $T_0 = a$ instead of $T_1 = a$, then:
  - Recursive rule
    $$T_{n+1} = rT_n, T_0 = a$$
  - General rule
    $$T_n = ar^n$$
- The interpretation of r as a percentage with respect to 1 is the **growth or decay** rate of an exponential growth or decay problem.

Geometric series and their applications

- The **geometric series** formula for a geometric sequence with a first term a and a common ratio $r > 0$ is given by
  $$S_n = \frac{a(1 - r^n)}{1 - r}$$
  where $r \neq 0, 1$.
- If $0 < r < 1$, then a **sum to infinity**, or an **infinite series** exists such that
  $$S_\infty = \frac{a}{1 - r}$$

Cumulative examination: Calculator-free

Total number of marks: 36   Reading time: 4 minutes   Working time: 36 minutes

1 (4 marks)

   a Show the use of Pascal's triangle to expand $(x + y)^4$. (2 marks)

   b Hence, or otherwise, determine the coefficient of the x^3 term in the expansion of $(x - 3)^4$. (2 marks)

2 (10 marks) As customers are walking into a local department store, they are asked the question *What are you shopping for today?* For a group of 200 customers, the results are shown below.

- 120 were shopping for clothing (C).
- 80 were shopping for electronics (E).
- 60 were shopping for décor (D).
- 30 were shopping for both clothing and electronics.
- 20 were shopping for both electronics and décor.
- 10 were shopping for clothing, electronics and décor.
- 25 were shopping for other items.

   a Construct and complete a Venn diagram to represent this situation. (2 marks)

   b Hence, calculate the probability that a randomly selected customer of this department store is

   i shopping exactly one of the three types of items (2 marks)

   ii shopping for electronics, given that they are also shopping for clothing (2 marks)

   iii not shopping for décor, given that they are shopping for clothing. (2 marks)

   c Use the relative frequencies to estimate the probability that the next two customers who walk into the store are both shopping for electronics. (2 marks)

3 (4 marks) Solve the following equations.

   a $x^2 - 14x + 48 = 0$ (1 mark)

   b $3(x + 7)^2 - 72 = 0$ (1 mark)

   c $x^2 - 5x + 2 = 0$ (2 marks)

4 (4 marks) Use the expansion of $\cos(x - y)$ and $\sin(x - y)$ to prove that $\cos\left(\dfrac{\pi}{2} - x\right) = \sin(x)$ and $\sin\left(\dfrac{\pi}{2} - x\right) = \cos(x)$.

5 (4 marks) Solve the following exponential equations for x.

   a $64^{x+3} \times 32 = 4 \times 8^{3x+1}$ (2 marks)

   b $\dfrac{27^{3x+2}}{9^{3-x}} = 243$ (2 marks)

6 (10 marks) Two sequences are defined according to the recursive rules below.

$A_{n+1} = A_n - 4, A_1 = a$

$B_n = rB_{n-1}, B_1 = 8$

It is known that $A_3 = B_3 = k$.

   a Determine a rule for a in terms of r. (3 marks)

   b Determine the possible values of a and r, if $a = 20r$. (5 marks)

   c Hence, determine the possible values of k. (2 marks)

Cumulative examination: Calculator-assumed

Total number of marks: 24   Reading time: 3 minutes   Working time: 24 minutes

1 (4 marks) Determine the equation of each of the following parabolas.

a (2 marks)

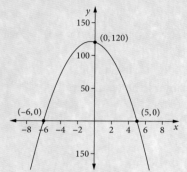

b (2 marks)

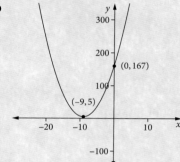

2 (2 marks) $f(x) = (x + 5)^3 - 4$ is dilated by a factor of 4 from the x-axis, reflected in the y-axis and translated 3 in the negative direction of the x-axis to give $g(x)$. What is the equation of $g(x)$?

3 (8 marks) The number of bacteria N in a colony after t days is given by the function $N(t) = N_0(a^t)$. A bacterial colony starts with 1000 bacteria and the number of bacteria doubles every day.

  a Find the function $N(t)$. (2 marks)

  b Find how many bacteria will be in the colony after 50 days. Write your answer in scientific notation, correct to two significant figures. (3 marks)

  c Find the number of days, correct to two significant figures, for the colony to grow to 10 times it's original size. (3 marks)

4 (10 marks) A town council is planning to build a new park, and has allocated a budget to plant trees and install infrastructure such as benches and a playground. The monthly plan outlines the following details:

- The number of trees to be planted initially is 20 and the number of trees in the park will increase by 2 trees per month thereafter for a year.
- The cost of installing infrastructure initially, I_0, is $1800 and the maintenance of this infrastructure decreases by 30% each subsequent month.

  a Write a recursive rule for the number of trees in the park in the form $T_{n+1} = T_n + d, T_0 = a$. (2 marks)

  b Calculate the total number of trees planted in the park by the end of the year. (2 marks)

  c Write a rule for the cost of the infrastructure maintenance in the nth month, I_n. (2 marks)

  d Calculate the total cost of the infrastructure and its maintenance by the end of the year. (2 marks)

  e Suppose the tree planting ceased at the end of the year, but the infrastructure maintenance continued indefinitely. Describe the long-term cost of maintaining this infrastructure according to this plan. (2 marks)

CHAPTER 8

DIFFERENTIAL CALCULUS

Syllabus coverage
Nelson MindTap chapter resources

8.1 Rates of change
Average rate of change
Tangents, secants and chords
Using CAS 1: Finding the rate of change using a tangent line

8.2 Instantaneous rates of change and the gradient function
Instantaneous rates of change
The gradient function
Using CAS 2: Finding the gradient function

8.3 Differentiation by first principles
Limits
Differentiation by first principles

8.4 Differentiating polynomial functions
Differentiating powers of x
Differentiating polynomial functions
Using CAS 3: Completing a term-by-term differentiation
Using CAS 4: Finding the derivative at a given point
Using CAS 5: Finding the function from the derivative

Examination question analysis
Chapter summary
Cumulative examination: Calculator-free
Cumulative examination: Calculator-assumed

Syllabus coverage

TOPIC 2.3: INTRODUCTION TO DIFFERENTIAL CALCULUS

Rates of change

2.3.1 interpret the difference quotient $\dfrac{f(x+h)-f(x)}{h}$ as the average rate of change of a function f

2.3.2 use the Leibniz notation δx and δy for changes or increments in the variables x and y

2.3.3 use the notation $\dfrac{\delta y}{\delta x}$ for the difference quotient $\dfrac{f(x+h)-f(x)}{h}$ where $y = f(x)$

2.3.4 interpret the ratios $\dfrac{f(x+h)-f(x)}{h}$ and $\dfrac{\delta y}{\delta x}$ as the slope or gradient of a chord or secant of the graph of $y = f(x)$

The concept of the derivative

2.3.5 examine the behaviour of the difference quotient $\dfrac{f(x+h)-f(x)}{h}$ as $h \to 0$ as an informal introduction to the concept of a limit

2.3.6 define the derivative $f'(x)$ as $\lim\limits_{h \to 0} \dfrac{f(x+h)-f(x)}{h}$

2.3.7 use the Leibniz notation for the derivative: $\dfrac{dy}{dx} = \lim\limits_{\delta x \to 0} \dfrac{\delta y}{\delta x}$ and the correspondence $\dfrac{dy}{dx} = f'(x)$ where $y = f(x)$

2.3.8 interpret the derivative as the instantaneous rate of change

2.3.9 interpret the derivative as the slope or gradient of a tangent line of the graph of $y = f(x)$

Computation of derivatives

2.3.10 estimate numerically the value of a derivative for simple power functions

2.3.11 examine examples of variable rates of change of non-linear functions

2.3.12 establish the formula $\dfrac{d}{dx}(x^n) = nx^{n-1}$ for non-negative integers n expanding $(x+h)^n$ or by factorising $(x+h)^n - x^n$

Properties of derivatives

2.3.13 understand the concept of the derivative as a function

2.3.14 identify and use linearity properties of the derivative

2.3.15 calculate derivatives of polynomials

Mathematics Methods ATAR Course Year 11 syllabus pp. 13–14 © SCSA

Video playlists (5):
- 8.1 Rates of change
- 8.2 Instantaneous rates of change and the gradient function
- 8.3 Differentiation by first principles
- 8.4 Differentiating polynomial functions
- **Examination question analysis** Differential calculus

Worksheets (3):
- 8.2 Instantaneous rates of change
- 8.4 Derivative of a sum of terms • Derivatives of linear products

To access resources above, visit
cengage.com.au/nelsonmindtap

8.1 Rates of change

Average rate of change

A **rate** indicates how one quantity changes as a result of a change in another quantity. We refer to these quantities as variables. For example, if there was a staircase and each step takes 2 seconds to climb, the rate of change is 2 seconds/step.

In each case, the rate of change is constant.

For variables x and y, with x as the independent variable and y as the dependent variable, the rate of change, $\dfrac{\delta y}{\delta x}$, is the rate of change in the dependent variable y with respect to the independent variable x, where the symbol δ represents change in a quantity.

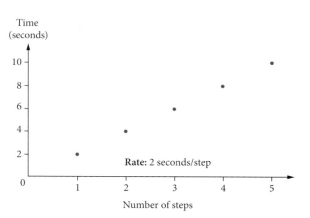

For a **linear function**, the rate of change is constant and represents the gradient of the straight line.

For 2 points $A(x_1, y_1)$ and $B(x_2, y_2)$ on the graph of a linear function, the rate of change between them is $\dfrac{\delta y}{\delta x} = \dfrac{y_2 - y_1}{x_2 - x_1}$.

For a non-linear function, $\dfrac{\delta y}{\delta x} = \dfrac{y_2 - y_1}{x_2 - x_1}$ is the **average rate of change** between the two points.

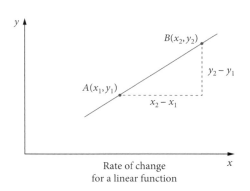

Rate of change for a linear function

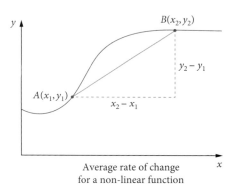

Average rate of change for a non-linear function

A rate of change can be positive, zero or negative, depending on the gradient of the straight line connecting the two points.

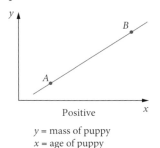

Positive

y = mass of puppy
x = age of puppy

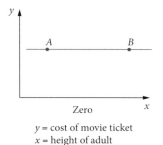

Zero

y = cost of movie ticket
x = height of adult

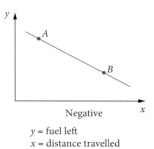

Negative

y = fuel left
x = distance travelled

> ### Rate of change
> The rate of change is the gradient of a straight-line segment.
>
> The rate of change can be positive, zero or negative.
>
> The rate of change between the points $A(x_1, y_1)$ and $B(x_2, y_2)$ is $\dfrac{\delta y}{\delta x} = \dfrac{y_2 - y_1}{x_2 - x_1}$.

WORKED EXAMPLE 1 — Finding constant rate of change

For each graph, calculate, where possible, the constant rate of change over the entire interval, also stating whether the rate of change is positive, zero or negative. If not possible, give a reason.

a
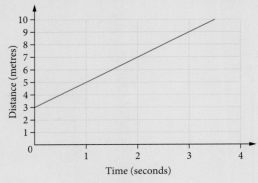

b
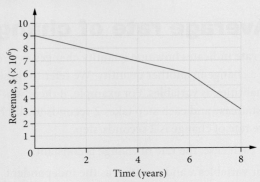

c
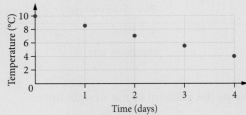

Steps	Working
a 1 Decide if the gradient is constant everywhere on the graph.	The line is increasing uniformly, so the gradient is constant and positive.
2 Calculate the gradient. Choose two points on the line, say $(0, 3)$ and $(3, 9)$, and calculate the gradient of the line.	$\dfrac{\delta y}{\delta x} = \dfrac{y_2 - y_1}{x_2 - x_1}$ $= \dfrac{9 - 3}{3 - 0}$ $= 2$
3 State the rate of change.	The rate of change is 2 metres/second.
b 1 Decide if the gradient is constant everywhere on the graph.	The gradient in the interval from 0 to 6 years is different from the gradient from 6 to 8 years, so the gradient is not constant.
2 State the rate of change.	Since the gradient is not constant over the whole interval, there is no constant rate of change.
c 1 Decide if the gradient is constant everywhere on the graph.	A line connecting the points is decreasing uniformly, so the gradient is constant and negative.
2 Use the points $(0, 10)$ and $(4, 4)$ to calculate the gradient of the line.	$\dfrac{\delta y}{\delta x} = \dfrac{y_2 - y_1}{x_2 - x_1}$ $= \dfrac{4 - 10}{4 - 0}$ $= -1.5$
3 State the rate of change.	The rate of change is -1.5°C/day or the temperature is decreasing at a rate of 1.5°C each day.

WORKED EXAMPLE 2 — Finding average rate of change

The position of a particle is described by the equation $x = 2t^3$, where x is in metres and t is in minutes. Find the average rate of change of the particle's position between $t = 1$ and $t = 3$.

Steps	Working
1 Work out the required x values.	When $t = 1$, $x = 2 \times 1^3 = 2$. When $t = 3$, $x = 2 \times 3^3 = 54$.
2 Write the coordinates.	$A(t_1, x_1) = (1, 2)$ $B(t_2, x_2) = (3, 54)$
3 Calculate the gradient.	$\dfrac{\delta x}{\delta t} = \dfrac{x_2 - x_1}{t_2 - t_1} = \dfrac{54 - 2}{3 - 1} = 26$
4 State the average rate of change.	The average rate of change between $t = 1$ and $t = 3$ is 26 metres/minute.

Tangents, secants and chords

A **tangent** is a straight line that touches a curve at one point. A straight line passing through two points on a curve is a **secant** and the interval (line segment) joining two points on a curve is a chord.

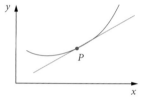
Tangent at point P

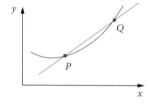
Secant passing through P and Q

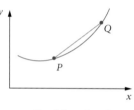
Chord from P to Q

The gradient of tangent lines can be used to approximate the rate of change of a function at a point.

WORKED EXAMPLE 3 — Finding the tangent from a graph

This graph shows the amount of water in a tank as it is being drained. Use a tangent to the graph to find the flow rate at 10 minutes.

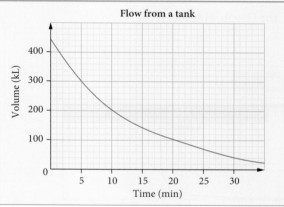

Steps	Working
1 Use a transparent ruler and a pencil to draw a tangent at 10 minutes.	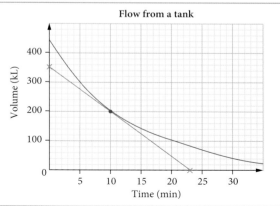

2 Find two points on the tangent.		$(0, 350)$ and $(23, 0)$ are on the tangent line.
3 Find the gradient of the tangent.		$m = \dfrac{y_2 - y_1}{x_2 - x_1}$
		$= \dfrac{0 - 350}{23 - 0}$
		$\approx -15 \text{ kL/min}$
4 Write the answer.		The water is flowing out at approximately $15\,\text{kL/min}$.

USING CAS 1 — Finding the rate of change using a tangent line

Graph $f(x) = \dfrac{1}{2 - x}$ and calculate the rate of change at $x = 4$ using the tangent.

ClassPad

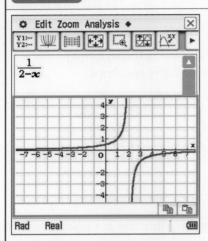

1 In **Main**, enter and highlight the expression.
2 Tap **Graph** and drag the expression down into the Graph window.

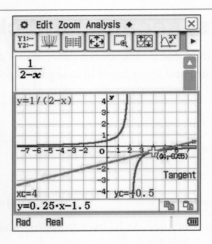

3 Tap **Analysis** > **Sketch** > **Tangent**.
4 Enter **4** then tap **OK** to display the tangent line when $x = 4$.
5 Press **EXE** to display the equation of the tangent line.

TI-Nspire

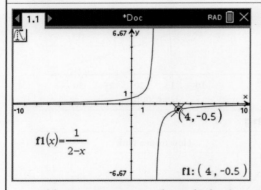

1 Add a **Graphs** page and graph the function.
2 Press **menu** > **Trace** > **Graph Trace**.
3 Enter **4** and press **enter** to jump to the point on the graph where $x = 4$.
4 Press **enter** to label the coordinates of this point.

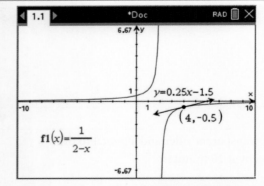

5 Press **menu** > **Geometry** > **Points & Lines** > **Tangent**.
6 Click on the point where $x = 4$ to display the tangent line at this point.
7 Press **enter** to display the equation of the tangent line.

The equation of the tangent line at $x = 4$ is $y = 0.25x - 1.5$.
Therefore, the rate of change is 0.25, which is the slope of the tangent line.

EXERCISE 8.1 Rates of change

ANSWERS p. 480

Mastery

1. **WORKED EXAMPLE 1** For each graph, calculate, where possible, the constant rate of change over the entire interval, also stating whether it is positive, zero or negative. If not possible, give a reason.

   a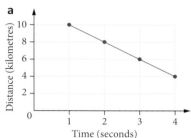
   b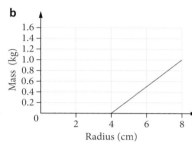
   c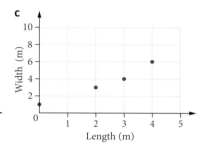

2. **WORKED EXAMPLE 2** The period, T seconds, of a pendulum of length L metres can be approximated by the rule $T = 2\sqrt{L}$.
   Find the average rate of change of the pendulum's period between $L = 4$ and $L = 9$.

3. **WORKED EXAMPLE 3** The graph displays the intensity of light (lumens) at a given distance from a candle source.
   Calculate the rate of change, correct to one decimal place, of the light intensity at a distance of

   a 2 m
   b 9 m

   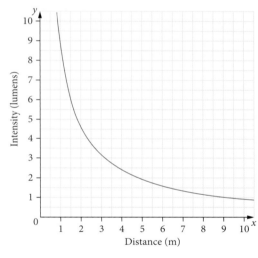

4. **Using CAS 1** Use CAS to sketch the graph of $f(x) = x^3 - 3x^2 - 4x + 12$ and calculate the rate of change using a tangent line at $x = 3.5$.

5. Copy the graph below and by drawing tangent lines, determine the rate of change at A, B and C shown in the graph.

   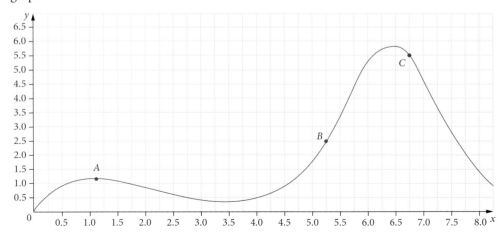

Chapter 8 | Differential calculus 389

Calculator-free

6 (2 marks) Determine the rate of change of this line.

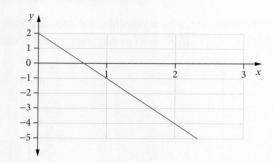

7 (5 marks) State whether each of the following statements is true or false.

a A negative rate of change means both variables decrease at the same time. (1 mark)

b For the equation $x = 3y$, the average rate of change of y with respect to x is $\dfrac{\delta x}{\delta y}$. (1 mark)

c The gradient of a line of best fit is the rate of change. (1 mark)

d If M is the independent variable and N is the dependent variable, its rate of change can be represented as $\dfrac{\delta N}{\delta M}$. (1 mark)

e For a linear graph, the rate of change is always positive. (1 mark)

8 (3 marks) For the function with the rule $f(x) = x^3 - 4x$, determine the average rate of change of $f(x)$ with respect to x for $1 \leq x \leq 3$.

9 (2 marks) The position x cm of a particle at time t seconds is described by $x = \dfrac{3}{4}(2t + 1)$. Determine the average rate of change of distance with respect to time for the first 4 seconds.

10 (5 marks) State whether each of the following statements is true or false.

a The tangent to any point on a graph intersects the graph at two points. (1 mark)
b The length of a chord cannot be greater than its corresponding secant. (1 mark)
c A chord on a curve cannot be a tangent elsewhere on the same curve. (1 mark)
d Two points on a curve cannot have a tangent that has the same gradient. (1 mark)
e The gradient of a secant cannot be negative. (1 mark)

11 (2 marks) State the rate of change at $(2, 10)$ on the graph.

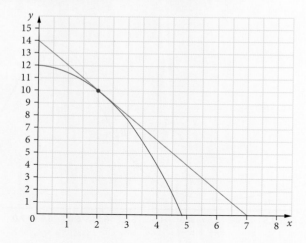

12 (2 marks) Approximate the rate of change at $x = 2$ on the graph below.

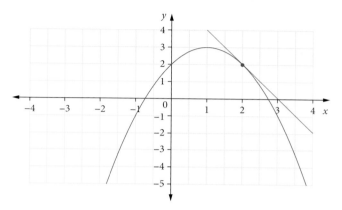

Calculator-assumed

13 (2 marks) Determine the average rate of change of $y = x - \frac{1}{2}x^2$ between $x = 4$ and $x = 8$.

14 (4 marks) A particle's position, P metres, at time t seconds can be described by $P = t^2$.
  a Find the average rate of change of the particle between $t = 1$ and $t = 2$. (1 mark)
  b Find the average rate of change of the particle between $t = 4$ and $t = 5$. (1 mark)
  c Find the average rate of change of the particle between $t = t_1$ and $t = t_2$. (2 marks)

15 (2 marks) After 1.5 minutes, an initially full leaking 4-litre bucket now has 3.4 litres of water remaining. Determine the average rate of change of the volume of the water.

16 (2 marks) At which point, P, Q, R, S or T on the graph below, is the rate of change approximately -0.4?

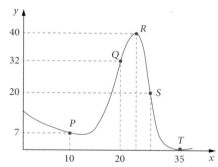

17 (2 marks) The points $P(16, 3)$ and $B(25, 0)$ are labelled on the diagram.

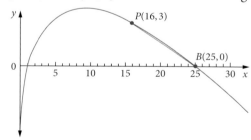

Find m, the gradient of the chord PB. (Exact value to be given.)

8.2 Instantaneous rates of change and the gradient function

Instantaneous rates of change

In the diagram on the right, point Q is moving towards point P, and $Q_1, Q_2, Q_3 \ldots$ represent the positions of Q as it approaches P.

The **instantaneous rate of change** at P is the gradient of the tangent line at P.

The gradients of the chords, $Q_1P, Q_2P, Q_3P \ldots$ are approximations for the instantaneous rate of change at P.

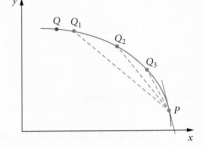

On the diagram on the right, consider the fixed point $P(x, f(x))$ and the variable point $Q(x+h, f(x+h))$. As Q approaches P, h approaches zero, thus the gradient of the curve at point P equals the gradient of the tangent line at point P.

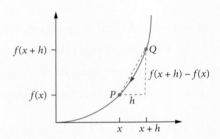

Instantaneous rates of change

For a small value of h, the rate of change at P can be approximated by using:

$$\text{gradient of the tangent} = \frac{\delta y}{\delta x} = \frac{f(x+h) - f(x)}{h}$$

WORKED EXAMPLE 4 — Approximating the rate of change

Find the approximate rate of change of $f(x) = x^3 - x^2 + 2x + 3$ between the points with x-coordinates $x = 2$ and $x = 2.1$.

Steps	Working
1 Work out the value of h.	Find the difference between the two x values. $h = 2.1 - 2 = 0.1$
2 Evaluate δy.	Find the difference between the two y values. $\delta y = f(x+h) - f(x)$ $= f(2.1) - f(2)$ $= 12.051 - 11$ $= 1.051$
3 Calculate $\dfrac{\delta y}{\delta x}$.	$\dfrac{\delta y}{\delta x} = \dfrac{f(x+h) - f(x)}{h} = \dfrac{1.051}{0.1} = 10.51$
4 State the answer.	The approximate rate of change is 10.51.

The value of the instantaneous rates of change can be determined by approaching the given value of x using a number of small h values.

WORKED EXAMPLE 5 — Finding the instantaneous rate of change at a point

a Use $h = 0.1, 0.01, 0.001$ to find approximations for the instantaneous rate of change of $f(x) = x^2$ at $x = 5$.
b Hence, state the instantaneous rate of change of $f(x)$ at $x = 5$.

Steps	Working
a Find each approximation using $\dfrac{\delta y}{\delta x} = \dfrac{f(x + h) - f(x)}{h}$.	$h = 0.1$ $\dfrac{f(x + h) - f(x)}{h} = \dfrac{(5 + 0.1)^2 - 5^2}{0.1} = 10.1$ $h = 0.01$ $\dfrac{f(x + h) - f(x)}{h} = \dfrac{(5 + 0.01)^2 - 5^2}{0.01} = 10.01$ $h = 0.001$ $\dfrac{f(x + h) - f(x)}{h} = \dfrac{(5 + 0.001)^2 - 5^2}{0.001} = 10.001$
b Deduce the instantaneous rate of change at the required value.	As h becomes smaller, the rates of change approach 10, so the instantaneous rate of change is 10.

The gradient function

The **gradient function** gives the rate of change, or the gradient of the tangent, for any value of x.

In the first graph below, the tangents at points A, B, C and D represent the instantaneous rates of change on the curve $y = x^2$ at each point. Notice that the gradient becomes steeper as we move from A to D.

This means that the rate of change of the gradient is increasing. Plotting each gradient, $\dfrac{\delta y}{\delta x}$, with respect to its x-coordinate produces a linear graph of gradient 2 (the second graph below).

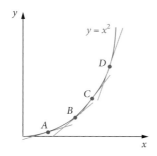
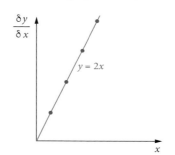

To confirm this, suppose we approach each point and calculate the rate of change of the function $y = f(x) = x^2$ using the x values 0, 1, 2, 3, 4 and $h = 0.02$.

The results for $\dfrac{\delta y}{\delta x} = \dfrac{f(x + h) - f(x)}{h}$ are summarised below.

$x = 0$, $\dfrac{\delta y}{\delta x} = \dfrac{(0 + 0.02)^2 - 0^2}{0.02} \approx 0$

$x = 1$, $\dfrac{\delta y}{\delta x} = \dfrac{(1 + 0.02)^2 - 1^2}{0.02} \approx 2$

$x = 2$, $\dfrac{\delta y}{\delta x} = \dfrac{(2 + 0.02)^2 - 2^2}{0.02} \approx 4$

$x = 3$, $\dfrac{\delta y}{\delta x} = \dfrac{(3 + 0.02)^2 - 3^2}{0.02} \approx 6$

$x = 4$, $\dfrac{\delta y}{\delta x} = \dfrac{(4 + 0.02)^2 - 4^2}{0.02} \approx 8$

> **Exam hack**
>
> The relationship between the x value and its rate of change may not be immediately obvious. Depending on h and the x values used, the rate of change may be a decimal requiring rounding before a relationship becomes apparent. Look for a general pattern in the values.

The results show that the rate of change is twice the value of the x-coordinate.

If we let $R(x)$ be the gradient function, then $R(x) = 2x$.

WORKED EXAMPLE 6 — Finding the gradient function using approximations

Use $h = 0.1$ to find the gradient function $R(x)$ of $f(x) = x^2 + x$ for the x values 1, 2 and 3.

Steps	Working
1 Obtain $\frac{\delta y}{\delta x}$ for each x value given.	Substitute each x value into $\frac{\delta y}{\delta x} = \frac{f(x + h) - f(x)}{h}$. $x = 1, \quad \frac{\delta y}{\delta x} = \frac{[(1.1)^2 + 1.1] - [1^2 + 1]}{0.1} \approx 3$ $x = 2, \quad \frac{\delta y}{\delta x} = \frac{[(2.1)^2 + 2.1] - [2^2 + 2]}{0.1} \approx 5$ $x = 3, \quad \frac{\delta y}{\delta x} = \frac{[(3.1)^2 + 3.1] - [3^2 + 3]}{0.1} \approx 7$
2 Identify a relationship between the x value used and the gradient.	It appears that the gradient is double the x value plus 1.
3 State the gradient function, $R(x)$.	$R(x) = 2x + 1$

Using CAS makes the computations in the previous worked example much easier.

USING CAS 2 — Finding the gradient function

Find approximate rates of change for the function $f(x) = 4x^2 - x + 5$ at various values of x to suggest a gradient function $g(x)$ for the rate of change of $f(x)$.

ClassPad

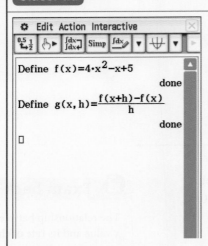

1 Define the functions **f(x)** and **g(x,h)** as shown above.

2 Find values of $g(x,h)$ for x from 0 to 5, using $h = 0.01$ or $h = -0.01$ or less.

Make a table of values. Approximate values of $g(x)$.

x	0	1	2	3	4	5
$g(x)$	−1	7	15	23	31	39

The function goes up in steps of 8. The function seems to be $g(x) = 8x - 1$.

TI-Nspire

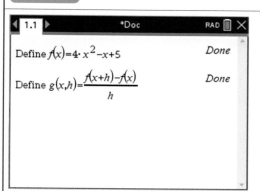

1 Define the functions **f(x)** and **g(x,h)** as shown above.

2 Find values of $g(x,h)$ for x from 0 to 5, using $h = 0.01$ or $h = -0.01$ or less.

Make a table of values. Approximate values of $g(x)$.

x	0	1	2	3	4	5
$g(x)$	−1	7	15	23	31	39

The function goes up in steps of 8. The function seems to be $g(x) = 8x - 1$.

EXERCISE 8.2 Instantaneous rates of change and the gradient function ANSWERS p. 480

Recap

1 The graph shows the speed of a particle as a function of time.

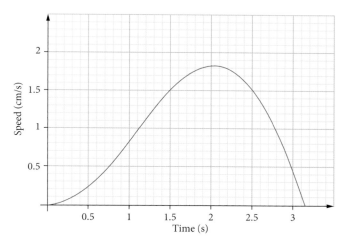

The average rate of change of the particle between 1.5 seconds and 2.5 seconds is closest to

A −3   B −1   C 0   D 1   E 3

2 For the function $y = \sqrt{x - 1}$, the average rate of change of y with respect to x from $x = 5$ to $x = 10$ is

A $\dfrac{1}{5}$ B $\dfrac{2}{5}$ C 1   D 3   E 5

▶ **Mastery**

3 [WORKED EXAMPLE 4] Find the approximate rate of change of $f(x) = x^3 + x^2 - 3x - 1$ between the points $x = 4$ and $x = 4.2$ using $\frac{\delta y}{\delta x} \approx \frac{f(x+h) - f(x)}{h}$.

4 [WORKED EXAMPLE 5] Using $h = 0.1, 0.01, 0.001, 0.0001$, state the approximations for the instantaneous rates of change of $f(x) = 2x^3 - 1$ at $x = -2$. Hence, state the instantaneous rate of change of $f(x)$ at $x = -2$.

5 [WORKED EXAMPLE 6] Using $h = 0.2$, find the gradient function $R(x)$ of $f(x) = 2x^2 - 1$ for the x values 0, 1, 2 and 3.

6 [Using CAS 2] Find approximate rates of change for the function $f(x) = 2x^2 + 2x - 7$ at various values of x (use $h = 0.1$) to suggest a gradient function $g(x)$ for the rate of change of $f(x)$.

Calculator-free

7 (3 marks) Using $h = 0.1$, find the approximate rates of change of $y = x^2 - x$ using $x = 1, 2, 3$.

8 (4 marks) Given $y = 3x^2$,
  a determine approximations for the instantaneous rates of change of y with respect to x between $x = 4$ and $x = 4 + h$, using $h = 0.1, 0.01, 0.001$ (3 marks)
  b what value will the approximation approach as the value of h approaches 0? (1 mark)

Calculator assumed

9 (4 marks) By trying various values of x, find approximate rates of change for $f(x) = 4x - 3x^2$ and use your results to determine the gradient function.

10 (4 marks) Find the approximate rates of change of $f(x) = 3x^2 + 2x + 2$ using $h = 0.1$ to suggest a gradient function $g(x)$ for the rate of change of $f(x)$.

11 (3 marks) For $y = 3 - 2x^3$, using $h = 1$ and by approaching $x = 10$, show that $\frac{\delta y}{\delta x}$ is -662.

12 (3 marks) For $y = -x^2$, using $h = 0.2$ and by approaching $x = 2$, show $\frac{\delta y}{\delta x}$ is -4.2.

13 (2 marks) Some approximations to the rate of change of a function are shown below.

$x = 0, \frac{\delta y}{\delta x} = 0.82 \quad x = 1, \frac{\delta y}{\delta x} = 6.89 \quad x = 2, \frac{\delta y}{\delta x} = 12.77 \quad x = 3, \frac{\delta y}{\delta x} = 18.49$

Determine an approximation for the gradient function in the form $y = mx + c$, where m and c are integers.

14 (4 marks) The height, $H(t)$ metres, of a toy rocket t seconds after launch is described by the function $H(t) = -t(t - 10)$.
  a Speed is the change in distance divided by the change in time. Estimate the speed of the rocket between 2 seconds and 4 seconds. (2 marks)
  b What is the rocket's rate of change of height between 3 and 3.01 seconds? (2 marks)

15 (4 marks)
  a Show that $\frac{\delta y}{\delta x} = \frac{f(x+h) - f(x)}{h} = 2ax + ah + b$ for $f(x) = ax^2 + bx + c$, where a, b, c are given constants. (3 marks)
  b If h approaches 0, what is the suggested gradient function, $R(x)$? (1 mark)

8.3 Differentiation by first principles

Limits

What happens to the function $f(x) = x + 1$ as $x \to 3$?

When we approach $x = 3$, we use any x value greater than 3, such as 3.1, 3.01 and 3.001, and find the corresponding values for $f(x)$: 4.1, 4.01 and 4.001. Or we can use any value less than 3, such as 2.9, 2.99, 2.999, and find the corresponding values for $f(x)$: 3.9, 3.99, 3.999.

As x approaches 3, $f(x)$ approaches $3 + 1 = 4$. This is written as $\lim\limits_{x \to 3} f(x) = 4$.

In general, we write the **limit**: $\lim\limits_{x \to a} f(x) = L$.

Differentiation by first principles

Suppose we want to find the instantaneous rate of change at point A shown on the below graph. We can use limits to see what happens to the gradient of the chord AB as B moves closer to A.

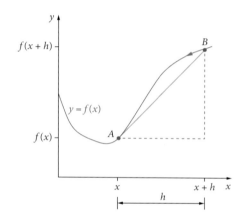

Points A and B lie on the curve $y = f(x)$ and have coordinates $A(x, f(x))$ and $B(x + h, f(x + h))$ respectively.

The gradient of the line connecting A to B is $\dfrac{f(x + h) - f(x)}{h}$.

As B approaches A, h becomes smaller. In the limiting case, we let h approach 0 and AB becomes the tangent to the curve at A.

We write $f'(x) = \lim\limits_{h \to 0} \dfrac{f(x + h) - f(x)}{h}$, where $f'(x)$ is the **gradient function** or the **derivative** of $f(x)$. This is a function that gives the gradient of the tangent (or instantaneous rate of change) at the point $(x, f(x))$. Finding the gradient function in this way is called **differentiation by first principles**.

$f'(x)$ is also written as $\dfrac{dy}{dx}$, meaning the derivative of y with respect to x. The d stands for 'difference', so the derivative can be interpreted as 'difference in y' over 'difference in x'.

WORKED EXAMPLE 7 — Using differentiation by first principles

Use differentiation by first principles to find the gradient function for $f(x) = 4x^2$.

Steps	Working
1 Write the function.	$f(x) = 4x^2$
2 Find an expression for $f(x+h)$. Expand and simplify.	$f(x+h) = 4(x+h)^2$ $= 4x^2 + 8xh + 4h^2$
3 Find an expression for $\dfrac{f(x+h) - f(x)}{h}$ and simplify.	$\dfrac{f(x+h) - f(x)}{h} = \dfrac{(4x^2 + 8xh + 4h^2) - 4x^2}{h}$ $= \dfrac{8xh + 4h^2}{h}$ $= 8x + 4h$
4 Find $\lim\limits_{h \to 0} \dfrac{f(x+h) - f(x)}{h}$.	$\lim\limits_{h \to 0} \dfrac{f(x+h) - f(x)}{h} = \lim\limits_{h \to 0}(8x + 4h)$ $= 8x + 4 \times 0$ $= 8x$
5 State the gradient function.	The derivative, or gradient function for $f(x) = 4x^2$ is $8x$. Hence, $f'(x) = 8x$.

WORKED EXAMPLE 8 — Finding the derivative at a point by first principles

Use differentiation by first principles to find $f'(x)$ for $f(x) = -x^3$, then evaluate $f'(-3)$.

Steps	Working
1 Write the function.	$f(x) = -x^3$
2 Find an expression for $f(x+h)$. Expand and simplify.	$f(x+h) = -(x+h)^3$ $= -x^3 - 3x^2h - 3xh^2 - h^3$
3 Find an expression for $\dfrac{f(x+h) - f(x)}{h}$ and simplify.	$\dfrac{f(x+h) - f(x)}{h} = \dfrac{(-x^3 - 3x^2h - 3xh^2 - h^3) - (-x^3)}{h}$ $= \dfrac{-3x^2h - 3xh^2 - h^3}{h}$ $= -3x^2 - 3xh - h^2$
4 Find $f'(x)$.	$f'(x) = \lim\limits_{h \to 0} \dfrac{f(x+h) - f(x)}{h}$ $= \lim\limits_{h \to 0}(-3x^2 - 3xh - h^2)$ $= -3x^2 - 3x \times 0 - 0^2$ $= -3x^2$
5 Evaluate $f'(x)$ using the given value.	$f'(-3) = -3 \times (-3)^2 = -27$

 Exam hack

When $f(x)$ is a polynomial, simplifying $\dfrac{f(x+h) - f(x)}{h}$ will result in $f(x)$ terms cancelling out with $f(x+h)$ terms.

If this does not happen, check your working.

EXERCISE 8.3 Differentiation by first principles

ANSWERS p. 480

Recap

1 Which formula shows the approximate rate of change of $f(x)$ at $x = 1$ with $h = 0.5$?

A $f'(x) \approx \dfrac{f(1.1) - f(0.5)}{0.5}$ **B** $f'(x) \approx \dfrac{f(1) - f(1.5)}{0.5}$ **C** $f'(x) \approx \dfrac{f(1.5) - f(1.1)}{0.5}$

D $f'(x) \approx \dfrac{f(1.5) - f(0.5)}{0.5}$ **E** $f'(x) \approx \dfrac{f(1.5) - f(1)}{0.5}$

2 Find approximate rates of change for the function $f(x) = x^2 - 3x$ at various values of x (use $h = 0.1$) to suggest a gradient function $g(x)$ for the rate of change of $f(x)$.

Mastery

3 **WORKED EXAMPLE 7** Use differentiation by first principles to find the gradient function for $f(x) = \dfrac{1}{3}x^3$.

4 **WORKED EXAMPLE 8** Use differentiation by first principles to evaluate $f'\left(\dfrac{2}{3}\right)$ for $f(x) = 6x$.

5 Use differentiation by first principles to evaluate $f'(x)$ for each of the functions below.

a $f(x) = 3x^2 + 2$ **b** $f(x) = -x^2 + x$ **c** $f(x) = \dfrac{2x + 3}{2}$

d $f(x) = 3 - 2x^2$ **e** $f(x) = (x + 1)^2$ **f** $f(x) = -x^3$

Calculator-free

6 (3 marks) Use differentiation by first principles to show that $f'(-1) = 2$ when $f(x) = -2 - x^2$.

7 (3 marks) Given $f(x) = \dfrac{x^2 - a^2}{x + a}$, use differentiation by first principles to show that $f'(x) = 1$ for all values of $x \neq -a$.

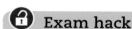

Exam hack

Check to see if you can simplify the function before applying first principles.

8 (2 marks) Use first principles to determine the gradient function of $f(x) = -7x^2$.

9 (2 marks) For the function $f(x) = 2 - 3x$, determine $\lim\limits_{h \to 0} \dfrac{f(x + h) - f(x)}{h}$.

10 (3 marks) Differentiate $f(x) = (2x + 3)^2$ by first principles.

Calculator-assumed

11 (2 marks) Use first principles to show that $f'(a) = 3a^2$ if $f(x) = x^3$.

12 (2 marks) Use differentiation by first principles to find the gradient function of the linear function $f(x) = mx + c$, where m and c are constants.

13 (2 marks) Use differentiation by first principles to find the derivative of the quadratic function $f(x) = ax^2 + bx + c$, where a, b and c are constants.

14 (3 marks) Use differentiation by first principles to find the gradient function of $f(x) = x^4$.

15 (4 marks) Use differentiation by first principles to show that for $f(x) = x^n$, $n = 1, 2, 3 \ldots$, $f'(x) = nx^{n-1}$.
Hint: $(x + h)^n = x^n + nx^{n-1}h + k_2 x^{n-2} h^2 + k_3 x^{n-3} h^3 + \ldots + h^n$, where n, k_2, k_3 … are constants.

8.4 Differentiating polynomial functions

Video playlist
Differentiating polynomial functions

Worksheets
Derivative of a sum of terms

Derivatives of linear products

Differentiating powers of x

When we differentiate powers of x by first principles, we obtain the derivatives as shown in the table.

Can you see a pattern in finding the derivative of $f(x) = x^n$, where $n = 1, 2, 3 \ldots$?

$f(x)$	$f'(x)$
x	1
x^2	$2x$
x^3	$3x^2$
x^4	$4x^3$

To differentiate any power of x, make the power become the coefficient of x and subtract 1 from the power.

> **The derivative of x^n**
>
> If $f(x) = x^n$, then $f'(x) = nx^{n-1}$.

For example, for $f(x) = x^{99}, f'(x) = 99x^{98}$.

What if there is a coefficient of x other than 1 involved, such as in $f(x) = 4x^2$?

For $f(x) = 4x^2, f'(x) = 8x$.

Notice that $f'(x) = 4 \times 2x^1 = 8x$

To differentiate $y = ax^n$, multiply the coefficient a by the power and subtract 1 from the power.

> **The derivative of ax^n**
>
> If $f(x) = ax^n$, then $f'(x) = anx^{n-1}$.

For example, the derivative of $f(x) = 3x^4$ is $f'(x) = 3 \times 4x^{4-1} = 12x^3$.

Furthermore, the derivative of a constant (number), c, is 0, because $c = c \times 1 = cx^0$, so its derivative is $c \times 0x^{-1} = 0$.

> **The derivative of a constant**
>
> If $f(x) = c$, then $f'(x) = 0$.

For example, the derivative of $f(x) = 7$ is $f'(x) = 0$.

This makes sense because if a function is a constant (number), such as $y = 7$, then its graph is a horizontal (flat) line, so its gradient is 0 at every point.

> The derivative of a constant is always 0.

Differentiating polynomial functions

We can differentiate polynomial functions by differentiating each term separately. This is called **term-by-term** differentiation.

For example, if $y = 4x^3 - x^2 + 5$ then $\dfrac{d}{dx}$ is another way of writing the derivative.

That is, $\dfrac{d}{dx}(4x^3 - x^2 + 5) = \dfrac{d}{dx}(4x^3) - \dfrac{d}{dx}(x^2) + \dfrac{d}{dx}(5) = 12x^2 - 2x$.

We can also differentiate functions that use variables other than x and y.

For example, if $r = 3t^2 + 2$, then $\dfrac{dr}{dt} = 6t$.

WORKED EXAMPLE 9 — Completing term-by-term differentiation

Differentiate each function.

a $f(x) = -x^8$

b $f(x) = 5x^4 - \dfrac{3}{2}x^2 + 1$

Steps	Working
a Differentiate using $f'(x) = anx^{n-1}$.	$f(x) = -x^8$ $f'(x) = -1 \times 8x^{8-1} = -8x^7$
b Differentiate each term separately using $f'(x) = anx^{n-1}$.	$f(x) = 5x^4 - \dfrac{3}{2}x^2 + 1$ $f'(x) = 5 \times 4x^{4-1} - \dfrac{3}{2} \times 2x^{2-1} + 0$ $= 20x^3 - 3x$

USING CAS 3 — Completing a term-by-term differentiation

Differentiate $f(x) = \dfrac{x^2}{2} - 3x + 2$.

ClassPad

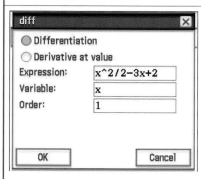
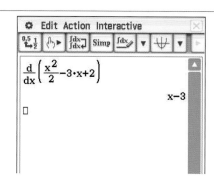

1 In **Main**, enter and highlight the expression $\dfrac{x^2}{2} - 3x + 2$.

2 Tap **Interactive > Calculation > diff**.

3 Tap **OK**.

4 The derivative of $f(x)$ will be displayed.

TI-Nspire

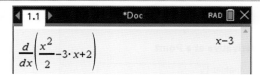

1 Press **menu > Calculus > Derivative**.

2 The derivative template will be displayed.

3 Enter **x** at the start of the template followed by the expression $\dfrac{x^2}{2} - 3x + 2$.

4 Press **enter** to display the derivative of $f(x)$.

The derivative of $f(x)$ is $x - 3$.

WORKED EXAMPLE 10 — Finding the derivative at a given point

Evaluate $f'(x)$ for the given value of x.

a $f(x) = 4x^2 - 2x + 1$, $f'\left(\dfrac{1}{2}\right)$

b $f(x) = (x+1)(x-3)$, $f'\left(-\dfrac{3}{2}\right)$

Steps	Working
a 1 Differentiate each term.	$f(x) = 4x^2 - 2x + 1$ $f'(x) = 8x - 2$
2 Evaluate $f'\left(\dfrac{1}{2}\right)$.	$f'\left(\dfrac{1}{2}\right) = 8 \times \dfrac{1}{2} - 2 = 2$
b 1 Expand and simplify $f'(x)$ before differentiating.	$f(x) = x^2 - 2x - 3$
2 Differentiate using $f'(x) = anx^{n-1}$.	$f'(x) = 2x - 2$
3 Evaluate $f'\left(-\dfrac{3}{2}\right)$.	$f'\left(-\dfrac{3}{2}\right) = 2\left(-\dfrac{3}{2}\right) - 2 = -5$

USING CAS 4 — Finding the derivative at a given point

Find $f'(2)$ for $f(x) = 5x^4$.

ClassPad

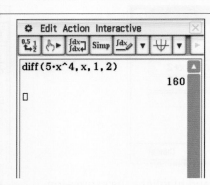

1. In **Main**, enter and highlight the expression **5x⁴**.
2. Tap **Interactive > Calculation > diff > Derivative at value**.
3. In the **Value:** field, enter **2**.
4. Tap **OK**. The value of $f'(2)$ will be displayed.

TI-Nspire

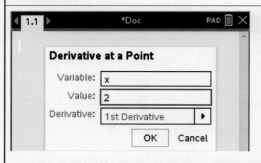
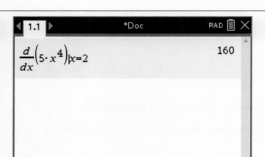

1. Press **menu > Calculus > Derivative at a Point**.
2. In the dialogue box, enter the value **2**.
3. In the derivative template, enter the expression **5x⁴**. The value of $f'(2)$ will be displayed.

WORKED EXAMPLE 11 — Finding a constant given the derivative at a given point

For the function $f(x) = x^4 + ax^2 + 5$, $f'(-2) = -44$. Find the value of the constant, a.

Steps	Working
1 Find the derivative of $f(x)$. Differentiate $f(x)$ term-by-term.	$f'(x) = 4x^3 + 2ax$
2 Form an equation.	$f'(-2) = 4(-2)^3 + 2a \times (-2)$ $= -32 - 4a$ $f'(-2) = -44 \Rightarrow -32 - 4a = -44$
3 Solve for the unknown.	$-32 - 4a = -44$ $-4a = -12$ $a = 3$

USING CAS 5 — Finding the function from the derivative

For the function $f(x) = ax^3 + bx^2 + x$, $f'(-1) = 15$ and $f'(2) = 9$. Find the value of the constants a and b.

ClassPad

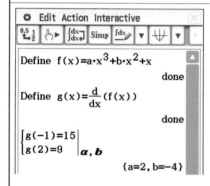

1. Define the function $f(x) = ax^3 + bx^2 + x$ using **Var** for variables. There is no need to insert multiplication signs when using **Var**.
2. Define $g(x)$ as the derivative of $f(x)$.
3. Open the **Keyboard > Math1** and insert the **simultaneous equations** template.
4. Enter the two derivative equations and the variables a, b using **Var** into the template as shown above.

TI-Nspire

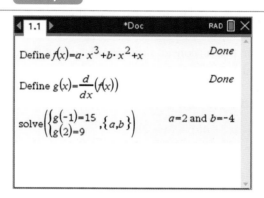

1. Define the function $f(x) = a \cdot x^3 + b \cdot x^2 + x$. Be sure to insert multiplication signs after the a and the b.
2. Define $g(x)$ as the derivative of $f(x)$.
3. Press **menu > Algebra > Solve System of Equations > Solve System of Equations**.
4. In the dialogue box, change the **Variables:** to a, b.
5. Enter the two derivative equations into the template as shown above.

$a = 2, b = -4$

Video Examination question analysis: Differential calculus

EXAMINATION QUESTION ANALYSIS

Calculator-free (12 marks)

A quadratic function is given by $f(x) = (ax + 3)(x + b)$, where a and b are constants.

a Use differentiation by first principles to show that the derivative of $f(x)$ is $f'(x) = 2ax + ab + 3$. (3 marks)

b Find the values of a and b given that $f'(-2) = 3$ and $f'(4) = 15$. (5 marks)

c Calculate $f'(-3.5)$. What can be concluded about the graph of $y = f(x)$ at $x = -3.5$? (2 marks)

d Calculate the value of k if $f'(k) = 3$. (2 marks)

Reading the question

- Read each question carefully to understand what method you should use to find the solution.
- Identify key words that suggest what approach is needed. For example, for part **a**, do not use the fast method of differentiation when differentiation by first principles is required.
- Part **b** relates to the values of the derivative function at the given points and not the value of the functions.

Thinking about the question

- Parts **a** and **b** are worth more than two marks, so working must be shown to attain full marks. For parts **c** and **d**, answer only is acceptable, though working out is recommended.
- In part **c**, consider the significance of the question in terms of the gradient of the tangent line at that point.

Worked solution (✓ = 1 mark)

a $f(x) = ax^2 + abx + 3x + 3b$
$= ax^2 + (ab + 3)x + 3b$ ✓

$$\frac{f(x+h) - f(x)}{h}$$

$$= \frac{[a(x+h)^2 + (ab+3)(x+h) + 3b] - [ax^2 + (ab+3)x + 3b]}{h}$$

$$= \frac{[ax^2 + 2axh + ah^2 + (ab+3)x + (ab+3)h + 3b] - [ax^2 + (ab+3)x + 3b]}{h}$$

$$= \frac{[2axh + ah^2 + (ab+3)h]}{h}$$

$= 2ax + ah + ab + 3$ ✓

$$\lim_{h \to 0} \frac{f(x+h) - f(x)}{h} = \lim_{h \to 0} 2ax + ah + ab + 3$$

$= 2ax + bx + 3$ ✓

b $f'(x) = 2ax + ab + 3$

$f'(-2) = 2a \times (-2) + ab + 3 = 3$ ✓

$f'(4) = 2a \times (4) + ab + 3 = 15$ ✓

Simplify to get

$-4a + ab = 0$

$8a + ab = 12$ ✓

Subtract the first equation from the second equation.

$8a - (-4a) + ab - ab = 12 - 0$

$12a = 12 \Rightarrow a = 1$ ✓

Substitute for a in $-4a + ab = 0$.

$-4 \times 1 + 1 \times b = 0 \Rightarrow b = 4$ ✓

c $f'(x) = 2ax + ab + 3$
$\quad\quad = 2x + 7$, using $a = 1$, $b = 4$
$f'(-3.5) = 2 \times (-3.5) + 7$
$\quad\quad\quad = 0$ ✓

The derivative is zero, which means the gradient of the tangent at $x = -3.5$ is zero. ✓

d $f'(x) = 2x + 7$
$\therefore f'(k) = 2k + 7$
$\quad\quad 3 = 2k + 7$ ✓
$\therefore k = -2$ ✓

EXERCISE 8.4 Differentiating polynomial functions

ANSWERS p. 480

Recap

1 If $f(x) = 3x^2 + x$, then $\dfrac{f(x+h) - f(x)}{h}$ is

A $3x + h + 1$ **B** $6x + 1$ **C** $3x + 3h$ **D** $6x + 3h + 1$ **E** $6x + h + 1$

2 A possible step in finding the gradient function of $f(x) = 1 - x^2$ is

A $\lim\limits_{h \to 0} \dfrac{(x+h)^2 + 1 - x^2}{h}$

B $\lim\limits_{h \to 0} -\dfrac{(x+h)^2 + x^2}{h}$

C $\lim\limits_{h \to 0} \dfrac{(x-h)^2 - x^2}{h}$

D $\lim\limits_{h \to 0} \dfrac{(x+h)^2 + 1 + x^2}{h}$

E $\lim\limits_{h \to 0} \dfrac{(x-h)^2 + x^2}{h}$

Mastery

3 **WORKED EXAMPLE 9** Differentiate each function with respect to x.

a $y = x^3 + x^2 + x$

b $y = 6x^7 + 10x^9 - 5x^{12} + 8x^6$

c $y = 14 - 2x^6 + 9x^8 + x^{11}$

d $y = 3x^8 - 5x^{12} + 31$

e $y = 6x^{12} + 11x^{13} - 24x^5$

f $y = 12x^3 + 7x^4 - 11x^6 + 38$

g $y = 4x^8 - 15x + 28x^2 + 14x^3$

h $y = 10 - 6x + 7x^2 + 14x^3 - 9x^4$

i $y = 15x^8 - 7x^5 + 4x^7 + 12x - 21$

j $y = 16x^2 - 9x^4 + 7x^3 + 18x - 32$

4 **Using CAS 3** Differentiate $y = \left(\dfrac{x^2}{2} - x^3\right)^2$ with respect to x.

5 **WORKED EXAMPLE 10** Calculate $f'(x)$ for the given value of x.

a $f(x) = -2x^2 + \dfrac{2}{3}x - 1$, $f'\left(\dfrac{1}{8}\right)$

b $f(x) = \dfrac{x^3 + x}{2x}$, $f'(25)$

> 🔒 **Exam hack**
>
> You may need to simplify a function before differentiating.

6 **Using CAS 4** Find the derivative of $y = 2x^2 - \dfrac{x}{2}$ and evaluate $\dfrac{dy}{dx}$ at $x = 6$.

7 WORKED EXAMPLE 11 For the function $f(x) = 3 + ax^2 - x^4$, $f'(-1) = -2$. Find the value of the constant, a.

8 Using CAS 5 Find the values of the constants a and b given that for the function $f(x) = (ax + b)^2$, $a + b > 0$, $f(1) = 16$ and $f'(1) = 40$.

Calculator-free

9 (10 marks) Find each derivative.

a $\dfrac{d}{dx}(3x^4 - 6x^2 + 18)$ (1 mark)

b $\dfrac{d}{du}(3u^4 - 6u^2 + 18)$ (1 mark)

c $\dfrac{d}{dm}(5m^6 + 8m^5 - 24m + 8)$ (1 mark)

d $\dfrac{d}{dy}(9y + 6y^5 - 4y^3 + 2y^7)$ (1 mark)

e $\dfrac{d}{dz}(3z^2 - 9z^3 + 38z - 5)$ (1 mark)

f $\dfrac{d}{dt}(5t^7 + 9t^4 - 16t)$ (1 mark)

g $\dfrac{d}{dr}(21 - 13r - 6r^2 + 9r^3)$ (1 mark)

h $\dfrac{d}{da}((a-2)(a+2)(3-a))$ (1 mark)

i $\dfrac{d}{dp}\left(\dfrac{8p - 3p^4}{2p}\right)$ (1 mark)

j $\dfrac{d}{dt}(8 + 4t^2)^2$ (1 mark)

10 (2 marks) Find the values of a and b in the function $f(x) = ax^2 + bx + 1$ given that $f(1) = 0$ and $f'(-1) = -10$.

11 (3 marks) Determine the derivative of the function $f(x) = \dfrac{5(x-2)(x^2 + 2x + 4)(x+3)}{x^2 + x - 6}$ by first simplifying $f(x)$.

12 (2 marks) Determine the derivative of $0.25x^{12}$.

13 (2 marks) For $f(x) = 1 + 2x - \dfrac{1}{4}x^8 - \dfrac{5}{18}x^9$, determine $f'(x)$.

Calculator-assumed

14 (2 marks) If $y = x^2 + 2x + 4$, determine the rate of change of y with respect to x at $x = k$.

15 (2 marks) If $y = 3x^2 + 7$ and $z = 9 - 4x$, determine the derivative of $y + z$.

16 (2 marks) If $f(x) = 3x - 5$ and $g(x) = 2x + 4$, determine the derivative of $f(x) \times g(x)$.

17 (2 marks) Find the derivative of $g(a) = (a - 6)^2$.

18 (3 marks) If $f(x) = \dfrac{1}{3}a^2x^3 + \dfrac{5}{2}ax^2 + 6x$ and $f'(-1) = 2$, determine the value(s) for the constant a.

Chapter summary

Average rate of change

- A rate of change occurs when a change in one variable produces a change in another variable.
- The **average rate of change** is the gradient of the straight line.
- The average rate of change $\dfrac{\delta y}{\delta x}$ between $A(x_1, y_1)$ and $B(x_2, y_2)$ is $\dfrac{\delta y}{\delta x} = \dfrac{y_2 - y_1}{x_2 - x_1}$.

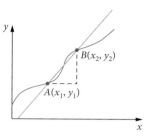

Instantaneous rates of change

- A straight line that touches a curve at one point is a **tangent**.
- If a straight line passes through two points on a curve, it is a **secant**.
- The interval (line segment) joining two points on a curve is a chord.

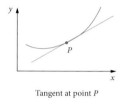 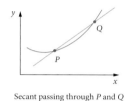 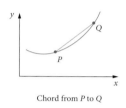

Tangent at point P Secant passing through P and Q Chord from P to Q

The gradient function

- The gradient of the tangent at a point on the curve is the **instantaneous rate of change**.
- An approximation $\dfrac{\delta y}{\delta x}$ for the instantaneous rate of change at point x is found using $\dfrac{\delta y}{\delta x} = \dfrac{f(x+h) - f(x)}{h}$.
- The **gradient function** gives the instantaneous rate of change (or gradient of the tangent) at any value of x.
- The rate of change describes how one variable changes in response to another variable.
- The instantaneous rate of change of a function for a particular value of x is measured by the gradient of the tangent to the graph of the function at the point $P(x, y)$.

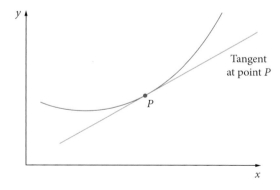

Differentiation by first principles

Finding the gradient function using the formula $f'(x) = \lim_{h \to 0} \dfrac{f(x+h) - f(x)}{h}$ is called **differentiation by first principles**, where $f'(x)$ or $\dfrac{dy}{dx}$ is the **derivative** of $f'(x)$. This is a function that gives the gradient of the tangent (or instantaneous rate of change) at the point $(x, f'(x))$.

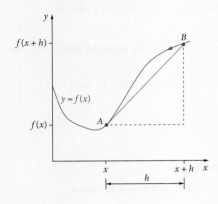

Differentiating polynomial functions

- If $f(x) = x^n$, then $f'(x) = nx^{n-1}$.
- If $f(x) = ax^n$, then $f'(x) = anx^{n-1}$.
- If $f(x) = c$, then $f'(x) = 0$.
- We can differentiate polynomial functions by differentiating each term separately. This is called **term-by-term differentiation**.

Cumulative examination: Calculator-free

Total number of marks: 31   Reading time: 4 minutes   Working time: 31 minutes

1 (4 marks) Christopher has five pairs of identical purple socks and three pairs of identical green socks. His socks are all randomly mixed in his drawer. He takes two individual socks at random from the drawer in the dark. Determine the probability that he obtains

  a two socks of different colours (2 marks)

  b a matching pair of socks. (2 marks)

2 (2 marks) Find the exact length of the arc described by the minute hand of a clock in 25 minutes if the minute hand is 8 cm.

3 (4 marks)

  a Evaluate

   i $16^{\frac{1}{2}}$ (1 mark)

   ii $125^{\frac{2}{3}}$ (1 mark)

  b Hence, find the value of $16^{\frac{1}{2}} \times 125^{\frac{2}{3}}$. (2 marks)

4 (9 marks) The first three terms of a geometric sequence with $r > 0$ are

  $T_1 = x + 2$
  $T_2 = x$
  $T_3 = 2x - 3$

  a Determine the value of x. (4 marks)

  b Hence, state the value of the common ratio, r. (2 marks)

  c Describe the behaviour of

   i T_n as $n \to \infty$ (1 mark)

   ii S_n, the sum of the first n terms, as $n \to \infty$. (2 marks)

5 (2 marks) The temperature, $T\,°C$, inside a greenhouse at time t hours after midday is given by the function $T(t) = \frac{1}{10}(t-1)(t-2)^2 + 5, 0 \le t \le 5$.

What is the average rate of change of the temperature in the greenhouse between 2 pm and 5 pm?

6 (3 marks) Differentiate $f(x) = \frac{1}{2}x^2 - x$ using first principles.

7 (3 marks) Determine the value of $f'(-1)$ if $f(x) = \frac{(2x+3)(x-3)}{2}$.

8 (4 marks) If $f(x) = \frac{(x-1)(x+a)}{2}$ and $f'(1) = 4$, determine the value of a.

Cumulative examination: Calculator-assumed

Total number of marks: 23   Reading time: 3 minutes   Working time: 23 minutes

1 (4 marks) Consider the following two-way table showing the relationship between two sets.

	B	B'	Total
A	y + 10	x	30
A'	2x	y	20
Total	30	20	50

   a Determine the values of x and y. (2 marks)

   b Describe the relationship between sets A and B. Justify your answer. (2 marks)

2 (3 marks) Consider the quadratic expression $y = x^2 - 6x - 11$.

   a Complete the square to rewrite the expression in the form $y = (x - h)^2 + k$. (2 marks)

   b Hence, solve the equation $x^2 - 6x + 11 = 0$. (1 mark)

3 (2 marks) If $f(x + 2) = x^2 + x - 7$, determine $f(x)$.

4 (4 marks)

   a Use $h = 0.1, 0.01, 0.001$ to find approximations for the instantaneous rates of change of $f(x) = 2x^3$ at $x = 2$. (3 marks)

   b Hence, state the instantaneous rate of change of $f(x)$ at $x = 4$. (1 mark)

5 (3 marks) Differentiate $f(x) = (3x^2 - 3)\left(2 - \dfrac{x}{2}\right)$ using first principles.

6 (3 marks) If $f(x) = ax^2 - bx + 3$ and $f'(k) = 3$, determine the value of k (in terms of a and b).

7 (4 marks) For the given the function $f(x) = ax^2 + bx$, $f(2) = 10$ and $f'(2) = 13$.
Write two equations and solve simultaneously to find the values of a and b.

CHAPTER 9
APPLICATIONS OF DIFFERENTIAL CALCULUS

Syllabus coverage
Nelson MindTap chapter resources

9.1 Equation of a tangent
 Using CAS 1: Finding the equation of a tangent at a given point
9.2 Straight line motion
9.3 Stationary points
 Gradient of a function
 Stationary points
 Using CAS 2: Finding turning points
 Stationary points of inflection
9.4 Curve sketching
 Key features of a graph
9.5 Optimisation problems
 Using CAS 3: Finding maximum and minimum points
 Solving optimisation problems
9.6 The anti-derivative
 Using CAS 4: Finding the anti-derivative
 Using CAS 5: Finding the anti-derivative and calculating the value of c
9.7 Applying the anti-derivative to straight line motion

Examination question analysis
Chapter summary
Cumulative examination: Calculator-free
Cumulative examination: Calculator-assumed

Syllabus coverage

TOPIC 2.3: INTRODUCTION TO DIFFERENTIAL CALCULUS

Applications of derivatives
- 2.3.16 determine instantaneous rates of change
- 2.3.17 determine the slope of a tangent and the equation of the tangent
- 2.3.18 construct and interpret position–time graphs with velocity as the slope of the tangent
- 2.3.19 recognise velocity as the first derivative of displacement with respect to time
- 2.3.20 sketch curves associated with simple polynomials, determine stationary points, and local and global maxima and minima, and examine behaviour as $x \to \infty$ and $x \to -\infty$
- 2.3.21 solve optimisation problems arising in a variety of contexts involving polynomials on finite interval domains

Anti-derivatives
- 2.3.22 calculate anti-derivatives of polynomial functions

Mathematics Methods ATAR Course Year 11 syllabus p. 14 © SCSA

Video playlists (8):
- 9.1 Equation of a tangent
- 9.2 Straight line motion
- 9.3 Stationary points
- 9.4 Curve sketching
- 9.5 Optimisation problems
- 9.6 The anti-derivative
- 9.7 Applying the anti-derivative to straight line motion
- **Examination question analysis** Applications of differential calculus

Worksheets (3):
- 9.3 Gradient of a function
- 9.4 Gradient functions • Sketching curves

Nelson MindTap

To access resources above, visit
cengage.com.au/nelsonmindtap

9.1 Equation of a tangent

We saw in the previous chapter that the derivative is a gradient function that gives the gradient of the tangent (or instantaneous rate of change) to the graph of $y = f(x)$ at the point $(x, f(x))$. This means that the derivative can be used to find the equation of the tangent to the graph at any point on the graph.

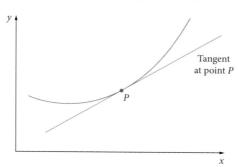

In chapter 3, we studied the equations of a straight line.

- The **gradient–intercept form** of a straight line with gradient m and y-intercept c is $y = mx + c$.
- The **point–gradient form** of a straight line with gradient m that passes through the point (x_1, y_1) is $y - y_1 = m(x - x_1)$.

Video playlist
Equation of a tangent

Equation of a tangent at a point

To find the **equation of the tangent** to the graph of $y = f(x)$ at the point (x_1, y_1):

1. Find $f'(x)$ by differentiation.
2. Find the gradient of the tangent by evaluating $f'(x_1)$.
3. Use the point to find the equation of the line, using either $y - y_1 = m(x - x_1)$ or $y = mx + c$.

WORKED EXAMPLE 1 Finding the equation of the tangent at a given point

Find the equation of the tangent to the curve $f(x) = x^2 + x + 1$ at the point $(1, 3)$.

Steps	Working
1 Find the gradient function by differentiation.	$f'(x) = 2x + 1$
2 Find the gradient of the tangent by substituting $x = 1$ in $f'(x)$.	$f'(1) = 2 \times 1 + 1 = 3$ The gradient is $m = 3$.
3 Find the equation of the tangent using $y - y_1 = m(x - x_1)$ or $y = mx + c$.	Substitute $m = 3$, $x_1 = 1$ and $y_1 = 3$: $y - 3 = 3(x - 1)$ $y - 3 = 3x - 3$ $y = 3x$ Substitute $m = 3$: $y = 3x + c$ To find the value of c, substitute $x = 1$ and $y = 3$. $3 = 3 \times 1 + c$ $c = 0$ $y = 3x + 0$ $y = 3x$

USING CAS 1 — Finding the equation of a tangent at a given point

Find the equation of the tangent to the curve $f(x) = x^2 - x + 3$ at the point $(1, 3)$.

ClassPad

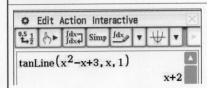

1. Enter and highlight the expression **$x^2 - x + 3$**.
2. Tap **Interactive > Calculation > line > tanLine**.
3. In the dialogue box **Point:** field, enter **1**.
4. Tap **OK**.

TI-Nspire

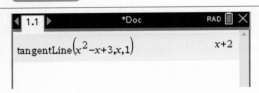

1. Press **menu > Calculus > Tangent Line**.
2. Enter **$x^2 - x + 3$, x, 1**.
3. Press **enter**.

The equation of the tangent line is $y = x + 2$.

WORKED EXAMPLE 2 — Finding points on a curve that have a given gradient

Find the point(s) on the graph of $f(x) = \dfrac{1}{3}x^3 - x^2 - 4x + 1$ where the tangent

a has a gradient of -1
b is parallel to the line $y = 4x + 1$.

Steps	Working
a 1 Find the gradient function by differentiation.	$f'(x) = x^2 - 2x - 4$
2 Solve $f'(x) = -1$.	$x^2 - 2x - 4 = -1$ $x^2 - 2x - 3 = 0$ $(x - 3)(x + 1) = 0$ $x = 3$ or $x = -1$
3 Find the corresponding y values.	$f(3) = \dfrac{1}{3}(3)^3 - 3^2 - 4(3) + 1 = -11$ $f(-1) = \dfrac{1}{3}(-1)^3 - (-1)^2 - 4(-1) + 1 = 3\dfrac{2}{3}$
4 State the coordinates required.	The gradient of the tangent is -1 at $(3, -11)$ and $\left(-1, 3\dfrac{2}{3}\right)$.
b 1 Determine the gradient value needed.	Parallel lines have the same gradient. The gradient of $y = 4x + 1$ is 4.
2 Solve $f'(x) = 4$.	$x^2 - 2x - 4 = 4$ $x^2 - 2x - 8 = 0$ $(x - 4)(x + 2) = 0$ $x = 4$ or $x = -2$
3 Find the corresponding y values.	$f(4) = \dfrac{1}{3}(4)^3 - 4^2 - 4(4) + 1 = -9\dfrac{2}{3}$ $f(-2) = \dfrac{1}{3}(-2)^3 - (-2)^2 - 4(-2) + 1 = 2\dfrac{1}{3}$
4 State the coordinates required.	The tangent is parallel to $y = 4x + 1$ at $\left(4, -9\dfrac{2}{3}\right)$ and $\left(-2, 2\dfrac{1}{3}\right)$.

EXERCISE 9.1 Equation of a tangent

ANSWERS p. 481

Mastery

1 **WORKED EXAMPLE 1** Find the equation of the tangent to the curve $f(x) = 6x^2 - x - 4$ at the point $(-1, 3)$.

2 **Using CAS 1** Find the equation of the tangent to the graph of $y = x^2 - 2x - 3$ at the point $(4, 0)$.

3 Find the equation of the tangent to the graph of each function at the given point.
   a $y = 5x^2 - 2x - 4$ at $x = 1$
   b $y = x - x^3 - 5$ at $x = 2$
   c $y = x^2 - x^3 + 3x + 2$ at $x = -1$

4 **WORKED EXAMPLE 2** Find the point(s) on the graph of $f(x) = x^3 - 3x^2 + 3x + 4$ where the tangent
   a has a gradient of 12
   b is parallel to the line $y = 3x$
   c is perpendicular to $y = -\dfrac{1}{27}x + 4$
   d is parallel to the x-axis.

Exam hack

Parallel to the x-axis means the gradient is zero.

5 Show that the tangent to the graph of $f(x) = -2x^4 + 2x^3 + 5x^2 - 4x - 4$ at $x = -1$ is parallel to the x-axis.

Calculator-free

6 (2 marks) Find the coordinates of the point where the gradient of the tangent to the parabola $y = 5x^2 - x$ is 2.

7 (3 marks) Show that $\dfrac{-1 \pm \sqrt{10}}{3}$ are the values of x where the tangent to the curve with equation $y = x^2(x + 1)$ is parallel to the line $y = 3x - 1$.

8 (2 marks) Determine the equation of the tangent to the curve of the function $f(x) = x - x^3$ at the point $(-2, 6)$.

9 (3 marks) Show that the tangent to the graph of the general quadratic function $f(x) = ax^2 + bx + c$, where a, b and c are constants, is parallel to the x-axis when $x = -\dfrac{b}{2a}$.

Calculator-assumed

10 (2 marks) Determine the gradient of the tangent to the graph of $f(x) = x - x^2$ at $x = 4$.

11 (2 marks) Find the equation of the tangent to the graph of $f(x) = (x - 4)(x + 6)$ at its positive x-intercept.

12 (2 marks) Show that the tangent to the graph of $f(x) = x^2 - 8x + 12$ at the point $(5, 11)$ is parallel to the line $y = 2x + 1$.

13 (3 marks) Find the coordinates of the point on the graph of $f(x) = 3x^2 - 2x + 3$ where the tangent is parallel to the x-axis.

14 (3 marks) The function $f(x) = x^3 + 3x^2 - 2$ has a tangent parallel to the x-axis at 2 points. Find the coordinates of these points.

9.2 Straight line motion

Video playlist
Straight line motion

Kinematics is a study of straight line motion using

- **displacement** = $x(t)$ = position of a particle at time t from a chosen origin
- **velocity** = $v(t)$ = velocity at time t, the rate of change of displacement.

Hence, velocity is the derivative of displacement.

Displacement is a 'signed distance' that can be positive or negative. For example, a negative displacement means the particle's position is to the left of the origin.

Velocity is a 'signed speed' that can be positive or negative. For example, a negative velocity means the particle is moving to the left.

Instantaneous velocity

$$\text{instantaneous velocity} = \frac{dx}{dt} = v$$

From the above equations, usually
- x is measured in metres
- t is measured in seconds
- velocity v is measured in m/s (metres per second).

For example, consider an ant that is travelling over the ground. The ant takes 20 seconds to travel 80 cm. This means the ant is travelling, on average, at $\frac{80}{20} = 4$ cm/s. This is considered the **average rate of change** of the ant's displacement with respect to time.

Average velocity and average speed

$$\text{average velocity} = \frac{x_2 - x_1}{t_2 - t_1}$$

$$\text{average speed} = \frac{\text{distance travelled}}{\text{time taken}}$$

 Exam hack

Note that although velocity can be positive or negative, speed is always positive.

If the ant is not travelling steadily, it is more likely to speed up and slow down when it meets food, so here the ant at a particular point may be travelling using the **instantaneous rate of change** at a particular time.

WORKED EXAMPLE 3 Differentiating for straight line motion

The displacement of a particle travelling in a straight line is given by $x(t) = 3t^3 - t$, with x in metres and t in seconds.

a Find an expression for $v(t)$.

b Find the instantaneous rate of change of the particle at $t = 3$.

Steps	Working
a Write the function for displacement, then use $\frac{dx}{dt}$ to find the velocity.	$x(t) = 3t^3 - t$ $\therefore v(t) = \frac{dx}{dt} = 9t^2 - 1$
b Substitute $t = 3$ into $v(t)$ to calculate $v(3)$.	$v(3) = 9(3)^2 - 1$ $= 80$ m/s

WORKED EXAMPLE 4 Graphical interpretation of straight line motion

The displacement of a particle undergoing straight line motion is graphed for $0 \leq t \leq 9$, where t is the time in seconds, and x is the displacement in metres.

a What is the initial displacement of the particle?
b Determine when the velocity equals zero.
c Determine the approximate speed of the particle at $t = 1$.

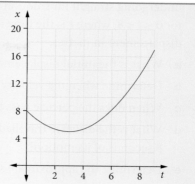

Steps	Working
a Read the x value off the graph when $t = 0$.	$x = 8$ m
b Determine the t value at the minimum point of the graph (gradient of the tangent is zero at that point).	$t = 3$ s
c Draw a tangent at $t = 1$ and find the approximate gradient of this line. (Note: the gradient of this line is negative, but speed must be positive).	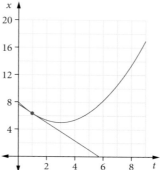 Note: the gradient of the tangent is closer to -1.3 so the approximate speed should be 1.3 m/s. It is only an approximation, so it could be left as is. Up to you. Speed is approximately 1.5 m/s.

EXERCISE 9.2 Straight line motion

ANSWERS p. 481

Recap

1 Determine the equation of the tangent line to the curve $y = x^3 - 2x + 3$ at $(2, 7)$.

2 Find the coordinates of the point(s) on the graph of $f(x) = (x - 3)^2(x + 3)$ where the tangent line is parallel to the x-axis.

Mastery

3 **WORKED EXAMPLE 3** The displacement of a particle travelling in a straight line is given by $x(t) = 2t^2 + 1$, where x is in metres and t is in seconds.
   a Find an expression for $v(t)$.
   b Find the velocity of the particle at $t = 3$.

4 **WORKED EXAMPLE 4** The displacement of a particle undergoing straight line motion is graphed on the right for $0 \le t \le 8$, where t is the time in seconds, and x is the displacement in metres.

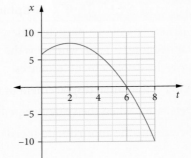

   a What is the initial displacement of the particle?
   b Determine when the velocity equals zero.
   c When does the particle pass through the origin?
   d What is the distance travelled by the particle for the first 8 seconds of its motion?
   e Determine the approximate speed of the particle at $t = 6$.

5 The displacement of a particle travelling in a straight line is given by $x(t) = -3t^3 + 2t$, where x is in metres and t is in seconds. Determine its velocity at $t = 1$.

Calculator-free

6 (5 marks) The displacement in centimetres after time t seconds of a particle moving in a straight line is given by $x = 2 - t - t^2$, $t \ge 0$.
   a Find the initial displacement. (1 mark)
   b Find when the particle will be at the origin. (1 mark)
   c Find the displacement at 2 seconds. (1 mark)
   d How far will the particle move in the first 2 seconds? (1 mark)
   e Find its velocity at 3 seconds. (1 mark)

7 (4 marks) The displacement of a particle is given by $s = t^3 - 4t^2 + 3t$, where s is in metres and t is in seconds.
   a Find the initial velocity. (2 marks)
   b Find the times when the particle will be at the origin. (2 marks)

Calculator-assumed

8 (4 marks) A projectile is fired into the air and its height in metres is given by $h = 40t - 5t^2 + 4$, where t is in seconds.
   a Find its initial height. (1 mark)
   b Find the initial velocity. (1 mark)
   c Find the height after 1 second. (1 mark)
   d What is the maximum height of the projectile? (1 mark)

9 (4 marks) The displacement of a particle is given by $x = t^3 - 9t$ cm, where t is in seconds.
   a Find an expression for its velocity, in m/s. (2 marks)
   b At what time is the velocity 3 cm/s? (2 marks)

10 (2 marks) The displacement of a particle travelling in a straight line is given by $x(t) = t^3 + 3t^2$, where x is in metres and t is in seconds.
   a Determine its velocity, in m/s. (1 mark)
   b Determine the velocity at $t = 1$. (1 mark)

Stationary points

Gradient of a function

The **gradient** of a function describes the direction of its graph. When a function is increasing, it is moving upwards to the right. When a function is decreasing, it is moving downwards to the right.

Consider the graph of a quartic function shown.

- The graph has a positive gradient for $-3 < x < -1$ and $x > 3$.
- The graph has a negative gradient for $x < -3$ and $-1 < x < 3$.
- The graph has a zero gradient at $x = -3$ and $x = 3$.

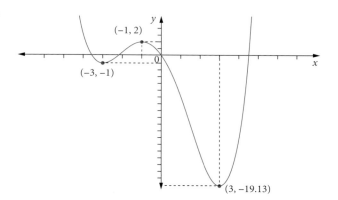

Gradients of a function

- A positive gradient points up: ↗
- A negative gradient points down: ↘
- A zero gradient is flat: →

WORKED EXAMPLE 5 | **Determining positive and negative gradients**

State the intervals over which the graph with the rule $y = 2x^2 - 2$ has a positive gradient and a negative gradient.

Steps	Working
1 Sketch the graph of $y = 2x^2 - 2$.	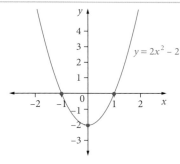
2 Find the x value of the turning point from the graph.	Turning point is at $x = 0$.
3 Identify ↗ or ↘ gradients.	Gradient is positive for $x > 0$. Gradient is negative for $x < 0$.

Stationary points

A **stationary point** on the graph of a function is where there is zero gradient, $f'(x) = 0$, such that the graph is flat, neither increasing nor decreasing. There are three types of stationary points: a **local minimum**, a **local maximum** (both called **turning points**) or a **stationary point of inflection**.

Turning points

A turning point is found where $f'(x) = 0$ and the sign of the gradient changes on either side of the stationary point.

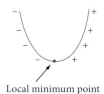
Local minimum point

Local maximum point

Consider the graph of $f(x) = 2x^3 - 6x$.

The turning points are found where $f'(x) = 0$.

$f'(x) = 6x^2 - 6 = 0$
$6(x - 1)(x + 1) = 0$
$\therefore x = 1$ or $x = -1$

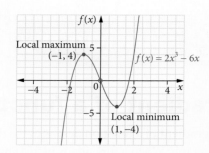

Substitute $x = 1$ and $x = -1$ into $f(x)$ to get the coordinates of the stationary points as $(1, -4)$ and $(-1, 4)$.

On either side of the point $(-1, 4)$, the gradient changes from positive to negative, giving a local maximum turning point.

On either side of the point $(1, -4)$, the gradient changes from negative to positive, giving a local minimum turning point.

WORKED EXAMPLE 6 Finding turning points

Find the coordinates of the turning points for the graph $f(x) = \dfrac{2}{3}x^3 + \dfrac{3}{2}x^2 - 2x$.

Steps	Working
1 Sketch the graph of $f(x) = \dfrac{2}{3}x^3 + \dfrac{3}{2}x^2 - 2x$.	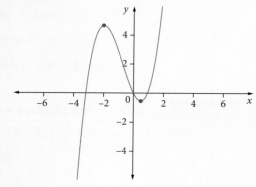
2 Find $f'(x)$ and solve for zero.	$f'(x) = 2x^2 + 3x - 2$ $f'(x) = 0$ for $x = -2$ and $x = \dfrac{1}{2}$.
3 Identify if these are turning points by checking whether the sign of the gradient changes on either side.	At $x = -2$, gradient changes from positive to negative. At $x = \dfrac{1}{2}$, gradient changes from negative to positive. Both are turning points.
4 Substitute x values into $f(x)$ for coordinates.	$f(-2) = \dfrac{14}{3}$ $\therefore \left(-2, \dfrac{14}{3}\right)$ is a local maximum turning point. $f\left(\dfrac{1}{2}\right) = -\dfrac{13}{24}$ $\therefore \left(\dfrac{1}{2}, -\dfrac{13}{24}\right)$ is a local minimum turning point.

USING CAS 2 Finding turning points

Confirm the coordinates of the turning points for the graph $f(x) = \dfrac{2}{3}x^3 + \dfrac{3}{2}x^2 - 2x$ from Worked example 6.

ClassPad

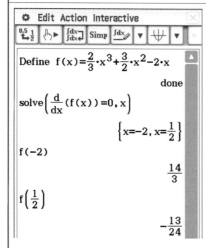

TI-Nspire

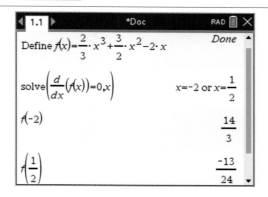

1 Define the function **f (x)**.
2 Set the **derivative of f (x)** equal to **0** and **solve** to find the x-coordinates of the turning points.
3 Substitute the solutions into **f (x)** to determine the corresponding y-coordinates.

The coordinates of the turning points are $\left(-2, \dfrac{14}{3}\right)$ and $\left(\dfrac{1}{2}, -\dfrac{13}{24}\right)$ and this confirms the coordinates from Worked example 6.

Stationary points of inflection

A **stationary point of inflection** is a flat 'bend' in the graph where $f'(x) = 0$ and the concavity of the graph changes: from concave down to concave up, or concave up to concave down. Also, the sign of the gradient stays *the same* on both sides of the stationary point. A stationary point is a point where the gradient of the graph is instantaneously zero.

Point of inflection

Concave down to concave up

Point of inflection

Concave up to concave down

Consider the graph of $f(x) = (x - 2)^3 - 27$.

$f'(x) = 3(x - 2)^2 = 0$ for stationary points.

Solving, we get $x = 2$.

Test for the gradient on either side of $x = 2$.

Substitute $x = 2$ into $f(x)$ to get the y-coordinate of the stationary point of inflection as $f(2) = -27$.

In the graph of $f(x) = (x - 2)^3 - 27$, the graph changes from concave down to concave up at $(2, -27)$ and the sign of the gradient stays *positive* on both sides.

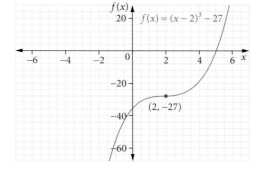

WORKED EXAMPLE 7 — Stationary points

Find the coordinates of any stationary points of inflection in the graph of $f(x) = x^4 + x^3 - 3x^2 - 5x - 2$.

Steps	Working
1 Sketch the graph of $f(x) = x^4 + x^3 - 3x^2 - 5x - 2$.	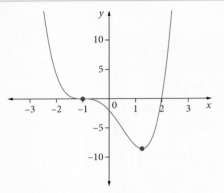
2 Find $f'(x)$ and solve for zero.	$f'(x) = 4x^3 + 3x^2 - 6x - 5 = 0$ $f'(x) = 0$ for $x = -1$ and $x = \dfrac{5}{4}$.
3 Check whether the sign of the gradient stays the same on both sides to identify the stationary point of inflection.	At $x = -1$, gradient is negative on both sides, so it is a stationary point of inflection. At $x = \dfrac{5}{4}$, gradient changes from negative to positive, so it is a turning point.
4 Find the coordinates of the stationary point of inflection.	$f(-1) = 0$ Stationary point of inflection is at $(-1, 0)$.

$f'(x)$ before the stationary point	$f'(x)$ at the stationary point	$f'(x)$ after the stationary point	Type of point
+	0	−	Local maximum point
−	0	+	Local minimum point
+	0	+	Stationary point of inflection
−	0	−	Stationary point of inflection

EXERCISE 9.3 Stationary points

ANSWERS p. 481

Recap

1. The displacement of a particle travelling in a straight line is given by $x(t) = 2t^2 + 1$, where x is in metres and t is in seconds.
   a. Find an expression for $v(t)$.
   b. Find the time when the velocity is zero.

2. The displacement of a particle travelling in a straight line is given by $x(t) = \dfrac{3t^2 - 2t}{4}$, where x is in metres and t is in seconds.
   a. Find an expression for $v(t)$.
   b. Initially, in which direction is the particle moving?
   c. When does the particle stop?

Mastery

3. **WORKED EXAMPLE 5** State the interval where the function $f(x) = 4x^3 - 12x^2 + 12x$ has a positive gradient.

4. **WORKED EXAMPLE 6** Find the coordinates of the turning points in the graph of $f(x) = x^3 + x^2 - 2$.

5. **Using CAS 2** Determine the turning point of the graph of $f(x) = 4x^2 - 16x + 2$ and state its nature.

6. **WORKED EXAMPLE 7** Find the coordinates of any stationary point of inflection in the graph of $f(x) = (x + 2)^3 - 2$.

7. Find any stationary points of each function.
   a. $f(x) = 2x^2 - 8x + 3$
   b. $f(x) = x^3 - x^2 - 8x + 10$

Calculator-free

8. (2 marks) Find any stationary points on the curve $y = (x - 2)^4$.

9. (2 marks) Determine the coordinates of the turning point of the graph $y = x^2 - 3x + 2$.

10. (2 marks) Determine the coordinates and nature of the turning point of the graph $y = -2(x + 1)^2 + 7$.

11. (2 marks) Determine the coordinates of the stationary point(s) of the graph $y = \dfrac{1}{3}x^3 + x^2 - 3x + 1$.

12. (2 marks) Determine the stationary point of inflection of the graph $y = 2(x + 1)^3 + 7$.

13. (2 marks) Determine the coordinates of the stationary point of inflection of the graph $y = -2(x + 1)^3 + 7$.

14. (2 marks) Determine the value of the stationary point of the graph of $y = 4m^2 - 16m + 1$.

15. (3 marks) The function $f(x) = 2x^2 + px + 7$ has a stationary point at $x = 3$. Find p.

Calculator-assumed

16 (4 marks) Determine the coordinates and nature of the stationary points of the function $f(x) = 2x^3 + 3x^2 - 36x + 7$.

17 (5 marks) For the graph $f(x) = x^4 + x^3 - 3x^2 - 5x - 1$, determine whether each of the statements below is true or false.

    a There is a local maximum. (1 mark)
    b There is a local minimum. (1 mark)
    c There is a stationary point of inflection. (1 mark)
    d There is a turning point at $(-1, 1)$. (1 mark)
    e There is a stationary point at $(-1, 1)$. (1 mark)

18 (2 marks) Let $f(x)$ be a function such that
- $f'(3) = 0$
- $f'(x) < 0$ when $x < 3$ and when $x > 3$.

Determine the type of stationary point, if there is any, at $x = 3$.

19 (3 marks) Consider $f(x) = x^2 + px + 1$. There is a stationary point on the graph of f when $x = -\dfrac{1}{4}$. Determine the value of p.

20 (2 marks) A cubic function has the rule $y = f(x)$. The graph of the derivative function $f'(x)$ crosses the x-axis at $(2, 0)$ and $(-3, 0)$. The maximum value of the derivative function is 10. Determine the value of x for which the graph of $y = f(x)$ has a local maximum.

21 (5 marks) State whether the following are true or false. A cubic function has

    a at least 1 stationary point (1 mark)
    b at most 2 stationary points (1 mark)
    c any number of stationary points (1 mark)
    d at least 2 stationary points (1 mark)
    e exactly 3 stationary points. (1 mark)

9.4 Curve sketching

By the end of Year 11, recognising and sketching a range of graphs is considered assumed knowledge for students. It is important to thoroughly understand how to deal with graphs of all types. Don't just rely on CAS for graphing.

Key features of a graph

When sketching a graph, it is important to identify the key features:

- general shape
- domain and range
- y-intercept
- x-intercept(s)
- stationary point(s)
- global and local maxima and minima
- behaviour as $x \to \pm\infty$.

WORKED EXAMPLE 8 Curve sketching

a Sketch the graph of $f(x) = -(x + 1)^2(x - 3)$ for $-1 \leq x \leq 4$, labelling key features.
b State the global maximum and minimum for the interval $-1 \leq x \leq 4$.
c State the behaviour of the function as $x \to \pm\infty$.

Steps	Working
a 1 Explore the general features of the graph.	Note from its equation that this is a cubic function with a leading coefficient of -1 and x-intercepts at -1 and 3.
2 Find the y-intercept.	$f(0) = -(0 + 1)^2(0 - 3) = 3$
3 Find the x-intercepts.	$-(x + 1)^2(x - 3) = 0$ $x = -1, x = 3$
4 Find $f'(x)$ and solve for zero for stationary points.	$f'(x) = -3x^2 + 2x + 5 = 0$ $x = -1, x = \dfrac{5}{3}$ $f(-1) = 0, f\left(\dfrac{5}{3}\right) = \dfrac{256}{27}$
5 Find the endpoint values of the domain.	$f(-1) = 0$ $f(4) = -25$
6 Sketch the graph of $f(x) = -(x + 1)^2(x - 3)$.	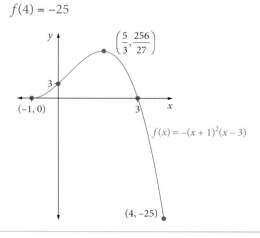

b Consider the highest and lowest points on this section of the graph.	global max. $y = \dfrac{256}{27}$ global min. $y = -25$
c Ignore the previous given interval, and now consider the graph for very large and very small values of x.	As $x \to \infty, f(x) \to -\infty$ As $x \to -\infty, f(x) \to \infty$ $(-1, 0)$ is a stationary point where gradient is 0. When $x < -1$, the graph increases and tends to infinity.

Sometimes, the turning points found on a function may not actually be the maximum or minimum point for the whole function being considered (called the **global maximum** or **minimum point**). We need to compare the turning points to the **interval endpoint values** to find the global maximum or minimum point for the given interval.

> **Exam hack**
>
> The range is not always the y values of the endpoints. You must also compare them with the y values of the local maximum and minimum points.

WORKED EXAMPLE 9 Finding the minimum value

Find the minimum value of the graph of $f(x) = (x - 1)^2(x + 3)$ for the interval $-2 \leq x \leq 3$.

Steps	Working
1 Explore the general features of the graph.	Note from its equation that this is a cubic function with a leading coefficient of 1 and x-intercepts at 1 and -3.
2 Find endpoint values.	$f(-2) = 9, f(3) = 24$
3 Find the y-intercept.	$f(0) = (0 - 1)^2(0 + 3) = 3$
4 Find $f'(x)$ and solve for zero, ignoring values of x outside of the given interval.	$f'(x) = 3x^2 + 2x - 5 = (3x + 5)(x - 1) = 0$ $\therefore x = 1, x = -\dfrac{5}{3}$ $f(1) = 0, f\left(-\dfrac{5}{3}\right) = \dfrac{256}{27}$ Minimum point is $(1, 0)$.
5 Compare endpoint minimum with local minimum.	Endpoint minimum is 9, so global minimum at $x = 1$.
6 Use CAS to sketch the graph to confirm.	

ClassPad

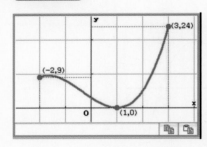

TI-Nspire

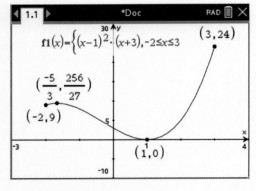

7 State the minimum value at $(1, 0)$.	Minimum value $= 0$

EXERCISE 9.4 Curve sketching

ANSWERS p. 482

Recap

1 The stationary point on the graph of $y = 2x^2 - 1$ is at

A $x = -1$ **B** $x = 0$ **C** $x = \dfrac{1}{2}$ **D** $x = 1$ **E** $x = 2$

2 The turning points of the graph of $y = x^3 - x^2 - x$ is/are at

A $(1, 0)$ and $\left(-\dfrac{1}{3}, 0\right)$ **B** $(1, -1)$ only   **C** $\left(-\dfrac{1}{3}, \dfrac{5}{27}\right)$ only

D $\left(-\dfrac{1}{3}, 0\right)$ only   **E** $\left(-\dfrac{1}{3}, \dfrac{5}{27}\right)$ and $(1, -1)$

Mastery

3 WORKED EXAMPLE 8

  a Sketch the graph of $g(x) = -\dfrac{1}{2}(x + 2)^2(x + 3)$ for $-4 \le x \le 0$, labelling key features.

  b State the global maximum and minimum for the interval $-4 \le x \le 0$.

  c State the behaviour of the function as $x \to \pm\infty$.

4 WORKED EXAMPLE 9   Find the maximum value of the graph of $f(x) = (x + 1)^2(x - 3)$ for the interval $-2 \le x \le 4$.

5 Sketch the graph of $f(x) = 2(x - 2)^2$ for $0 \le x \le 5$, labelling key features.

6 Sketch the graph of $f(x) = -\dfrac{3}{2}(x - 2)^2(x + 1)^2$ for $-1 \le x \le 3$, labelling key features.

Calculator-free

7 (8 marks) The graph of $g(x)$ is shown.

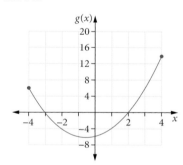

Use the graph to estimate the

  a coordinates of the stationary point   (1 mark)

  b y value at $x = -4$ (1 mark)

  c y value at $x = 1$ (1 mark)

  d intervals where the function has a positive gradient   (1 mark)

  e intervals where the function has a negative gradient   (1 mark)

  f rate of change of the function at $x = 1$ (1 mark)

  g range   (1 mark)

  h global maximum.   (1 mark)

8 (7 marks) The graph of $p(x)$ is shown.

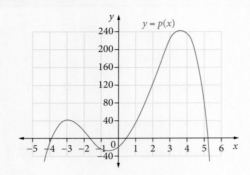

Use the graph to estimate the

a coordinates of the stationary points (3 marks)

b y value at $x = 1$ (1 mark)

c y value at $x = 2$ (1 mark)

d intervals where the rate of change of the function is positive (1 mark)

e intervals where the rate of change of the function is negative. (1 mark)

9 (4 marks) Consider the function $f(x) = 3x^2 - x^3$ for $-1 \leq x \leq 3$.

a Find the coordinates of the stationary points of the function. (2 marks)

b Sketch the graph of f. Label any endpoints with their coordinates. (2 marks)

10 (4 marks) Find any stationary points on the curve $y = 2x^3 - 9x^2 - 24x + 30$ and sketch its graph.

11 (4 marks) Determine any stationary points on the curve $y = x^3 + 6x^2 - 7$ and sketch the curve.

12 (4 marks) Find the stationary points on the curve $y = (x - 4)(x + 2)^2$ and hence sketch the curve.

Calculator-assumed

13 (5 marks) For the function $f(x) = -\frac{1}{2}(x + 1)^2$ state whether each of the statements below is true or false.

a $f(x)$ has a range which includes all real values. (1 mark)

b $f(x)$ has a local maximum at $x = 1$. (1 mark)

c $f(x)$ has a local minimum at $x = -1$. (1 mark)

d $f(x)$ has a local maximum at $x = -1$. (1 mark)

e $f(x)$ has exactly one stationary point. (1 mark)

14 (3 marks) Determine the maximum point of the graph of $f(x) = -\frac{1}{3}(x + 1)^3(x - 1)$.

15 (5 marks) For the function $f(x) = -\frac{1}{2}(x + 1)^2(x - 2)$ state whether each of the statements below is true or false.

a $f(x)$ has a range which includes all real values. (1 mark)

b $f(x)$ has a local maximum at $x = 2$. (1 mark)

c $f(x)$ has a local minimum at $x = -1$. (1 mark)

d $f(x)$ has a local maximum at $x = -1$. (1 mark)

e $f(x)$ has an x-intercept at $x = 1$. (1 mark)

16 (3 marks) Determine the local maximum point of the graph of $f(x) = -\frac{1}{2}(x + 1)^2(x - 2)$.

9.5 Optimisation problems

Differentiation techniques are used to solve problems that ask for maximum or minimum points of a graph, or the maximum or minimum values of a function. Such problems are quite common in exams, often appearing more than once and in a range of contexts, question types and length.

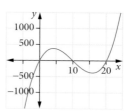

Consider the graph of $f(x) = x(20 - x)(10 - x)$.

If we differentiate and solve for zero, we find the stationary points of the graph. This question could apply to a common real-life problem with the height of liquid in a container written as $H(t) = t(20 - t)(10 - t)$, where $H(t)$ is the height of liquid in cm, at $t > 0$.

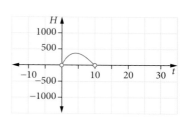

In reality, the function $H(t)$ will only exist for H, $0 < t < 10$.

The maximum height can be found by finding the local maximum of $H(t)$.

Solving $H'(t) = 0$, gives the local maximum turning point $(4.2, 384.9)$ correct to one decimal place.

The maximum height equals $384.9\,\text{cm}$.

Video playlist
Optimisation problems

WORKED EXAMPLE 10 — Finding maximum and minimum points

Find the coordinates of the local maximum and minimum points in the graph of
$f(x) = \dfrac{1}{5}(x - 1)^2(x + 4)$ for $-5 \le x \le 2$.

Steps	**Working**
1 Find $f'(x)$ and solve for zero.	$f'(x) = \dfrac{3x^2 + 4x - 7}{5} = \dfrac{1}{5}(3x + 7)(x - 1) = 0$
	$\therefore x = -\dfrac{7}{3},\ x = 1$
2 Consider the shape of the graph to decide whether each point is a maximum or minimum. Note from its equation that this is a cubic function with a leading coefficient of $\dfrac{1}{5}$ and x-intercepts at 1 and -4.	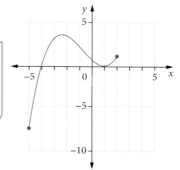 Local maximum is at $x = -\dfrac{7}{3}$. Local minimum is at $x = 1$.
3 Find the y values of the turning points.	$f\!\left(-\dfrac{7}{3}\right) = \dfrac{100}{27},\ f(1) = 0$ Local maximum is at $\left(-\dfrac{7}{3}, \dfrac{100}{27}\right)$. Local minimum is at $(1, 0)$.

Chapter 9 | Applications of differential calculus

USING CAS 3 — Finding maximum and minimum points

Confirm the local maximum and minimum points in the graph of $f(x) = \frac{1}{5}(x-1)^2(x+4)$ from Worked example 10.

ClassPad

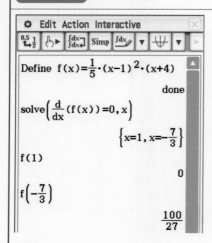
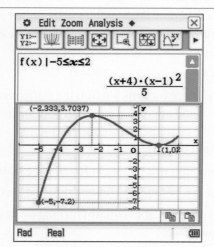

1. Define the function **f(x)**.
2. Set the **derivative of f(x)** equal to **0** and **solve** to find the x-coordinates of the turning points.
3. Substitute the solutions into **f(x)** to determine the corresponding y-coordinates.
4. To visualise these solutions, graph **f(x)** over the domain **[−5, 2]**.
5. Adjust the window settings to suit.
6. Tap **Interactive > G-Solve > Max** to find the local maximum.
7. Tap **Interactive > G-Solve > Min** to find the local minimum.

TI-Nspire

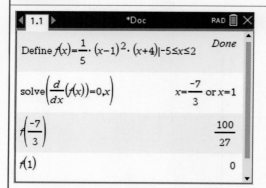
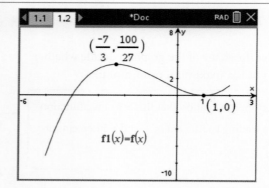

1. Define the function **f(x)**.
2. Set the **derivative of f(x)** equal to **0** and **solve** to find the x-coordinates of the turning points.
3. Substitute the solutions into **f(x)** to determine the corresponding y-coordinates.
4. To visualise these solutions, add a **Graphs** page and graph **f(x)**.
5. Adjust the window settings to suit.
6. To find the exact coordinates of the turning points, press **menu > Geometry > Points & Lines > Point On**.
7. Click to add 2 points onto the curve then press **esc** to remove the **Point On** tool.
8. Click on the x-coordinates of each point and enter the values as shown above.

The local maximum is $\left(-\frac{7}{3}, \frac{100}{27}\right)$. The local minimum is $(1, 0)$.

Solving optimisation problems

WORKED EXAMPLE 11 Solving optimisation problems

A rectangular sheet of tin, measuring 4 m by 2 m, has square corners of length x units cut out at each corner. These corners are folded up to make a storage box. Find the maximum possible volume, in m³, correct to one decimal place, for this storage box.

Steps	Working
1 Sketch a diagram.	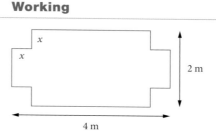
2 Set up an equation to describe the volume of the box.	length = $4 - 2x$, width = $2 - 2x$, height = x $V(x) = x(4 - 2x)(2 - 2x)$
3 Identify the domain for this problem.	x is the length of a square corner so it must be positive and less than half of the width 2 m. The domain is $0 < x < 1$.
4 Find $V'(x)$ and solve for zero.	$V(x) = x(4 - 2x)(2 - 2x) = 4x^3 - 12x^2 + 8x$ $V'(x) = 12x^2 - 24x + 8 = 0$ $\therefore x = \dfrac{3 \pm \sqrt{3}}{3}$ $\dfrac{3 - \sqrt{3}}{3} \approx 0.4227$ for domain $0 < x < 1$
5 Test whether this is a maximum point by using CAS to sketch a graph or by checking whether the sign of $V'(x)$ changes from positive to negative.	The point is a maximum. 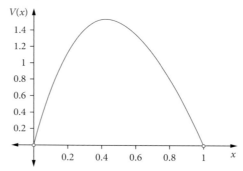
6 Substitute this value of x into $V(x)$ to find the maximum volume of the storage box.	$V(0.4227) = 1.5396$ Maximum possible volume is about 1.54 m^3.

> 🔓 **Exam hack**
>
> For maximum and minimum problems, re-read the question to make sure that you have answered it. One common mistake students make in exams is to forget to answer the actual question, even though all the working is correct.

EXERCISE 9.5 Optimisation problems

ANSWERS p. 483

Recap

1. Find the maximum value of the function $y = -4x(x - 1)$.

2. Sketch the graph of $f(x) = -3(x + 2)(x - 2)^2 + 2$ for $0 \leq x \leq 5$, labelling key features.

Mastery

3. **WORKED EXAMPLE 10** Find the coordinates of the local maximum and minimum points in the graph of $f(x) = (x - 1)(x + 3)^2$.

4. **Using CAS 3** Determine the local minimum point on the graph of $f(x) = 2(x + 1)^2(x + 2)$ in the interval $-3 \leq x \leq 3$.

5. **WORKED EXAMPLE 11** Zoe has a rectangular piece of cardboard that is 16 cm long and 8 cm wide. Zoe cuts squares of side length x centimetres from each of the corners of the cardboard, and turns up the sides to form an open box. Determine the value of x for which the volume of the box is a maximum.

6. A rectangular field shares one side with an existing paddock and so this requires no fence. There is only 1000 m of fencing material to fence the remaining sides. Find the maximum possible area of the field.

7. The perimeter of a rectangle is 60 m and its length is x m. Show that the area of the rectangle is given by the equation $A = 30x - x^2$. Hence, find the greatest area of the rectangle.

Calculator-free

8. (3 marks) Find two numbers whose sum is 28 and whose product is a maximum.

9. (3 marks) Find the greatest volume of a rectangular box with a square base if the sum of the height and the side of the base must not exceed 24 cm.

10. (3 marks) The cost, in dollars per hour, of a bike ride is given by the formula $C = x^2 - 15x + 70$, where x is the distance travelled in kilometres. Find the distance that gives the lowest cost.

11. (3 marks) The height, in metres, of a ball is given by the equation $h = 16t - 4t^2$, where t is the time in seconds. Find when the ball will reach its greatest height, and what the greatest height will be.

Calculator-assumed

12. (6 marks) A box is made from an 80 cm by 30 cm rectangular cardboard by cutting out 4 equal squares of size x cm from each corner. The edges are turned up to make an open box.

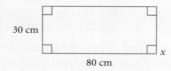

   a Show that the volume of the box is given by the equation $V = 4x^3 - 220x^2 + 2400x$. (2 marks)
   b Find the value of x that gives the box its greatest volume. (2 marks)
   c Find the maximum volume of the box, correct to one decimal place. (2 marks)

13 (3 marks) A welder wants to design an open metal container having a square base and a surface area of 108 square metres. What dimensions will produce a container with maximum volume?

14 (3 marks) In an experiment, it is found that the temperature (°C) of an object is approximated by the function $f(t) = 40 - 24t + 9t^2 - t^3$, where t is the time in minutes after the start of the experiment, given that $\frac{3}{2} \leq t \leq 6$. Find the highest and lowest temperatures of the object during the given time interval.

15 (2 marks) Determine the local maximum point on the graph of $f(x) = -(x-1)(x+3)^2$.

16 (2 marks) Determine the global minimum point on the graph of $f(x) = 2(x+1)^2(x+2)$ for $-3 \leq x \leq 3$.

17 (4 marks) A right-angled triangle, OAB, is formed by the horizontal axis and the point $A(k, 9 - (k+1)^2)$, where $0 \leq k \leq 2$, on the parabola $y = 9 - (x+1)^2$, as shown below. Find the maximum area of the triangle OAB to two decimal places.

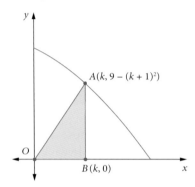

18 (10 marks) A solid block in the shape of a rectangular prism has a base of width x cm. The length of the base is two-and-a-half times the width of the base.

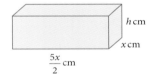

The block has a total surface area of 6480 square centimetres.

a Show that if the height of the block is h cm, $h = \dfrac{6480 - 5x^2}{7x}$. (2 marks)

b Show that the volume, V cm^3, of the block is given by $V(x) = \dfrac{5x(6480 - 5x^2)}{14}$. (3 marks)

c Find $\dfrac{dV}{dx}$. (2 marks)

d Find the exact values of x and h if the block is to have maximum volume. (3 marks)

▶ **19** (9 marks) A train is travelling at a constant speed of w km/h along a straight level track from M toward Q. The train will travel along a section of track $MNPQ$.

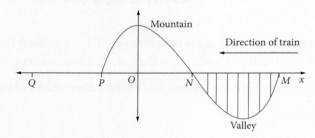

Section MN passes along a bridge over a valley. Section NP passes through a tunnel in a mountain. Section PQ is 6.2 km long.

From M to P, the curve of the valley and the mountain, directly below and above the train track, is modelled by the graph of $y = \dfrac{1}{200}(ax^3 + bx^2 + c)$ where a, b and c are real numbers. All measurements are in kilometres.

The curve defined from M to P passes through $N(2, 0)$. The gradient of the curve at N is -0.06 and the curve has a turning point at $x = 4$.

From this information write down three simultaneous equations in a, b and c.

9.6 The anti-derivative

Video playlist
The anti-derivative

The **anti-derivative** of a function $f(x)$ is the function $F(x)$ whose derivative is $f(x)$. In other words, $F'(x) = f(x)$. The process of finding a function from its derivative is called **anti-differentiation**, the reverse process of differentiation.

When we differentiate a **constant** term (number), we get 0, so when anti-differentiating we must include '$+ c$' in the answer, where c stands for any real number.

The general anti-derivative, or the **integral**, of $2x$ is $x^2 + c$, where c is a constant.

In algebraic terms, this is written as $\int (2x)\,dx = x^2 + c$.

This is read as 'the integral of $2x$ with respect to x is $x^2 + c$'.

Another name for the process of anti-differentiation is **integration**.

The anti-derivative (or integral) of ax^n

If $\dfrac{dy}{dx} = ax^n$, then $y = \int (ax^n)\,dx = a\int (x^n)\,dx = \dfrac{ax^{n+1}}{n+1} + c$, $n \neq -1$.

Consider the function $f(x) = 5 + 4x + 9x^2$.

To anti-differentiate, we 'add 1 to the power, and divide by the new power'.

The term 5 becomes $5x$.

The term $4x$ becomes $2x^2$.

The term $9x^2$ becomes $3x^3$.

$$\int (5 + 4x + 9x^2)\,dx = \dfrac{5x^1}{1} + \dfrac{4x^2}{2} + \dfrac{9x^3}{3} + c$$

So the anti-derivative of the expression $5 + 4x + 9x^2$ becomes $5x + 2x^2 + 3x^3 + c$.

WORKED EXAMPLE 12 | Finding the anti-derivative

Find the anti-derivative of the function
$f(x) = 2x^2 - 4x + 5$.

> **Exam hack**
>
> You can always check your answer by differentiating it to see whether you get the original function.

Steps	Working
1 Write the function as a derivative.	$\dfrac{dy}{dx} = 2x^2 - 4x + 5$
2 Integrate each term using the formula $\int ax^n \, dx = \dfrac{ax^{n+1}}{n+1} + c$ to find y.	$y = \dfrac{2x^3}{3} - \dfrac{4x^2}{2} + 5x$ $\therefore y = \dfrac{2x^3}{3} - 2x^2 + 5x + c$

We can also use CAS to anti-differentiate.

USING CAS 4 | Finding the anti-derivative

Find the anti-derivative of $3x^2 + 2x - 4$.

ClassPad

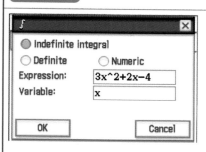
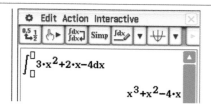

1. Enter and highlight the expression.
2. Tap **Interactive** > **Calculation** > $\int$.
3. In the dialogue box, tap **OK**.

4. The anti-derivative will be displayed.

TI-Nspire

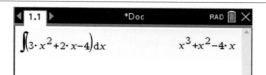

1. Press **menu** > **Calculus** > **Integral**.
2. The **Integral** template will be displayed.
   Alternatively, press **template** to access the integral template.

3. Enter the expression, including the **x** at the end as shown above.
4. Tap **enter** to display the anti-derivative.

The anti-derivative of $3x^2 + 2x - 4$ is $x^3 + x^2 - 4x + c$.

> CAS does not give you the $+ c$. Remember to include $+ c$ in your answers.

Note that CAS does not give you the + c.

To find the value of c, we need extra information.

For example, if $\dfrac{dy}{dx} = 3x^2 + 2x^2 + 4x + 1$,

we anti-differentiate and get $y = \dfrac{3}{4}x^4 + \dfrac{2}{3}x^3 + 2x^2 + x + c$.

If we are also told that when $x = 1$, $y = 2$, we substitute to get

$$2 = \dfrac{3}{4}(1)^4 + \dfrac{2}{3}(1)^3 + 2(1)^2 + (1) + c$$

$$\therefore 2 = \dfrac{3}{4} + \dfrac{2}{3} + 2 + 1 + c$$

giving $c = -\dfrac{29}{12}$.

So in this case, $y = \dfrac{3}{4}x^4 + \dfrac{2}{3}x^3 + 2x^2 + x - \dfrac{29}{12}$.

WORKED EXAMPLE 13   Applying the anti-derivative

Find y if $\dfrac{dy}{dx} = 3x^2 + x + 3$ and $x = 1$ when $y = 0$.

Steps	Working
1 Write the expression as a derivative.	$\dfrac{dy}{dx} = 3x^2 + x + 3$
2 Integrate each term and simplify.	$y = \dfrac{3x^3}{3} + \dfrac{x^2}{2} + 3x + c$ $\therefore y = x^3 + \dfrac{x^2}{2} + 3x + c$
3 Find the value of c by substituting $x = 1$ when $y = 0$.	$0 = 1^3 + \dfrac{1^2}{2} + 3(1) + c$ $0 = \dfrac{9}{2} + c$ $\therefore c = -\dfrac{9}{2}$
4 State the answer including the value of c.	$y = x^3 + \dfrac{x^2}{2} + 3x - \dfrac{9}{2}$

USING CAS 5 — Finding the anti-derivative and calculating the value of c

Find the anti-derivative of $f'(x) = 6x^2 + 3x - 4$, given that $f(0) = 2$.

ClassPad

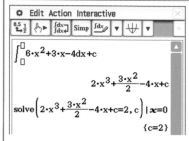

1. Enter and highlight the expression.
2. Tap **Interactive > Calculation > ∫**.
3. Enter **+c** at the end of the integral as shown above.
4. Set the anti-derivative expression equal to 2 and solve for **c**. Include the condition that $x = 0$.

TI-Nspire

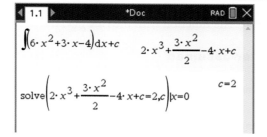

1. Press **menu > Calculus > Integrate**.
2. Enter the expression, including the **x**.
3. Enter **+c** at the end of the integral as shown above.
4. Set the anti-derivative expression equal to 2 and solve for **c**. Include the condition that $x = 0$.

$$f(x) = 2x^3 + \frac{3x^2}{2} - 4x + 2$$

EXERCISE 9.6 The anti-derivative

ANSWERS p. 483

Recap

1. Find two numbers whose difference is 4 and whose product is a minimum.

2. The height of a stone thrown vertically upwards is given by $h = 64t - 16t^2$, where h is the height in metres and t is the time in seconds. Find the maximum height reached by the stone.

Mastery

3. **WORKED EXAMPLE 12** Find the anti-derivative of the function $f(x) = 3x^3 + 2x^2$.

4. **Using CAS 4** Determine the anti-derivative of the function $y = -2(3x + 2)^2$.

5. **WORKED EXAMPLE 13** Find the anti-derivative of $f'(x) = x^2 + 3x + 7$, given that $f(1) = 2$.

6. **Using CAS 5** Find the anti-derivative of $\frac{dy}{dx} = x^2 - x$, given that $y = 0$ when $x = 1$.

Calculator-free

7. (8 marks) Find the anti-derivative of

   a. x^2 (1 mark)
   b. $x^3 + x^2$ (1 mark)
   c. $4x^3 + 2x - 4$ (1 mark)
   d. $3x^2 + 5x + 7$ (1 mark)
   e. $\frac{3x - 4}{2}$ (1 mark)
   f. $(x - 2)(2x + 3)$ (1 mark)
   g. $(1 - 2x)^2$ (1 mark)
   h. $(x^2 + x + 1)^2$ (1 mark)

   Exam hack
   Functions may need to be expanded and/or simplified prior to anti-differentiating.

8 (3 marks) The derivative with respect to x of the function $f(x)$ is $f'(x) = \dfrac{1}{2} - \dfrac{3x-2}{2}$. Given that $f(2) = 0$, find $f(x)$ in terms of x.

9 (2 marks) Find the anti-derivative of $(x-1)^3$ with respect to x.

10 (3 marks) The function with the rule $g(x)$ has derivative of $g'(x) = 3x^2 + 1$. Given that $g(1) = 2$, find $g(x)$.

Calculator-assumed

11 (2 marks) Determine the anti-derivative of the function $y = 8x^3 + 2x^2 - 7$.

12 (3 marks) Find the anti-derivative of $f'(x) = 3x^3 + 2x^2$ if $x = 1$ when $y = 0$.

13 (3 marks) If $f'(x) = 2x + b$ (where b is a constant), $f(2) = 1$ and $f(-1) = -2$, determine the value of b.

Video playlist
Applying the anti-derivative to straight line motion

9.7 Applying the anti-derivative to straight line motion

Kinematics is a study of motion using displacement (x) and velocity (v), all in terms of time. We know that $v = \dfrac{dx}{dt}$.

Conversely, displacement $= x = \int v(t)\,dt$.

WORKED EXAMPLE 14 — Anti-differentiating for straight line motion

The velocity of a particle travelling in a straight line is given by $v(t) = 2t$.

a Find an expression for the displacement $x(t)$ if $t = 3$ when $x = 0$.
b Find the displacement at $t = 4$ seconds.

Steps	Working
a 1 State the rule for velocity.	$v(t) = 2t$
2 Use displacement $= x = \int v(t)\,dt$.	$x = \int 2(t)\,dt$ $\therefore x = t^2 + c$
3 Find c using $t = 3$ when $v = 0$.	Solve $0 = 3^2 + c$ $\therefore c = -9$
4 State the displacement function.	$x = t^2 - 9$
b Substitute $t = 4$ into displacement function.	$x = 4^2 - 9 = 7$ The particle is at $x = 7$ at $t = 3$.

EXAMINATION QUESTION ANALYSIS

Calculator-free (10 marks)

There are two toy cars racing on a model car track, both starting at the origin. It is known that Car A has a displacement given by $x(t) = \dfrac{t^3}{3} - \dfrac{5t^2}{2} + 6t$ metres, and Car B has a velocity given by $v(t) = \dfrac{4t - 3}{2}$ metres/second.

a What is the displacement of Car A at 3 seconds? (2 marks)
b When does Car A momentarily stop? (3 marks)
c Determine the displacement of Car B at 5 seconds. (2 marks)
d Are both cars heading in the same direction initially? Explain your answer. (3 marks)

Reading the question

- Take note that both cars start at the origin.
- Take note that different information is given about the two cars. One is about displacement, and one is about velocity.
- As this question is calculator free, questions will need to be solved manually.

Thinking about the question

- Part **a** is straightforward as you already have the displacement equation.
- Part **b** requires finding the velocity, and then when velocity is zero.
- Part **c** requires displacement, so you will need to anti-differentiate to find this. Remember to consider the value for the constant.
- For part **d**, to find the direction the cars are heading, velocity will need to be found for both cars. Remember, initial velocity means when $t = 0$.

Worked solution ($\checkmark$ = 1 mark)

a $x(3) = \dfrac{3^3}{3} - \dfrac{5(3)^2}{2} + 6(3)$ ✓

$x(3) = 9 - \dfrac{45}{2} + 18$

$x(3) = \dfrac{9}{2}$ m ✓

b $v(t) = t^2 - 5t + 6$ ✓

Let $v(t) = 0$.
$0 = (t - 3)(t - 2)$
$\therefore t = 3, 2$ ✓

Car A stops momentarily at 3 seconds and 2 seconds. ✓

c $x(t) = t^2 - \dfrac{3t}{2}$ ✓

$\therefore x(5) = 25 - \dfrac{15}{2} = \dfrac{35}{2}$ m ✓

d Car A: $v(0) = 6$ m/s ✓

Car B: $v(0) = -\dfrac{3}{2}$ m/s ✓

The **different signs for velocity at $t = 0$** indicate the cars are heading in **different directions**. ✓

EXERCISE 9.7 Applying the anti-derivative to straight line motion

ANSWERS p. 483

Recap

1. Find the anti-derivative of $10x^4 + 4x^3 - 3x^2 - 2x$.

2. $2x^2 + 3x - 2$ has the anti-derivative

   A $6x + 3 + c$

   B $6x + 3x - 2$

   C $\dfrac{2x^3}{3} + \dfrac{3x^2}{2} - 2x + c$

   D $9x$

   E $2x^2 + 3x + c$

Mastery

3. **WORKED EXAMPLE 14** The velocity, in m/s, of a particle is expressed by $v(t) = t^2 + 3t$. Find the displacement, x metres, given that $x(0) = 2$.

4. The velocity of a toy car, starting at the origin, is 20 m/s and at $t = 1$ its displacement is 10 m. Determine an expression for the displacement.

5. The velocity of an object is given by $v = 6t + 4$ cm/s. The particle starts at 3 m to the left of the origin. Find its displacement after 5 seconds.

Calculator-free

6. (4 marks) The velocity of a particle is given by $v = 8t^3 - 3t^2$ cm/s. If the particle is initially at the origin, find

   a its displacement after 3 seconds (2 marks)

   b when it will be at the origin again. (2 marks)

7. (4 marks) The velocity (in cm/s) of an object is given by $v = t^2(t^3 + 1)^2$ and the object is initially 2 centimetres to the right of the origin. Find the displacement of the object after 2 seconds.

8. (4 marks) A particle has a velocity given by $\dfrac{dx}{dt} = 2t - 3$ m/s. After 3 seconds, the particle is at the origin. Find its displacement after 7 seconds.

9. (4 marks) The velocity of a particle is given by $v = 4 - 3t$ cm/s. If the particle is 5 centimetres from the origin after 2 seconds, find the displacement after 7 seconds.

Calculator-assumed

10. (4 marks) The velocity of a particle is given by $v = 3t^2 - 5$ cm/s. If the particle is initially at the origin, find its displacement after 3 seconds.

11. (4 marks) A toy car is travelling with a velocity expressed by $v(t) = 2t + 1$ m/s. If the toy car started at the origin, determine its position after travelling for 5 seconds.

12. (4 marks) A car is travelling with velocity expressed by $v(t) = 2t + 1$ m/s. If the car's displacement at $t = 1$ is 40 m, determine an expression for displacement.

Chapter summary

Equation of a tangent

- To find the equation of the tangent to the graph of $y = f(x)$ at the point (x_1, y_1):
  1. Differentiate $f(x)$ to get $f'(x_1)$.
  2. Find the gradient of the tangent by evaluating $f'(x_1)$.
  3. Use the point to find the equation of the line.
- The equation of the tangent line of gradient m is found using:
  - the gradient–intercept form $y = mx + c$, where c is the y-intercept
  - the point–gradient form $y - y_1 = m(x - x_1)$, where (x_1, y_1) is a point on the line.

Gradient of a function

- A positive gradient points up: ↗
- A negative gradient points down: ↘
- A zero gradient is flat ⟶

Stationary points

- A **turning point** is found where $f'(x) = 0$ and the sign of the gradient changes on either side of the stationary point.

Local minimum point

Local maximum point

- A **stationary point of inflection** is a flat 'bend' in the graph where the concavity changes and the sign of the gradient stays *the same* on both sides of the stationary point.

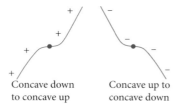
Concave down to concave up

Concave up to concave down

Maximum and minimum points

- A turning point is also called a local maximum or minimum point.

Curve sketching

Identify key features of a graph when sketching:
- general shape
- domain and range
- y-intercept
- x-intercept(s)
- stationary point(s)
- global and local maxima and minima
- behaviour as $x \to \pm\infty$
- To find the **global maximum** or **minimum point** of the graph of a function over an interval, we need to compare the interval endpoint values to the local maximum and minimum points.

Optimisation problems

Techniques of finding maximum and minimum points of graphs will help solve maxima and minima problems.

The anti-derivative

- The **anti-derivative** of a function $f(x)$ is the function $F(x)$ whose derivative is $f(x)$. In other words, $F'(x) = f(x)$. The process of finding a function from its derivative is called **anti-differentiation** or **integration**.
- If $\dfrac{dy}{dx} = ax^n$, then $y = \int ax^n dx = \dfrac{ax^{n+1}}{n+1} + c, n \neq -1$.
- Notations for an **anti-derivative or integral** of a function are $F(x)$ and $\int f(x)dx$.
- If we are given extra information about the anti-derivative function, then we can find the value of c.

Straight line motion

Kinematics is the study of motion using displacement (x) and velocity (v), all in terms of time (t).

- velocity $= v = \dfrac{dx}{dt}$
- displacement $= x = \int v(t) dt$

Cumulative examination: Calculator-free

Total number of marks: 25   Reading time: 3 minutes   Working time: 25 minutes

1 (3 marks) In a group of 100 people, it is found that 42 own at least one Android device, 78 own at least one Apple device and 6 own neither an Android nor an Apple device. Determine the number of people in the group who own at least one Android and at least one Apple device.

2 (4 marks) Sketch the parabola for each of the following equations. Label the turning point with its coordinates and any x- and y-intercepts.
  a $y = -(x + 3)^2 - 4$ (2 marks)
  b $y = x^2 + 11x + 28$ (2 marks)

3 (2 marks) If $x - a$ is a factor of $8x^3 - 14x^2 - a^2x$, where a can be any real value except for zero, determine the value of a.

4 (3 marks) Simplify $\dfrac{(3a^4b^5)^2 \times 5a^{-1}b^{-14}}{(6a^3b^2)^2}$, expressing your answer with positive indices.

5 (4 marks) Let $f(x) = 2x^2 - 1$.
  a Show use of first principles to determine the derivative of $f(x)$. (2 marks)
  b Calculate the instantaneous rate of change of f at $x = -1$. (2 marks)

6 (3 marks)
  a Anti-differentiate $y' = 2x^3 - x$, with respect to x. (1 mark)
  b Evaluate $f'(4)$, where $f(x) = \dfrac{2x^2 - 3x}{x}$. (2 marks)

7 (3 marks) Find $f(x)$ given that $f(3) = 4$ and $f'(x) = \dfrac{8x}{3} - 9x^2 + 1$.

8 (3 marks) Sketch the curve of $f(x) = x^2 - 4x + 4$, labelling all key features.

Cumulative examination: Calculator-assumed

Total number of marks: 39   Reading time: 4 minutes   Working time: 39 minutes

1 (7 marks) Australia's population by the top 10 countries of birth in 2021 is shown in the table, rounded to the nearest thousand.

Country of birth	Population ('000)
England	967
India	710
China	596
New Zealand	560
Philippines	310
Vietnam	268
South Africa	202
Malaysia	172
Italy	172
Sri Lanka	146
Total overseas-born	7 502
Total Australian-born	18 236

Source: ABS (Australian Bureau of Statistics) ABS website 2021

Use relative frequencies to estimate the probability, correct to four decimal places, that a randomly selected Australian resident:

a has England as their country of birth (1 mark)

b has Vietnam as their country of birth (1 mark)

c was Australian-born (1 mark)

d that was born overseas has China as their country of birth (2 marks)

e that was born overseas has New Zealand as their country of birth. (2 marks)

2 (3 marks) Determine the size of the smallest angle of a triangle whose sides are 3.2 cm, 4.4 cm and 5.3 cm long.

3 (11 marks) Two university social clubs launch a membership campaign to try to increase their membership numbers. The promotions teams of each club propose the following models for membership increases over the next six months.

Month	Current	1	2	3	4	5	6
Club A	215	220	225	230	235	240	245
Club B	180						240

a Describe the proposed model for Club A's membership over the next six months. (2 marks)

The president of Club B suggests that knowing a membership goal for month 6 is not enough information to know how the memberships are going to increase.

Let A_n be the number of members in Club A n months after the campaign launch.

Let B_n be the number of members in Club B n months after the campaign launch.

  b Write a general rule for Club B's membership if the membership model is

    i an arithmetic sequence (3 marks)

    ii a geometric sequence. (3 marks)

  c Assuming that Club B's membership model is a geometric sequence and Club A's membership continues in same way after the first six months, compare the predicted membership numbers for each club after 12 months. (3 marks)

4 (2 marks) Let $f'(x) = 3x^2 - 2x$ such that $f(4) = 0$. Determine $f(x)$.

5 (3 marks) A rectangular sheet of cardboard has a length of 80 cm and a width of 50 cm. Squares of side length x cm are cut from each of the corners, as shown in the diagram below. A rectangular box with an open top is then constructed. Determine the value of x when the volume of the box is a maximum.

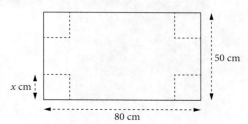

6 (2 marks) If a particle has a velocity of $v(t) = \dfrac{1}{2}(3t - 5)$ and $x(3) = -2$, determine an expression for its displacement.

7 (4 marks) A particle moves so that its distance, x metres, from a fixed point after t seconds is given by $x(t) = 16 + 48t - t^3$. Determine when and where the particle stops momentarily.

8 (7 marks) A piece of wire, 3 metres long, is used to make the 9 edges of the frame of a wedge (see diagram). The height and length of the wedge are $5x$ cm and $12x$ cm respectively.

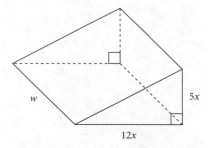

  a Given that L is the length of the hypotenuse of the cross-section, show that $L = 13x$. (1 mark)

  b Write an equation for the width of the wedge (w) in terms of x (in centimetres). (1 mark)

  c Show that the volume of the wedge (V) is $V = 600x^2(6 - x)$ cm^2. (2 marks)

  d Use a calculus technique to determine the dimensions of the frame that will maximise the volume of the wedge. (3 marks)

Answers

CHAPTER 1

EXERCISE 1.1

1 a 5   **b** 2   **c** $C = \{3, 6, 9\}$

2 a $\overline{A} = \{1, 4, 6, 8, 9, 10\}$
  b $A \cap C = \{3\}$
  c $A \cup B = \{2, 3, 4, 5, 6, 7, 8, 10\}$
  d $C \cap \overline{A \cup B} = \{9\}$
  e $A \cap B \cap C = \{\ \} = \varnothing$

3 a

	F	F'	Total
H	25	40	65
H'	35	25	60
Total	60	65	125

  b 25, meaning that 25 students like neither hot food nor fizzy drinks.
  c It represents the students who don't like both hot food and fizzy drinks.

4 $n(\overline{A \cup B}) = 10$

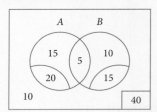

5 a

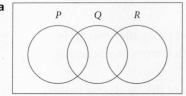

  b

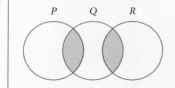

6 110

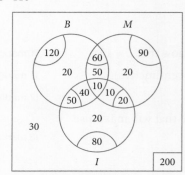

7 a 8
  b $\{1, 7, 11, 13\}$
  c $\{15\}$
  d This is because the lowest common multiple of 2, 3 and 5 is 30, which is not in the universal set.

8 78

9 a $80 - x - y$ **b** 25

10 28

EXERCISE 1.2

1 D   **2** C

3 a 40 320   **b** 6720   **c** 56

4 a 10   **b** 35
  c 4   **d** 21
  e 70   **f** 11
  g 66   **h** 100
  i 780   **j** 105

5 a 1.3077×10^{12} **b** 210   **c** 105

6 a 1540   **b** 319 770

7 2 827 440

8 a 36   **b** 7   **c** 21

9 a 455   **b** 70   **c** 350

10 a 56   **b** 1   **c** 10   **d** 55

11 a 646 646   **b** 17 920   **c** 1001   **d** 37 128

EXERCISE 1.3

1 E   **2** D

3 a 1, 6, 15, 20, 15, 6, 1   **b** $r = 1$ **c** 64

4 a 729
  b $x^3 + 6x^2 + 12x + 8$
  c $x^3 - 9x^2 + 27x - 27$
  d $27x^3 - 27x^2 + 9x - 1$

5 $81x^4 + 540x^3 + 1350x^2 + 1500x + 625$

6 a 216   **b** 25
  c -32 **d** $-\dfrac{15}{16}$

7 a 128
  b $x^7 + 14x^6 + 84x^5 + 280x^4 + 560x^3 + 672x^2 + 448x + 128$

8 a 5 terms   **b** -8

9 Proof: see worked solutions

CUMULATIVE EXAMINATION: CALCULATOR-FREE

1 a $\{e, i\}$
  b i 5   **ii** 14

2 a 10   **b** 210   **c** 4

3 a $16x^4 - 32x^3 + 24x^2 - 8x + 1$
  b 64   **c** 30

CUMULATIVE EXAMINATION: CALCULATOR-ASSUMED

1 a 2 895 620 **b** 1 167 600 **c** 2 448 040

2 a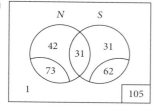

or

	S	S'	Total
N	31	42	73
N'	31	1	32
Total	62	43	105

b The number of students who liked the food at North Café or South Café or both.

c 1

CHAPTER 2

EXERCISE 2.1

1 a $\frac{4}{7}$ **b** $\frac{2}{7}$ **c** $\frac{4}{7}$

2 a $\frac{15}{20}$ **b** $\frac{11}{20}$ **c** 1

3 a experimental: $\frac{22}{30} = 0.73$, theoretical: $\frac{1}{2} = 0.5$
Experimental probability is $\frac{7}{30}$ greater than the theoretical probability.

b experimental: $\frac{11}{20} = 0.55$, theoretical: $\frac{10}{20} = 0.5$
Experimental probability is $\frac{1}{20}$ greater than the theoretical probability.

4 a 0.53 **b** 0.4 **c** 0.57

5 a $\frac{21}{30}$ **b** $\frac{26}{30}$ **c** $\frac{16}{30}$

6 1 red, 1 blue, 1 yellow, 4 green, 5 white

EXERCISE 2.2

1 E **2** C

3

	B	B'	Total
A	0.32	0.35	0.67
A'	0.22	0.11	0.33
Total	0.54	0.46	1

$P(A') = 0.33$

4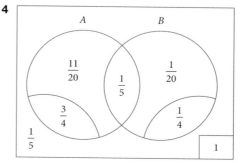

$P(A' \cap B') = \frac{1}{5}$

5 $\frac{7}{48}$

6 a mutually exclusive **b** neither

7 $\frac{19}{20}$

8 a 0 **b** 1 **c** $\frac{1}{20}$

9 $\frac{40}{200}$

10 0.87

11 a $\frac{5}{24}$ **b** $\frac{1}{3}$

12 $\frac{2}{5}$

13 a 0.35 **b** 0.05

14 a 0.16 **b** 0.32

15 a $\frac{3}{52}$ **b** $\frac{16}{52}$

16 a

	Disease	No disease	Total
Positive	0.28	0.02	0.3
Negative	0.01	0.69	0.7
Total	0.29	0.71	1

b **i** 0.01 **ii** 0.99

17 a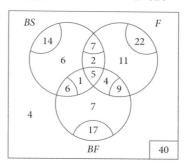

b **i** $\frac{7}{40}$ **ii** $\frac{1}{40}$ **iii** $\frac{24}{40}$

EXERCISE 2.3

1 C   **2** A

3 a

	1	2	3	4	5
1	(1, 1)	(1, 2)	(1, 3)	(1, 4)	(1, 5)
2	(2, 1)	(2, 2)	(2, 3)	(2, 4)	(2, 5)
3	(3, 1)	(3, 2)	(3, 3)	(3, 4)	(3, 5)
4	(4, 1)	(4, 2)	(4, 3)	(4, 4)	(4, 5)
5	(5, 1)	(5, 2)	(5, 3)	(5, 4)	(5, 5)

b i $\dfrac{2}{25}$ **ii** $\dfrac{16}{25}$

4 $\dfrac{4}{10}$

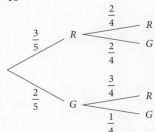

5 a

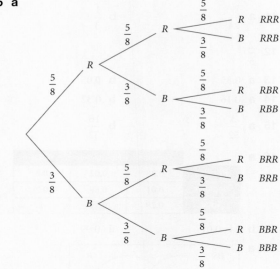

b $\dfrac{485}{512}$

6 a $\dfrac{216}{1000}$ **b** $\dfrac{210}{720}$ **c** $\dfrac{40}{72}$

7 $\dfrac{14}{36}$

8 a

	1	2	3	4	5	6
1	(1,1)	(1, 2)	(1, 3)	(1, 4)	(1, 5)	(1, 6)
2	(2,1)	(2, 2)	(2, 3)	(2, 4)	(2, 5)	(2, 6)
3	(3,1)	(3, 2)	(3, 3)	(3, 4)	(3, 5)	(3, 6)
4	(4,1)	(4, 2)	(4, 3)	(4, 4)	(4, 5)	(4, 6)

b i $\dfrac{6}{24}$ **ii** $\dfrac{12}{24}$ **iii** $\dfrac{18}{24}$

9 a

	1	2	3	4	5	6
1	1	$\dfrac{1}{2}$	$\dfrac{1}{3}$	$\dfrac{1}{4}$	$\dfrac{1}{5}$	$\dfrac{1}{6}$
2	2	1	$\dfrac{2}{3}$	$\dfrac{1}{2}$	$\dfrac{2}{5}$	$\dfrac{1}{3}$
3	3	$\dfrac{3}{2}$	1	$\dfrac{3}{4}$	$\dfrac{3}{5}$	$\dfrac{1}{2}$
4	4	2	$\dfrac{4}{3}$	1	$\dfrac{4}{5}$	$\dfrac{2}{3}$
5	5	$\dfrac{5}{2}$	$\dfrac{5}{3}$	$\dfrac{5}{4}$	1	$\dfrac{5}{6}$
6	6	3	2	$\dfrac{3}{2}$	$\dfrac{6}{5}$	1

b i $\dfrac{3}{36}$ **ii** $\dfrac{2}{36}$ **iii** $\dfrac{15}{36}$ **iv** $\dfrac{14}{36}$

10 a $\dfrac{8}{30}$ **b** $\dfrac{14}{30}$ **c** $\dfrac{18}{30}$

11 0.4545

12 a 0.1775   **b** 0.54   **c** 0.8825

EXERCISE 2.4

1 B   **2** D

3 0.3023   **4** 0.9964

5 $1 - \dfrac{\binom{15}{4}}{\binom{20}{4}}$ **6** 0.3571

7 a 0.3297   **b** 0.0769   **c** 0.8462

8 a $\dfrac{2}{16}$ **b** $\dfrac{8}{16}$

9 a 0.4286   **b** 0.4286   **c** 0.5

EXERCISE 2.5

1 C   **2** B

3 a conditional probability, due to the conditional language 'given that'

b simple probability, as no conditional language used

4 $\dfrac{4}{7}$

5 a

	B	B'	Total
A	0.5	0.2	0.7
A'	0.05	0.25	0.3
Total	0.55	0.45	1

$P(A' \mid B') = \dfrac{5}{9}$

b

	B	B'	Total
A	$\dfrac{7}{24}$	$\dfrac{1}{8}$	$\dfrac{10}{24}$
A'	$\dfrac{5}{12}$	$\dfrac{4}{24}$	$\dfrac{14}{24}$
Total	$\dfrac{17}{24}$	$\dfrac{7}{24}$	1

$P(A \mid B') = \dfrac{3}{7}$

6 $\dfrac{22}{33}$

7 a

	3	4	5	6
3	(3, 3)	(3, 4)	(3, 5)	(3, 6)
4	(4, 3)	(4, 4)	(4, 5)	(4, 6)
5	(5, 3)	(5, 4)	(5, 5)	(5, 6)
6	(6, 3)	(6, 4)	(6, 5)	(6, 6)

  b $\frac{4}{8}$ **c** $\frac{3}{10}$

8 0.8571

9 a 0.0625   **b** 0.399

10 0.3499

11 a

	B	B'	Total
A	0.35	0.09	0.44
A'	0.25	0.31	0.56
Total	0.6	0.4	1

  b i $\frac{7}{12}$ **ii** $\frac{9}{40}$ **iii** $\frac{31}{40}$
  iv $\frac{35}{44}$ **v** $\frac{25}{56}$

12 a

	S	S'	Total
C	30	60	90
C'	30	30	60
Total	60	90	150

A Venn diagram is also appropriate.

  b i $\frac{1}{3}$ **ii** $\frac{2}{3}$
  c $\frac{59}{149}$

13 a

	1	2	3	4
1	–	(1, 2)	(1, 3)	(1, 4)
2	(2, 1)	–	(2, 3)	(2, 4)
3	(3, 1)	(3, 2)	–	(3, 4)
4	(4, 1)	(4, 2)	(4, 3)	–

  b $\frac{1}{4}$

14 a $\frac{13}{24}$ **b** $\frac{9}{13}$

15 a $\frac{2}{5}p + \frac{1}{5}$
  b i $\frac{1-p}{2p+1}$ **ii** $\frac{7}{16}$

16 a $\frac{25}{100}$ **b** $\frac{97}{100}$ **c** $\frac{3}{100}$
  d $\frac{16}{36}$ **e** $\frac{61}{64}$

EXERCISE 2.6

1 C   **2** A

3 $P(A \cap B) = 0.15$, $P(A) = 0.25$.
$P(B \mid A) = \frac{0.15}{0.25} = 0.6 = P(B)$

4 0.94   **5** $\frac{27}{256}$

6 If A and B are independent, then A is also independent of B'. $P(A \mid B') = P(A) = 0.56$.

7 a $\frac{1}{4}$ **b** $\frac{1}{12}$ **c** $\frac{5}{6}$

8 a If $p = 0$, then $P(A' \cap B) = -\frac{1}{8} < 0$ and probability cannot be negative.
  b $p = \frac{3}{8}$

9 0.42   **10** 0.096

11 0.1678   **12** 192 : 175

EXERCISE 2.7

1 D   **2** false

3 a

	L	L'	Total
R	50	12	62
R'	40	198	238
Total	90	210	300

  b i 0.1667 **ii** 0.1333 **iii** 0.5556

4 a

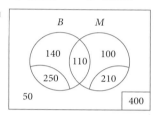

  b i 0.5238 **ii** 0.6667 **iii** 0.56

5 a i 0.3333 **ii** 0.8065
  b $P(L) = 0.3$, $P(L \mid R) = 0.8065$
  $0.8065 - 0.3 = 0.5065 > 0.05$
  $P(L) \ne P(L \mid R)$ and so there is insufficient evidence to suggest that studying literature is independent of whether a student enjoys reading books.

6 $P(M') = 0.475$, $P(M' \mid B) = 0.56$
$0.56 - 0.475 = 0085 > 0.05$
Given that $P(M') \ne P(M' \mid B)$, there is sufficient evidence to suggest that not enjoying mountainous destinations is not independent of enjoying destinations with a beach. Therefore, there is validity to the travel agent's claim.

7 a

	T	T'	Total
B	150	170	320
B'	100	80	180
Total	250	250	500

  b i 0.64 **ii** 0.6 **iii** 0.4444
  c $P(B) = 0.64$, $P(B \mid T) = 0.6$
  $0.64 - 0.6 = 0.04 < 0.05$
  $P(B) \approx P(B \mid T)$ and so it does appear that liking the bus as a mode of transport may be independent of liking the train as a mode of transport.

8

	A	A'	Total
T	0.63	0.09	0.72
T'	0.07	0.21	0.28
Total	0.7	0.3	1

$P(T) = 0.72$, $P(T \mid A) = 0.9$
$0.9 - 0.72 = 0.18 > 0.05$
Given that $P(T) \ne P(T \mid A)$, there is insufficient evidence to suggest that Marnie arriving on time to school is not independent of her setting her alarm. Therefore, there is validity to the teacher's claim.

9

	W	W'	Total
F	0.18	0.12	0.3
F'	0.07	0.63	0.7
Total	0.25	0.75	1

$P(W) = 0.25$, $P(W|F) = 0.6$

$0.6 - 0.25 = 0.35 > 0.05$

Given that $P(W) \neq P(W|F)$, there is insufficient evidence to suggest that the team winning is not independent of them scoring first. Therefore, there is validity to the manager's claim.

CUMULATIVE EXAMINATION: CALCULATOR-FREE

1 $n = 5$

2 a 10

b Prime numbers are positive and so $n(P \cap N) = 0$.

c $\{1, 9\}$

d i $\dfrac{1}{10}$ **ii** $\dfrac{1}{2}$

3 256   **4** 0.3

5 a $\dfrac{1}{15}$ **b** $\dfrac{1}{3}$

CUMULATIVE EXAMINATION: CALCULATOR-ASSUMED

1 a 45   **b** 45   **c** 45

d 50   **e** 25

2 a i 120   **ii** 85

b 1   **c** $\dfrac{1}{36}$

3 a 0.036   **b** 0.36

4 a If A is independent of B, then A is also independent of B' and so $P(A|B') = P(A) = 0.35$.

b $P(A) = 0.1077$

5

	G	G'	Total
F	156	116	272
F'	174	66	240
Total	330	182	512

$P(F) = 0.5313$, $P(F|G) = 0.4727$

$0.5313 - 0.4727 = 0.0585 > 0.05$

Given that $P(F) \neq P(F|G)$, the organisation's claim could be supported as there is sufficient evidence to suggest that exercising more than four times a week is not independent of whether someone owns a gym membership.

CHAPTER 3

EXERCISE 3.1

1 a -3 **b** 1   **c** $-\dfrac{1}{4}$

d $\dfrac{2}{11}$ **e** $-\dfrac{1}{17}$ **f** 0

2 a 5   **b** -1.5 **c** $\dfrac{5}{8}$ **d** $-\dfrac{9}{7}$

3 a 3   **b** $-\dfrac{7}{4}$ **c** $\dfrac{31}{19}$

d 1   **e** $-\dfrac{33}{8}$ **f** $-\dfrac{3}{8}$

4 a x-intercept: $\left(\dfrac{7}{2}, 0\right)$ y-intercept: $(0, -7)$

b x-intercept: $(24, 0)$ y-intercept: $(0, 8)$

c x-intercept: $(8, 0)$ y-intercept: $\left(0, -\dfrac{8}{5}\right)$

d x-intercept: $(-9, 0)$ y-intercept: $(0, 4)$

e x-intercept: $(\pi, 0)$ y-intercept: $(0, -\pi)$

f x-intercept: $\left(-\dfrac{1}{4}, 0\right)$ y-intercept: $(0, 17)$

5 a 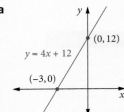   **b**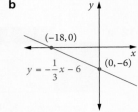

c 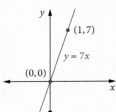   **d**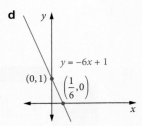

e 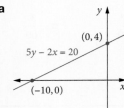   **f**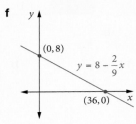

6 a 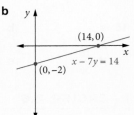   **b**

c 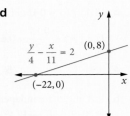   **d**

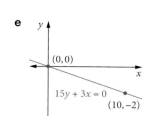
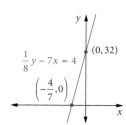
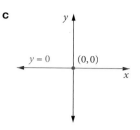
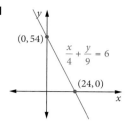

7 a 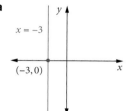   **b** 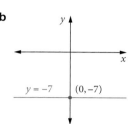   **e** 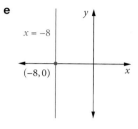   **f**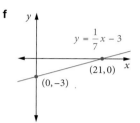

c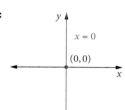

d

10 a -11 **b** $-\dfrac{29}{13}$

11 24 units^2

12 a 5   **b** $\dfrac{1}{14}$

13 a 711 km   **b** 8 hours and 20 minutes

c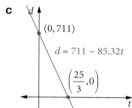

e

f

8 a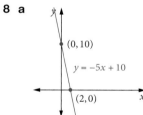

b

d 85.32 is the average speed of the car, in km/h, when driving. The value is negative because the distance remaining (d) is decreasing.

EXERCISE 3.2

1 $-\dfrac{2}{3}$

2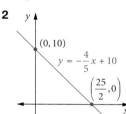

c

9 a

b

3 a $y = 4x - 3$ **b** $y = 4x - 3$

c Most find $y - y_1 = m(x - x_1)$ to be more efficient.

4 a $y = -x - 1$ **b** $y = -\dfrac{3}{4}x + \dfrac{7}{2}$ **c** $y = \dfrac{5}{2}x - \dfrac{19}{2}$

5 a $y = 2x + 1$ **b** $y = -\dfrac{1}{3}x - \dfrac{2}{9}$ **c** $y = 2x - 4$

6 a $y = -2x + 1$ **b** $y = -\dfrac{x}{3} + 2$ **c** $y = -\dfrac{x}{5} + \dfrac{33}{5}$

7 a $y = -2x + 1$ **b** $y = \dfrac{3}{2}x + 3$ **c** $y = 14$

d $y = -\dfrac{1}{4}x + 7$ **e** $y = \dfrac{2}{3}x$ **f** $x = -9$

8 a $y = \frac{x}{2} + 4$ b $y = \frac{x}{3} - 2$

c $y = -\frac{5}{3}x + \frac{19}{3}$

9 a $y = \frac{5}{3}x - \frac{17}{3}$ b $y = -\frac{x}{2} - 3$

c $y = -3x + 11$

10 a $y = -\frac{1}{3}x - 1$ b $y = 5$

c $y = -\frac{3}{4}x - 3$ d $y = \frac{3}{4}x + \frac{25}{4}$

e $x = -5$ f $y = -\frac{4}{5}x - \frac{6}{5}$

g $y = -7$ h $y = -x + 11$

i $y = -6$ j $y = \frac{1}{3}x - \frac{5}{3}$

k $y = \frac{3x}{2} - 3$ l $y = -5x - 12$

11 a L_2 and L_4 b L_3 and L_6

12 a $120   b $110   c $E = 110t + 120$

13 Proof: see worked solutions

14 $m = \frac{1}{3}$

15 Proof: see worked solutions

EXERCISE 3.3

1 A   2 $y = -\frac{1}{6}x - 1$

3 a $4x + 6$ b $-8x + 12$

c $35 - 7x$ d $x^2 - 8x$

e $9x - 106$ f $2x^2 - 6x$

4 a $x^2 + 5x + 6$ b $x^2 + x - 20$

c $x^2 - 14x + 48$ d $x^2 - 10x - 11$

e $2x^2 - 3x + 41$ f $-5x - 6$

5 a $6x^2 + 13x + 6$ b $4x^2 + 8x - 5$

c $6x^2 + 41x - 7$ d $15x^2 - 17x + 4$

e $x^2 + 7x - 72$ f $-x^2 - 2x + 49$

6 a $x^2 + 14x + 49$ b $x^2 - 16x + 64$

c $4x^2 + 12x + 9$ d $9x^2 - 30x + 25$

e $5x^2 - 6x + 2$ f $-4pq$

g $x^2 + x + \frac{1}{4}$ h $x^2 - 5x + \frac{25}{4}$

7 a $x^2 - 9$ b $x^2 - 49$

c $x^2 - 5$ d $4x^2 - 1$

e $16x^2 - 3$ f $64x^2 - \frac{1}{81}$

8 a $15x^2 - 19x + 13$ b $52x^2 - 78x - 56$

9 a $m^2 - 10mn + 25n^2$ b $6x + 3$

c $49p^2 - \frac{1}{16}$ d $4x^2 - 20x - 24$

e $6t^2 - 39t + 63$ f $-16x^2 + 48x - 36$

g $2x^2 - 3xy + 11x - 6y + 14$

h $x^2y^2 - 121v^2$

10 a expansion of $(x + 4)^2$

b $(x - 5)^2$ c $(3x + 1)^2$

d not a perfect square expansion

11 a $t = 36$

b $p = 49$

c $q = 70$ or $q = -70$

12 a Proof: see worked solutions   b $x^3 - 27$

13 a $x^2 - 5x + 7x - 35 = x^2 + 2x - 35$

b $288 \, \text{m}^2$

c $0.012 \, \text{m}^2$

d $100 = x^2 + 2x - 35$ or $100 = (x - 5)(x + 7)$

EXERCISE 3.4

1 $(0, -25)$ and $(30, 0)$

2 a $20x^2 - 4x$ b $x^2 - 16x + 64$

c $x^2 - 9$ d $4x^2 + 28x + 49$

3 a $2(x + 8)$ b $4(x - 3)$

c $9b(1 - 4b)$ d $-2(2xy - 5x - 4)$

e $12x^2(1 - 2x)$ f $7(x^2 - 2x + 8)$

g $5d(3y - 8d + 1)$ h $-6xyz(x - 5y)$

4 a $(x + 6)^2$ b $(x + 9)^2$

c $(x - 4)^2$ d $(x - 11)^2$

e $(7x - 1)^2$ f $(3x - 5)^2$

g $8(x - 3)^2$ h $2(4x - 1)^2$

5 a $(x - 5)(x + 5)$ b $(x - 7)(x + 7)$

c $(t - 6)(t + 6)$ d $(x - \sqrt{6})(x + \sqrt{6})$

e $(p - \sqrt{7})(p + \sqrt{7})$ f $(x - 7)(x + 11)$

g $(x - 7 - \sqrt{5})(x - 7 + \sqrt{5})$

h $3(2u - 3p^3)(2u + 3p^3)$

6 a $(x - 4)(x + 1)$ b $(x - 5)(x - 2)$

c $(x - 8)(x + 6)$ d $(3x + 2)(x - 5)$

e $3(x - 2)^2$ f $(4p + q)(p - 2q)$

g $4(x^2 + 1)(x - 4)$ h $a(ax - 2)(x + 3)(x - 3)$

7 a $(x + 4)(x - 4)$ b $3(x + 1)^2$

c $(x - 8)(x + 7)$ d $(x - 3)(2x + 4)$

e $(2x - 7)^2$ f $-24x(1 - 3y)$

g $(6m - 1)(6m + 1)$ h $x^2 + 9$

i $(p - 7)(p + 5)$

j $(u - 7 - \sqrt{13})(u - 7 + \sqrt{13})$

k $(5x - 2)(x + 8)$

l $2(x - 3a)(x + 3a)$

8 a $5(4x - 9)^2$

b $3(x - pq)(x + pq)$

c $-7(x - 10)(x + 6)$

9 a $(x - 2)^2$ b $-44x$

10 9

11 a Proof: see worked solutions   b $11 \, \text{mm}$

EXERCISE 3.5

1 a 10   b -6 c -11

2 $(x - 8)^2$

3 a 1, 12; 2, 6; 3, 4; −1, −12; −2, −6; −3, −4
   b 1, 20; 2, 10; 4, 5; −1, −20; −2, −10; −4, −5
   c 1, −15; 3, −5; −1, 15; −3, 5
   d 1, 24; 2, 12; 3, 8; 4, 6; −1, −24; −2, −12; −3, −8; −4, −6
   e 1, −30; 2, −15; 3, −10; 5, −6; −1, 30; −2, 15; −3, 10; −5, 6
   f 1, −18; 2, −9; 3, −6; −1, 18; −2, 9; −3, 6
   g 1, 72; 2, 36; 3, 24; 4, 18; 6, 12; 8, 9;
       −1, −72; −2, −36; −3, −24; −4, −18; −6, −12; −8, −9
   h 1, −21; 3, −7; −1, 21; −3, 7
   i 1, 16; 2, 8; 4, 4; −1, −16; −2, −8; −4, −4
   j 1, −5; −1, 5
   k 1, −28; 2, −14; 4, −7; −1, 28; −2, 14; −4, 7
   l 1, 45; 5, 9; −1, −45; −5, −9

4 a $(x+2)(x+3)$ **b** $(x+4)(x+7)$
  c $(x+2)(x+12)$ **d** $(x-4)(x-6)$
  e $(x+9)(x-1)$ **f** $(x+5)(x-8)$
  g $(x-3)(x+20)$ **h** $(x-4)(x-8)$
  i $(x+5)(x-7)$

5 a $2(x-2)(x+9)$ **b** $4(x-5)(x-12)$
  c $-5(x+15)(x-1)$ **d** $-3(x-8)(x+3)$

6 a $(2x+1)(x+1)$ **b** $(x-3)(3x-4)$
  c $(2x-1)(x+10)$ **d** $(5x-6)(x-1)$
  e $(x+2)(4x-5)$ **f** $(3x-2)(2x-5)$

7 a $x = -3, 1$ **b** $x = 1, -8$ **c** $x = 0, 6$
  d $-\frac{1}{4}, 2$ **e** $x = -\frac{2}{3}, -5$ **f** $x = \frac{4}{7}, \frac{1}{9}$

8 a $x = -1, -6$ **b** $x = -9, 7$ **c** $x = 2, 10$
  d $x = -5, 6$ **e** $x = 3, 4$ **f** $x = -1, 8$

9 a $x = -\frac{1}{3}, 4$ **b** $x = -5, -\frac{3}{2}$ **c** $x = -\frac{1}{6}, 2$

10 a $x = -25, -3$ **b** $x = -3$ **c** $x = -8, 12$
   d $x = -9, 9$ **e** $x = 3, 15$ **f** $x = 0, 2$
   g $x = -7, 6$ **h** $x = -5, \frac{1}{4}$ **i** $x = -12, 3$
   j $x = -\sqrt{5}, \sqrt{5}$ **k** $x = -11, 3$ **l** $x = -\frac{1}{2}, \frac{3}{2}$

11 a Proof: see worked solutions
   b $x = 6$, length = 12 m, width = 3 m

12 a $x + 2$ **b** $\frac{12}{x}$
   c $\frac{12}{x+2}$ **d** mango $2, melon $4

13 a 2 m
   b 1.2 seconds, 2 seconds
   c 3.2 seconds

14 a 210 dots
   b i $n(n+1) = 72$
     ii $n^2 + n - 72 = 0$
     iii the 8th triangle

EXERCISE 3.6

1 a $(x+3)^2$ **b** $(x-9)^2$ **c** $4(x-5)^2$

2 a $x = -2, 8$ **b** $x = -\frac{3}{2}, 9$ **c** $x = -\frac{3}{2}, 0$

3 a $(x+3)^2 + 8$ **b** $(x+4)^2 - 2$
  c $(x-1)^2 - \frac{7}{2}$ **d** $\left(x+\frac{1}{2}\right)^2 - \frac{5}{4}$
  e $\left(x-\frac{5}{2}\right)^2 + \frac{3}{4}$ **f** $\left(x-\frac{3}{2}\right)^2 - \frac{53}{4}$

4 a $2(x-5)^2 + 4$ **b** $3(x-1)^2 - 10$
  c $-2(x+9)^2 - 2$ **d** $-4(x-4)^2 - 6$
  e $2\left(x-\frac{7}{2}\right)^2 - \frac{5}{2}$ **f** $-5\left(x+\frac{3}{2}\right)^2 - \frac{11}{4}$

5 a $x = 6, 18$
  b $x = -6, -4$
  c no real solutions
  d $x = -\frac{29}{2}, \frac{7}{2}$
  e $x = 4 - \sqrt{7}, 4 + \sqrt{7}$
  f $x = -\frac{5}{2} + 2\sqrt{6}, -\frac{5}{2} - 2\sqrt{6}$

6 a $x = -3 - \sqrt{19}, -3 + \sqrt{19}$
  b no real solutions
  c $x = -\frac{7}{2} - \frac{\sqrt{61}}{2}, -\frac{7}{2} + \frac{\sqrt{61}}{2}$
  d $x = -1, 11$
  e no real solutions
  f $x = -\frac{5}{2} - \frac{3\sqrt{5}}{2}, -\frac{5}{2} + \frac{3\sqrt{5}}{2}$

7 a $\left(x+\frac{b}{2}\right)^2 - \left(\frac{b^2}{4} - c\right) = 0$ **b** $b^2 - 4c < 0$

8 a $-2\left(x - \frac{5}{2}\right)^2 + \frac{15}{2}$ **b** $x = \frac{5}{2} - \frac{\sqrt{15}}{2}, \frac{5}{2} + \frac{\sqrt{15}}{2}$

EXERCISE 3.7

1 a $(x-5)^2 - 22$ **b** $x = 5 - \sqrt{22}, 5 + \sqrt{22}$

2 −9

3 a i $a = 1, b = 4, c = -11$
     ii $x = -2 - \sqrt{15}, -2 + \sqrt{15}$
  b i $a = 4, b = -5, c = -3$
     ii $x = \frac{5 - \sqrt{73}}{8}, \frac{5 + \sqrt{73}}{8}$
  c i $a = -3, b = 6, c = 2$
     ii $x = 1 - \frac{\sqrt{15}}{3}, 1 + \frac{\sqrt{15}}{3}$
  d i $a = -5, b = 10, c = -3$
     ii $x = 1 - \frac{\sqrt{10}}{5}, 1 + \frac{\sqrt{10}}{5}$
  e i $a = -1, b = -1, c = 11$
     ii $x = \frac{-1 - 3\sqrt{5}}{2}, \frac{-1 + 3\sqrt{5}}{2}$

f i $a = 2, b = 5, c = -3$
   ii $x = -3, \dfrac{1}{2}$

g i $a = -4, b = -6, c = 1$
   ii $x = \dfrac{-3 - \sqrt{13}}{4}, \dfrac{-3 + \sqrt{13}}{4}$

h i $a = 2, b = 7, c = -4$
   ii $x = -4, \dfrac{1}{2}$

i i $a = 11, b = 9, c = -1$
   ii $x = \dfrac{-9 - 5\sqrt{5}}{22}, \dfrac{-9 + 5\sqrt{5}}{22}$

4 a $x = -5 - 3\sqrt{2}, -5 + 3\sqrt{2}$

  b $x = -\dfrac{1}{6}, 2$

  c $x = \dfrac{9 - \sqrt{129}}{6}, \dfrac{9 + \sqrt{129}}{6}$

5 a i 65   **ii** two solutions
  b i 25   **ii** two solutions
  c i -119 **ii** no solutions
  d i 0   **ii** one solution
  e i -47 **ii** no solutions
  f i -80 **ii** no solutions
  g i 0   **ii** one solution
  h i 104   **ii** two solutions
  i i -104 **ii** no solutions

6 a $k = \dfrac{49}{4}$ **b** $p < \dfrac{9}{4}$ **c** $m < -\dfrac{1}{42}$

7 a $q = -2\sqrt{30}, 2\sqrt{30}$
  b $r < -\dfrac{\sqrt{5}}{2} \cup r > \dfrac{\sqrt{5}}{2}$
  c $-20 - 4\sqrt{30} < d < -20 + 4\sqrt{30}$

8 a $x = \dfrac{3 \pm \sqrt{9 + 4t}}{2}$
  b $x = \dfrac{3 - \sqrt{13}}{2}, \dfrac{3 + \sqrt{13}}{2}$
  c $t = 4$

9 a $p^2 - 12p + 36$
  b Proof: see worked solutions

EXERCISE 3.8

1 $x = \dfrac{-7 - \sqrt{93}}{2}, \dfrac{-7 + \sqrt{93}}{2}$

2 a -24
  b There are no solutions because $\Delta < 0$.

3 a i $(7, 5)$ **ii** minimum **iii** $x = 7$
  b i $(-7, 32)$ **ii** minimum **iii** $x = -7$
  c i $(11, -12)$ **ii** minimum **iii** $x = 11$
  d i $(-5, -8)$ **ii** maximum **iii** $x = -5$
  e i $(3, 4)$ **ii** minimum **iii** $x = 3$
  f i $(-50, -26)$ **ii** maximum **iii** $x = -50$

4 a $x = -13, 1$
  b $x = 2, 8$
  c $x = 11 - \dfrac{2\sqrt{3}}{3}, 11 + \dfrac{2\sqrt{3}}{3}$
  d no x-intercepts
  e $x = -3, 7$
  f $x = 10 - \dfrac{3\sqrt{5}}{5}, 10 + \dfrac{3\sqrt{5}}{5}$

5 a i $(-3, -25)$, minimum
   ii -16
   iii $x = -8, 2$
   iv
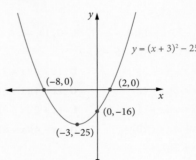

  b i $(-1, -32)$, minimum
   ii -30
   iii $x = -5, 3$
   iv
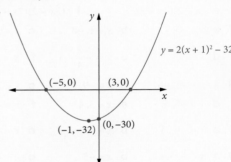

  c i $(6, 100)$, maximum
   ii 64
   iii $x = -4, 16$
   iv
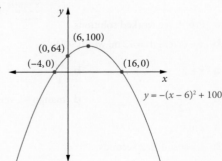

d **i** $(-1, 17)$, maximum
  ii 13
  iii $x = -1 - \frac{\sqrt{17}}{2}, -1 + \frac{\sqrt{17}}{2}$
  iv
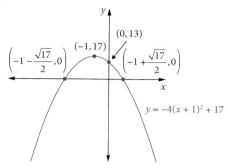

e **i** $(5, -8)$, maximum
  ii -58
  iii no x-intercepts
  iv
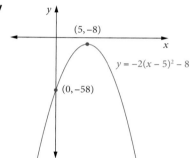

f **i** $(1, -6)$, minimum
  ii -2
  iii $x = 1 - \frac{\sqrt{6}}{2}, 1 + \frac{\sqrt{6}}{2}$
  iv
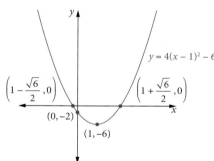

g **i** $(5, 45)$, minimum
  ii 170
  iii no x-intercepts
  iv
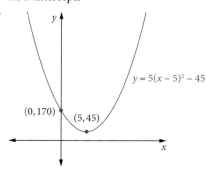

h **i** $(4, -72)$, minimum
  ii -40
  iii $x = -2, 10$
  iv
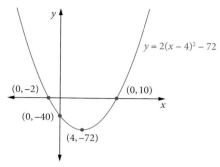

i **i** $(-3, 7)$, maximum
  ii -47
  iii $x = -3 - \frac{\sqrt{42}}{6}, -3 + \frac{\sqrt{42}}{6}$
  iv
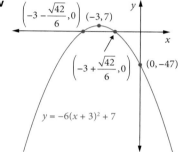

6 a **i** $y = (x + 1)^2 - 4$
  ii
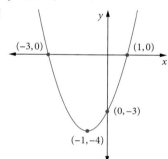

  b **i** $y = -(x + 5)^2 + 9$
  ii
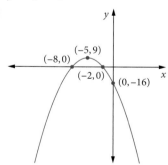

c **i** $y = 3(x+3)^2 - 432$
   ii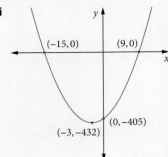

d **i** $y = (x-5)^2 + 49$
   ii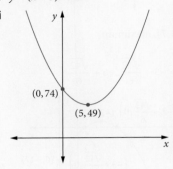

e **i** $y = -(x-3)^2 + 10$
   ii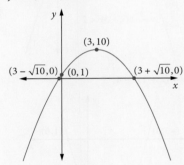

f **i** $y = 2(x-4)^2 - 12$
   ii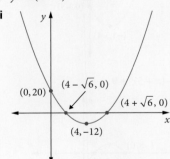

7 a $h = 12 - b$

b $A = \dfrac{1}{2}b(12-b)$

c $A = -\dfrac{1}{2}(b-6)^2 + 18$

d Maximum area is $18\,\text{cm}^2$ when the base is 6 cm.

8 a $x = h \pm \sqrt{-\dfrac{k}{a}}$

b $\dfrac{k}{a} > 0$

c $k = 0$

9 a $y = -(x-13)^2 + 144$

b 24 metres

c 144 metres

d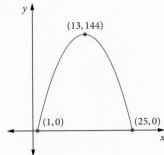

EXERCISE 3.9

1 $(-1, 14)$, maximum

2 $x = -3 - 2\sqrt{5}, -3 + 2\sqrt{5}$

3 a **i** -75 **ii** $x = -5, 15$
   iii 5 **iv** -100
   v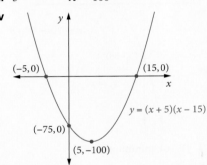

b **i** 32 **ii** $x = -4, 8$
   iii 2 **iv** 36
   v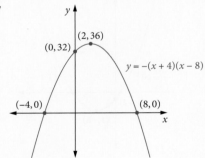

c **i** -12 **ii** $x = -3, 4$
   iii $\dfrac{1}{2}$ **iv** $-\dfrac{49}{4}$
   v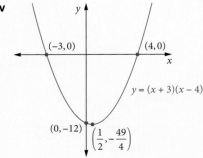

d **i** −15 **ii** $x = -3, \dfrac{5}{2}$
   iii $-\dfrac{1}{4}$ **iv** $-\dfrac{121}{8}$
   v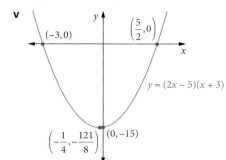

e **i** 72 **ii** $x = -3, 12$
   iii $\dfrac{9}{2}$ **iv** $\dfrac{225}{2}$
   v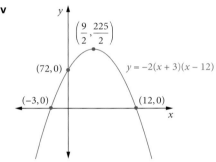

f **i** 84 **ii** $x = 14, 2$
   iii 8 **iv** −108
   v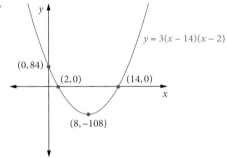

4 a **i** −8 **ii** $x = -4, 2$
   iii −1 **iv** −9
   v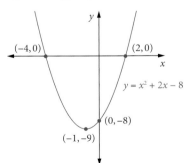

b **i** 36 **ii** $x = -6, 6$
   iii 0 **iv** 36
   v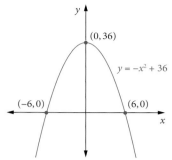

c **i** −11 **ii** no x-intercepts
   iii −1 **iv** −7
   v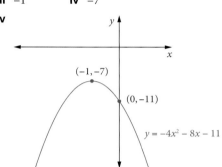

d **i** −32 **ii** $x = -8, -4$
   iii −6 **iv** 4
   v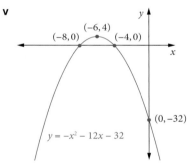

e **i** 0 **ii** $x = -3, 0$
   iii $-\dfrac{3}{2}$ **iv** $-\dfrac{9}{2}$
   v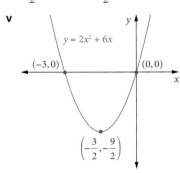

f i −63   **ii** $x = -3, 7$
   iii 2   **iv** −75
   v

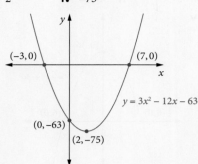

5 a i −2   **ii** $x = 3 - \sqrt{11}, 3 + \sqrt{11}$
   iii 3   **iv** −11
   v

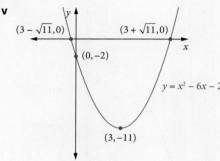

b i 8   **ii** $x = -2 + 2\sqrt{3}, -2 - 2\sqrt{3}$
   iii −2   **iv** 12
   v

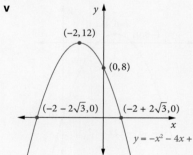

c i −10   **ii** $x = -3 + \sqrt{19}, -3 - \sqrt{19}$
   iii −3   **iv** −19
   v

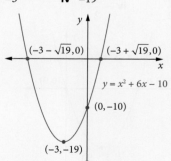

d i 1   **ii** no x-intercepts
   iii $-\dfrac{3}{8}$ **iv** $\dfrac{7}{16}$
   v

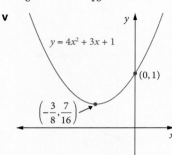

e i 4   **ii** $x = 9 - \sqrt{85}, 9 + \sqrt{85}$
   iii 9   **iv** 85
   v

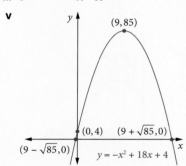

f i 1   **ii** $x = \dfrac{1}{6} - \dfrac{\sqrt{13}}{6}, \dfrac{1}{6} + \dfrac{\sqrt{13}}{6}$
   iii $\dfrac{11}{6}$ **iv** $\dfrac{13}{12}$
   v

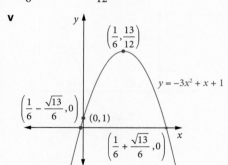

6 a

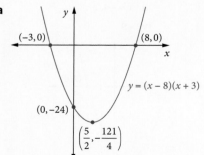

b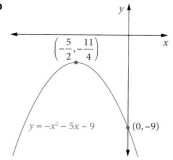

c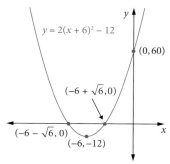

d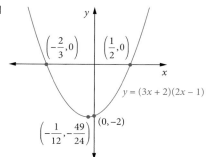

e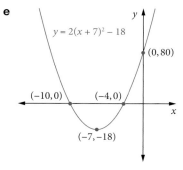

f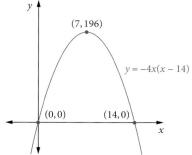

g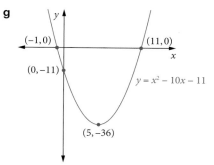

h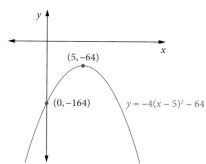

i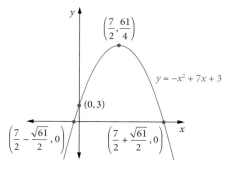

j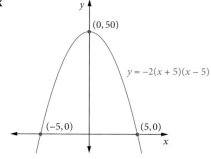

k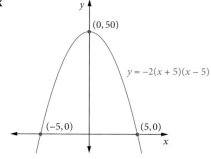

l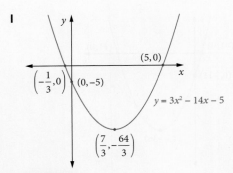

7 a ClassPad

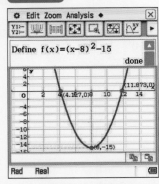

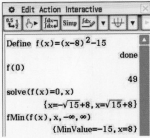

TI-Nspire

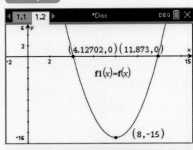

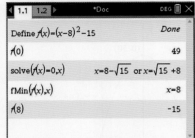

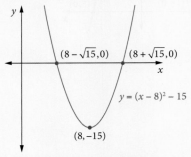

b ClassPad

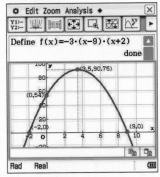

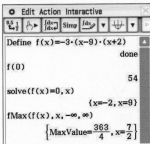

TI-Nspire

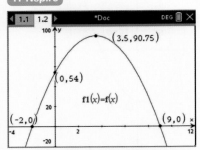

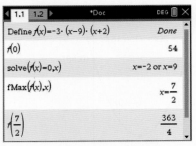

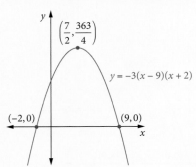

c ClassPad

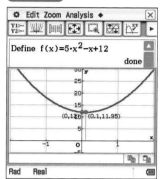

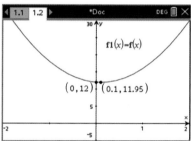

TI-Nspire

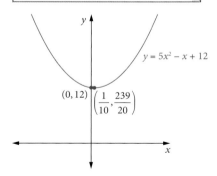

8 a $y = 2x^2 - 4x - 70$
  b $2(x-1)^2 - 72$
  c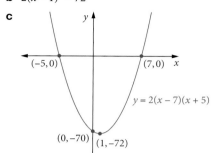

9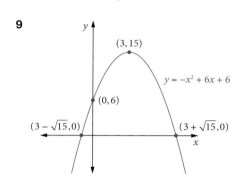

10 a 23 m   b 3.07 m
   c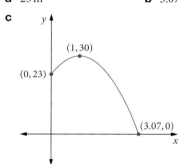

11 a $(2250, 3200)$ b 8516 m
   c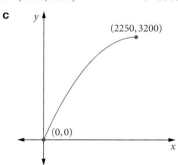

EXERCISE 3.10

1 $(5, 17)$

2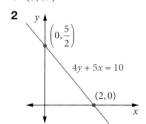

3 a $(-1,-1)$

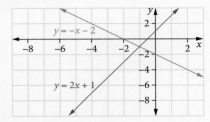

b $(-5,-1)$

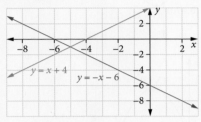

c $(4,2)$

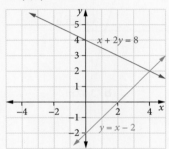

4 a $x=2, y=3$ **b** $x=4, y=1$
c $x=5, y=0$ **d** $x=4, y=-3$
e $x=\frac{25}{4}, y=\frac{15}{4}$ **f** $x=-9, y=-2$

5 a $x=-8, y=4$ **b** $x=-6, y=-2$
c $x=-10, y=4$ **d** $x=10, y=-3$
e $x=-3, y=-7$ **f** $x=-6, y=1$

6 a one point of intersection
b no points of intersection
c infinite points of intersection
d infinite points of intersection
e one point of intersection
f one point of intersection

7 a $(-2,3), (6,27)$ **b** $(-2,-10), (2,-14)$
c $(1,-7)$ **d** $(2,-9)$
e $(-2,4), (0,2)$ **f** $(-6,38), (2,-2)$

8 a $x=\frac{20}{63}, y=\frac{19}{21}$

b $a=-1, b=0, c=-\frac{1}{3}, d=-\frac{1}{3}$

or $a=\frac{3}{2}, b=\frac{5}{4}, c=\frac{1}{2}, d=\frac{1}{2}$

9 $(-3.65, 16.94), (1.65, 1.06)$
10 any value of m except -7 or 7
11 $m=-2$ and $k=-2$
12 a $-2 < k < 6$ **b** $k < -2 \cup k > 6$ **c** $k=-2, 6$

EXERCISE 3.11

1 $(-2,-49)$ **2** $y = \left(x - \frac{3}{2}\right)^2 - \frac{1}{4}$

3 a i 1
   ii k: the y value of the turning point or a point on the graph
 b i 1
   ii a: a point on the graph
 c i 2
   ii a, b: two points on the graph
 d i 3
   ii a: a point on the graph
      u, v: two x-intercepts
 e i 3
   ii a, b, c: three points on the graph
 f i 3 (turning point is considered two pieces of information)
   ii a: a point on the graph
      h, k: the turning point

4 a $y = (x-2)(x-5)$
 b $y = (x+7)(x-9)$
 c $y = -2x(x-10)$
 d $y = -\frac{1}{5}(2x+11)(x-4)$

5 a $y = (x-6)^2 - 7$ **b** $y = -(x-2)^2 + 22$
 c $y = 3(x+4)^2 - 1$ **d** $y = -\frac{4}{5}(x+10)^2 + 20$

6 a $y = x^2 + 5x + 6$ **b** $y = -x^2 - 2x - 9$
 c $y = 2x^2 + 5x + 5$ **d** $y = -3x^2 - x - 10$

7 a i $y = a(x-h)^2 + k$
   ii $y = -(x-5)^2 + 11$
 b i $y = a(x-u)(x-v)$
   ii $y = -\frac{1}{2}(x+4)(x+1)$
 c i $y = ax^2 + bx + c$ **ii** $y = 4x^2 + 5$
 d i $y = a(x-h)^2 + k$ **ii** $y = 2(x+3)^2 + 6$
 e i $y = ax^2 + bx + c$ **ii** $y = -x^2 - 4x + 4$
 f i $y = a(x-u)(x-v)$ **ii** $y = 3(x+2)(x-6)$

8 a $a=1, s=1, u=-2, t=3, v=5$
 b $y = 3x^2 + x - 10$

9 a $y = \frac{1}{(4-m)^2}(x-m)^2$ **b** one

10 a $y = 2x^2 - 3x + 1$
  b $y = 7x^2 - 13x + 56$

11 a $32 = a - b + c$
    $98 = 4a + 2b + c$
    $b^2 - 4ac = 0$
  b i 0
    ii $32 = a(1+h)^2$
       $98 = a(2-h)^2$
  c $y = 2(x+5)^2$ or $y = 2x^2 + 20x + 50$

EXERCISE 3.12

1 $y = -6(x+5)(3x+4)$ **2** $(6, -3)$

3 a $A = x(x+28)$
  b width: 12 m, length: 40 m

4 $\dfrac{121}{10}$ m³

5 a $A = xy, P = x + 2y$ **b** $A = \dfrac{x(28-x)}{2}$
  c 98 m², when $x = 14$

6 a $\dfrac{9}{4}$ m below ground at $\dfrac{9}{2}$ s

  b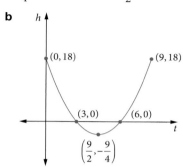

  c 3 seconds

7 a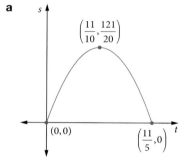

  b 6.05 m **c** 1.1 s **d** 1.6 s

8 a $r^2(4 - \pi) = 5$ **b** 2.41 cm

9 a $a = \dfrac{3}{10}, h = -100, k = 0$ **b** 1%

10 384 000 km

11 a $200 - 4x^2$
  b The corners won't be cut if $x = 0$. The corners will be too big if $x \geq 5$.
  c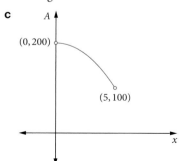

  d 3 cm

12 a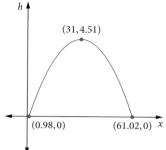

  b 0.98 m **c** 4.51 m **d** 2.51 m; yes

CUMULATIVE EXAMINATION: CALCULATOR-FREE

1 a i 0.2 **ii** 0.28
  b If complementary, $P(A) + P(B) = 1$
    $x + 0.1 + 0.6 = 1$
    $x = 0.3$
  c Proof: see worked solutions

2 a $x = \dfrac{3}{4}, y = \dfrac{5}{4}$
  b perpendicular; $m_1 = 3, m_2 = -\dfrac{1}{3}; m_1 m_2 = -1$.
  c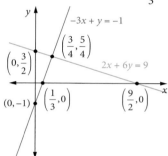

3 a $(x-16)(x+2)$ **b** $x = -2, 16$

4 $x = \dfrac{11 - 9\sqrt{2}}{2}, \dfrac{11 + 9\sqrt{2}}{2}$

5 $k = -\dfrac{25}{2}$

6 a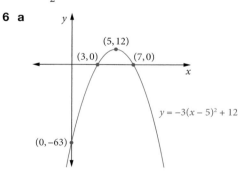

  b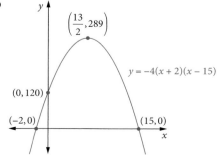

7 a $y = -\dfrac{7}{32}(x+4)(x-8)$

b $3x^2 - x - 14$

8 $(5, 3)$ and $\left(-\dfrac{1}{5}, \dfrac{101}{25}\right)$

CUMULATIVE EXAMINATION: CALCULATOR-ASSUMED

1 a

	Apple	Other	Total
≤ 25 years old	250	120	370
> 25 years old	180	130	310
Total	430	250	680

b i 0.6324 **ii** 0.6757 **iii** 0.4194

c P(Apple) = 0.6324, P(Apple | ≤ 25) = 0.6757
0.6757 − 0.6324 = 0.0433 < 0.05

Given that P(Apple) ≈ P(Apple | ≤ 25), there is sufficient evidence to suggest a customer buying an Apple phone is independent of their age group and so the analyst's claim appears to be valid.

2 a $p(1-p) + (1-p)(1.2-p)$ when $p = 0.9$
$= 0.9(0.1) + 0.1(0.3)$
$= 0.12$

b $0.9^3 = 0.729$

c Proof: see worked solutions

3 $f = \dfrac{9}{5}c + 32$

4

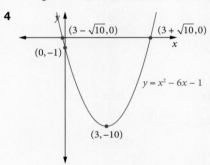

5 $-6 - 2\sqrt{6} < b < -6 + 2\sqrt{6}$

6 a $P = -2(x^2 - 2x - 25)$

b 3 units

c $P = 52$ units, when $x = 1$

7 a $-32 = 4a - 2b + c$
$-5 = a + b + c$
$-32 = 16a + 4b + c$

b $y = -3x^2 + 6x - 8$

8 a 32 cm **b** 12.8 m **c** 2.3 m **d** 10 cm

9 a $y = \left(x + \dfrac{p}{2}\right)^2 - \left(\dfrac{p^2}{4} + 7\right)$

b $x = \dfrac{-p \pm \sqrt{p^2 + 28}}{2}$

c $y = \left(x + \dfrac{p}{2} - \dfrac{\sqrt{p^2 + 28}}{2}\right)\left(x + \dfrac{p}{2} + \dfrac{\sqrt{p^2 + 28}}{2}\right)$

CHAPTER 4

EXERCISE 4.1

1 a not polynomial **b** not polynomial
c polynomial **d** polynomial
e not polynomial **f** polynomial

2 a 5 **b** 4 **c** −3 **d** 0

3 $a = 46, b = -28$

4 a i 4 **ii** 0
b i −2 **ii** 2 **iii** 14
c 9 **d** $x = \pm 5$
e $2x + 2h - 9$ **f** $x^2 + 2$

5 a i 2 **ii** 8
b i 34 **ii** −8

6

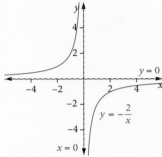

7

(graph: $x = -2$, $y = 0$, $y = \dfrac{1}{x+2}$)

8 They have the same general shape, but $y = x^5$ will be steeper.

9 a $a = 0$ **b** $b = 10$ **c** $c = -6$
d $a = -1, b \neq 7$ **e** $a = 4$

10 $A = 6, B = -2, C = -10$

11 $b = 29$

12 $m = 3, n = -2, p = 13$

13 When $x \to 0$, $y \to \pm\infty$ and when $x \to \pm\infty$, $y \to 0$.

EXERCISE 4.2

1 −36

2 a polynomial **b** not polynomial
c not polynomial **d** polynomial

3 a i Proof: see worked solutions **ii** $x = -2, -4, 3$
b i Proof: see worked solutions **ii** $x = -1, 3, -3$

4 $(x-3)(x+2)(x-10)$

5 a $x = 6$ **b** $x = -8$
c $x = 1$ **d** $x = \dfrac{4}{3}$

6 a $x = -4, 1, 3$ **b** $x = -3, 2, 3$
c $x = -2, 2$ **d** $x = 1, 2$
7 a $(x - 2)$ **b** $(x + 1)$
c $(x - 2)$ **d** $(x - 2)$
8 -3 and 0
9 a $(x + 2)(x^2 - 4x + 3)$ **b** $-2, 1, 3$
10 a $(x + 1)(x - 2)(x - 6)$ **b** $(x + 2)(x - 3)(x + 3)$
c $(x + 2)(2x + 1)(x - 5)$ **d** $(x - 2)(2x + 1)(x + 3)$
11 $-2, 0$
12 a $f(0) = p(0)^3 + q(0)^2 + r(0) + s = 7$, so $s = 7$.
b $D(4, 0), F(7, 0)$

EXERCISE 4.3

1 $P(x) = (x - 3)(x - 2)(x + 2)$

2 $x = 5$

3 a 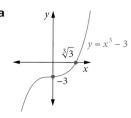 **b**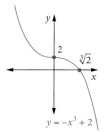

c 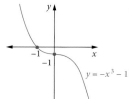 **d**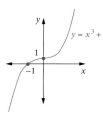

e 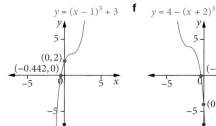 **f**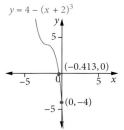

g 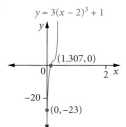 **h**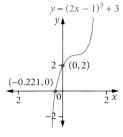

i **j**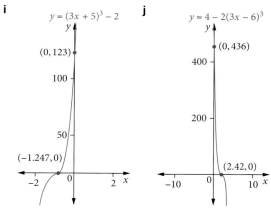

4 a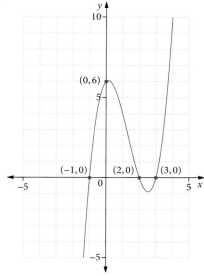

b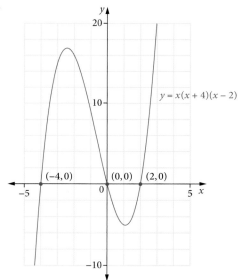

c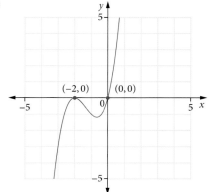

d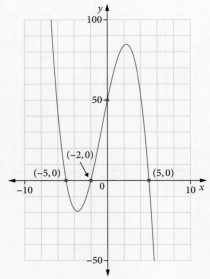

e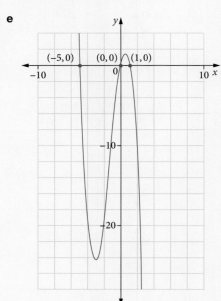

f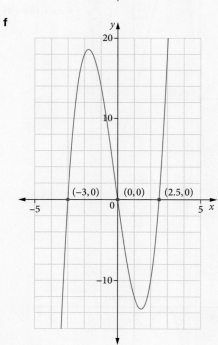

5 $y = (x + 2)(x - 1)(x - 4)$

6 a $y = 2(x + 2)(x - 1)(x - 3)$
  b $y = -1(x + 1)(x - 1)(x - 2)$
  c $y = 1(x + 2)(x - 3)^2$

7 $a = 3, b = 3$

8 a no **b** yes **c** no **d** no **e** yes

9 a no **b** no **c** yes **d** yes **e** no

10 $x + 8$ and $x + 3$

11 a $0.5 \, \text{cm}^2$

  b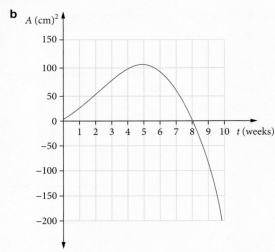

  c after $t = 8.006$ **d** 85.32% **e** 93.31%

12 $y = 2x^3 - 4x^2 - 22x + 24$

EXERCISE 4.4

1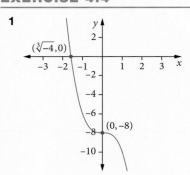

2 $y = -2(x + 5)(x + 1)^2$

3 a $\{(-1, 5), (4, 3), (6, 3), (7, 5)\}$, domain: $\{-1, 4, 6, 7\}$, range: $\{3, 5\}$, function, many-to-one

  b $\{(2, 7), (2, 3), (5, 3), (5, -2), (8, -2), (8, 4)\}$, domain: $\{2, 5, 8\}$, range: $\{-2, 3, 4, 7\}$, not a function

  c $\{(4, 1), (-6, -4), (6, 2), (3, -1)\}$, domain: $\{-6, 3, 4, 6\}$, range: $\{-4, -1, 1, 2\}$, function, one-to-one

  d $\{(7, 5), (-4, 3), (3, 3), (3, 6)\}$, domain: $\{-4, 3, 7\}$, range: $\{3, 5, 6\}$, not a function

4 a not a function   **b** not a function
  c function   **d** function

5 a not a function   **b** function
  c function   **d** not a function
  e not a function   **f** function
  g function   **h** function

6 a domain: all real values, $x \neq -5$; range: all real values, $y \neq 0$
  b domain: all real values; range: all real values
  c domain: all real values, $x \neq 2, -2$; range: $y \leq -\frac{1}{4}$ or $y > 0$
  d domain: all real values; range: $0 < y \leq 1$
  e domain: $x \leq -4$ or $x \geq 4$; range: $y \geq 0$
  f domain: $x \geq 0$; range: $y \geq 0$
  g domain: $x \leq -2$ or $x \geq 3$; range: $y \geq 0$
  h domain: $-3 \leq x \leq 5$; range: $0 \leq y \leq 4$
  i domain: $x < 7$; range: $y > 0$

7 $f(x) \geq -16$

8 a $-14 \leq y \leq 6.25$ **b** $-20.25 \leq y < 22$
  c $5.25 \leq y \leq 21$ **d** $-\frac{4}{11} \leq y \leq -\frac{1}{9}$

9 a $-2 < x \leq 2$ **b** $1 < x \leq 6$
  c $-\frac{4}{3} \leq x < \frac{1}{3}$ **d** $0.5 \leq x < 2.75$

10 a function, many-to-one **b** not a function
   c not a function **d** function, one-to-one

11 a function **b** not a function
   c not a function **d** function

12 $x \geq -3$

13 $x < 3$

14 $0 < x \leq 9$

15 a function **b** not a function
   c not a function **d** not a function
   e not a function **f** function
   g function **h** function
   i not a function **j** function

16 a $y \leq 25$ **b** $y \geq -1$
   c $y \leq 9$ **d** all real values except 0
   e all positive real values **f** $-4 \leq y \leq 4$
   g $-17 < y < 23$ **h** $-21 < y \leq 18$
   i $y \geq 4$ **j** $-16 \leq y \leq 20$

17 domain: $x \leq -2$ or $x \geq 2$, range: $y \geq 0$

18 $-2 < x \leq 6$

EXERCISE 4.5

1 D **2** B

3 a
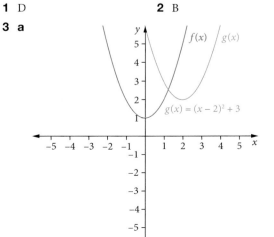

b
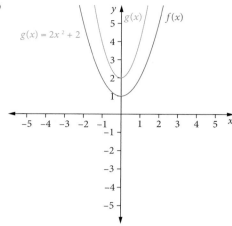

c
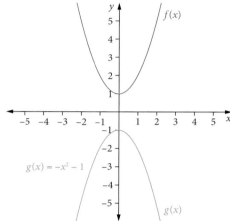

4 a $A(-8, -2), B(6, 22), C(-6, 19)$
  b $A(-10, 2), B(18, -2), C(-6, -1.5)$
  c $A(21, 2), B(0, -6), C(18, -5)$
  d $A(1, 12), B(8, -12), C(2, -9)$
  e $A\left(1, -\frac{11}{3}\right), B(8, -1), C\left(2, -\frac{4}{3}\right)$

5 a
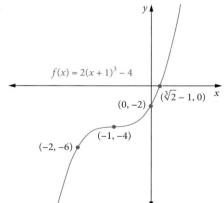

b
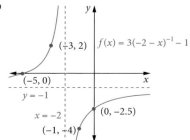

c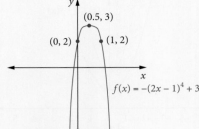

d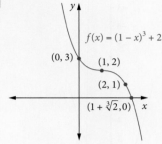

e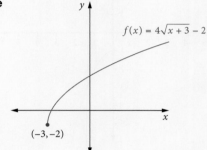

f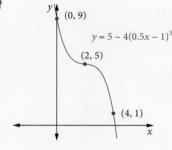

g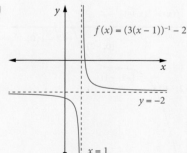

h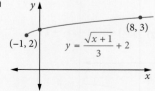

i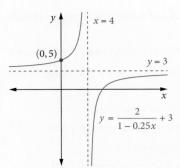

j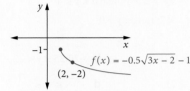

k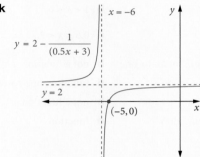

l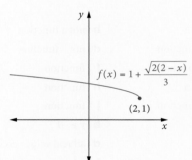

6 a $g(x) = -3(x-4)^2 - 1$ **b** $g(x) = 1 - 4(x+3)^4$

   c $g(x) = 3 - \dfrac{1}{3(x-2)^2}$ **d** $g(x) = -2(4x+1)^2 - 1$

   e $g(x) = 3\sqrt[3]{4-x} - 9$ **f** $g(x) = 3(2x+1)^{-1} - 6$

   g $g(x) = 3(4-x)^3 + 2.5$ **h** $g(x) = \sqrt{x} + 9$

7 a Reflection in the *y*-axis, dilation by a factor of 2 from the *x*-axis, translation 3 up and 2 to the right; $g(x) = 2\sqrt{2-x} + 3$.

   b Reflection in an axis, dilation by a factor of 2 from the *x*-axis, translation 1 to the left and 1 down; $g(x) = -2(x+1)^3 - 1$.

   c Reflection in an axis, translation 1 to the left and 1 down; $g(x) = -(x+1)^3 - 1$.

   d Reflection in an axis, dilation by a factor of 2 from the *x*-axis, translation 2 to the right and 4 up; $g(x) = 4 - 2\sqrt{(x-2)}$.

   e Reflection in the *x*-axis, dilation by a factor of 0.5 from the *x*-axis, translation 3 to the left and 2 down; $g(x) = 2(x+3)^2$.

f Dilation by a factor of 0.25 from the x-axis, dilation of a factor of 0.5 from the y-axis; $g(x) = 0.5(2x + 3)^4$.

g Reflection in the x-axis, translation 2 to the right and 3 down; $g(x) = \dfrac{2}{x-3} - 6$.

h Reflection in both axes, dilation by a factor of 0.25 from the x-axis, a translation of 4 to the left and a translation 10 up; $g(x) = -0.25\sqrt{x+6} + 9$.

i Reflection in the y-axis, dilation from the y-axis by a factor of 2 and translation of a factor of 1 to the right, dilation from the x-axis by a factor of 0.125 and translation of 2.125 up.

j Reflection in the x-axis and dilation from the x-axis by a factor of 0.5, translation of 0.5 down, dilation by a factor of 0.25 from the y-axis and translation of 2 to the left; $g(x) = \dfrac{1}{x+1} - 2$.

8 a
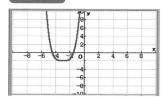

b
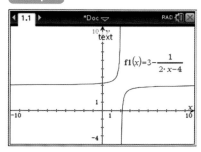

c
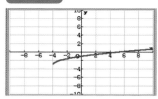

d
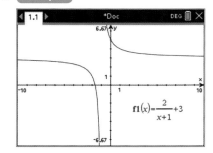

e
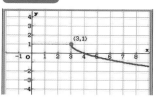

f
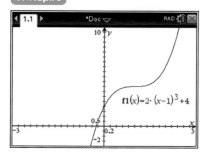

9 a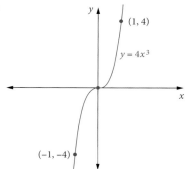

b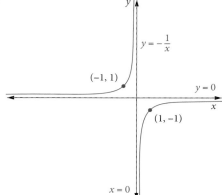

c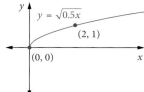

d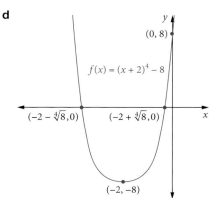

e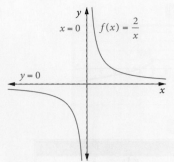

f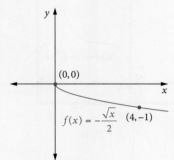

g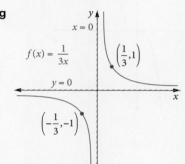

h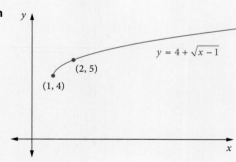

i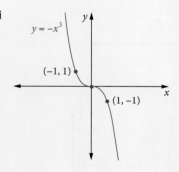

j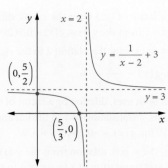

k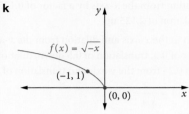

l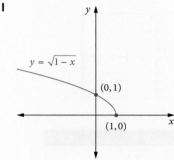

10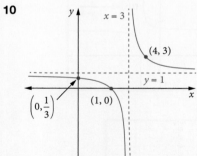

11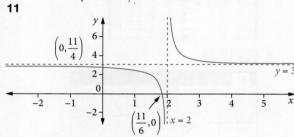

12 $(4, 7)$

13 $f(x) = -\sqrt{x - 5}$

14 $-2, -3, 4$

15 Dilation by a factor of 4 from the y-axis and a translation 5 to the left.

16 $f(x) = x^2 + 2$

EXERCISE 4.6

1 Translation of 1 unit to the left followed by a reflection in the *x*-axis.

2 $y = -2(x + 3)^2 - 1$

3
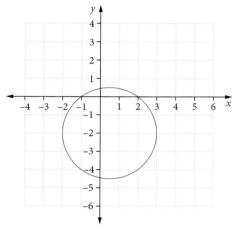

4 centre: $(1, -2)$ and radius $\sqrt{13}$

5
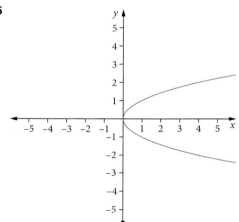

6 a $(3, -4), 9$

b $\left(-\dfrac{7}{2}, 5\right), 4\sqrt{3}$

c $(x - 5)^2 + (y + 3)^2 = 49$

d $(x + 2)^2 + (y - 4)^2 = 16$

e $\left(x - 7\dfrac{1}{2}\right)^2 + \left(y - 4\dfrac{1}{2}\right)^2 = 41$

7 a An example is $(4, -2)$ and $(4, 2)$.

b An example is $(0, 6)$ and $(0, -6)$.

8 $a = 3, b = 12$

9 $(x + 4)^2 + (y - 3)^2 = 49$

EXERCISE 4.7

1 $(x + 4)^2 + (y - 2)^2 = 100$

2
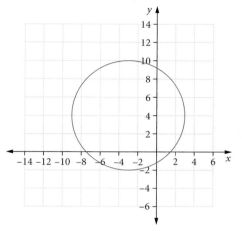

3 a $xy = 30$ **b** 3 diaries   **c** $7.50

4 a The faster you are travelling, the less times it will take to reach the destination.

b 180 km   **c** 1 hour and 48 minutes

d 30 km/h. There may have been roadworks or a traffic jam.

5 a $xy = 24, a = 2, b = 3$ **b** $xy = -18, a = 2, b = -36$

6 10 people

7 a approx. 4 hours and 22 minutes

b The assumption is that all the hoses are pumping water into the pool at the same rate.

CUMULATIVE EXAMINATION: CALCULATOR-FREE

1 a 32   **b** 7

2 a i p^3 **ii** $3p^2(1 - p)$

b $p = \dfrac{3}{4}$

3 a Proof: see worked solutions

b
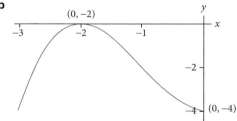

4 $x < -4$ or $x > 4$

5
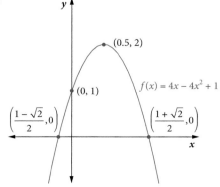

6 $(-4, 0)$ and $(0, 4)$

CUMULATIVE EXAMINATION: CALCULATOR-ASSUMED

1 a $y = -2x + 10$ **b** $y = -\frac{1}{3}x + \frac{1}{3}$

2 $y = (x-2)^2 - 3$ **3** $f(x) = x^2 - 4x + 6$

4 $\frac{1}{2}$ **5** $x \le -1$ or $x \ge 6$

6 $(-4, 12)$ **7** $(x-2)^2 + (y-3)^2 = 16$

CHAPTER 5

EXERCISE 5.1

1 a $x = 6.359$ **b** $c = 10.723$ **c** $x = 13.347$
2 a $40°$ **b** $35°$ **c** $38°$
3 $y = -x - 1$
4 a F **b** T **c** F
  d F **e** T
5 a $\frac{5}{3}$ **b** $\frac{3}{\sqrt{34}}$ **c** $\frac{5}{\sqrt{34}}$
6 $20.98\,m$ **7** $23.58°$
8 $120°$ **9** $11\,m$
10 AYZ $51.34°$, AZW $59.04°$
11 Proof: see worked solutions
12 $1.74\,km$

EXERCISE 5.2

1 D **2** $y = 1.73x + 3.20$
3 a $a = 13.7, b = 13.4$ **b** $p = 13.4, q = 10.9$
  c $h = 17.2, k = 13.9$ **d** $c = 34.3, d = 21.1$
  e $m = 11.8, n = 6.6$ **f** $x = 38.5, y = 51.0$
4 a $Y = 39°, x = 17\,m, X = 76°$
  b $p = 49\,cm, r = 85\,cm, R = 92°$
5 a $41°$ **b** $32°$ **c** $54°$
  d $38°$ **e** $27°$ **f** $17°$
6 $\theta = 26.2°$.
  $180° - 26.2° = 153.8°$. However, $153.8° + 32° > 180°$, so not a possible answer, therefore only one solution.
7 $88\,cm$
8 a $\theta = 22°$ **b** $\theta = 53°$ **c** $\theta = 96°$
  d $\theta = 78°$ **e** $\theta = 109°$ **f** $\theta = 134°$
9 $169.14\,cm^2$
10 a false **b** true **c** false **d** false **e** true
11 a false **b** false **c** true **d** true **e** false
12 $121.42°$ and $58.58°$ **13** $369.3\,m^2$
14 a $x = 17$ **b** $x = 6$
15 $76\,mm$ **16** $7\,m$
17 $36.3°$ and $26.4°$
18 $31.38\,cm$, to 2 decimal places
19 a $20.9\,cm$ **b** $23.1\,cm$
20 $\theta = 111.4°$
21 a $166.2\,km$ **b** $187.3°T$
22 a $21.21\,cm$ **b** $30.97°$ **c** $30.97°$ **d** $40.32°$

EXERCISE 5.3

1 A **2** B
3 a $36°$ **b** $120°$ **c** $225°$
  d $210°$ **e** $540°$ **f** $140°$
  g $240°$ **h** $420°$ **i** $20°$
  j $50°$ **k** $62.5°$ **l** $44.0°$
4 a 0.98 **b** 1.19 **c** 2.22
  d 5.04 **e** 5.45
5 a $4\pi\,cm$ **b** $\pi\,m$ **c** $\frac{25\pi}{3}\,cm$
  d $\frac{\pi}{2}\,cm$ **e** $\frac{7\pi}{4}\,mm$ **f** $0.65\,m$
  g $3.92\,cm$ **h** $6.91\,mm$
6 a $12.5\,cm$ **b** $5.0\,cm$
7 a $10.47\,cm$ **b** $26.18\,cm^2$ **c** $15.35\,cm^2$
8 a $\frac{3\pi}{4}$ **b** $\frac{\pi}{6}$ **c** $\frac{5\pi}{6}$ **d** $\frac{4\pi}{3}$
  e $\frac{5\pi}{3}$ **f** $\frac{7\pi}{20}$ **g** $\frac{\pi}{12}$ **h** $\frac{5\pi}{2}$
  i $\frac{5\pi}{4}$ **j** $\frac{2\pi}{3}$
9 2
10 a minor **b** major **c** minor **d** major
11 a 1.75 **b** 0.30 **c** 1.5
  d 1.56 **e** 1.12 **f** 0.67
  g 0.63 **h** 2
12 a $b = 2.309$ **b** $e = 6.467$ **c** $k = 7.071$
13 $160°$ **14** $29\,cm$
15 $3.91\,mm^2$ **16** $\frac{45\pi}{4}\,cm$ **17** 56%

EXERCISE 5.4

1 a $\pi\,cm$ **b** $3\pi\,cm^2$ **c** $3\pi - 9\,cm^2$
2 a $2.09\,cm$ **b** $1.05\,cm^2$ **c** $0.61\,cm^2$
3 a $\frac{3}{5}$ **b** $-\frac{4}{3}$
4 a $-\cos(80°)$ **b** $-\tan\left(\frac{\pi}{7}\right)$ **c** $\sin(3°)$
5 a $\sin(80°)$ **b** $\cos\left(\frac{\pi}{4}\right)$ **c** $-\tan\left(\frac{\pi}{6}\right)$
6 Proof: see worked solutions
7 a $-\frac{12}{13}$ **b** $-\frac{5}{13}$
8 $-\frac{24}{7}$
9 a false **b** false **c** false
  d true **e** false
10 $29°$
11 $-\frac{1}{\tan(\theta)}$ **12** $\frac{4\pi}{3}, \frac{5\pi}{3}$
13 Proof: see worked solutions

EXERCISE 5.5

1. B
2. D
3. a $-\dfrac{1}{2}$
   b $-\dfrac{\sqrt{3}}{2}$
   c $-\dfrac{1}{\sqrt{3}}$ (or $-\dfrac{\sqrt{3}}{3}$)
   d $\dfrac{1}{\sqrt{2}}$
   e $-\dfrac{1}{\sqrt{2}}$
   f $-\dfrac{\sqrt{3}}{2}$
   g undefined
   h $\dfrac{\sqrt{3}}{2}$
   i 0
4. $\dfrac{-\sqrt{3}-1}{2\sqrt{2}}$
5. $x = \dfrac{\pi}{3}$
6. $\dfrac{\sqrt{3}-1}{\sqrt{3}+1} = 2 - \sqrt{3}$
7. a $\dfrac{\pi}{6}$ b $\dfrac{\pi}{4}$ c $\dfrac{\pi}{3}$
8. $\dfrac{1}{3}$ 9. 1   10. 0
11. a false   b true   c false
    d false   e false
12. a $-330°, -210°, 30°, 150°$
    b $\pm\dfrac{\pi}{6}$ c $-\dfrac{\pi}{3}, \dfrac{2\pi}{3}$
13. $30°, 150°$
14. a $\dfrac{117}{125}$ b $\dfrac{4}{5}$
15. a $\cos 75°$ b $\sin(-3A)$ or $-\sin(3A)$

EXERCISE 5.6

1. C   2. 1
3. a amplitude 4, period 2π
   b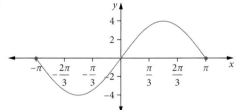
4. a $y = 2\sin\left(\dfrac{x}{2}\right)$ b $y = 2\sin\left(x - \dfrac{2\pi}{3}\right)$
5. a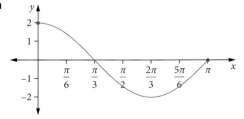
   b maximum at $(0, 2)$, minimum at $\left(\dfrac{2\pi}{3}, -2\right)$
6. range is $-3 \le y \le 3$, period = 6
7. $p = 4$ 8. $p = 2$
9. 10   10. $f(x) = 3\sin(20\pi x)$

11. $y = 3\cos\left(x - \dfrac{\pi}{2}\right)$ 12. 3π
13. Amplitude changes from 1 to 2, period changes from 2π to 2.
14. π
15.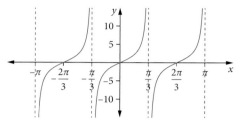
16.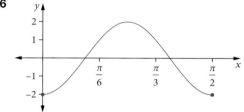

EXERCISE 5.7

1. Amplitude is 4 and period is $\dfrac{2\pi}{3}$.
2. $y = -3\cos(3x)$
3. a 13
   b 43 cm below the rest position
   c 6.14 s
4. a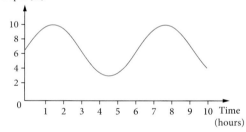
   b minimum 3 m, maximum 10 m
   c 3.42 hours after midday, which is 3:25 pm
   d 6.06 m
5. a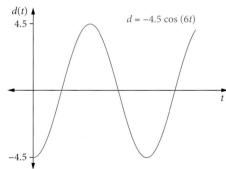
   b The mass is suspended 4.5 cm below the unstretched position.
   c 0.52 s   d -4.3 cm

6 maximum 35°C at 6 am, 2 pm
minimum 15°C at 10 am, 6 pm

7 16 s **8** 2

9 2.4 m **10** 0.4 m

11 $\dfrac{\pi}{10}$ **12** 2 km

13 a Period is 1.795 s, number of swings is 33.
   b 4.97 cm **c** 1.17 s

14 a 4 cm **b** 9 cycles
   c 24 cm **d** $\dfrac{1}{3}$ milliseconds

15 a $r = 20, a = 80$
   b $y = 4\cos(80t) + 20$ in $0 < t < 2$

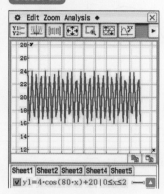

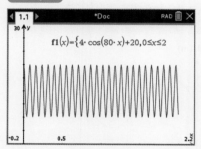

c 25 **d** 19.3 cm above G

CUMULATIVE EXAMINATION: CALCULATOR-FREE

1 a 20 **b** 210 **c** 16

2 a $y = -9x + 12$
   b

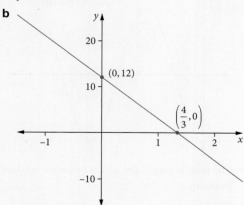

3

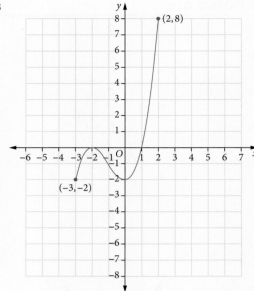

4 $\tan\theta$

5 a 6 m **b** $(14 + 4\sqrt{3})$ m

6 a F **b** F **c** T
   d T **e** F **f** F

7 $\left(\dfrac{\pi}{3} - 1\right)\text{cm}^2$ or $\left(\dfrac{\pi - 3}{3}\text{cm}^2\right)$

CUMULATIVE EXAMINATION: CALCULATOR-ASSUMED

1 a

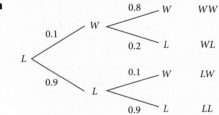

b i 0.08 **ii** 0.11 **iii** 0.1

2 $\dfrac{1}{2}$ **3** 8.01 cm, 25.92°

4 a period = 6, amplitude = 400
   b max = 1600, min 800
   c 1000

5 a minimum 8 m, maximum 118 m
   b 30 min
   c 28.80 m

6 a 435.84 m
   b 5576.35 m^2

CHAPTER 6

EXERCISE 6.1

1. **a** a^3 **b** x^{20} **c** a^{24}
   d $40a^9b^5$ **e** $\dfrac{m^5n}{27}$ **f** $8a^{18}b^{12}$
2. **a** $200x^{26}y^{25}$ **b** $20x^{12}y^{17}$ **c** $4a^8b^{11}$
3. **a** 2^{41} **b** 3^7
   c $5^{12} \times 3^{14}$ **d** $2^{24} \times 3$
4. **a** $\dfrac{1}{m^7}$ **b** $\dfrac{10}{3x^5y^4}$
   c $\dfrac{1}{27a^{18}b^4}$ **d** $\dfrac{5x^2}{6y^4}$
5. **a** $3^{5n-2} \times 2^{14n+4}$ **b** $5^{n-5} \times 2^{7n-1}$ **c** $2^{16-7n} \times 3^{8-6n}$
6. **a** 2^{15x} **b** 3^{-7x} **c** 2^{90x}
7. **a** $\dfrac{20}{x^5y^6}$ **b** $\dfrac{xy^7}{2}$
8. **a** $64a^2b^2$ **b** $64b^{10}$
9. **a** **i** $2^2 \times 3^3$ **ii** 2×3^3 **iii** $2^4 \times 3^{11}$
   b $2^{5-4n} \times 3^{9-4n}$
10. $a = 11n - 14, b = 4n - 4, c = 3n - 4$

EXERCISE 6.2

1. **a** m^6 **b** x^{45} **c** $16a^{28}$
2. **a** 3^{11x} **b** 2^{11x} **c** 5^{12x}
3. **a** **i** $\sqrt{x}$ **ii** $\sqrt[3]{m}$ **iii** $\sqrt[5]{x^6}$
   b **i** $x^{\frac{1}{6}}$ **ii** $a^{\frac{1}{2}}b^{\frac{3}{4}}$ **iii** $p^{\frac{2}{7}}$
4. **a** 5 **b** 2 **c** 8 **d** $\dfrac{1}{8}$
   e 8 **f** $\dfrac{1}{27}$ **g** 32 **h** $\dfrac{8}{125}$
5. **a** $x^{\frac{8}{15}}$ **b** $x^{\frac{23}{6}}$ **c** $x^{-\frac{5}{6}}y^{\frac{1}{2}}$
   d $14x^{\frac{7}{4}}y^{\frac{11}{4}}$ **e** $x^{\frac{5}{6}}y^{\frac{5}{6}}$ **f** $9x^{-\frac{11}{6}}y^{-\frac{5}{2}}$
6. **a** $n = 12$ **b** $n = 0$ **c** $a = -\dfrac{1}{12}, b = -\dfrac{1}{4}$
7. **a** **i** 8 **ii** 3 **b** 72
8. **a** $a = \dfrac{7}{4}, b = -\dfrac{5}{2}$ **b** $x^{-\frac{3}{4}}$
9. **a** Proof: see worked solutions **b** 1

EXERCISE 6.3

1. **a** $\sqrt{n^5}$ **b** $a^{\frac{7}{3}}$
2. $x^{\frac{37}{12}}$

3. **a** $x = \dfrac{5}{4}$ **b** $x = \dfrac{9}{4}$ **c** $x = \dfrac{2}{3}$ **d** $x = -\dfrac{2}{3}$
4. **a** $x = \dfrac{15}{11}$ **b** $x = 4$ **c** $x = -\dfrac{3}{19}$ **d** $x = 6$
5. **a** $x = \dfrac{5}{3}$ **b** $x = -2$ **c** $x = \dfrac{1}{9}$ **d** $x = 2$
6. **a** $x = 2, 3$ **b** $x = 0, 4$ **c** $x = 1, 3$ **d** $x = 2$
7. **a** $x = -2, 1$ **b** $x = -2, 0$ **c** $x = -1, 2$ **d** $x = -3$
8. **a** $x = \dfrac{3}{2}$ **b** $x = -6$ **c** $x = -\dfrac{3}{8}$ **d** $x = \dfrac{8}{11}$
9. **a** $x = \dfrac{3}{7}$ **b** $x = 6$ **c** $x = \dfrac{10}{9}$ **d** $x = -2$
10. **a** $x = 0.77$ **b** $x = 1.77$ **c** $x = 13.23$

EXERCISE 6.4

1. **a** $x = \dfrac{3}{2}$ **b** $x = -\dfrac{2}{3}$
2. $x = -2$
3. **a** The y-intercept is 8.5, the graph is increasing with increasing gradient, and is always positive. The horizontal asymptote is $y = 0$. The coordinates of the endpoints are $(-3, 0.544)$ and $(3, 132.813)$, to three decimal places.

ClassPad

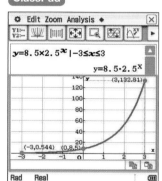

TI-Nspire

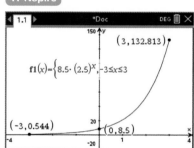

b The y-intercept is 16, the graph is increasing with increasing gradient, and is always positive. The horizontal asymptote is $y = 0$. The coordinates of the endpoints are $(-5, 0.5)$ and $(5, 512)$.

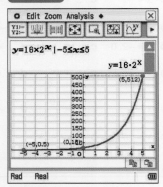

4 a y-intercept $(0, -2)$, asymptote $y = -3$

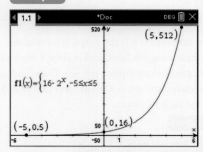

b y-intercept $(0, 0)$, asymptote $y = -1$

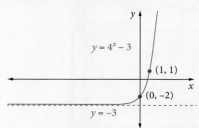

c y-intercept $\left(0, \dfrac{26}{25}\right)$, asymptote $y = 1$

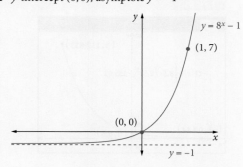

5 a D   **b** C   **c** A
6 $b = 3, c = 1, y = 2^{(x-1)} + 3$
7 $b = -10, c = -4$
8 $b = -4, c = -2$
9 $g(x) = 4^{x-5} - 3$
10 a   **i** $y = -2$ **ii** $(0, 0)$
   b 1 unit left and 2 units down
11 a $-3 = 3^{(-4-c)} + b$ and $-9 = 3^{(-5-c)} + b$
   b $b = -12, c = -6$ **c** $y = -12$
12 a $32 = a^5$
      $a^5 = 2^5$
      $a = 2$
   b $g(x) = 2^{x+3} + 4$ **c** $g(4) = 132$

EXERCISE 6.5

1 $a = 3$
2 asymptote $y = -8$, y-intercept $(0, 8)$
3 a 7.4   **b** 13   **c** 0.36
   d 1800   **e** 21   **f** 72
   g 9.8
4 a   **i** 4.36×10^5 **ii** 8.2×10^{-5} **iii** 1.35×10
   b   **i** 5560   **ii** 0.000 86   **iii** 0.40
5 a   **i** 45   **ii** 1.15
   b   **i** 1100   **ii** 0.82
6 a $A(t) = 3(1.05)^t$
   b   **i** $3.47 \, \text{m}^2$ **ii** $4.89 \, \text{m}^2$ **iii** $7.96 \, \text{m}^2$
   c 105 days
7 a $V(t) = 24\,900(0.85)^t$
   b   **i** \$15 291.71   **ii** \$11 048.26   **iii** \$4902.17
   c 19.8 years
8 a $T(t) = 20 + 480(0.9)^t$
   b   **i** 303.4°C   **ii** 187.4°C   **iii** 54.46°C
   c 2.22 min
9 a $p(m) = 3(2)^m$ **b** 768
10 a 6191   **b** 4%
11 a \$10 000   **b** 20%   **c** \$8000
12 a $V(t) = 3000(1.08)^t$
   b   **i** \$3779.14   **ii** \$6476.77   **iii** \$20 545.43
   c 9.01 years
13 \$160.10   **14** 1.68×10^7 **15** 43 872

CUMULATIVE EXAMINATION: CALCULATOR-FREE

1 a   **i** {3, 4}   **ii** {3, 4, 5, 6, 7, 8}
     iii {4}   **iv** {5, 6}
   b 6   **c** 12
2 a $(x - 11)(x + 3)$
   b $3(2x - 5)^2$
   c $(9x - \sqrt{10}y^2)(9x + \sqrt{10}y^2)$

3

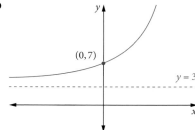

4 Period is 4, amplitude is 2 and range is $-2 \geq y \geq 2$.

5 a 15   b $x = -1$

6 a $\dfrac{5a^{10}b^{13}}{3}$ b $2^{n-7}3^{7n-1}$

7 a   i $(0, 7)$ ii $y = 3$
   b

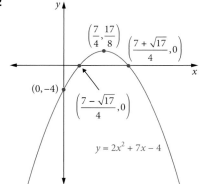

CUMULATIVE EXAMINATION: CALCULATOR-ASSUMED

1 a 2300   b 420
   c   i 0.1583   ii 0.5043   iii 0.3621

2

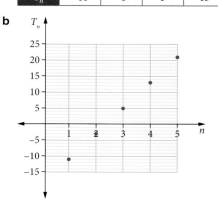

3 a 25°C
   b 76.68°C
   c 2.12 minutes
   d 7.71 minutes
   e 222°C

4 a $1 = 2^{9-c} + b$ and $13 = 2^{11-c} + b$
   b $b = -3, c = 7$ c $y = -3$

5 a $a = 3$ b $g(x) = 3^{x-4} + 3$
   c 30   d $n + 3$

CHAPTER 7

EXERCISE 7.1

1 a It is an arithmetic sequence with a first term of 9 and a common difference of 3.
   b It is an arithmetic sequence with a first term of 68 and a common difference of -4.

2 a 11, 8, 5, 2, -1 b -4, 5, 14, 23, 32

3 a $T_{n+1} = T_n + 2, T_1 = -5$ b $F_{n+1} = F_n + 0.9, F_1 = 6.1$

4 $T_{n+1} = T_n + 3, T_1 = -2$

5 $x = 1; T_{n+1} = T_n - 1, T_1 = 7$

6 a $A_n = 3n + 97$ b $T_n = 54 - 4n$

7 a 100, 95, 90, 85, 80   b 17, 28, 39, 50, 61

8 a

n	1	2	3	4	5
T_n	-11	-3	5	13	21

   b

9 a $T_{n+1} = T_n + 7, T_1 = 2$ b $T_n = 7n - 5$

10 a 3, 7, 11, 15, 19   b $T_n = 4n - 1$
    c   i $T_{50} = 199$ ii $n = 7$

11 a $A_8 = a + 7d$ and $A_{12} = a + 11d$
    b $d = 4, a = -3; A_{n+1} = A_n + 4, A_1 = -3$
    c Proof: see worked solutions

12 a $t_6 = -5$ b $n = 29, t_{29} = -51$

EXERCISE 7.2

1 D   2 A

3 a $S_n = 4150 + 30n$
   b $S_{26} = \$4930$
   c 29 weeks

4 a The population of the bacteria decreases by 5000 cells each hour.
   b After 58 hours, the population is at 10 000.

5 a $R_{n+1} = R_n + 200, R_1 = 400$
   b $R_n = 200n + 200$
   c $R_{10} = 2200$ metres
   d day 24

6 a $T_3 = 92°C$
   b $T_n = 98 - 2n$
     $26 = 98 - 2n$
     $2n = 72$
     $n = 36$

7 a $T_{n+1} = T_n + 55$, $T_1 = 80$
b $T_{10} = 575$
c $n = \dfrac{710}{27} \approx 26.296$. $T_{26} = 1455$, $T_{27} = 1510$. They should sell tickets for 27 days.

8 a $M_n = 540 - 40n$
b **i** $M_{10} = 140$
ii $n = \dfrac{54}{4} = 13.5$. $M_{13} = 20$, $M_{14} = -20$. It is expected to generate sales for approximately 13 months.

9 a $A_n = 8000 + 140n$
b $A_4 = \$8560$
c $A_{28} = \$11\,920$, $A_{29} = \$12\,060$. Therefore, within 28 months.

EXERCISE 7.3

1 B **2** D
3 a −19 **b** −55 **c** 4
4 a 190 **b** 1845
5 1545
6 a $3.57 **b** $1575
7 a $T_{n+1} = T_n + 3$, $T_1 = 4$ **b** $T_{15} = 46$
c i $S_4 = 34$ **ii** $S_{10} = 175$
8 a $B_1 = a = 3$, $d = 6$ **b** $S_n = 3n^2$
c $S_{10} = 300$ **d** $n = 20$
9 a $S_1 = -1$, $S_2 = 2$, $S_3 = 9$
b $T_1 = -1$, $T_2 = 3$, $T_3 = 7$
c $T_n = 4n - 5$; $T_{100} = 395$
d Proof: see worked solutions
10 a $R_{n+1} = R_n - 5$, $R_1 = 110$
110, 105, 100, 95, 90
b $S_n = \dfrac{n}{2}(225 - 5n)$
c 1125
d $R_{22} = 5$, $R_{23} = 0$.
No more blocks added after the 22nd row.
$S_{22} = 1265$.

EXERCISE 7.4

1 E **2** C
3 a It is a geometric sequence with the first term of −2 and a common ratio of 4.
b It is a geometric sequence with the first term of 160 and a common ratio of $\dfrac{1}{4}$.
4 a 999, 333, 111, 37, $\dfrac{37}{3}$
b −0.2, −2, −20, −200, −2000
5 $T_{n+1} = 5T_n$, $T_1 = 15$
6 $T_{n+1} = \dfrac{2}{3}T_n$, $T_1 = 540$
7 $x = \dfrac{3}{2}$; $T_{n+1} = 2T_n$, $T_1 = \dfrac{11}{2}$

8 a $T_n = 9\left(\dfrac{1}{3}\right)^{n-1}$
b $B_n = -2 \times 6^{n-1}$
c $T_n = \sqrt{2}(\sqrt{2})^{n-1} = (\sqrt{2})^n$
9 a $T_{30} = -1.304 \times 10^{-6}$ **b** $T_{30} = 6.863 \times 10^{10}$

10 a

n	1	2	3	4	5
T_n	$\dfrac{2}{3}$	2	6	18	54

b (graph of T_n vs n)

11 a $T_{n+1} = \dfrac{1}{2}T_n$, $T_1 = 10$
b $T_n = 10\left(\dfrac{1}{2}\right)^{n-1}$

12 a $T_5 = \dfrac{81}{2}$
b $r = \dfrac{3}{2}$, $a = 96$; $T_{n+1} = \dfrac{3}{2}T_n$, $T_1 = 96$

13 a $A_n \to 0$
b $a = 6$, $A_{20} = 5.162 \times 10^{-9}$

14 a i Proof: see worked solutions
ii $d = 12$, $a = -18 \Rightarrow X_1 = -18$, $X_2 = -6$, $X_3 = 6$
$r = 2$, $b = 3 \Rightarrow Y_1 = 3$, $Y_2 = 6$, $Y_3 = 12$
b $3n - 1 = 2 \times 3^{n-1}$ has the solutions $n = 1$ and $n = 1.54$. Since n must be a positive integer, there is only one solution for n.

EXERCISE 7.5

1 A **2** B
3 a $A_{n+1} = 1.036A_n$, $A_0 = 40\,000$; $A_1 = \$41\,440$
b $A_{10} = \$56\,971.49$
c $B_n = 40\,000 \times (1.006)^n$; $B_{60} = \$57\,271.54$. The balance of the loan is approximately $300.05 more.

4 a $P_n = 1800 \times 0.6^n$
b $P_3 = \$388.80$
c $P_4 = 233.28 < 300$ and so it is expected to be written off after 4 years.

5 a $a = 3000$, $r = 1.15$
b $a = 30$, $r = 0.98$
c $a = 26$, $r = 1.15$

6 a The population of the city as of January 2020 is 120 000.
b The growth rate of the population is 2% every year.
c 161 504 **d** 36 years

7 a 1.2599; the bacteria are growing at a rate of 25.99% per hour.
  b 8000   **c** after 10 hours
8 a 479 g   **b** 13.51 years
9 a $A_{n+1} = 1.01A_n$, $A_0 = 76\,000$; $B_{n+1} = 1.02B_n$, $B_0 = 72\,500$
  b $A_{10} = \$83\,951.28$, $B_{10} = \$88\,377.10$; Employer B offers \$4425.81 more after 10 years.
  c She should choose employer B, as $B_n > A_n$ for $0 \leq n \leq 5$.

EXERCISE 7.6

1 C   **2** C
3 a 77.5   **b** −43   **c** 374.4
4 a 2441.406   **b** 3375
5 a S_∞ exists as $0 < r < 1$; $S_\infty = 180$
  b S_∞ does not exist as $r > 1$; $S_5 = 372.08$
  c S_∞ exists as $0 < r < 1$; $S_\infty = 720$
6 a $a = 9, r = \dfrac{1}{2}$ **b** $S_n = 18\left(1 - \dfrac{1}{2^n}\right)$
  c Given that $0 < r < 1$, a sum to infinity exists; $S_\infty = 18$.
7 a \$25, \$22.50, \$20.25, \$18.23, \$16.40
  b $S_3 = \$67.75$
  c Given that $0 < r < 1$, a limiting value of S_n exists. $S_\infty = 250$, and so they can expect to pay no more than \$250.
8 a $r = \dfrac{1}{4}$ **b** $16, 4, 1, \dfrac{1}{4}, \dfrac{1}{16}$
  c Proof: see worked solutions
  d 21
  e As $n \to \infty$, $S_n \to \dfrac{64}{3}$
9 a $r = \dfrac{2}{5}$
  b $S_n = \dfrac{25\left(1 - \left(\dfrac{2}{5}\right)^n\right)}{3}$
  c $S_8 = 8.33$
10 a $C_4 = \$66.55$
  b $S_n = 500(1.1^n - 1)$
  c $S_8 = \$571.79$
  d $S_{11} = 926.56$, $S_{12} = 1069.21$
  Minimum number of sales required is 12 in a month.
11 a $B_n = 4096\left(\dfrac{1}{4}\right)^{n-1}$
  b 7 rows
  c $S_7 = 5461$ blocks
12 a 9 m, 5.4 m, 3.24 m, 1.94 m
  b $15 + 2S_4 = 54.17$ m
  c $S_\infty = 37.5$. The total vertical distance will be the initial 15 m plus twice the sum to infinity, $15 + 2S_\infty = 80$ m.

CUMULATIVE EXAMINATION: CALCULATOR-FREE

1 a When $n = 4$, 1 4 6 4 1: $x^4 + 4x^3y + 6x^2y^2 + 4xy^3 + y^4$
  b $4y = 4(-3) = -12$
2 a

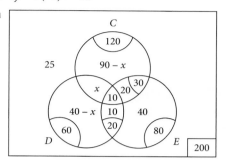

Let $n(C \cap D \cap E') = x$.
$$80 + 90 - x + x + 40 - x + 25 = 200$$
$$235 - x = 200$$
$$x = 35$$

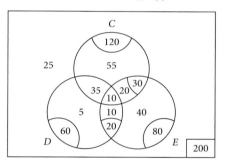

  b i $\dfrac{100}{200}$ or 0.5   ii $\dfrac{30}{120}$ or 0.25   iii $\dfrac{75}{120}$
  c $\left(\dfrac{80}{200}\right)^2 = \left(\dfrac{2}{5}\right)^2 = \dfrac{4}{25}$ or 0.16
3 a $x = 6, x = 8$
  b $-7 + 2\sqrt{6}, -7 - 2\sqrt{6}$
  c $x = \dfrac{5 - \sqrt{17}}{2}, x = \dfrac{5 + \sqrt{17}}{2}$
4 Proof: see worked solutions
5 a $x = 6$
  b $x = \dfrac{5}{11}$
6 a $a = 8r^2 + 8$
  b $a = 40, r = 2$ or $a = 10, r = \dfrac{1}{2}$
  c $k = 32$ or $k = 2$

CUMULATIVE EXAMINATION: CALCULATOR-ASSUMED

1 a $y = -4(x - 5)(x + 6)$
  b $y = 2(x + 9)^2 + 5$
2 $g(x) = 4(2 - x)^3 - 16$
3 a $N(t) = 1000(2^t)$
  b 1.1×10^{18}
  c 3.3 days

4 a $T_{n+1} = T_n + 2$, $T_0 = 20$
  b $T_{12} = 20 + 12(2) = 44$ trees
  c $I_n = 1800(0.7)^n$
  d $S_{12} = \$5941.87$
  e $S_\infty = \dfrac{1800}{1 - 0.7} = 6000$. The total long-term cost for maintaining the park infrastructure according to this plan is $6000.

CHAPTER 8

EXERCISE 8.1

1 a $\dfrac{10 - 4}{1 - 4} = -2$ km/s

  b $\dfrac{1 - 0}{8 - 4} = 0.25$ kg/cm

  c Not possible because the points when connected do not give a straight line.

2 0.4 s/m

3 a -2.1 lumens/m   **b** -0.1 lumens/m

4

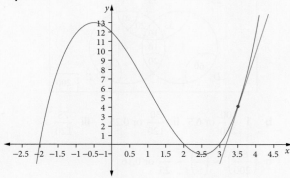

Gradient at $x = 3.5$ is 11.75.

5 $A = 0$, $B = 3$, $C = -2$

6 -3

7 a false **b** false **c** true **d** true **e** false

8 9

9 1.5 cm/s

10 a false **b** true **c** true **d** false **e** false

11 -2 **12** -2 **13** -5

14 a 3 m/s   **b** 9 m/s   **c** $t_1 + t_2$

15 $-\dfrac{2}{5}$ litres/minute **16** P **17** $m = -\dfrac{1}{3}$

EXERCISE 8.2

1 C   **2** A   **3** 55.64

4 $h = 0.1$, $\dfrac{\delta y}{\delta x} = 22.82$

  $h = 0.01$, $\dfrac{\delta y}{\delta x} = 23.8802$

  $h = 0.001$, $\dfrac{\delta y}{\delta x} = 23.988$

  $h = 0.0001$, $\dfrac{\delta y}{\delta x} = \dfrac{23.9988}{n}$

  $f'(-2) = 24$

5 $x = 0$ $R(x) = 0.4$
  $x = 1$ $R(x) = 4.4$
  $x = 2$ $R(x) = 8.4$
  $x = 3$ $R(x) = 12.4$
  In general, $R(x) \approx 4x$.

6 $4x + 2$ **7** $1.1, 3.1, 5.1$

8 a $h = 0.1$, $\dfrac{\delta y}{\delta x} = 24.3$

  $h = 0.01$, $\dfrac{\delta y}{\delta x} = 24.03$

  $h = 0.001$, $\dfrac{\delta y}{\delta x} = 24.003$

  b 24

9 $y = 4 - 6x$ **10** $g(x) = 6x + 2$

11 $\dfrac{\delta y}{\delta x} = \dfrac{f(11) - f(10)}{1} = -662$

12 $\dfrac{\delta y}{\delta x} = \dfrac{f(2.2) - f(2)}{0.2} = -4.2$

13 $y = 6x + 1$

14 a 4 m/s   **b** 3.99 m/s

15 a Proof: see worked solutions
  b $R(x) = 2ax + b$

EXERCISE 8.3

1 E   **2** $g(x) = 2x - 3$

3 $f'(x) = x^2$ **4** $f'(x) = 6$, $f'\left(\dfrac{2}{3}\right) = 6$

5 a $f'(x) = 6x$ **b** $f'(x) = -2x + 1$
  c $f'(x) = 1$ **d** $f'(x) = -4x$
  e $f'(x) = 2x + 2$ **f** $f'(x) = -3x^2$

6 $f'(x) = -2x$, $f'(-1) = 2$

7 Proof: see worked solutions

8 $-14x$ **9** -3

10 $f'(x) = 8x + 12$ **11** $f'(x) = 3x^2$, $f'(a) = 3a^2$

12 $f'(x) = m$ **13** $f'(x) = 2ax + b$

14 $4x^3$

15 Proof: see worked solutions

EXERCISE 8.4

1 D   **2** C

3 a $3x^2 + 2x + 1$
  b $42x^6 + 90x^8 - 60x^{11} + 48x^5$
  c $-12x^5 + 72x^7 + 11x^{10}$
  d $24x^7 - 60x^{11}$
  e $72x^{11} + 143x^{12} - 120x^4$
  f $36x^2 + 28x^3 - 66x^5$
  g $32x^7 - 15 + 56x + 42x^2$
  h $-6 + 14x + 42x^2 - 36x^3$
  i $120x^7 - 35x^4 + 28x^6 + 12$
  j $32x - 36x^3 + 21x^2 + 18$

4 $\dfrac{dy}{dx} = 6x^5 - 5x^4 + x^3$

5 a $\frac{1}{6}$ b 25

6 $y = 4x - \frac{1}{2}$ 7 $a = 3$

$\frac{dy}{dx} = 23.5$

8 $a = 5, b = -1$

9 a $12x^3 - 12x$ b $12u^3 - 12u$
   c $30m^5 + 40m^4 - 24$ d $9 + 30y^4 - 12y^2 + 14y^6$
   e $6z - 27z^2 + 38$ f $35t^6 + 36t^3 - 16$
   g $27r^2 - 13 - 12r$ h $-3a^2 + 6a + 4$
   i $-\frac{9p^2}{2}$ j $64t^3 + 108t$

10 $a = 3, b = -4$ 11 $f'(x) = 10x + 10$

12 $3x^{11}$ 13 $2 - 2x^7 - 2.5x^8$

14 $2k + 2$ 15 $6x - 4$

16 $2(6x + 1)$ 17 $2a - 12$

18 $a = 1, a = 4$

CUMULATIVE EXAMINATION: CALCULATOR-FREE

1 a 0.5   b 0.5

2 $\frac{20\pi}{3}$ cm

3 a i 4   ii 25
   b 100

4 a $x = 2$ b $r = \frac{1}{2}$
   c i $T_n \to 0$ ii $S_\infty = 8$

5 1.2°C/hour   6 $f'(x) = x - 1$

7 $f'(-1) = -3.5$ 8 $a = 7$

CUMULATIVE EXAMINATION: CALCULATOR-ASSUMED

1 a $x = 0, y = 20$
   b If $x = 0$, then only $A \cap B$ and $\overline{A \cup B}$ are not empty sets. If $n(A) = 30, n(B) = 30$ and $n(A \cap B) = 30$, then $A = B$.

2 a $y = (x - 3)^2 - 20$
   b $x = 3 + 2\sqrt{5}, x = 3 - 2\sqrt{5}$

3 $f(x) = x^2 - 3x - 5$

4 a $h = 0.1, \frac{\delta y}{\delta x} = 25.22$
     $h = 0.01, \frac{\delta y}{\delta x} = 24.120$
     $h = 0.001, \frac{\delta y}{\delta x} = 24.012$
   b 24

5 $f'(x) = -4.5x^2 + 12x + 1.5$

6 $k = \frac{b+3}{2a}$

7 $a = 4, b = -3$

CHAPTER 9

EXERCISE 9.1

1 $y = -13x - 10$ 2 $y = 6x - 19$

3 a $y = 8x - 9$ b $y = -11x + 11$
   c $y = -2x - 1$

4 a $(-1, -3), (3, 13)$ b $(0, 4), (2, 6)$
   c $(-2, -22), (4, 32)$ d $(1, 5)$

5 Proof: see worked solutions

6 $\left(\frac{3}{10}, \frac{3}{20}\right)$

7 Proof: see worked solutions

8 $y = -11x - 16$

9 Proof: see worked solutions

10 -7

11 $y = 10x - 40$

12 Proof: see worked solutions

13 $\left(\frac{1}{3}, \frac{8}{3}\right)$

14 $(0, -2)$ and $(-2, 2)$

EXERCISE 9.2

1 $y = 10x - 13$ 2 $(-1, 32), (3, 0)$

3 a $v(t) = 4t$ b 12 m/s

4 a 6 m   b 2 s
   c 6 s   d 20 m
   e 4 m/s

5 -7 m/s

6 a $x(0) = 2$ cm   b $t = 1$
   c $x(2) = -4$ cm   d 6 cm
   e $v(3) = -7$ cm/s

7 a $v(0) = 3$ m/s   b $t = 0, t = 1, t = 3$

8 a $h(0) = 4$ m   b $v(0) = 40$ m/s
   c $h(1) = 39$ m   d at $t = 4, h(4) = 84$ m

9 a $v(3) = 18$ cm/s   b $t = 2$ seconds

10 a $3t^2 + 6t$ b 9 m/s

EXERCISE 9.3

1 a $v(t) = 4t$ b $t = 0$

2 a $v(t) = \frac{1}{4}(6t - 2)$ b to the left of the origin
   c $t = \frac{1}{3}$

3 everywhere except at $x = 1$

4 $(0, -2)$ and $\left(-\frac{2}{3}, -\frac{50}{27}\right)$

5 minimum at $(2, -14)$

6 $(-2, -2)$

7 a $(2,-5)$ **b** $(2,-2)$ and $\left(-\dfrac{4}{3}, \dfrac{446}{27}\right)$

8 $(2,0)$ **9** $\left(\dfrac{3}{2}, -\dfrac{1}{4}\right)$

10 maximum at $(-1,7)$ **11** $(-3,10)$ and $\left(1, -\dfrac{2}{3}\right)$

12 $x = -1$ **13** $(-1,7)$

14 $m = 2$ **15** $p = -12$

16 maximum at $(-3, 88)$ and minimum at $(2, -37)$

17 a false   **b** true   **c** true   **d** false   **e** true

18 stationary point of inflection

19 $p = \dfrac{1}{2}$ **20** $x = 2$

21 a true   **b** true   **c** false   **d** false   **e** false

EXERCISE 9.4

1 B   **2** E

3 a

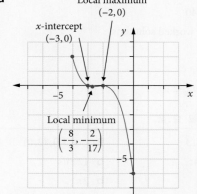

b maximum is 2 and minimum is -6

c as $x \to \infty, f(x) \to -\infty$ as $x \to -\infty, f(x) \to \infty$

4 25

5

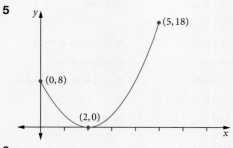

6

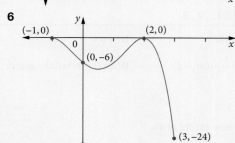

7 a $(-0.5, -6)$ **b** $y = 6$

c $y = -4$ **d** $-0.5 < x \le 4$

e $-4 \le x < -0.5$ **f** ≈ 4

g $-6 \le y \le 14$ **h** approx. 13

8 a $(-3, 40), (-0.5, -25)$ and $(3.5, 240)$

b $y = 40$ **c** $y = 120$

d $x < -3$ and $-0.5 < x < 3.5$

e $-3 < x < -0.5$ and $x > 3.5$

9 a $(0, 0)$ and $(2, 4)$

b

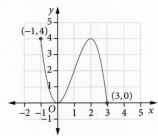

10 $(-1, 43)$ and $(4, -82)$

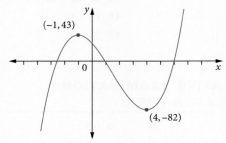

11 $(-4, 25)$ and $(0, -7)$

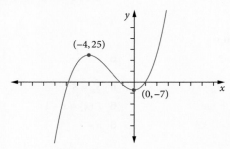

12 $(-2, 0)$ and $(2, -32)$

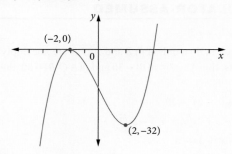

13 a false   **b** false   **c** false   **d** true   **e** true

14 $\left(0.5, \dfrac{9}{16}\right)$

15 a true   **b** false   **c** true   **d** false   **e** false

16 $(1, 2)$

EXERCISE 9.5

1 1

2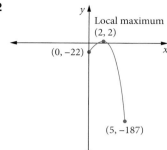

3 maximum = $(-3, 0)$ and minimum = $\left(-\dfrac{1}{3}, -\dfrac{256}{27}\right)$

4 $x = -1$

5 1.69 cm

6 125 000 m^2

7 225 m^2

8 14, 14

9 2048 cm^3

10 7.5 km

11 $t = 2$ s, $h = 16$ m

12 a Proof: see worked solutions

b $x = \dfrac{20}{3}$

c 7407.4 cm^3

13 6 m by 6 m by 3 m

14 highest = 24°C, lowest = 4°C

15 $\left(-\dfrac{1}{3}, \dfrac{256}{27}\right)$

16 $(-3, -8)$

17 2.52 units^2

18 a Proof: see worked solutions

b Proof: see worked solutions

c $\dfrac{dv}{dx} = -\dfrac{75}{14}x^2 + \dfrac{16\,200}{7}$

d $x = 12\sqrt{3}$, $h = \dfrac{120\sqrt{3}}{7}$

19 $\dfrac{1}{200}(8a + 4b + c) = 0$

$\dfrac{1}{200}(12a + 4b) = -0.06$

$\dfrac{1}{200}(48a + 8b) = 0$

EXERCISE 9.6

1 -2 and 2

2 64 metres

3 $\dfrac{3}{4}x^4 + \dfrac{2}{3}x^3 + c$

4 $-6x^3 - 12x^2 - 8x + c$

5 $f(x) = \dfrac{1}{3}x^3 + \dfrac{3}{2}x^2 + 7x - \dfrac{41}{6}$

6 $\dfrac{1}{3}x^3 - \dfrac{1}{2}x^2 + \dfrac{1}{6}$

7 a $\dfrac{x^3}{3} + c$

b $\dfrac{x^4}{4} + \dfrac{x^3}{3} + c$

c $x^4 + x^2 - 4x + c$

d $x^3 + \dfrac{5x^2}{2} + 7x + c$

e $\dfrac{3x^2}{4} - 2x + c$

f $\dfrac{2x^3}{3} - \dfrac{x^2}{2} - 6x + c$

g $x - 2x^2 + \dfrac{4x^3}{3} + c$

h $\dfrac{x^5}{5} + \dfrac{x^4}{2} + x^3 + x^2 + x + c$

8 $f(x) = \dfrac{-3x^2}{4} + \dfrac{3x}{2}$

9 $f(x) = \dfrac{x^4}{4} - x^3 + \dfrac{3x^2}{2} - x + c$

10 $g(x) = x^3 + x$

11 $y = 2x^4 + \dfrac{2}{3}x^3 - 7x + c$

12 $f(x) = \dfrac{3}{4}x^4 + \dfrac{2}{3}x^3 - \dfrac{17}{12}$

13 $b = 0$

EXERCISE 9.7

1 $2x^5 + x^4 - x^3 - x^2 + c$

2 C

3 $x(t) = \dfrac{1}{3}t^3 + \dfrac{3}{2}t^2 + 2$

4 $x = 20t - 10$

5 92 m

6 a 135 m

b $t = 0.5$ s

7 $\dfrac{746}{9}$ m

8 28 m

9 -42.5 cm

10 12 cm

11 30 m

12 $x(t) = t^2 + t + 38$

CUMULATIVE EXAMINATION: CALCULATOR-FREE

1 26

2 a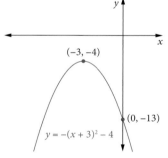

b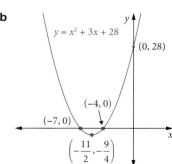

3 $a = 2$

4 $\dfrac{5a}{4b^8}$

5 a $4x$

b $f'(-1) = -4$

6 a $y = \dfrac{x^4}{2} - \dfrac{x^2}{2} + c$ **b** $f'(4) = 2$

7 $f(x) = \dfrac{4x^2}{3} - 3x^3 + x + 70$

8

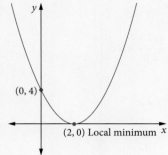

CUMULATIVE EXAMINATION: CALCULATOR-ASSUMED

1 a 0.0376   **b** 0.0104   **c** 0.7085
  d 0.0794   **e** 0.0746

2 37.1°

3 a It is an arithmetic sequence with an initial term of 215 and a common difference of 5.
  b i $B_n = 180 + 10n$ **ii** $B_n = 180(1.0491)^n$
  c $A_{12} = 275$, $B_{12} = 320$

Club B can expect to have 45 more members than Club A after 12 months.

4 $f(x) = x^3 - x^2 - 48$ **5** 10 cm

6 $x(t) = \dfrac{3t^2}{4} - \dfrac{5t}{2} - \dfrac{5}{4}$ **7** 4 seconds, 144 metres

8 a $L = \sqrt{(12x)^2 + (5x)^2} = 13x$
  b $w = 100 - 20x$
  c $V = 30x^2(100 - 20x) = 600x^2(5 - x)$
  d width = $\dfrac{100}{3}$ cm, height = $\dfrac{50}{3}$ cm, length = 40 cm

Glossary and index

addition principle The principle used to calculate combinations or probabilities when the word 'or' is used. (p. 17)

alternating sequence A geometric sequence with a common ratio $r < 0$. (p. 354)

ambiguous case A situation where 2 solutions for an unknown angle are possible. Sometimes, both the acute and obtuse angles are correct, but at other times only 1 angle is correct for that triangle. It depends on the size of the given angle. (p. 245)

amplitude The vertical distance from the centre of a trigonometric function to its maximum and minimum points. The amplitude of $y = \sin(x)$ and $y = \cos(x)$ is 1. (p. 273)

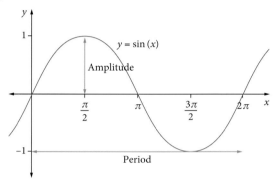

angle of inclination The angle a straight line makes with the positive x-axis. (p. 240)

angle sum and difference identities
$\sin(A \pm B) = \sin A \cos B \pm \cos A \sin B$
$\cos(A \pm B) = \cos A \cos B \mp \sin A \sin B$
$\tan(A \pm B) = \dfrac{\tan A \pm \tan B}{1 \mp \tan A \tan B}$ (p. 270)

anti-derivative (or integral) The opposite of the derivative. The anti-derivative of $f(x)$ is a function $F(x)$ whose derivative is $f(x)$: $F'(x) = f(x)$. (p. 434)

anti-differentiation (or integration) The opposite of differentiation; that is, the process of finding the anti-derivative or integral. (p. 434)

arc Part of the circumference of a circle. (p. 253)

arithmetic sequence A sequence of numbers that have a common difference between consecutive terms. (p. 333)

arithmetic series The sum of an arithmetic sequence. (p. 349)

array An $m \times n$ dimensional table representing the m possible outcomes of the first stage of an experiment and the n possible outcomes of the second stage of an experiment. (p. 48)

ASTC rule An acronym (All Stations To Claremont) to remember the quadrants where sine, cosine and tangent ratios are positive. (p. 262)

asymptote A line towards which a graph gets closer but never touches. In the graph below, the lines $x = 2$ and $y = 4$ are asymptotes. (p. 192)

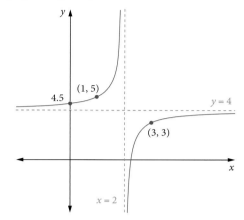

average rate of change The rate of change between point $A(x_1, y_1)$ and point $B(x_2, y_2)$ on the graph of a function, given by the gradient $m = \dfrac{y_2 - y_1}{x_2 - x_1}$ of the straight line connecting the two points. (p. 385)

axis of symmetry A line about which a graph is reflected equally on both sides. For a parabola, this is a vertical line halfway between the x-intercepts, given by $x = -\dfrac{b}{2a}$. (p. 142)

binomial coefficients The set of values
$\left\{ \binom{n}{0}, \binom{n}{1}, \binom{n}{2} \cdots \binom{n}{n} \right\}$ in the binomial theorem. (p. 23)

binomial expression An algebraic expression with two terms, such as $x + 4$ or $3x^2 - 6$. (p. 22)

binomial product Two or more binomial expressions multiplied together; for example, $(3x + 2)(x - 1)$. (p. 110)

binomial theorem The generalised expansion of $(x + y)^n$ for integers $n \geq 0$. (p. 23)

cardinality (of a set) The size of a set; that is, the number of elements in the set. (p. 3)

chord The interval (line segment) joining 2 points on a curve. (p. 254)

circumference The perimeter, or distance around the outside, of a circle. (p. 253)

co-domain The set of values that the range of a function or relation can fall in. (p. 206)

coefficient The number multiplying a variable. For example, in $x^3 - 3x^2 + 6x$, the coefficient of x^2 is -3. (p. 90)

combination An unordered selection of objects. (p. 13)

common difference A constant addition or subtraction between terms of a sequence. (p. 333)

common ratio A constant multiplication factor between the consecutive terms of a geometric sequence. (p. 354)

complement (of a set) The set of elements that are not in a given set. (p. 3)

completing the square A method allowing the transposition of a quadratic expression from general form to turning point form by rewriting part of the expression in the form $a^2 + 2ab + b^2 = (a+b)^2$ or $a^2 - 2ab + b^2 = (a-b)^2$. (p. 130)

compound event An event as a result of a multi-staged probability experiment. (p. 50)

conditional probability The probability of an event occurring given that another event has occurred. (p. 58)

constant A fixed value. For example, in $x^2 - 5x + 7$ the constant term is 7. (p. 189)

convergent When a sequence or series tends to a specific value. (p. 360)

cosine rule In any triangle ABC, $a^2 = b^2 + c^2 - 2bc\cos(A)$. (p. 245)

cubic function A polynomial of degree 3, whose equation has the general form $y = ax^3 + bx^2 + cx + d$. (p. 199)

cubic polynomial A polynomial of degree 3; for example, $P(x) = -2x^3 + x^2 - 10x - 7$. (p. 189)

De Morgan's laws The equivalences that relate complements, intersections and unions of sets: $\overline{A \cap B} = \overline{A} \cup \overline{B}$ and $\overline{A \cup B} = \overline{A} \cap \overline{B}$. (p. 6)

decreasing sequence A sequence such that the next term is always smaller than the previous term for any value of n. (p. 333)

degree of a polynomial The highest power on the variable in a polynomial. For example, $x^3 - 3x^2 + 6x$ has degree 3. (pp. 89, 189)

dependent variable A variable that is calculated from another variable according to a rule or equation. The variable on the vertical scale of a graph. For $y = f(x)$, y is the dependent variable. (p. 206)

derivative For the function $y = f(x)$, the derivative of y, $\frac{dy}{dx}$ or $f'(x)$, is the function of the instantaneous rate of change of $f(x)$; also called the gradient function. (p. 397)

difference of perfect squares The expression $a^2 - b^2$. It can be factorised to $(a-b)(a+b)$. (p. 113)

differentiation The process of finding the derivative of a function. (p. 400)

differentiation by first principles Finding the derivative of a function using the formula $\frac{dy}{dx} = f'(x) = \lim_{h \to 0} \frac{f(x+h) - f(x)}{h}$. (p. 397)

dilation Stretching or squashing (compressing) of a graph, either from the x-axis (vertically) or y-axis (horizontally). (p. 215)

discontinuity A break in the graph. (p. 192)

discriminant, Δ $b^2 - 4ac$. Represented by the symbol Δ. The part of the quadratic formula under the square root. For the equation $ax^2 + bx + c = 0$, when $\Delta > 0$ there are two distinct real solutions; when $\Delta = 0$, there is one solution; when $\Delta < 0$, there are no real solutions. (p. 136)

disjoint sets Two sets that do not have an intersection. (p. 5)

displacement The 'signed distance' from the origin of a moving object, represented by the function $x(t)$, where displacement is a function of time. (p. 416)

divergent When a sequence or series does not tend towards a specific value, but instead increases or decreases indefinitely. (p. 360)

domain The set of values of the independent variable x in a relation or function. (p. 206)

element A member of a set. (p. 3)

elimination method An algebraic method for solving simultaneous linear equations by adding or subtracting equations to eliminate one of the variables. One or more equations may first need to be multiplied by a constant to make elimination possible. (p. 158)

empty set The set that contains no elements. (p. 5)

equating coefficients (of identical polynomials) Applying the rule that if
$P(x) = a_n x^n + a_{n-1} x^{n-1} + a_{n-2} x^{n-2} + \ldots + a_1 x^1 + a_0$ and
$Q(x) = b_n x^n + b_{n-1} x^{n-1} + b_{n-2} x^{n-2} + \ldots + b_1 x^1 + b_0$ are equal for all values of x, then $a_n = b_n$, $a_{n-1} = b_{n-1}$, $a_{n-2} = b_{n-2}$, and so on up to $a_0 = b_0$. (p. 190)

event An outcome or set of outcomes in a probability experiment. (p. 37)

exact values The trigonometric ratios of $30°\left(\frac{\pi}{6}\right)$, $45°\left(\frac{\pi}{4}\right)$, $60°\left(\frac{\pi}{3}\right)$ and their multiples, which can be expressed as exact rational or surd values such as $\frac{1}{2}$ or $\frac{\sqrt{3}}{2}$. (p. 267)

expanding Expanding involves multiplying each term inside brackets by the factor outside the brackets. In the process, all brackets are removed. It is the reverse process of factorising. (p. 110)

experimental probability The relative frequency of an event in an observed experiment being used as an estimate for the likelihood of the event occurring. (p. 39)

explicit rule See **general rule**.

exponent Another name for power; for example, in 2^5, the exponent is 5. (p. 301)

exponential decay A decrease that is happening quickly at first, then gradually more slowly, according to the function $A(t) = A_0(a^t)$, where A is the quantity, A_0 is the initial value, a is the growth factor ($0 < a < 1$) and t is the time; a context that can be modelled by a geometric sequence with $a > 0$ and a common ratio, $0 < r < 1$. (p. 320)

exponential equation An equation involving a^x where the variable is in the exponent (power). (p. 310)

exponential function A function of the form $y = a^x$ where the variable is in the exponent (power). (p. 313)

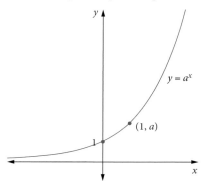

exponential growth An increasing function of the form $A(t) = A_0(a^t)$, where A is the quantity, A_0 is the initial value, a is the growth factor ($a > 1$) and t is the time; a context that can be modelled by a geometric sequence with $a > 0$ and a common ratio $r > 1$. (p. 320)

factorial The operation whereby a positive integer is multiplied by all positive integers smaller than it. (p. 13)

factorising The reverse process of expanding. It involves rewriting an expression as a product of factors. (p. 110)

finite sum The sum of the first n consecutive terms of a sequence. (p. 347)

first term The initial term of a sequence, a. (p. 333)

FOIL method A method for expanding $(a + b)(c + d)$ to $ac + ad + dc + bd$. Stands for First, Outer, Inner, Last. (p. 111)

function A relation such that each value in the domain has only one ordered pair; each x value matches with only one y value. (p. 206)

general form (of a linear equation) $ax + by + c = 0$, where a, b and c are whole numbers. (p. 93)

general form (of a quadratic function) $y = ax^2 + bx + c$ (p. 140)

general rule A rule that describes how to obtain the nth term in a sequence. (p. 336)

geometric sequence A sequence of numbers that have a common ratio between consecutive terms. (p. 354)

geometric series The sum of a geometric sequence. (p. 367)

global maximum/minimum point The maximum/minimum point of a function over a given interval. It is either a local maximum/minimum point or one of the endpoints of the interval. For the function graphed over the given interval below, the global maximum point is $(1.6667, 9.4815)$ and the global minimum point is $(4, -25)$. (p. 426)

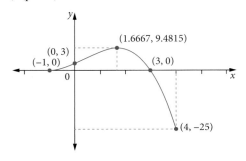

gradient The slope of a line, given by $m = \dfrac{\text{rise}}{\text{run}} = \dfrac{y_2 - y_1}{x_2 - x_1}$. For a curve, the gradient is the slope of the tangent to the curve at a particular point. (p. 89)

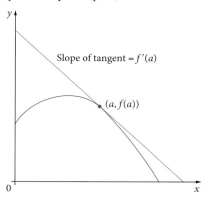

gradient function *See* **derivative**.

gradient–intercept form (of a linear function) $y = mx + c$, where m is the gradient and c is the y-intercept. (p. 89)

growth rate The value of r interpreted as a percentage in contexts of exponential growth and decay. (p. 364)

highest common factor The highest number that can be divided exactly into every term. (p. 116)

horizontal asymptote A horizontal line to which the graph of a function tends as x approaches infinity. (p. 192)

hyperbola A discontinuous graph with equation of the form $y = \dfrac{a}{x - b}$. (p. 192)

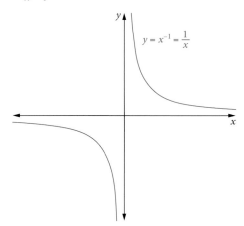

implied domain *See* **natural domain**.

increasing sequence A sequence such that the next term is always larger than the previous term for any value of n. (p. 333)

independent events Two events that do not influence the likelihood of the other occurring. (p. 67)

independent variable A variable that is used to calculate the value of another variable. The variable on the horizontal scale of a graph. For $y = f(x)$, x is the independent variable. (p. 206)

index (or exponent) Another name for power; for example, 5 is the index in 2^5. (p. 301)

infinite geometric series *See* **sum to infinity**.

instantaneous rate of change The rate of change at a point (x_1, y_1) on the graph of a function $f(x)$, given by the gradient of the tangent to the graph at that point, which is the derivative of the function at that point, $f'(x_1)$. (p. 392) See **gradient** for diagram.

integral (or anti-derivative) The opposite of the derivative. The anti-derivative of $f(x)$ is a function $F(x)$ whose derivative is $f(x)$: $F'(x) = f(x)$. (p. 434)

integration (or anti-differentiation) The opposite of differentiation, the process of finding the anti-derivative or integral. (p. 434)

intercept form (of a linear function) $\frac{x}{a} + \frac{y}{b} = 1$ (p. 93)

intercept form (of a quadratic function) $y = a(x - u)(x - v)$. (p. 140)

intersection (of sets) The set of elements that are contained in both given sets. It contains the overlapping elements of the sets. (p. 3)

inverse proportion Where x increases in value, y decreases proportionally, or vice-versa. (p. 227)

leading coefficient (of a polynomial) The coefficient of the term of a polynomial that contains the highest power. For example, $x^3 - 3x^2 + 6x$ has a leading coefficient of 1 (the coefficient of x^3). (p. 189)

leading term (of a polynomial) The term of a polynomial that contains the highest power. For example, $x^3 - 3x^2 + 6x$ has a leading term of x^3. (p. 189)

like terms Two or more terms with the same variables raised to the same powers. (p. 110)

limit The value of the function $f(x)$ at $x = a$ as x gets infinitely closer to a. (p. 397)

linear decay A context that can be modelled by an arithmetic sequence with a negative common difference. (p. 343)

linear function An equation that can be expressed in the form $y = mx + c$. The degree is 1. The graph of a linear function is a straight line. (p. 385)

linear growth A context that can be modelled by an arithmetic sequence with a positive common difference. (p. 343)

linear polynomial A polynomial of degree 1; for example $P(x) = -2x$. (p. 189)

linear relationship See **linear function**.

major arc, sector or segment An arc, sector or segment that covers more than half of a circle. (p. 254)

minor arc, sector or segment An arc, sector or segment that covers less than half of a circle. (p. 254)

monic polynomial A polynomial whose leading coefficient is 1; for example, $x^3 - 3x^2 + 6x$. (p. 189)

monic quadratic trinomial A quadratic trinomial whose leading coefficient is 1. The general form is $x^2 + bx + c$. (p. 122)

multiplication principle The principle used to calculate combinations or probabilities when the word 'and' is used. (p. 17)

mutually exclusive events Two events in a sample space such that they cannot both occur at the same time, $P(A \cap B) = 0$. (p. 43)

natural domain The largest set of real numbers for which a function rule is defined. Also called implied domain. (p. 209)

null factor law If $mn = 0$, then $m = 0$ or $n = 0$ or $m = n = 0$. (p. 126)

one-to-one function A function such that each value in the range has only one ordered pair (x, y); each y value matches with only one x value. (p. 208)

oscillating sequence See **alternating sequence**.

outcome An element of the sample space of a probability experiment. (p. 37)

pairwise disjoint When any pairing of two sets (from three sets) are disjoint sets. (p. 9)

parabola The graph of a quadratic function. Key features include any axes intercepts and the turning point. (p. 140)

parallel lines Two lines are parallel if and only if they have the same gradient. They never meet and they are equidistant. (p. 104)

parameter A variable within a function that is neither the input nor the output of the function. (p. 138)

partial sum See **finite sum**.

Pascal's triangle A triangular arrangement of numbers showing the number of combinations possible when r objects are selected from n distinct objects. (p. 20)

perfect square $(a + b)^2$ or $(a - b)^2$. $(a + b)^2$ expands to $a^2 + 2ab + b^2$, and $(a - b)^2$ expands to $a^2 - 2ab + b^2$. (p. 112)

period The horizontal distance a trigonometric function moves before repeating itself. The period of $y = \sin(x)$ and $y = \cos(x)$ is 2π. (p. 273) See **amplitude** for diagram.

permutation An arrangement or an ordered selection of objects. (p. 13)

perpendicular lines Two lines that intersect at a right angle (90°). The gradients are negative reciprocals of each other: $m_1 m_2 = -1$. (p. 105)

phase change A horizontal translation of the graph of a trigonometric function. (p. 273)

point–gradient form (of a linear function) $y - y_1 = m(x - x_1)$, where m is the gradient and the line passes through the point (x_1, y_1). (p. 93)

point of inflection A point where a function changes concavity. (p. 199)

polynomial An expression involving a sum of terms of the form $P(x) = a_n x^n + a_{n-1} x^{n-1} + a_{n-2} x^{n-2} + \ldots + a_1 x^1 + a_0$, where n is a positive whole number and $a_0, a_1, a_2 \ldots a_{n-1}, a_n$ are real numbers and $a_n \neq 0$. (p. 189)

power function A function of the form x^n, where n is a real number. (p. 192)

probability The mathematics of chance; the numerical quantity between 0 and 1 inclusive, given to the likelihood of an event occurring. (p. 37)

product The result of multiplication; for example, 27 is the product of 3 and 9 and $5x$ is the product of 5 and x. (p. 110)

proper subset A set contained within another set, not including the set itself. (p. 21)

Pythagorean identity The trigonometric identity $\sin^2 \theta + \cos^2 \theta = 1$. (p. 261)

quadrant One-quarter of the number plane, bounded by the x- and y-axes. For example, quadrant 1 is the region enclosed by the positive x-axis and the positive y-axis. (p. 244)

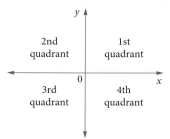

quadratic See **quadratic relationship**.

quadratic equation Any equation that can be rearranged into the form $ax^2 + bx + c = 0$. (p. 122)

quadratic expression An algebraic expression where the degree is 2; for example, $-5x^2 + 1$. (p. 122)

quadratic formula The formula $x = \dfrac{-b \pm \sqrt{b^2 - 4ac}}{2a}$ that gives the solutions of the quadratic equation $ax^2 + bx + c = 0$. (p. 134)

quadratic function A function that can be expressed in the form $y = ax^2 + bx + c$. The degree is 2. The graph of a quadratic function is a parabola. (p. 140)

quadratic polynomial A polynomial of degree 2; for example $P(x) = x^2 - 10x - 7$. (p. 189)

quadratic relationship See **quadratic function**.

quadratic trinomial A quadratic expression with 3 terms. The general form is $ax^2 + bx + c$. (p. 122)

quartic polynomial A polynomial of degree 4; for example $P(x) = 2x^4 + x^2 - 10x - 7$. (p. 189)

radian A measure of angle defined as arc lengths around the circumference of a circle: π radians = 180°. (p. 253)

radius The distance from the centre of a circle to the circumference. (p. 253)

range The set of values of the dependent variable y in a relation or function $y = f(x)$. (p. 206)

recursive rule A rule that describes how to obtain the next term in a sequence given the previous term. (p. 333)

reference triangles Special right-angled triangles with angles of 30°, 45° and 60° that are used to find exact values of trigonometric functions. (p. 267)

reflection Mirror-image or 'flipping' of a graph so that it is back-to-front or bottom-to-top. (p. 216)

relation A set of ordered pairs (x, y). (p. 206)

relative frequency The number of times an event occurs out of the total number of trials. (p. 39)

repeated solution A equation's solution that occurs multiple times due to a repeated linear factor. For example, $x = 5$ in $(x - 5)^2 = 0$. Graphically, a repeated solution occurs when a stationary point is also an x-intercept. (p. 167)

replacement The context of a staged probability experiment whereby the sample space is returned to the original set after a stage of the experiment. (p. 49)

root The solution to $f(x) = 0$. It is the x-intercepts of the function. (p. 135)

sample space The set of all possible outcomes of a probability experiment. (p. 37)

scientific notation A value written as a number between 1 and 10 multiplied by a power of 10. (p. 321)

secant A straight line passing through two points on a curve. (p. 387)

sector A region of a circle cut off by 2 radii. (p. 254)

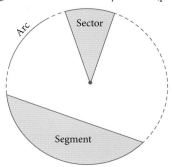

segment A region of a circle cut off by a chord. (p. 254)

sequence A pattern of numbers. (p. 333)

series The sum of a sequence. (p. 349)

set A group of objects, such as numbers. (p. 3)

set operations The ways in which sets relate to other sets. (p. 3)

significant figures (rounding to a number of) A method of rounding involving all the non-zero digits of a number plus the zeros that are included between them, or that are final zeros and signify accuracy. (p. 320)

simultaneous equations A set of equations that are all satisfied by the same values. Graphically, the solution is the point of intersection between the graphs of the equations. Also known as a **system of equations**. (p. 157)

sine rule In any triangle ABC, $\dfrac{a}{\sin A} = \dfrac{b}{\sin B} = \dfrac{c}{\sin C}$. (p. 243)

square root function A function of the form $y = \sqrt{x}$. (p. 194)

stationary point A point on a graph where the gradient equals 0. The graph is flat at the stationary point, neither increasing nor decreasing. $\dfrac{dy}{dx} = f'(x) = 0$. A stationary point is either a turning point (local maximum or minimum point) or a stationary point of inflection. (p. 419)

stationary point of inflection A stationary point on a graph where the concavity changes but the sign of the gradient stays the same on both sides. (p. 419)

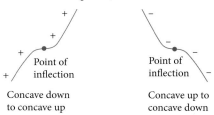

steady state The limiting value of a sequence as $n \to \infty$. (p. 360)

subset A set contained within another set, including the set itself. (p. 21)

substitution method An algebraic method of solving simultaneous equations by making one of the variables, x or y, the subject of one equation and then substituting it into the other equation. (p. 158)

sum to infinity The limiting value of a geometric series when $0 < r < 1$. (p. 371)

symmetry of independence If an event A is independent of an event B, then B is also independent of A. (p. 70)

tangent A straight line that touches a curve at only one point. (p. 387) See **gradient** for diagram.

term (algebra) A part of an expression separated by + or − operations. For example, the expression $5x + \frac{2}{3} - 10y^2$ has three terms. (p. 110)

term (sequences) A number in a sequence. (p. 333)

theoretical probability The likelihood of an event occurring without having to conduct an experiment. (p. 39)

transformation A change to a function and its graph through dilation, reflection or translation. (p. 214)

translation Shifting of a graph in the direction of the x-axis (horizontally) or y-axis (vertically). (p. 214)

tree diagram A display showing each possible outcome of every stage of a probability experiment, and the corresponding probabilities. (p. 48)

trigonometric function A trigonometric function such as $y = \sin(x)$, $y = \cos(x)$ or $y = \tan(x)$. (p. 239)

turning point A stationary point on a graph where the sign of the gradient changes on either side. If it changes from negative (decreasing) to positive (increasing), it is a local minimum point. If it changes from positive (increasing) to negative (decreasing), it is a local maximum point. On the parabola $y = a(x - h)^2 + k$, the coordinates of the turning point are (h, k). (p. 142)

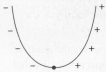 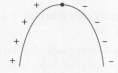

Local minimum point   Local maximum point

turning point form A quadratic function expressed in the form $y = a(x - h)^2 + k$. The turning point of the parabola is (h, k). (p. 140)

two-way table A table that shows the size of the sets and the possible intersections. (p. 5)

union (of sets) The set of elements formed by combining both given sets. It contains the elements that are in one or the other or both sets. (p. 3)

unit circle A circle of radius 1 unit on the number plane, used to define the trigonometric function of values of angles of any size. (p. 260)

universal set The set of all possible elements. (p. 3)

velocity The 'signed speed' of a moving object, represented by the function $v = \frac{dx}{dt}$, the rate of change of the displacement, x. Conversely, $x = \int v(t)\,dt$. (p. 416)

Venn diagram A diagram that shows all possible relationships that exist between two (or more) sets. (p. 5)

vertical asymptote A vertical line to which the graph of a function tends as y approaches infinity. (p. 192)

***x*-intercept** The point of a graph that intersects the x-axis. Found by letting $y = 0$ and solving for x. (p. 89)

***y*-intercept** The point of a graph that intersects the y-axis. Found by letting $x = 0$ and solving for y. (p. 89)